# Modeling of Magnetic Particle Suspensions for Simulations

# Modeling of Magnetic Particle Suspensions for Simulations

**Akira Satoh**

Department of Machine Intelligence and System Engineering
Akita Prefectural University
Yuri-Honjo, Japan

CRC Press is an imprint of the
Taylor & Francis Group, an **informa** business
A SCIENCE PUBLISHERS BOOK

CRC Press
Taylor & Francis Group
6000 Broken Sound Parkway NW, Suite 300
Boca Raton, FL 33487-2742

First issued in paperback 2020



ISBN-13: 978-1-4987-4091-3 (hbk)
ISBN-13: 978-0-367-78262-7 (pbk)



**Library of Congress Cataloging-in-Publication Data**

Names: Satō, Akira, 1958-
Title: Modeling of magnetic particle suspensions for simulations / Akira Satoh, Department of Machine Intelligence and System Engineering, Akita Prefectural University.
Description: Boca Raton, FL : CRC Press, [2016] | "A science publishers book." | Includes bibliographical references and index.
Identifiers: LCCN 2016036652 | ISBN 9781498740913 (hardback : alk. paper)
Subjects: LCSH: Magnetic materials--Mathematical models.
Classification: LCC TK454.4.M3 S28 2016 | DDC 620.1/1297--dc23
LC record available at https://lccn.loc.gov/2016036652

**Visit the Taylor & Francis Web site at**
**http://www.taylorandfrancis.com**

**and the CRC Press Web site at**
**http://www.crcpress.com**

# Preface

A suspension composed of magnetic fine particles may frequently be used as an intermediary material where the ultimate goal is to generate a more complex material; a typical application along this line is in the production of high density recording material. A more direct application of a magnetic particle suspension may be found in mechanical dampers and actuators, which make use of the magneto-rheological effect that is controlled in a desirable manner, by an external magnetic field. There are currently numerous researchers vigorously studying these applications in the fields of magnetic materials and fluid engineering that have made an enormous contribution through the innovation of related commercial products. The present era may request researchers to open up as yet unknown new fields for the application of magnetic particles and suspensions. One of the new pioneering fields is certainly the biomedical engineering field, where for instance, many researchers have been investigating the development of a drug delivery system using magnetic composite materials. Modern synthesis technology enables one to design and generate magnetic-particle-based composite materials with an eye to their application in the fields of biomedical and fluid engineering.

In the above-mentioned research fields, particle-based simulation methods such as Monte Carlo, molecular dynamics and Brownian dynamics play a vitally important role for investigating the related physical phenomena from a microscopic point of view. This is mainly because of the difficulty in employing a theoretical or an experimental approach in a microscopic investigation of a multi-particle system in which multi-body interactions induce a complex aggregate microstructure that is significantly influenced by a wide variety of factors and situations.

From this background, the main objective of the present book is to highlight the modeling of magnetic particles with differing shapes and magnetic properties, from the viewpoint of providing graduate students and young researchers with information regarding both the theoretical aspects and actual techniques for the treatment of magnetic particles in particle-based simulations. We focus here on the more common and useful techniques for modeling magnetic particles from a simulation point of view, but we avoid those for specifically narrow topic areas where a special simulation technique may be required. Nowadays, academic research fields are narrowly subdivided into many specialist research fields and researchers are required to obtain and update mountains of information including the related theories, experimental methods, simulation techniques and the important results obtained from these various approaches. The present book is intended to provide a sound foundation for readers who wish to take up the challenge of particle-based simulation research in a pioneering research field where the standard simulation methods are not directly applicable.

As a general rule, specialized knowledge concerning the modeling and simulation of a magnetic particle suspension will be acquired in an effective manner under the supervision of an expert. The present book is written to play such a role for readers who wish to develop the skill of modeling magnetic particles and develop a computer simulation program using their own ability. It is therefore a self-learning book for gradient students and young researchers who wish to investigate the related physical phenomena from a microscopic point of view. The present book treats a series of important common procedures necessary for investigating a physical phenomenon by means of simulation, i.e., modeling of magnetic particles, expressions for interactions between particles, formalization of basic equations, particle-based simulation methods, strategy of simulations, and quantities describing system characteristics. Moreover, two academic-oriented simulation programs of Monte Carlo for cube-like particles and Brownian dynamics for disk-like particles are shown to provide readers with an aid for acquiring the skill of developing a simulation code. From the support obtained from using this book, readers are expected to be able to sufficiently advance their present skill in preparation for tackling any challenging physical problem they may encounter in the future.

We focus on the following subjects in the present book. Non-spherical particles such as the rod-like, spheroidal, disk-like and cube-like particles are addressed, and the related expressions regarding forces, torques, energies and diffusion coefficients are derived and relevant mathematical expressions are shown. Characteristics of particle-particle interaction energies are discussed with reference to curves plotted in figures to facilitate straightforward understanding. Modeling of the magnetic properties of non-spherical particles is focused on, based on the point dipole model and plus-minus charges model, which are commonly and widely used in particle-based simulations. Since the interaction energy or force due to the overlap between two repulsive surface layers is in general not known as an analytical expression in the case of non-spherical particles, we discuss a modeling method for which the known theory for two spherical particles is applicable to the case of non-spherical particles. This approach with a simplified model may be regarded as a useful first approximation because the main hurdle in developing a particle-based simulation code (program) may doubtless be in treating the complex interaction of the overlap of the repulsive layers. The assessment of the particle overlap becomes more difficult for the more complex shaped particles, such as the cube-like particles, in both the form of the mathematical analysis and actual development of the simulation code. Hence, in this book, we discuss the analysis of the particle overlap in detail with regard to non-spherical particles including the cube-like particles. For the particle-based simulation methods, we concentrate on Brownian dynamics and lattice Boltzmann as well as the usual powerful methods of Monte Carlo and molecular dynamics. In addition, the stochastic rotation dynamics (multi-particle collision dynamics) is briefly summarized. The lattice Boltzmann, stochastic rotation dynamics and dissipative particle dynamics methods can simulate both the particle motion and the ambient flow field simultaneously, and therefore, these methods may become more important in addressing actual physical problems from the various application fields. The author believes that the two sample simulation programs shown in the present book will help readers to overcome some of the more difficult hurdles and therefore be able to develop an academic-oriented simulation program without supervision. The numerous detailed descriptions are intended as an aid for readers to understand perfectly the simulation programs in a straightforward manner.

The author acknowledges Dr. Geoff N. Coverdale, for the useful advice and help provided in accomplishing the present book. Also, the author appreciates the work put in by Ms. Aya Saitoh in the preparation of the digital manuscripts.

Finally, the author strongly hopes that young researchers will pioneer a new research subject related to magnetic particle-based materials by means of simulation techniques, based on the theoretical foundation which the present book may offer to readers.

*"Goddess of fortune smiles upon a researcher with no cloudy heart."*

**Akira Satoh**
Kisarazu City, Chiba Prefecture, Japan
April 2016

# Contents

CHAPTER 1

# General Remarks

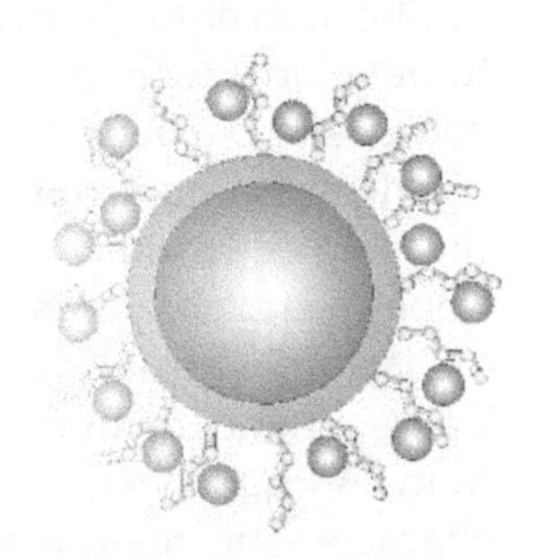

## 1.1 Application fields of magnetic particle suspensions

Magnetic particle suspensions have an important potential for application and therefore a variety of studies on these functional fluids have been conducted in various fields including the traditional fluid engineering field and the recent bioengineering field where there is interest in drug delivery systems. In the field of magnetic recording material, which is a typical application field for magnetic particles, high-density recording materials [1, 2] and optical units [3–7], as shown in Fig. 1.1, have been actively studied. In these applications, suspensions composed of magnetic particles function as an intermediary medium to obtain an ultimate goal material. In another typical application field of fluid engineering, as shown in Fig. 1.2, the magneto-rheological effect, which implies that the apparent viscosity of magnetic suspensions varies due to the strength of an external magnetic and flow fields, has been applied in order to develop mechanical actuators and dampers [8, 9]. Functional particles responding to an external magnetic field, which may be developed by coating materials with magnetic particles, have a feasible potential for innovative application [10]. This kind of composite particle is designed to be controlled by an applied magnetic field whilst a non-magnetic material part of the composite particle has been synthesized to possess another specific function. In this application, the latter function constitutes the main function of the composite particle and the former function is used for transporting the particle to a specific site. Currently these types of magnetic composite particles are regarded as a key material in their application to the biomedical engineering field [11–13]. Hence, there have been vigorous attempts to synthesize multi-functionalized composite particles that can include medicines as a goal in the application to a drug delivery system [14–27], as shown in Fig. 1.3. From a suspension physics engineering point of view, the application of these higher functionalized magnetic particles to a drug delivery system is a significantly hopeful application field. From an academic point of view, it is a challenging subject to develop a technology employing a gradient magnetic field for guiding composite particles to the site of a specific cancer cell and to administer the drug agents from the particle [28–36]. Moreover, their application to the field of resource engineering and environmental engineering is also hopeful and challenging [37–39]. In these applications the precious metals or harmful substances dissolved in sea water are captured and recovered by magnetic composite particles using a gradient applied magnetic field. As clearly demonstrated in these application examples, using the characteristic feature that their physical properties and behavior in a

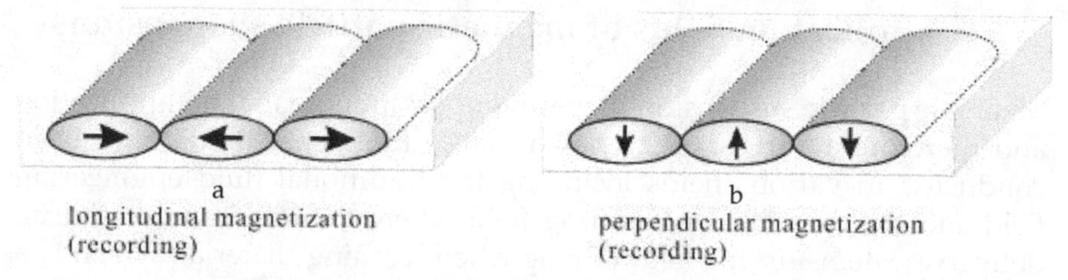


**Figure 1.1.** Recording materials: (a) longitudinal magnetization and (b) perpendicular magnetization.

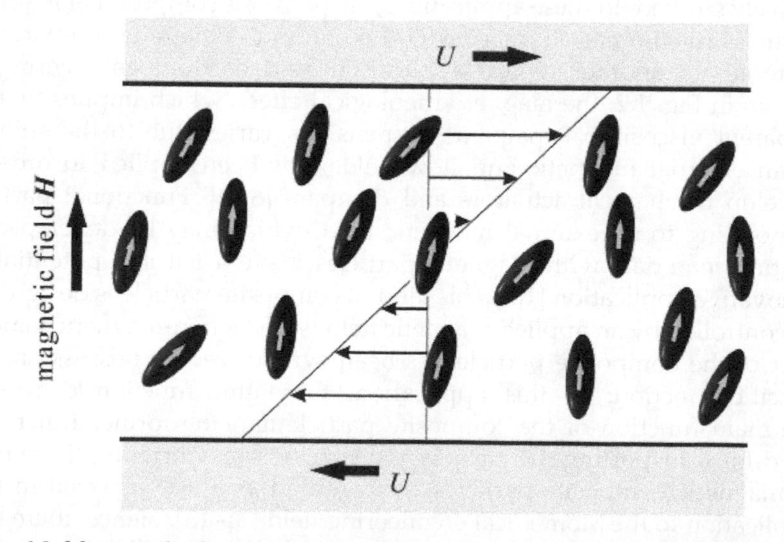


**Figure 1.2.** Magneto-rheological effect due to the orientational characteristics of magnetic particles in a flow field under an applied magnetic field.

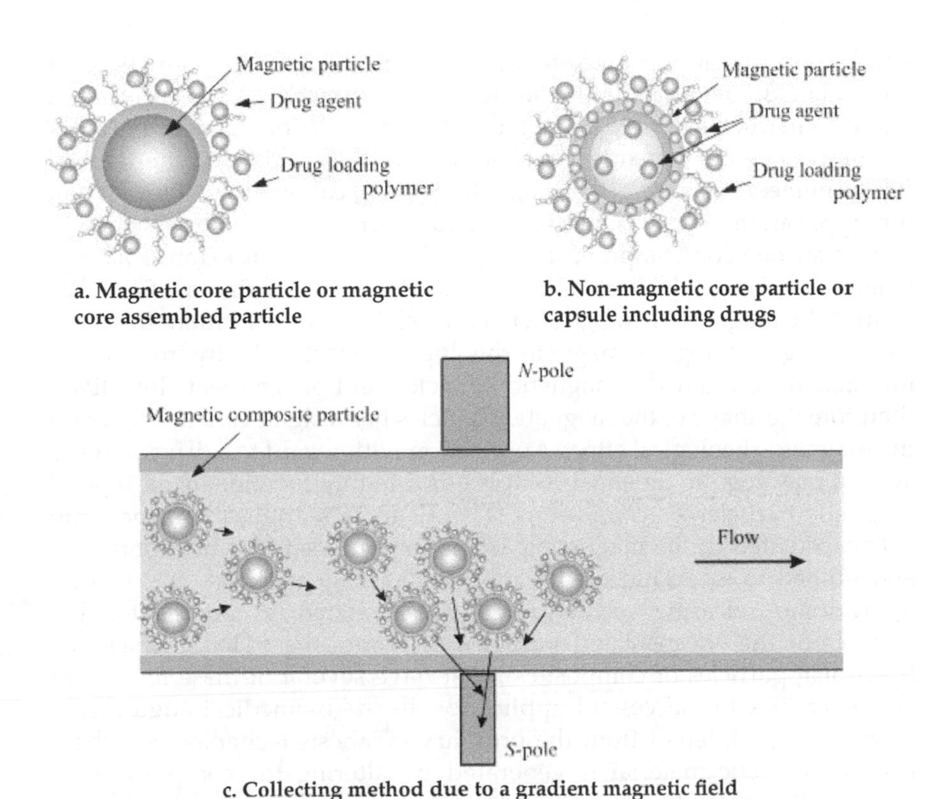


**Figure 1.3.** Application to the drug delivery system: (a) and (b) composite functional particles, and (c) a collecting method due to a gradient magnetic field.

flow field can be controlled by an external non-uniform magnetic field, functional suspensions composed of new composite magnetic particles offer significant potential for expanding their application in various fields.

## 1.2 Multi-functionalized magnetic particles

New magnetic particles (magnetic materials) such as magnetic recording materials have actively been developed by combining different kinds of molecules by many researchers [40], and representative magnetic recording materials are the FePt particles [1, 2]. In addition to the ordinary $Fe_2O_3$ or $Fe_3O_4$-based particles, there have been vigorous attempts to generate multi-functionalized magnetic particles in a variety of application fields. Such particles are synthesized by coating non-magnetic base particles with smaller magnetic particles or another magnetic coating material, which formulates a composite material made of both magnetic and non-magnetic materials with the desired characteristics for application [10, 41]. Hence, the desired behavioral characteristics of magnetic particles or composites are dependent on the field of application. In the field of fluid engineering, where the magneto-rheological effect is mainly used for applications such as mechanical dampers and actuators, we will need a suspension composed of magnetic particles that exhibits a large magneto-rheological effect and also need a mechanism to effectively control the magneto-rheological effect [8, 9]. In this application, the main factor for governing the magneto-rheological effect is the hydrodynamic interaction between the magnetic particles and an ambient flow field, therefore the shape of the magnetic particles has a significant influence on the magneto-rheological effect. Application with regard to a drug delivery system may require the most complex and highly functionalized type of magnetic particles or composites [11–13]. In this application, the drug is encapsulated in the magnetic particle or composite for transportation and guided to a specific site by an external magnetic field, and further appropriate releasing technology is necessitated to administer the drug from the encapsulated particles or composites. Hence, magnetic functional particles or composites must have several of these important characteristics for successful application in the biomedical engineering field [33–36]. Different from the ordinary synthesis technologies, where a new magnetic material is generated by altering the composition of the constituting materials (molecules), the composite magnetic particles themselves are generated by a design for obtaining the most desired characteristics as functional particles [41], as shown in Fig. 1.4, and so may lead to the appearance of characteristics that cannot be expected from the ordinary development approach. The concept of this developing approach corresponds to certain kinds of design engineering regarding mechanical machines. Hence, in addition to clarifying the physical

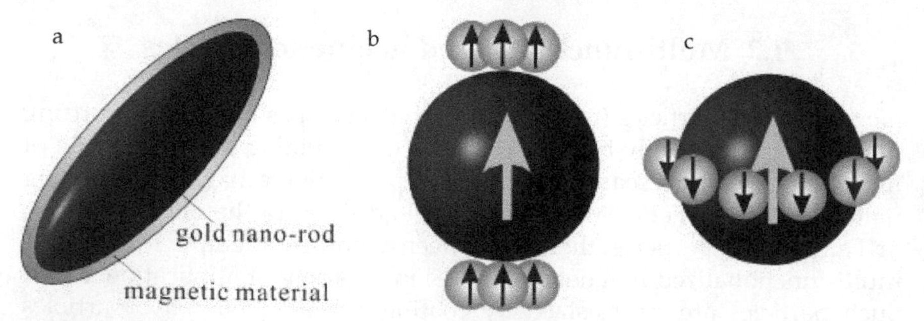


**Figure 1.4.** Various types of magnetic composite particles.

phenomena of magnetic particle suspensions, molecular simulations have an important role in providing inspiration for the feasibility of new types of magnetic particles or composites and for stimulating the motivation for experimental researchers to generate these materials.

## 1.3 General magnetic characteristics of magnetic particles

There is a long and rich history regarding the studies of magnetism and the magnetization theory of magnetic materials in the field of magnetic material science. The studies regarding the theory of magnetism teach us that the magnetization of magnetic materials has a strong relationship with the magnetic domain structure of the materials [42, 43]: readers are recommended to refer to standard textbooks of magnetism and magnetic material if they have an interest in understanding in detail the mechanism of the magnetization process. Here we only briefly address the magnetic characteristics of magnetic particles that may be useful for synthesizing new magnetic particle suspensions.

In order to perform simulations of a magnetic particle suspension, it is necessary to clarify the magnetic properties of the dispersed magnetic particles and to idealize them in the form of a mathematical model. Since calculation of the particle-particle interactions is a significantly time-consuming task for a multi-particle system, it is desirable, from a simulation point of view, to simplify the magnetization model as much as possible. Hence, we first discuss the general magnetic characteristics of magnetic particles, which may be a valuable aid in understanding how the magnetization model for magnetic particles is simplified and idealized in order to perform simulations in a reasonable computation time. The following discussion regarding the magnetic characteristics and the modeling is a foundation that enables one to develop more complex magnetization models.

Magnetic or magnetizable particles smaller than nano-size generally tend to exhibit superparamagnetism, which implies that the magnetic moment of the magnetic particle has a freedom to rotate independent of the particle body [42], as shown in Fig. 1.5(a). This is an example of the fact that the magnetic characteristics exhibited by magnetic particles are strongly dependent on their material properties, size and shape [42–44]. Increase in the size of magnetic particles from nano-size makes characteristics of the magnetization more complex, and these characteristics have a strong relationship with the structure of the magnetic domains in the particle body. A variety of factors, such as material properties and particle shape, significantly influence the magnitude and direction of the magnetization within the particle body [42–44], as shown in Figs. 1.5(b) and (c).

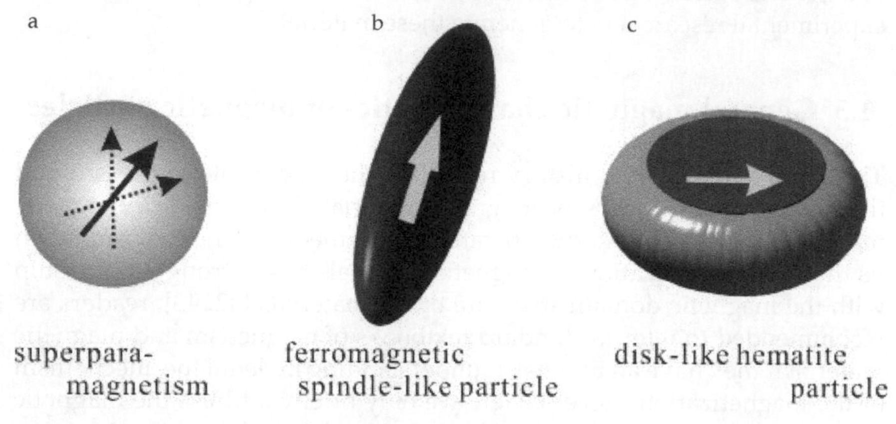


**Figure 1.5.** Magnetization dependent on the shape and size.

Before we discuss the theory of magnetism, it should be noted that in treating the theory of magnetism and magnetic characteristics, it is very important to understand which unit system is used, since there are three unit systems for expressing magnetic quantities and data in the commercial and academic societies. The CGS unit system is commonly used in the commercial world and the SI unit system is more frequently used in textbooks. Moreover, there are two unit systems for the SI unit, i.e., the Sommerfeld and Kennelly systems. In the present book, we employ the Sommerfeld unit system, and the unit for each magnetic quantity that is usually used for expressing equations and magnetic characteristics is shown in Table 1.1. In the Sommerfeld unit system, the unit of magnetization corresponds to the unit of magnetic field strength.

Magnetic characteristics of particles are described by the magnetization strength. In general, the magnitude of the magnetization of a magnetic particle, $M$, is governed by the magnetic field strength, $H$. However, as shown in Fig. 1.6, a demagnetizing field appears within the particle body, and therefore the induced magnetic field $H^{(ind)}$ will make a contribution to the magnetization strength of the particles. Hence, the magnetic field strength inside the body, $H^{(ind)}$, is obtained by reducing the applied magnetic field $H$ by the demagnetizing field $H^{(demag)}$ as $H^{(ind)} = H - H^{(demag)}$. The demagnetization field is dependent on the particle shape and is expressed as $H^{(demag)} = N_d M(H^{(ind)})$, where $N_d$ is the demagnetizing factor as for instance, $N_d = 1/3$ in the case of spherical particles [42, 43]. Employing the notation

**Table 1.1.** Sommerfeld unit system used in magnetism.

| Magnetic Quantities | $B = \mu_0 (H + M)$ (Sommerfeld) |
|---|---|
| Magnetic field strength $H$ | [A/m] |
| Magnetization strength $M$ | [A/m] |
| Magnetic flux density $B$ | [T] (=[Wb/m$^2$]) |
| Permeability of free space $\mu_0$ | $\mu_0 = 4\pi \times 10^{-7}$ [H/m] (=[Wb/(A·m)] |
| Magnetic charge $q$ | [A·m] |
| Magnetic moment $m$ | [A·m$^2$] |
| Potential energy $U$ | $U = -\mu_0 \boldsymbol{m} \cdot \boldsymbol{H}$ [J] (=[Wb·A]) |
| Torque $T$ | $\boldsymbol{T} = \mu_0 \boldsymbol{m} \times \boldsymbol{H}$ [N·m] (=[Wb·A]) |
| Magnetic field induced by magnetic charge $H$ | $\boldsymbol{H} = \frac{q}{4\pi r^2} \cdot \frac{\boldsymbol{r}}{r}$ [A/m] |
| Magnetic force acting between two magnetic charges $F$ | $\boldsymbol{F} = \frac{\mu_0 q q'}{4\pi r^2} \cdot \frac{\boldsymbol{r}}{r}$ [N] (=[Wb·A/m]) |
| Magnetic interaction between two magnetic moments, $U$ | $U = \frac{\mu_0}{4\pi r^3}\{\boldsymbol{m}_1 \cdot \boldsymbol{m}_2 - \frac{3}{r^2}(\boldsymbol{m}_1 \cdot \boldsymbol{r})(\boldsymbol{m}_2 \cdot \boldsymbol{r})\}$ [J] (=[Wb·A]) |
| Combined units: [H] = [Wb/A], [T] = [Wb/m$^2$], [J] = [N·m] Equivalent units: [N] = [Wb·A/m] | |

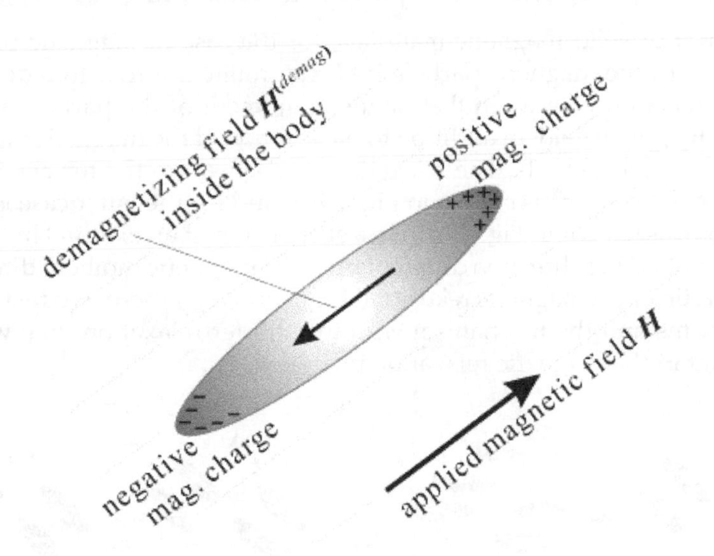


**Figure 1.6.** External magnetic field $\boldsymbol{H}$ and demagnetizing field $\boldsymbol{H}^{(demag)}$.

$\chi(H^{(ind)})$ for the magnetic susceptibility function, the magnetization $M$ is expressed as $M = \chi H^{(ind)}$. The magnetization is therefore as a function of the applied magnetic field strength $H$,

$$M = \frac{\chi}{1 + N_d \chi} H \tag{1.1}$$

If the shape of a magnetic particle is specified, the demagnetization factor may be evaluated. Moreover, if the magnetic susceptibility $\chi$ is known as a function of the applied field strength, the magnetization $M$ can be evaluated from Eq. (1.1). In general, the dependence of $\chi$ on the field strength will be experimentally determined. The magnetization is a measure expressed per volume of the material of interest, and therefore the magnetization and the volume of the particle body determine the magnetic moment $m$ as

$$m = V_p M \tag{1.2}$$

in which $V_p$ is the volume of the particle. It should be noted that the above expressions may be valid in the situation where the particle is uniformly magnetized and the magnetic moment of the particle is aligned with the applied magnetic field. In general, the magnetization $\boldsymbol{M}$ and the magnetic moment $\boldsymbol{m}$ are vector quantities and depend on both the strength and the direction of the magnetic field, expressed in vector form as $\boldsymbol{m} = V_p \boldsymbol{M}$.

In contrast to solid magnetic materials, for the case of magnetic particle suspensions, the magnetic particle itself can rotate in order to reduce the interaction energy between the magnetic moment of the particle and an applied magnetic field, in addition to the rotation of the magnetic moment within the particle body. The relaxation mechanism in the former case is Brownian relaxation, as shown in Fig. 1.7(a) and for the latter case is Néel relaxation, as shown in Fig. 1.7(b). In a suspension, if an external magnetic field is applied in a direction different from the magnetic moment direction of the particle, the magnetic relaxation is governed by these two relaxation mechanisms and the mechanism with the shorter relaxation time will be dominant in the magnetic relaxation phenomenon.

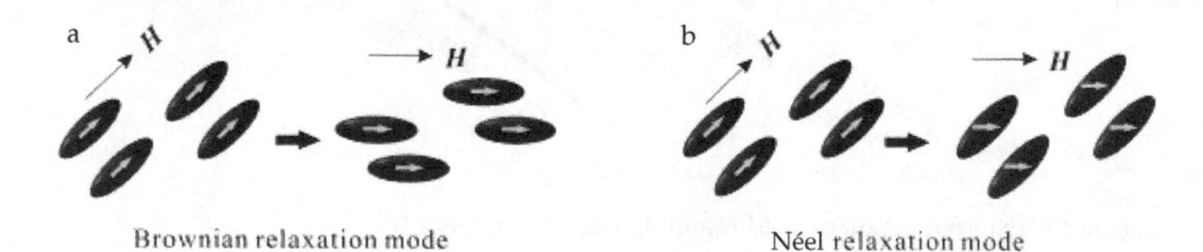


**Figure 1.7.** Two mechanisms for relaxation of the magnetic moment direction: (a) Brownian relaxation mode and (b) Néel relaxation mode.

The Brownian relaxation time $\tau_B$, for instance, for a spherical particle is expressed as [45]

$$\tau_B = \frac{3\eta V_h}{k_B T} \tag{1.3}$$

in which $\eta$ is the viscosity of the base liquid, $k_B$ is Boltzmann's constant, $T$ is the liquid temperature, and $V_h$ is the hydrodynamic volume of the particle. For the case of magnetic particles with uniaxial anisotropy, the Néel relaxation time $\tau_N$ is given by [46]

$$\tau_N = \tau_0 \exp(K_a V_c / k_B T) \tag{1.4}$$

in which $\tau_0$ is a constant that is generally in the order of $10^{-9}$ s, $V_c$ is the magnetic core volume of the particle, and $K_a$ is the effective anisotropy constant that depends on the magnetic material of the particle. The quantity $K_a V_c$ is the activation energy which is the effective energy barrier for change in the orientation of the magnetic moment. In order for the magnetic moment to align with the field direction, if the Néel relaxation time is much shorter than the Brownian relaxation time, the magnetic moment will rotate within the particle with no rotation of the particle body, otherwise the particle will rotate as a whole without experiencing the Néel relaxation mode.

In contrast to standard ferromagnetic particles, as shown in Fig. 1.8(a), hematite particles have a weaker magnetization in comparison with magnetite particles [47–56], so that if the particle diameter can be controlled in the synthesis process, it is relatively straightforward to obtain a stable dispersion composed of hematite particles. Spindle-like hematite particles exhibit the characteristic feature of being magnetized in the short axis direction [54, 55], so that raft-like clusters are formed in the magnetic field direction [57], as shown in Fig. 1.8(b). Another characteristic feature of a spindle-like particle dispersion is that the contribution to the viscosity arising from magnetic properties may become negative under certain conditions in an applied magnetic field. This negative magneto-rheological effect was predicted from the theory based on the orientational distribution function [58–60] and verified by an experiment using a cone-plate type rheometer [61]. It is clear from this example that different types of magnetic particle dispersions may exhibit different and unique characteristics when subjected to various situations of an applied magnetic field and a flow field.

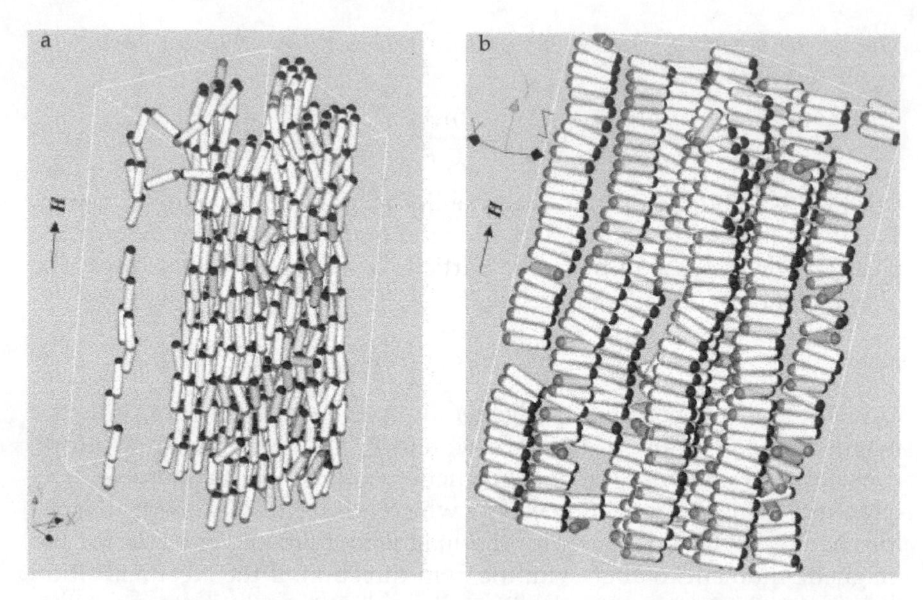


**Figure 1.8.** Different types of clusters formed for different magnetic properties of rod-like particles in an external magnetic field: (a) ferromagnetic particles and (b) hematite particles.

## 1.4 Modeling of magnetic characteristics of fine particles from a simulation point of view

From the viewpoint of simulating magnetic suspensions, it is necessary to idealize real magnetic particles by employing simpler particle shapes and simpler magnetization characteristics. This is due to the large amount of computation time that is required in calculating the interaction energies and forces acting between magnetic particles in the various particle-based simulation techniques. Hence it is a logically reasonable procedure to use a simpler particle model for the shape and the magnetization characteristics in order to obtain the physical properties of the dispersion as a first approximation, and then to proceed to the development of a more sophisticated particle model to understand experimental data, such as magneto-rheological characteristics, at a much deeper level. In the following, we briefly address the magnetic particle models that may be used for simulating magnetic particle suspensions, whilst several important particle models are treated in more detail in the following chapters.

The simplest magnetic particle model employs a spherical particle that has a point dipole moment with a constant magnitude at its center. The magnetic moment direction is fixed to the particle body and the magnitude of the magnetic moment is assumed to be constant and independent of an applied magnetic field [62, 63]. This implies that the relaxation motion of the magnetic moment is performed only by the Brownian relaxation mechanism. In order to improve this model, the magnitude of the magnetic moment may be treated as a function of the magnetic field strength, for instance, the magnitude of the magnetic moment may be assumed to be proportional to the magnetic field strength as in the linear magnetization model [64–66].

For non-spherical magnetic particles, there may be rod-like, plate-like, disk-like, and cube-like geometries, shown in Fig. 1.9, that have unique magnetic characteristics. Rod-like particles are frequently idealized as a prolate spheroid, spherocylinder, or plain cylinder and in each case there is a choice regarding several magnetization models that may be employed. Similar to the previous example, the simplest magnetization model is the point dipole model at the particle center, and the magnitude of the magnetization is assumed to be constant [67–69], as shown in Fig. 1.10(a). Again, the next simplest model is the linear magnetization model where the magnetization is in proportion to the magnetic field strength [64, 65], as shown in Fig. 1.10(b). The nonlinear magnetization model where the magnetization is as a nonlinear function of the magnetic field strength may be employed for expressing more complex magnetic characteristics [64, 65]. For certain rod-like magnetic particles, a more complex magnetization model may be necessary in order to approach realistic magnetic characteristics. For instance, point magnetic charges or point dipole moments may be located at the center of each of the hemisphere caps in the case of the spherocylinder particle [42, 70–72], as shown in Figs. 1.10(d) and (e). Moreover, magnetic rod-like particles may be modeled as a linear sphere-connected particle model where each constituent sphere has a point dipole moment at its center, as shown in Fig. 1.10(f). Progress to a more realistic magnetization model may employ magnetic point dipoles that are distributed within the particle body, and therefore this approach is frequently suitable as a magnetization model for magnetic particles with complex shape [40, 73, 74]. In general, additional complexity in the particle shape and the magnetization characteristics leads to a computationally expensive particle-based simulation, and therefore such complex models are ordinarily applied in only a significantly limited number of cases.

From a simulation point of view, the most complex magnetic particles have such characteristics that the relaxation of the magnetic moment in an external magnetic field is governed by both the mechanisms of Brownian and Néel relaxation, that is, the Brownian relaxation time is of the same

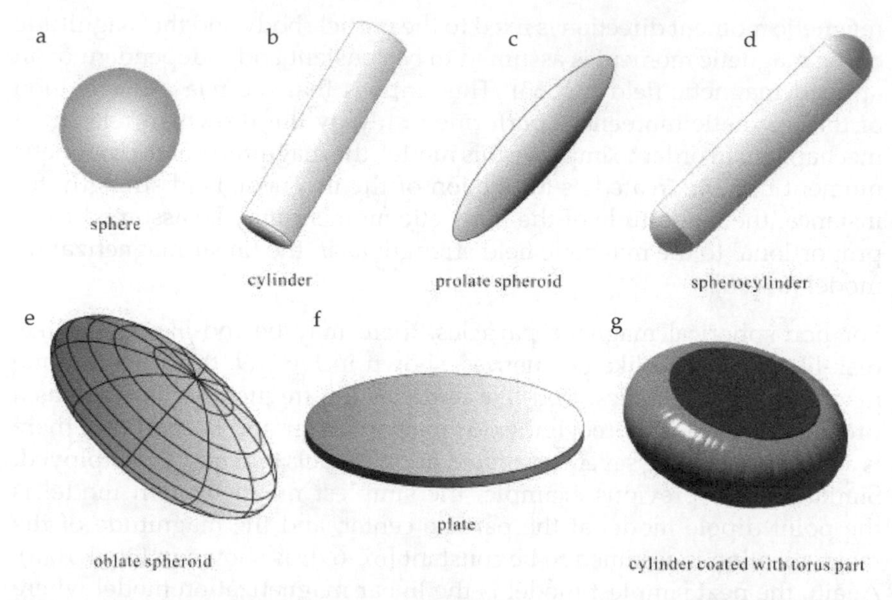


**Figure 1.9.** Particle models with various simple shapes for particle-based simulations.

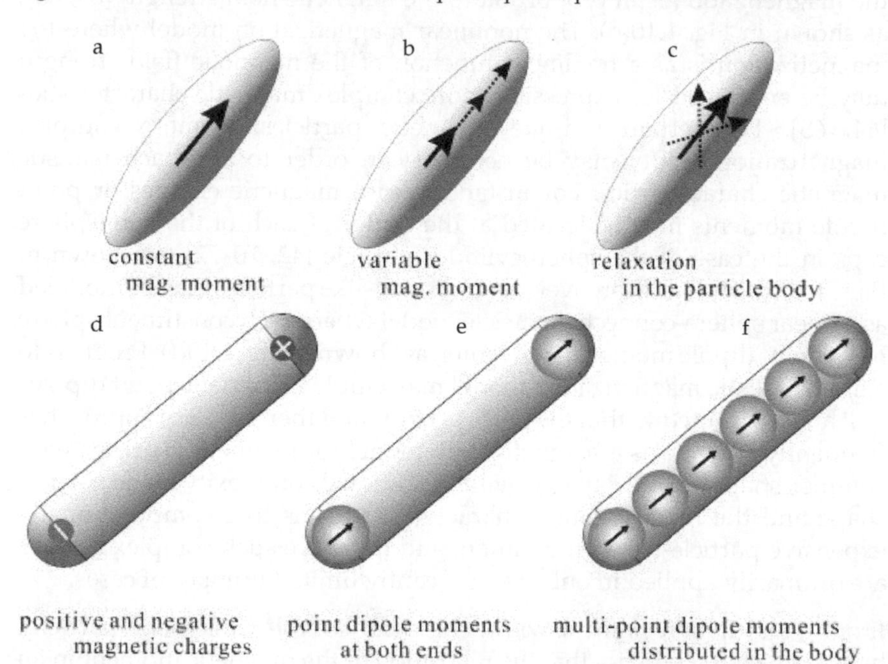


**Figure 1.10.** Various magnetization models.

order as the Néel relaxation time [75–77]. In this situation, it is necessary to simulate both the Brownian motion of the particle body and the relaxation of the magnetic moment within the particle body simultaneously. As a further improvement of the model, the magnetic characteristics of each particle may be evaluated from the Maxwell equations [78] and then combined with the translational and rotational motion of the particles at each simulation step. Unfortunately, although these improved magnetization models can explain the experimental observations more accurately, computation time will become much longer and frequently unrealistic for obtaining reasonable results. Therefore, the following approach may be a reasonable and significant procedure from the viewpoint of simulations. In conducting a simulation, a simpler magnetization model is initially employed as a first approximation and then it is successively refined until it is finally able to reproduce experimental results with quantitatively good agreement. Simulations using such refined magnetization models will be able to investigate physical phenomena at a detailed and accurate level. This step-by-step approach toward a difficult phenomenon may be the shortest path to achieve an understanding of the substantial characteristics of the phenomenon of interest.

As noted above, in the Brownian relaxation mode the entire particle rotates to align with the applied magnetic field direction, whilst in the Néel relaxation mode the magnetic moment itself aligns to the field direction within the body with a certain relaxation time. This behavior regarding the relation between the interaction of the magnetic moment and an applied magnetic field depends strongly on the material properties and the size of the particle. In the present book, we do not address the specific material properties of the magnetic particles but rather treat general magnetic particle models that may be applicable to various physical problems regarding a variety of magnetic suspensions. Hence, here we focus on the simpler shape and simpler magnetization models in which the magnitude of the magnetic moment is constant and the direction of the magnetic moment is fixed within the particle body, that is, the Néel relaxation of the magnetic moment is not taken into account.

## 1.5 Related physical phenomena

Surface modification technology plays an important role for successful synthesis of the magnetic composite particles used in drug delivery systems. Moreover, depending on the synthesis method, effects such as diffusion and sedimentation phenomena in the gravity field may have a significant influence on the final quality of these functional particles.

Also, the sedimentation behavior of magnetic composite particles has a significant relationship with the drug concentration on the surface properties of biological membranes [79]. Further, it is necessary to develop the technology for controlling self-assembled layers, aggregation structures and the orientational features of magnetic particles on a material surface after sedimentation by means of an applied magnetic field. Therefore, in developing functional composite particles for the magnetic drug delivery system, it is clear that physical phenomena such as aggregation phenomena, orientional distribution and the phase change of magnetic particles on a material surface have a significant relationship with the synthesis procedure. Hence, it is important to investigate the dependence of the magnetic particle behavior at a material surface in regard to various factors such as the magnetic field and the magnetic particle-particle interaction strength [79–83]. In addition, we are required to investigate the sedimentation characteristics of magnetic particles in the gravity field and also the integration of the structures on a material surface after sedimentation, which are dependent on various factors including the magnetic field strength [84].

Current advance in the technology for synthesizing magnetic composite particles [41] enables one to develop a magnetic particle suspension with specific magnetic characteristics for function in various circumstances of application. This advancement in synthesis technology stimulates studies regarding the application of magnetic functionalized particles in both the bioengineering field and the fluid engineering field. For instance, in the bioengineering field, fluid mechanics researchers have been attempting to develop a technology employing magnetic particles for transporting drugs to a specific site in the human body by means of a gradient external magnetic field [28–32]. In the fluid engineering field, the main targets for the application of magnetic dispersions are mechanical dampers and actuators that make use of the magneto-rheological effect.

In order to achieve successful applications, it is necessary to clarify aggregation phenomena and the phase change of the internal structure of magnetic particles in an applied magnetic field and a flow field [85], as shown in Table 1.2. Since magnetic particles are expected to aggregate in an applied magnetic field, we must investigate the relationship between the aggregate structure and the magneto-rheology in order to develop a technique for controlling these phenomena in an effective manner. In the application of the drug delivery system which enables one to administer magnetic drug-loading particles to a specific site, the main target is to develop a technique for manipulating the transport of composite particles using a gradient magnetic field generated by designing the locations of a

**Table 1.2.** Representative physical phenomena related to magnetic particle suspensions.

(1) Surface modification
(2) Diffusion and self-assembled layer on a material surface
(3) Sedimentation in the gravity field
(4) Aggregate structures in thermodynamic equilibrium
    Internal structures of aggregates, Phase change, Orientational distribution
(5) Flow problems of magnetic particles
    Magneto-rheological properties, Particle motion, Coupling of particle motion and ambient flow field

plurality of magnets [28–32]. Such a controlling technique for magnetic particles in terms of a non-uniform magnetic field is also applicable to the field of resource and environmental engineering in order to capture precious metals or hazardous substances dissolved in sea water. In this method, magnetic composite particles designed to adsorb precious metals or hazardous substances are captured by a non-uniform magnetic field which is similarly generated by an array of magnets [37–39].

## 1.6 Particle-based simulation methods

In order to investigate the behavior of various types of magnetic suspensions, simulation techniques at a microscopic level are proven to be indispensable. The contribution from experimental approaches is also important and helpful for developing the modeling of the magnetic particles which are used in simulations. For the aid of readers' understanding of the images of particle-based simulation methods, several snapshots of the particle configurations are shown in Fig. 1.11, which were obtained from a variety of simulations.

Monte Carlo method is a powerful microscopic technique for simulating particle suspensions in thermodynamic equilibrium [86]. This method is applicable to almost all types of magnetic particles with spherical, spheroidal, disk-like, or cube-like shape. The great advantage of this method arises from the fact that Monte Carlo is a stochastic approach and microscopic states of a system are generated according to the stochastic theory and not according to the basic equations of motion. Hence, if the interaction energy between particles can be expressed in mathematical form, Monte Carlo simulations are generally executable irrespective of the shape of particles. However, for the more complex particle shapes, the theoretical analysis of particle overlap, which is indispensable for developing a simulation program, becomes more difficult to implement. The Monte Carlo method may still be useful if the interaction between

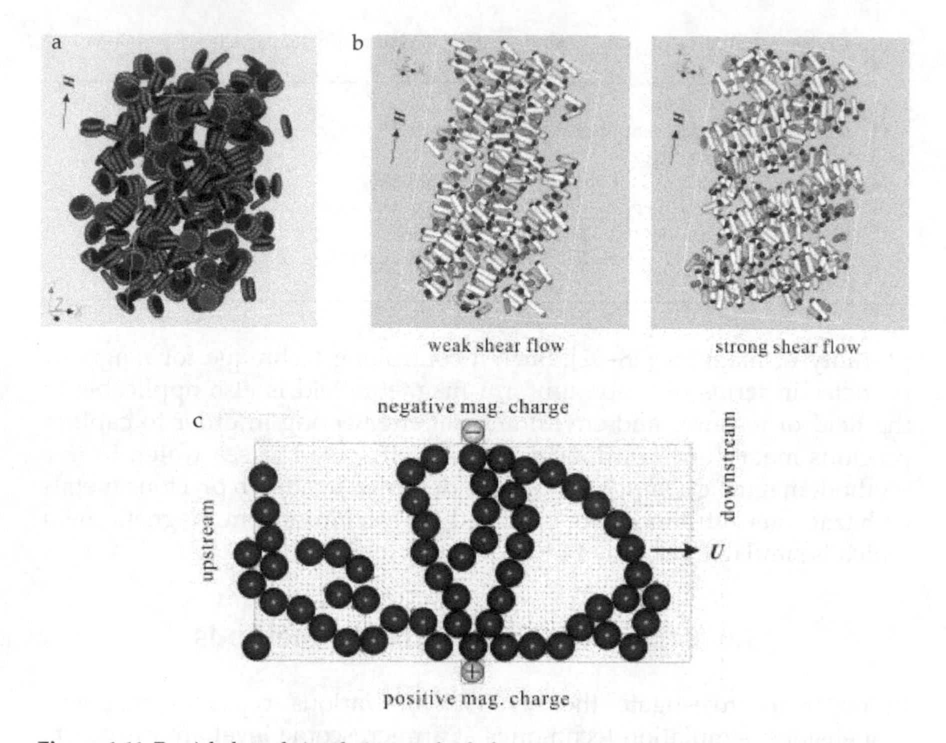


**Figure 1.11.** Particle-based simulation methods for magnetic particle suspensions: (a) Monte Carlo simulation for thermodynamic equilibrium, (b) Brownian dynamics simulation in a simple shear flow, and (c) lattice Boltzmann simulation for simulating the particle motion and the flow field simultaneously.

the soft steric or electric repulsive layers is not available or formulated because Monte Carlo simulations may still be performed for a solid particle system if the soft repulsive layers can be neglected. Solutions for this simplified system will be valid as a first approximate solution and certainly able to advance the understanding of physical phenomena as the first step. In contrast, molecular dynamics, Stokesian dynamics and Brownian dynamics simulations cannot generally be performed if the soft repulsive layers are neglected because these simulation methods make use of the equations of motion [86, 87]. It must be emphasized, however, that the Monte Carlo method is only suitable for a system in thermodynamic equilibrium and not appropriate for a dynamic situation such as the flow problem of particles.

In order to investigate the dynamic properties of a magnetic particle suspension, molecular dynamics and Brownian dynamics methods are usually employed. Molecular dynamics method is suitable for a suspension of particles larger than micron-order where the Brownian motion is negligible. This method is appropriate to investigate the influence of aggregate structures on the rheological characteristics of a suspension in a linear or simple shear flow field. If magnetic particles are smaller than micron-order, we need a simulation method that can activate the translational and rotational Brownian motion of the dispersed magnetic particles at a physically reasonable level, leading to the use of the Brownian dynamics method. In this method only the particle motion may be simulated, for instance, in a given linear flow or simple shear flow field, and therefore it may be limited in application because the particle motion and the ambient flow field cannot be solved simultaneously. In molecular dynamics and Brownian dynamics methods, the effect of the ambient fluid is taken into account as a friction term in the basic equation of the particle motion; the flow field is not solved in the simulation of the particle motion. Situations that require the multi-body hydrodynamic interactions between particles to be taken into account may be treated for a spherical particle suspension by both the methods of Stokesian dynamics [88, 89] and Brownian dynamics [86, 87]. However, in the case of a non-spherical particle suspension, the treatment of the multi-body hydrodynamic interactions is significantly more difficult. Hence, there is a need for a simulation method that is able to simultaneously take account of particle motion, the ambient flow field and multi-body hydrodynamic interactions between the dispersed particles. A hopeful simulation approach for this objective is the lattice Boltzmann method. In this method the ambient fluid is modeled by virtual fluid particles and the flow field may be obtained by simulating the virtual fluid particles and the dispersed particles simultaneously. Moreover, the multi-body hydrodynamic interactions between particles are reproduced indirectly through the interactions between dispersed and virtual fluid particles [90–93]. Also in this method, a special technique for activating particle Brownian motion may be combined with the ordinary lattice Boltzmann method [94, 95]. The concept of virtual fluid particles may be a key development idea for new particle-based simulation methods in which the multi-body hydrodynamic interactions between the dispersed particles are able to be taken into account and both the particle motion and the ambient flow field are able to be simulated simultaneously [96, 97].

Since there are various strategies in performing simulations at a microscopic level in order to obtain reliable results, in the following chapters we will discuss simulation techniques and the modeling of magnetic particles. Discussion will include important topics for treating the motion of magnetic particles in a suspension such as the criterion for particle overlap, which may be a serious hurdle to the development of a successful and reliable computational code.

## Bibliography

[1] Harrell, J. W., Kang, S., Jia, Z., Nikles, D. E., Chantrell, R. W. and Satoh, A. 2005. Model for the easy-axis alignment of chemically synthesized L10 FePe nanoparticles. Appl. Phys. Lett. 87: 202–208.

[2] Verdes, C., Chantrell, R. W., Satoh, A., Harrell, J. W. and Nikles, D. 2006. Self-organization, orientation and magnetic properties of FePt nanoparticle arrays. J. Magn. Magn. Mater. 304: 27–31.

[3] Iwayama, Y., Yamanaka, J., Takiguchi, Y., Takasaka, M., Ito, K., Shinohara, T., Sawada, T. and Yonese, M. 2003. Optically tunable gelled photonic crystal covering almost the entire visible light wavelength region. Langmuir. 19: 977–980.

[4] Reese, C. E., Guerrero, C. D., Weissman, J. M., Lee, K. and Asher, S. A. 2000. Synthesis of highly charged, monodisperse polystyrene colloidal particles for the fabrication of photonic crystals. J. Colloid Interface Sci. 232: 76–80.

[5] Mine, E., Hirose, M., Nagao, D., Kobayashi, Y. and Konno, M. 2005. Synthesis of submicrometer-sized titania spherical particles with a sol-gel method and their application to colloidal photonic crystals. J. Colloid Interface Sci. 291: 162–168.

[6] Furumi, S. and Sakka, Y. 2006. Chiroptical properties induced in chiral photonic-bandgap liquid crystals leading to a highly efficient laser-feedback effect. Adv. Mater. 18: 775–780.

[7] Furumi, S. and Sakka, Y. 2006. Circularly polarized laser emission induced by supramolecular chirality in cholesteric liquid crystals. J. Nanosci. Nanotech. 6: 1819–1822.

[8] Bullough, W. A. (ed.). 1996. Electro-Rheological Fluids, Magneto-Rheological Suspensions and Associated Technology, World Scientific, Singapore.

[9] Wereley, N. M. (ed.). 2013. Magnetorheology: Advances and Applications, Royal Society of Chemistry, London.

[10] Hellenthal, C., Ahmed, W., Kooij, E. S., Van Silfhout, A., Poelsema, B. and Zandvliet, H. J. W. 2012. Tuning the dipole-directed assembly of core-shell nickel-coated gold nanorods. J. Nanoparticle Research. 14: 1–14.

[11] Häfeli, U., Schütt, W., Teller, J. and Zborowski, M. (eds.). 1997. Scientific and Clinical Applications of Magnetic Carriers, Springer, Berlin.

[12] Kuznetsov, A. A., Filippov, V. I., Kuznetsov, O. A., Gerlivanov, V. G., Dobrinsky, E. K. and Malashin, S. I. 1999. New ferro-carbon adsorbents for magnetically guided transport of anti-cancer drugs. J. Magn. Magn. Mater. 194: 22–30.

[13] Weingart, J., Vabbilisetty, P. and Sun, X. 2013. Membrane mimetic surface functionalization of nanoparticles: Methods and applications. Adv. Colloid Interface Sci. 197-198: 68–84.

[14] Zhang, J. L., Srivastava, R. S. and Misra, R. D. K. 2007. Core-shell magnetite nanoparticles surface encapsulated with smart stimuli-responsive polymer: Synthesis, characterization, and LCST of viable drug-targeting delivery system. Langmuir. 23: 6342–6351.

[15] Liu, T., Hu, S., Hu, S., Tsai, S. and Chen, S. 2007. Preparation and characterization of thermal-sensitive ferrofluids for drug delivery application. J. Magn. Magn. Mater. 310: 2850–2852.

[16] Yang, J., Lee, J., Kang, J., Lee, K., Suh, J., Yoon, H., Huh, Y. and Haam, S. 2008. Hollow silica nanocontainers as drug delivery vehicles. Langmuir. 24: 3417–3421.

[17] Balakrishnan, S., Bonder, M. J. and Hadjipanayis, G. C. 2009. Particle size effect on phase and magnetic properties of polymer-coated magnetic nanoparticles. J. Magn. Magn. Mater. 321: 117–122.

[18] Abdalla, M. O., Aneja, R., Dean, D., Rangari, V., Russell, A., Jaynes, J., Yates, C. and Turner, T. 2010. Synthesis and characterization of noscapine loaded magnetic polymeric nanoparticles. J. Magn. Magn. Mater. 322: 190–196.

[19] Tomitaka, A., Koshi, T., Hatsugai, S., Yamada, T. and Takemura, Y. 2011. Magnetic characterization of surface-coated magnetic nanoparticles for biomedical application. J. Magn. Magn. Mater. 323: 1398–1403.

[20] Cao, Z., Yue, X., Li, X. and Dai, Z. 2013. Stabilized magnetic cerasomes for drug delivery. Langmuir. 29: 14976–14983.

[21] Tomitaka, A., Ueda, K., Yamada, T. and Takemura, Y. 2012. Heat dissipation and magnetic properties of surface-coated Fe3O4 nanoparticles for biomedical applications. J. Magn. Magn. Mater. 324: 3437–3442.

[22] Wang, N., Guan, Y., Yang, L., Jia, L., Wei, X., Liu, H. and Guo, C. 2013. Magnetic nanoparticles (MNPs) covalently coated by PEO–PPO–PEO block copolymer for drug delivery. J. Colloid Interface Sci. 395: 50–57.

[23] Zhang, X., Xue, L., Wang, J., Liu, Q., Liu, J., Gao, Z. and Yang, W. 2013. Effects of surface modification on the properties of magnetic nanoparticles/PLA composite drug carriers and *in vitro* controlled release study. Colloids Surf. A. 431: 80–86.

[24] Liu, G., Hu, D., Chen, M., Wang, C. and Wu, L. 2013. Multifunctional PNIPAM/Fe3O4–ZnS hybrid hollow spheres: Synthesis, characterization, and properties. J. Colloid Interface Sci. 397: 73–79.

[25] Jankiewicz, B. J., Jamiola, D., Choma, J. and Jaroniec, M. 2012. Silica–metal core–shell nanostructures. Advances in J. Colloid Interface Sci. 170: 28–47.

[26] Ambrogio, M. W., Frasconi, M., Yilmaz, M. D. and Chen, X. 2013. New methods for improved characterization of silica nanoparticle-based drug delivery systems. Langmuir. 29: 15386–15393.

[27] Richert, H., Surzhenko, O., Wangemann, S., Heinrich, J. and Gömert, P. 2005. Development of a magnetic capsule as a drug release system for future applications in the human GI tract. J. Magn. Magn. Mater. 293: 497–500.

[28] Mishima, F., Fujimoto, S., Takeda, S., Izumi, Y. and Nishijima, S. 2007. Development of control system for magnetically targeted drug delivery. J. Magn. Magn. Mater. 310: 2883–2885.

[29] Cao, Q., Han, X. and Li, L. 2011. Enhancement of the efficiency of magnetic targeting for drug delivery: Development and evaluation of magnet system. J. Magn. Magn. Mater. 323: 1919–1924.

[30] Takeda, S., Mishima, F., Fujimoto, S., Izumi, Y. and Nishijima, S. 2007. Development of magnetically targeted drug delivery system using superconducting magnet. J. Magn. Magn. Mater. 311: 367–371.

[31] Shapiro, B. 2009. Towards dynamic control of magnetic fields to focus magnetic carriers to targets deep inside the body. J. Magn. Magn. Mater. 321: 1594–1599.

[32] Vargha-Butler, E. I., Foldvari, M. and Mezei, M. 1989. Study of the sedimentation behaviour of liposomal drug delivery system. Colloids Surf. 42: 375–389.

[33] Müller-Schulte, D. and Schmitz-Rode, T. 2006. Thermosensitive magnetic polymer particles as contactless controllable drug carriers. J. Magn. Magn. Mater. 302: 267–271.

[34] Kim, D., Nikles, D. E., Johnson, D. T. and Brazel, C. S. 2008. Heat generation of aqueously dispersed CoFe2O4 nanoparticles as heating agents for magnetically activated drug delivery and hyperthermia. J. Magn. Magn. Mater. 320: 2390–2396.

[35] Shamim, N., Hidajat, L., Hong, K. and Uddin, M. S. 2006. Thermosensitive-polymer-coated magnetic nanoparticles: Adsorption and desorption of bovine serum albumin. J. Colloid Interface Sci. 304: 1–8.

[36] Liu, T., Hu, S., Liu, K., Shaiu, R., Liu, D. and Chen, S. 2008. Instantaneous drug delivery of magnetic/thermally sensitive nanospheres by a high-frequency magnetic field. Langmuir. 24: 13306–13311.
[37] Girginova, P. I., Daniel-da-Silva, A. L., Lopes, C. B., Figueira, P., Otero, M., Amaral, V. S., Pereira, E. and Trindade, T. 2010. Silica coated magnetite particles for magnetic removal of $Hg2^{+}$ from water. J. Colloid Interface Sci. 345: 234–240.
[38] Lan, S., Wu, X., Li, L., Li, M., Guo, F. and Gan, S. 2013. Synthesis and characterization of hyaluronic acid-supported magnetic microspheres for copper ions removal. Colloids Surf. A. 425: 42–50.
[39] Bruce, I. J. and Sen, T. 2005. Surface modification of magnetic nanoparticles with alkoxysilanes and their application in magnetic bioseparations. Langmuir. 21: 7029–7035.
[40] Jorgensen, F. 1996. The Complete Handbook of Magnetic Recording. 4th ed., McGraw-Hill, New York.
[41] Erb, R. M., Son, H. S., Samanta, B., Rotello, V. M. and Yellen, B. B. 2009. Magnetic assembly of colloidal superstructures with multipole symmetry. Nature Lett. 457: 999–1002.
[42] Chikazumi, S. 1997. Physics of Ferromagnetism. 2nd ed., Oxford Science Publications, London.
[43] Jiles, D. 1991. Introduction to Magnetism and Magnetic Materials, Chapman and Hall, London.
[44] Gubin, S. P. (ed.). 2009. Magnetic Nanoparticles, Wiley-VCH, Weinheim.
[45] Frenkel, J. 1955. Kinetic Theory of Liquids, Dover, New York.
[46] Néel, L. 1955. Some theoretical aspects of rock-magnetism. Advan. Phys. 4: 191–243.
[47] Matijevic, E. and Scheiner, P. 1978. Ferric hydrous oxide sols: III. Preparation of uniform particles by hydrolysis of Fe(III)-chloride, -nitrate, and -perchlorate solutions. J. Colloid Interface Sci. 63: 509–524.
[48] Muench, G. J., Arajs, S. and Matijević, E. 1981. Magnetic properties of monodispersed submicromic α-Fe2O3 particles. J. Appl. Phys. 52: 2493–2495.
[49] Ozaki, M., Kratohvil, S. and Matijevic, E. 1984. Formation of monodispersed spindle-type hematite particles. J. Colloid Interface Sci. 102: 146–151.
[50] Ozaki, M., Ookoshi, N. and Matijević, E. 1990. Preparation and magnetic properties of uniform hematite platelets. J. Colloid Interface Sci. 137: 546–549.
[51] Li, J., Qin, Y., Kou, X. and Huang, J. 2004. The microstructure and magnetic properties of Ni anoplatelets. Nanotech. 15: 982–986.
[52] Sugimoto, T., Itoh, H. and Mochida, T. 1998. Shape control of monodisperse hematite particles by organic additives in the gel-sol system. J. Colloid Interface Sci. 205: 42–52.
[53] Raming, T. P., Winnubst, A. J. A., Van Kats, C. M. and Philipse, A. P. 2002. The synthesis and magnetic properties of nanosized hematite (α-Fe2O3) particles. J. Colloid Interface Sci. 249: 346–350.
[54] Ozaki, M., Senna, M., Koishi, M. and Honda, H. 2000. Particles of specific functions. pp. 662–682. In: T. Sugimoto (ed.). Fine Particles. Marcel Dekker, New York.
[55] Van der Beek, D., Reich, H., Van der Schoot, P., Dijkstra, M., Schilling, T., Vink, R., Schmidt, M., Van Roij, R. and Lekkerkerker, H. 2006. Isotropic-nematic interface and wetting in suspensions of colloidal platelets. Phys. Rev. Lett. 97: 087801.
[56] Aoshima, M., Ozaki, M. and Satoh, A. 2012. Structural analysis of self-assembled lattice structures composed of cubic hematite particles. J. Phys. Chem. 116: 17862–17871.
[57] Satoh, A. 2008. Three-dimensional Monte Carlo simulations of internal aggregate structure in a colloidal dispersion composed of rod-like particles with magnetic moment normal to the particle axis. J. Colloid Interface Sci. 318: 68–81.

[58] Satoh, A. and Ozaki, M. 2006. Transport coefficients and orientational distributions of spheroidal particles with magnetic moment normal to the particle axis: Analysis for an applied magnetic field normal to the shear plane. J. Colloid Interface Sci. 298: 957–966.
[59] Satoh, A. and Sakuda, Y. 2007. Negative viscosity due to magnetic properties of a non-dilute suspension composed of ferromagnetic rod-like particles with magnetic moment normal to the particle axis. Molec. Phys. 105: 3145–3153.
[60] Satoh, A. 2013. Influence of the spin Brownian motion on the negative magneto-rheological effect in a rod-like hematite particle suspension. Molec. Phys. 111: 1042–1052.
[61] Sakuda, Y., Aoshima, M. and Satoh, A. 2012. Negative magneto-rheological effect of a dispersion composed of spindle-like hematite particles. Molec. Phys. 110: 1429–1435.
[62] Piet, D. L., Straube, A. V., Snezhko, A. and Aranson, I. S. 2013. Model of dynamic self-assembly in ferromagnetic suspensions at liquid interfaces. Phys. Rev. E. 88: 033024.
[63] Streekumari, A. and Ilg, P. 2013. Slow relaxation in structure-forming ferrofluids. Phys. Rev. E. 88: 042315.
[64] Horák, D., Babič, M., Macková, H. and Beneš, M. J. 2007. Preparation and properties of magnetic nano- and microsized particles for biological and environmental separations. J. Separation Sci. 30: 1751–1772.
[65] De Vicente, J., Segovia-Gutiérrez, J. P., Andablo-Reyes, E., Vereda, F. and Hidalgo-Álvarez, R. 2009. Dynamic rheology of sphere- and rod-based magnetorheological fluids. J. Chem. Phys. 131: 194902.
[66] Ivanov, A. S. and Pshenichnikov, A. F. 2014. Vortex flows induced by drop-like aggregate drift in magnetic fluids. Phys. Fluids. 26: 012002.
[67] Alvarez, C. E. and Klapp, S. H. L. 2013. Translational and rotational dynamics in suspensions of magnetic nanorods. Soft Matter. 9: 8761–8770.
[68] Sanchez, J. H. and Rinaldi, C. 2009. Rotational Brownian dynamics simulations of non-interacting magnetized ellipsoidal particles in d.c. and a.c. magnetic fields. J. Magn. Magn. Mater. 321: 2985–2991.
[69] Gil-Villegas, A., Jackson, G. and McGrother, S. C. 1998. Computer simulation of dipolar liquid crystals. J. Mol. Liquids. 76: 171–181.
[70] Kuzhir, P., Magnet, C., Bossis, G., Meunier, A. and Bashtovoi, V. 2011. Rotational diffusion may govern the rheology of magnetic suspensions. J. Rheology. 55: 1297–1318.
[71] Dussi, S., Rovigatti, L. and Sciortino, F. 2013. On the gas–liquid phase separation and the self-assembly of charged soft dumbbells. Molec. Phys. 111: 3608–3617.
[72] Williamson, D. C., Thacker, N. A. and Williams, S. R. 2005. Effects of intramolecular dipolar coupling on the isotropic-nematic phase transition of a hard spherocylinder fluid. Phys. Rev. E. 71: 021702.
[73] Kwaadgras, B. W., Van Roij, R. and Dijkstra, M. 2014. Self-consistent electric field-induced dipole interaction of colloidal spheres, cubes, rods, and dumbbells. J. Chem. Phys. 140: 154901.
[74] Alvarez, C. E. and Klapp, S. H. L. 2012. Percolation and orientational ordering in systems of magnetic nanorods. Soft Matter. 8: 7480–7489.
[75] Puri, I. K. and Ganguly, R. 2014. Particle transport in therapeutic magnetic fields. Ann. Rev. Fluid Mech. 46: 407–440.
[76] Kimura, T., Okamoto, S. and Uemura, T. 2007. Magnetic alignment of magnetically isotropic diamagnetic rod-like particle in modulated magnetic field. Jpn. J. App. Phys. 46: 586–588.
[77] Rosensweig, R. E. 1985. Ferrohydrodynamics, Cambridge University Press, Cambridge.

[78] Lopez-Lopez, M. T., Kuzhir, P., Caballero-Hernandz, J., Rodriguez-Arco, L., Duran, J. D. G. and Bossis, G. 2012. Yield stress in magnetorheological suspensions near the limit of maximum-packing fraction. J. Reology. 56: 1209–1224.
[79] Vargha-Butler, E. I., Foldvari, M. and Mezei, M. 1989. Study of the sedimentation behaviour of liposomal drug delivery system. Colloids Surf. 42: 375–389.
[80] Aoshima, M., Satoh, A. and Chantrell, R. W. 2008. Influence of perpendicular external magnetic field on microstructures of monolayer composed of ferromagnetic particles: Analysis by means of quasi-two-dimensional Monte Carlo simulation. J. Colloid Interface Sci. 323: 158–168.
[81] Satoh, A., Sakuda, Y. and Katayama, Y. 2009. Two-dimensional Monte Carlo simulations of aggregation phenomena in ferromagnetic colloidal dispersion composed of rod-like hematite particles. Inter. J. Emerg. Multidiscip. Fluid Sci. 1: 127–139.
[82] Satoh, A. and Sakuda, Y. 2009. Quasi-2D Monte Carlo simulations of a colloidal dispersion composed of magnetic plate-like particles with magnetic moment normal to the particle axis. Molec. Phys. 107: 1621–1627.
[83] Satoh, A. and Sakuda, Y. 2010. Quasi-2D Monte Carlo simulations of phase transitions and internal aggregate structures in a highly dense suspension composed of magnetic plate-like particles. Molec. Phys. 108: 2105–2113.
[84] Hayasaka, R. and Satoh, A. 2008. Brownian dynamics simulations of sedimentation phenomena of ferromagnetic spherical particles in a colloidal dispersion. Proceedings of 2008 ASME International Mechanical Engineering Congress and Exposition. pp. 125–130, Boston, U.S.A., October 31–November 6.
[85] Satoh, A. 2003. Introduction to Molecular-Microsimulation of Colloidal Dispersions, Elsevier, Amsterdam.
[86] Allen, M. P. and Tildesley, D. J. 1987. Computer Simulation of Liquids, Clarendon Press, Oxford.
[87] Satoh, A. 2010. Introduction to Practice of Molecular Simulation: Molecular Dynamics, Monte Carlo, Brownian Dynamics, Lattice Boltzmann and Dissipative Particle Dynamics, Elsevier Insights, Amsterdam.
[88] Bossis, G. and Brady, J. F. 1984. Dynamic simulation of sheared suspensions. I. General method. J. Chem. Phys. 80: 5141–5154.
[89] Brady, J. F. and Bossis, G. 1985. The rheology of concentrated suspensions of spheres in simple shear flow by numerical simulation. J. Fluid Mech. 155: 105–129.
[90] Succi, S. 2001. The Lattice Boltzmann Equation for Fluid Dynamics and Beyond, Clarendon Press, Oxford.
[91] Rothman, D. H. and Zaleski, S. 1997. Lattice-Gas Cellular Automata, Simple Models of Complex Hydrodynamics, Cambridge University Press, Cambridge.
[92] Rivet, J. -P. and Boon, J. P. 2001. Lattice Gas Hydrodynamics, Cambridge Univ. Press, Cambridge.
[93] Chopard, B. and Droz, M. 1998. Cellular Automata Modeling of Physical Systems, Cambridge University Press, Cambridge.
[94] Ladd, A. J. C. 1994. Numerical simulations of particulate suspensions via a discretized Boltzmann equation. Part 1. Theoretical foundation. J. Fluid Mech. 271: 285–309.
[95] Ladd, A. J. C. and Verberg, R. 2001. Lattice-Boltzmann simulations of particle-fluid suspensions. J. Stat. Phys. 104: 1191–1251.
[96] Satoh, A. 2012. On the method of activating the Brownian motion for application of the lattice Boltzmann method to magnetic particle dispersions. Molec. Phys. 110: 1–15.
[97] Kim, E., Stratford, K., Camp, P. J. and Cates, M. E. 2009. Hydrodynamic interactions in colloidal ferrofluids: A lattice Boltzmann study. J. Phys. Chem. B 113: 3681–3693.

CHAPTER 2

# Forces, Energies and Torques Acting on Magnetic Particles

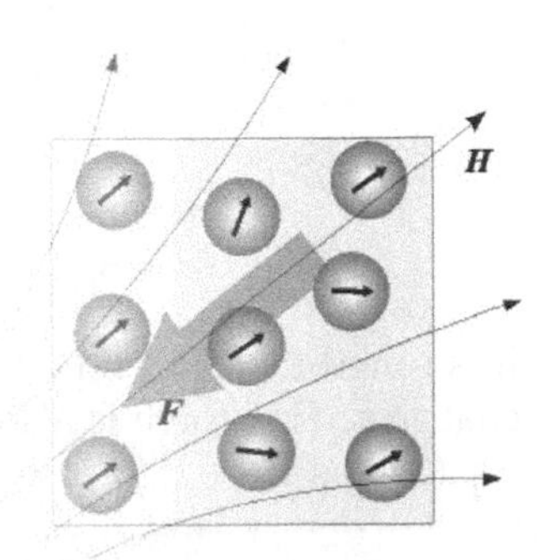


## 2.1 Similarity to electrostatic expressions

In the following, we show expressions for forces, energies, torques and other physical quantities related to interactions between magnetic particles and interactions with an external magnetic field. Although a single independent magnetic charge is not found to exist, the concept of this magnetic charge, similar to electric charge, is quite useful to derive expressions for magnetic forces, torques and energies acting between two magnetic particles [1]. Hence, we employ the concept of single positive and negative magnetic charges in a derivation procedure similar to that for an electrically charged particle system in order to derive the expressions below.

A magnetic charge $q$ induces a magnetic field at the relative position $\boldsymbol{r}$ from the magnetic charge:

$$\boldsymbol{H}^{(ind)} = \frac{q}{4\pi r^2} \cdot \frac{\boldsymbol{r}}{r} \tag{2.1}$$

in which $r$ is the magnitude of the vector $\boldsymbol{r}$, expressed as $r = |\boldsymbol{r}|$. If a single magnetic charge $q$ is placed in an applied field $\boldsymbol{H}$, a force $\boldsymbol{F}$ acts at the center of the magnetic charge due to an interaction between the charge and the field as

$$\boldsymbol{F} = \mu_0 q \boldsymbol{H} \tag{2.2}$$

Hence, a force $\boldsymbol{F}$ and an interaction energy $U$ acting between a magnetic charge $q$ and another magnetic charge $q'$ at the relative position $\boldsymbol{r}$ from the first charge, are expressed from Eqs. (2.1) and (2.2), respectively, as

$$\boldsymbol{F} = \mu_0 q' \boldsymbol{H}^{(ind)} = \frac{\mu_0 q q'}{4\pi r^2} \cdot \frac{\boldsymbol{r}}{r} \tag{2.3}$$

$$U = \frac{\mu_0 q q'}{4\pi r} \tag{2.4}$$

It is noted that $\boldsymbol{F}$ in Eq. (2.3) is a force acting on the charge $q'$ by the first charge $q$. This force is also derived from the relationship $\boldsymbol{F} = -\nabla U$ using Eq. (2.4) as

$$\boldsymbol{F} = -\nabla U = -\left( \boldsymbol{i}\frac{\partial U}{\partial x} + \boldsymbol{j}\frac{\partial U}{\partial y} + \boldsymbol{k}\frac{\partial U}{\partial z} \right)$$

$$= -\left( \boldsymbol{i}\frac{dU}{dr} \cdot \frac{\partial r}{\partial x} + \boldsymbol{j}\frac{dU}{dr} \cdot \frac{\partial r}{\partial y} + \boldsymbol{k}\frac{dU}{dr} \cdot \frac{\partial r}{\partial z} \right) = \frac{\mu_0 q q'}{4\pi r^2} \cdot \frac{\boldsymbol{r}}{r} \tag{2.5}$$

In this derivation, $\boldsymbol{r} = (x, y, z)$ and the relationship of $(\partial r/\partial x, \partial r/\partial y, \partial r/\partial z) = (x/r, y/r, z/r)$ have been used under the assumption that the magnetic charges are constant. The vectors $(\boldsymbol{i}, \boldsymbol{j}, \boldsymbol{k})$ are the unit vectors in each axis direction.

Next we consider an idealized rod-like particle that has a positive and a negative point magnetic charge at each end of the magnet between distance $l_0$, as shown in Fig. 2.1. The origin of an orthogonal coordinate system is taken at the center of the magnet. If the orientation of the magnet is described by $\boldsymbol{e}$, then the positions of the positive and the negative magnetic charge, $\boldsymbol{r}^+$ and $\boldsymbol{r}^-$, are expressed respectively as

$$\boldsymbol{r}^+ = (l_0/2)\boldsymbol{e}, \boldsymbol{r}^- = -(l_0/2)\boldsymbol{e} \tag{2.6}$$

This particle magnet is placed in a uniform magnetic field $\boldsymbol{H}$ and the force $\boldsymbol{F}$ acting on the magnet is evaluated by summing the forces $\boldsymbol{F}^+$ and $\boldsymbol{F}^-$ acting on the positive and negative charges as

$$\boldsymbol{F} = \boldsymbol{F}^+ + \boldsymbol{F}^- = \mu_0 q\boldsymbol{H} + \mu_0(-q)\boldsymbol{H} = 0 \tag{2.7}$$

It is seen from this equation that in the situation of a uniform magnetic field, no force acts on the magnet, which in this case is a rod-like particle, with a positive and a negative magnetic charge of the same magnitude

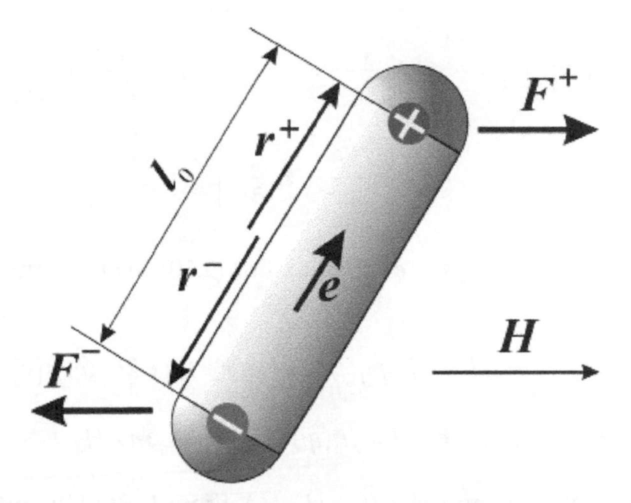


**Figure 2.1.** Magnet (magnetic rod-like particle) with positive and negative magnetic charges at both the ends in a uniform external magnetic field.

at the ends of the body. On the other hand, due to its inclination in the magnetic field, a torque $\boldsymbol{T}$ generally acts on the magnet. It is defined by the outer product of the position vector and the force, $(l_0\boldsymbol{e}/2) \times (\mu_0 q\boldsymbol{H})$, and therefore the torque is evaluated as

$$\boldsymbol{T} = \boldsymbol{r}^+ \times \boldsymbol{F}^+ + \boldsymbol{r}^- \times \boldsymbol{F}^- = (l_0\boldsymbol{e}/2) \times (\mu_0 q\boldsymbol{H}) + (-l_0\boldsymbol{e}/2) \times (-\mu_0 q\boldsymbol{H}) = \mu_0(ql_0\boldsymbol{e}) \times \boldsymbol{H} = \mu_0\boldsymbol{m} \times \boldsymbol{H} \tag{2.8}$$

in which the magnetic moment $\boldsymbol{m}$ is defined by $\boldsymbol{m} = ql_0\boldsymbol{e}$.

An interaction energy $U$ of the magnet $\boldsymbol{m} = (ql_0\boldsymbol{e})$ with a uniform applied magnetic field $\boldsymbol{H}$ is obtained by evaluating the work done in placing each magnetic charge from an infinity point along the opposite direction to the magnetic field direction. The difference in the path of positive and negative magnetic charges along the magnetic field direction is $l_0\boldsymbol{e} \cdot (\boldsymbol{H}/H)$, so that the interaction energy is obtained as

$$U = \mu_0 qH(-l_0\boldsymbol{e}/2)\cdot(\boldsymbol{H}/H) - \mu_0 qH(+l_0\boldsymbol{e}/2)\cdot(\boldsymbol{H}/H) = -\mu_0(ql_0\boldsymbol{e})\cdot\boldsymbol{H} = -\mu_0\boldsymbol{m}\cdot\boldsymbol{H} \tag{2.9}$$

in which $H$ is the magnitude of the magnetic field as $H = |\boldsymbol{H}|$. The expression in Eq. (2.8) is also obtained from the definition of a torque using Eq. (2.9). The definition of the torque is expressed as

$$T = -\left( e \times \frac{\partial}{\partial e} \right) U = -i \left( e_y \frac{\partial}{\partial e_z} - e_z \frac{\partial}{\partial e_y} \right) U - j \left( e_z \frac{\partial}{\partial e_x} - e_x \frac{\partial}{\partial e_z} \right) U$$

$$- k \left( e_x \frac{\partial}{\partial e_y} - e_y \frac{\partial}{\partial e_x} \right) U \tag{2.10}$$

Substitution of $U = -(\mu_0 q l_0)\, \boldsymbol{e} \cdot \boldsymbol{H}$ into Eq. (2.10) leads to the following equation:

$$\boldsymbol{T} = \boldsymbol{i}(\mu_0 q l_0)\left(e_y H_z - e_z H_y\right) + \boldsymbol{j}(\mu_0 q l_0)\left(e_z H_x - e_x H_z\right) + \boldsymbol{k}(\mu_0 q l_0)\left(e_x H_y - e_y H_x\right)$$

$$= (\mu_0 q l_0)\boldsymbol{e} \times \boldsymbol{H} = \mu_0 (q l_0 \boldsymbol{e}) \times \boldsymbol{H} = \mu_0\, \boldsymbol{m} \times \boldsymbol{H} \tag{2.11}$$

in which $\boldsymbol{e} = (e_x, e_y, e_z)$ and $\boldsymbol{H} = (H_x, H_y, H_z)$. From Eqs. (2.8) and (2.9), it is seen that the magnetic moment of the magnet or the magnetic particle tends to incline in the magnetic field direction.

As already pointed out, no force acts on the magnet in a uniform magnetic field. However, if an external magnetic field is not uniform but is as a function of the position, a body force will act on the magnetic body. We now consider the body force acting on the magnet with a constant positive and negative magnetic charge at each end in a non-uniform magnetic field. If the position of the center of the magnetic body is denoted by $\boldsymbol{r}$, then the positions of the positive and the negative magnetic charge, $\boldsymbol{r}^+$ and $\boldsymbol{r}^-$, are expressed, respectively, as

$$\boldsymbol{r}^+ = \boldsymbol{r} + (l_0/2)\boldsymbol{e},\ \boldsymbol{r}^- = \boldsymbol{r} - (l_0/2)\boldsymbol{e} \tag{2.12}$$

The magnetic fields at the positions $\boldsymbol{r}^+$ and $\boldsymbol{r}^-$ are approximately expressed as

$$\begin{aligned} \boldsymbol{H}(\boldsymbol{r}^+) &= \boldsymbol{H}(\boldsymbol{r} + (l_0/2)\boldsymbol{e}) \approx \boldsymbol{H}(\boldsymbol{r}) + (l_0/2)\boldsymbol{e} \cdot \nabla \boldsymbol{H}, \\ \boldsymbol{H}(\boldsymbol{r}^-) &= \boldsymbol{H}(\boldsymbol{r} - (l_0/2)\boldsymbol{e}) \approx \boldsymbol{H}(\boldsymbol{r}) - (l_0/2)\boldsymbol{e} \cdot \nabla \boldsymbol{H} \end{aligned} \tag{2.13}$$

Hence, according to the procedure similar to Eq. (2.7), the force acting on the magnetic body is obtained as

$$\boldsymbol{F} = \boldsymbol{F}^+ + \boldsymbol{F}^- = \mu_0 q(\boldsymbol{H}(\boldsymbol{r}) + (l_0/2)\boldsymbol{e}) \cdot \nabla \boldsymbol{H}) + \mu_0(-q)(\boldsymbol{H}(\boldsymbol{r}) - (l_0/2)\boldsymbol{e}) \cdot \nabla \boldsymbol{H})$$
$$= \mu_0 (q l_0\, \boldsymbol{e}) \cdot \nabla \boldsymbol{H} = \mu_0 \boldsymbol{m} \cdot \nabla \boldsymbol{H} = \mu_0 (\boldsymbol{m} \cdot \nabla)\boldsymbol{H} \tag{2.14}$$

## 2.2 Magnetic particle-particle and particle-field interactions

In this section, we consider the interaction between two magnetic particles and also between a magnetic particle and a magnetic field. We address a magnetic body with a constant positive and negative magnetic charge at each end. As before, if the position of the center of the magnetic body is taken as the origin of an orthogonal coordinate system, the positions of the positive and the negative magnetic charge, $\boldsymbol{r}^+$ and $\boldsymbol{r}^-$, are expressed in Eq. (2.6). These positive and negative magnetic charges induce a magnetic field $\boldsymbol{H}$ at an arbitrary position $\boldsymbol{r}$ as

$$\boldsymbol{H}(\boldsymbol{r}) = \frac{q}{4\pi}\cdot\frac{\boldsymbol{r}-\boldsymbol{r}^+}{\left|\boldsymbol{r}-\boldsymbol{r}^+\right|^3} - \frac{q}{4\pi}\cdot\frac{\boldsymbol{r}-\boldsymbol{r}^-}{\left|\boldsymbol{r}-\boldsymbol{r}^-\right|^3} = \frac{q}{4\pi}\left\{\frac{\boldsymbol{r}-(l_0/2)\boldsymbol{e}}{\left|\boldsymbol{r}-(l_0/2)\boldsymbol{e}\right|^3} - \frac{\boldsymbol{r}+(l_0/2)\boldsymbol{e}}{\left|\boldsymbol{r}+(l_0/2)\boldsymbol{e}\right|^3}\right\} \tag{2.15}$$

If $r\ (=|\boldsymbol{r}|)$ is much larger than $l_0$, then the denominator is approximately expressed as

$$\frac{1}{\left|\boldsymbol{r}-(l_0/2)\boldsymbol{e}\right|^3} = \left\{r^2 - l_0(\boldsymbol{r}\cdot\boldsymbol{e}) + l_0^2/4\right\}^{-3/2} \approx \frac{1}{r^3}\left\{1+\frac{3l_0}{2r^2}(\boldsymbol{r}\cdot\boldsymbol{e})\right\} \tag{2.16}$$

A similar approximate expression is obtained for the second term on the right-hand side in Eq. (2.15). Substitution of these approximate expressions into Eq. (2.15) leads to

$$\boldsymbol{H}(\boldsymbol{r}) \approx \frac{q}{4\pi}\left\{(\boldsymbol{r}-(l_0/2)\boldsymbol{e})(\frac{1}{r^3}(1+(3l_0/(2r^2))(\boldsymbol{r}\cdot\boldsymbol{e}))) \right.$$

$$\left. -(\boldsymbol{r}+(l_0/2)\boldsymbol{e})(\frac{1}{r^3}(1-(3l_0/(2r^2))(\boldsymbol{r}\cdot\boldsymbol{e})))\right\}$$

$$= \frac{1}{4\pi r^3}\left\{-(ql_0)\boldsymbol{e} + \frac{3}{r^2}(ql_0)(\boldsymbol{r}\cdot\boldsymbol{e})\boldsymbol{r}\right\} = \frac{1}{4\pi r^3}\left\{-\boldsymbol{m} + \frac{3}{r^2}(\boldsymbol{r}\cdot\boldsymbol{m})\boldsymbol{r}\right\} \tag{2.17}$$

Hence, if subscript 1 is attached to the quantities of the present magnetic body, an interaction energy $U_{12}$ with another magnetic body, subscript 2, with the magnetic moment $\boldsymbol{m}_2$ at the position $\boldsymbol{r}$ is obtained from Eqs. (2.9) and (2.17) as

$$U_{12} = -\mu_0 \boldsymbol{m}_2 \cdot \boldsymbol{H}(\boldsymbol{r}) = \frac{\mu_0}{4\pi r^3}\left\{\boldsymbol{m}_1 \cdot \boldsymbol{m}_2 - \frac{3}{r^2}(\boldsymbol{r}\cdot\boldsymbol{m}_1)(\boldsymbol{r}\cdot\boldsymbol{m}_2)\right\} \tag{2.18}$$

It is noted that this equation is valid under the situation where $r$ is much larger than $l_0$.

If the assumption of $r$ being much larger than $l_0$ is not satisfied, the interaction energy $U_{ij}$ between the two magnetic bodies $i$ and $j$, shown in Fig. 2.2, is expressed from Eq. (2.4) as

$$U_{ij} = \frac{\mu_0 q^2}{4\pi}\left\{\frac{1}{|\boldsymbol{r}_i^+ - \boldsymbol{r}_j^+|} - \frac{1}{|\boldsymbol{r}_i^+ - \boldsymbol{r}_j^-|} - \frac{1}{|\boldsymbol{r}_i^- - \boldsymbol{r}_j^+|} + \frac{1}{|\boldsymbol{r}_i^- - \boldsymbol{r}_j^-|}\right\}$$

$$= \frac{\mu_0 q^2}{4\pi}\left\{\frac{1}{|\boldsymbol{r}_{ij} + l_0\boldsymbol{e}_{ij}/2|} - \frac{1}{|\boldsymbol{r}_{ij} + l_0(\boldsymbol{e}_i + \boldsymbol{e}_j)/2|} - \frac{1}{|\boldsymbol{r}_{ij} - l_0(\boldsymbol{e}_i + \boldsymbol{e}_j)/2|} + \frac{1}{|\boldsymbol{r}_{ij} - l_0\boldsymbol{e}_{ij}/2|}\right\} \tag{2.19}$$

In this equation, the following notation has been employed. The center position and the orientation of particle $i$ are denoted by $\boldsymbol{r}_i$ and $\boldsymbol{e}_i$, respectively, the position vectors of the positive and negative magnetic charges are denoted by $\boldsymbol{r}_i^+ = \boldsymbol{r}_i + (l_0/2)\boldsymbol{e}_i$ and $\boldsymbol{r}_i^- = \boldsymbol{r}_i - (l_0/2)\boldsymbol{e}_i$, and $\boldsymbol{r}_{ij} = \boldsymbol{r}_i - \boldsymbol{r}_j$

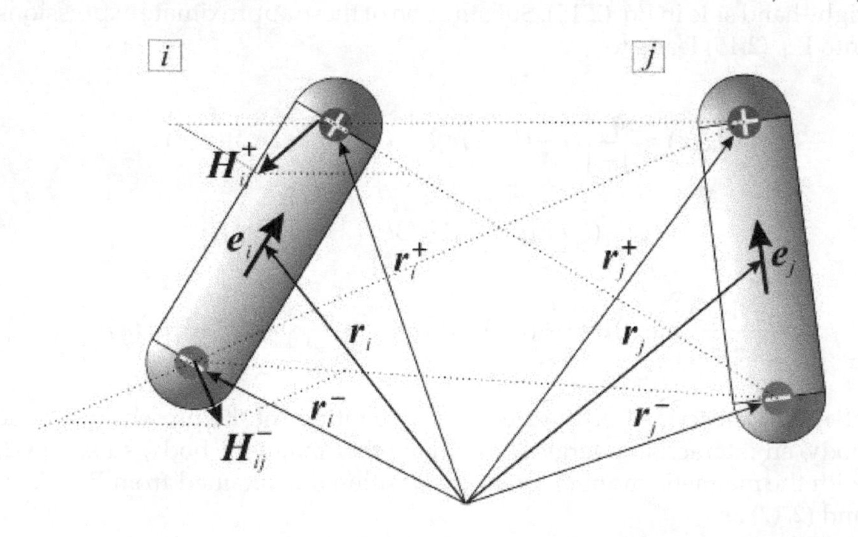


**Figure 2.2.** Interaction between two magnets (magnetic rod-like particles) with positive and negative magnetic charges at both the ends.

and $\boldsymbol{e}_{ij} = \boldsymbol{e}_i - \boldsymbol{e}_j$; similar notation is used for particle $j$. It is noted that the energy expression in Eq. (2.19) is obtained by summing the interaction energies for the four pairs of magnetic charges, i.e., the positive-positive, positive-negative, negative-positive, and negative-negative pairs.

Next we evaluate the force and torque acting on particle $i$ by particle $j$ for the magnetic charge model shown in Fig. 2.2. The magnetic field $\boldsymbol{H}_{ij}^{+}$ induced by particle $j$ at the position of the positive charge of particle $i$ is derived from Eq. (2.1) as

$$\boldsymbol{H}_{ij}^{+} = \frac{q}{4\pi}\cdot\frac{\boldsymbol{r}_i^{+} - \boldsymbol{r}_j^{+}}{\left|\boldsymbol{r}_i^{+} - \boldsymbol{r}_j^{+}\right|^3} - \frac{q}{4\pi}\cdot\frac{\boldsymbol{r}_i^{+} - \boldsymbol{r}_j^{-}}{\left|\boldsymbol{r}_i^{+} - \boldsymbol{r}_j^{-}\right|^3}$$

$$= \frac{q}{4\pi}\left\{\frac{\boldsymbol{r}_{ij} + l_0\boldsymbol{e}_{ij}/2}{\left|\boldsymbol{r}_{ij} + l_0\boldsymbol{e}_{ij}/2\right|^3} - \frac{\boldsymbol{r}_{ij} + l_0(\boldsymbol{e}_i + \boldsymbol{e}_j)/2}{\left|\boldsymbol{r}_{ij} + l_0(\boldsymbol{e}_i + \boldsymbol{e}_j)/2\right|^3}\right\} \tag{2.20}$$

The magnetic field $\boldsymbol{H}_{ij}^{-}$ induced at the position of the negative charge is obtained similarly. The force $\boldsymbol{F}_{ij}^{+}$ and $\boldsymbol{F}_{ij}^{-}$ and the torque $\boldsymbol{T}_{ij}^{+}$ and $\boldsymbol{T}_{ij}^{-}$ exerted by particle $j$ at the position of the positive and negative charge of particle $i$ are derived as

$$\boldsymbol{F}_{ij}^{+} = \mu_0 q\boldsymbol{H}_{ij}^{+},\quad \boldsymbol{F}_{ij}^{-} = -\mu_0 q\boldsymbol{H}_{ij}^{-},\quad \boldsymbol{T}_{ij}^{+} = (l_0/2)\boldsymbol{e}_i \times \boldsymbol{F}_{ij}^{+},\quad \boldsymbol{T}_{ij}^{-} = -(l_0/2)\boldsymbol{e}_i \times \boldsymbol{F}_{ij}^{-} \tag{2.21}$$

Hence, the force $\boldsymbol{F}_{ij}$ and torque $\boldsymbol{T}_{ij}$ exerted by particle $j$ are evaluated as

$$\boldsymbol{F}_{ij} = \boldsymbol{F}_{ij}^{+} + \boldsymbol{F}_{ij}^{-},\quad \boldsymbol{T}_{ij} = \boldsymbol{T}_{ij}^{+} + \boldsymbol{T}_{ij}^{-} \tag{2.22}$$

The action-reaction theorem yields $\boldsymbol{F}_{ji}^{(m)} = -\boldsymbol{F}_{ij}^{(m)}$.

It is simple to show that Eq. (2.19) reduces to Eq. (2.18) by means of a Maclaurin expansion if the separation $r_{ij}$ is much larger than $l_0$. The following formula is used for reforming each term on the right-hand side in Eq. (2.19).

$$(1+x)^{-1/2} = 1 - \frac{1}{2}x + \frac{(-1/2)(-3/2)}{2!}x^2 + O(x^3) \tag{2.23}$$

in which $x$ has been assumed to be much smaller than unity. From this Maclaurin expansion, the first term on the right-hand side in Eq. (2.19) can be expressed as

$$\frac{1}{|\boldsymbol{r}_i^+ - \boldsymbol{r}_j^+|} = \frac{1}{r_{ij}}\left[1 - \frac{l_0}{2r_{ij}^2}(\boldsymbol{r}_{ij}\cdot\boldsymbol{e}_{ij}) - \frac{l_0^2}{8r_{ij}^2}(\boldsymbol{e}_{ij}\cdot\boldsymbol{e}_{ij})\right.$$

$$\left. + \frac{3}{8}\left\{\frac{l_0^2}{r_{ij}^4}(\boldsymbol{r}_{ij}\cdot\boldsymbol{e}_{ij})^2 + \frac{l_0^3}{2r_{ij}^4}(\boldsymbol{e}_{ij}\cdot\boldsymbol{e}_{ij})(\boldsymbol{r}_{ij}\cdot\boldsymbol{e}_{ij})\right\} + O((l_0/r_{ij})^4)\right] \tag{2.24}$$

Similar equations are derived for the other terms in Eq. (2.19), and consequently the following relationships are obtained by neglecting higher order infinitesimal terms:

$$\frac{1}{|\boldsymbol{r}_i^+ - \boldsymbol{r}_j^+|} + \frac{1}{|\boldsymbol{r}_i^- - \boldsymbol{r}_j^-|} = \frac{1}{r_{ij}}\left[2 - \frac{l_0^2}{4r_{ij}^2}(\boldsymbol{e}_{ij}\cdot\boldsymbol{e}_{ij}) + \frac{3l_0^2}{4r_{ij}^4}(\boldsymbol{r}_{ij}\cdot\boldsymbol{e}_{ij})^2\right] \tag{2.25}$$

$$\frac{1}{|\boldsymbol{r}_i^+ - \boldsymbol{r}_j^-|} + \frac{1}{|\boldsymbol{r}_i^- - \boldsymbol{r}_j^+|} = \frac{1}{r_{ij}}\left[2 - \frac{l_0^2}{4r_{ij}^2}|\boldsymbol{e}_i + \boldsymbol{e}_j|^2 + \frac{3l_0^2}{4r_{ij}^4}(\boldsymbol{r}_{ij}\cdot(\boldsymbol{e}_i + \boldsymbol{e}_j))^2\right] \tag{2.26}$$

Substitution of Eqs. (2.25) and (2.26) into Eq. (2.19) and consideration of $\boldsymbol{m}_i = ql_0\boldsymbol{e}_i$ and $\boldsymbol{m}_j = ql_0\boldsymbol{e}_j$ leads to Eq. (2.18), that is,

$$U_{ij} = \frac{\mu_0}{4\pi r_{ij}^3}\left\{q^2l_0^2(\boldsymbol{e}_i\cdot\boldsymbol{e}_j) - \frac{3q^2l_0^2}{r_{ij}^2}(\boldsymbol{r}_{ij}\cdot\boldsymbol{e}_i)(\boldsymbol{r}_{ij}\cdot\boldsymbol{e}_j)\right\}$$

$$= \frac{\mu_0}{4\pi r_{ij}^3}\left\{\boldsymbol{m}_i\cdot\boldsymbol{m}_j - \frac{3}{r_{ij}^2}(\boldsymbol{r}_{ij}\cdot\boldsymbol{m}_i)(\boldsymbol{r}_{ij}\cdot\boldsymbol{m}_j)\right\} \tag{2.27}$$

Next, we derive expressions for the force $\boldsymbol{F}_{ij}$ and torque $\boldsymbol{T}_{ij}$ acting on particle $i$ by particle $j$ for the situation of the separation being much larger than $l_0$. A procedure similar to the derivation of Eq. (2.27) will be significantly complex, and therefore we employ another approach based on Eqs. (2.5) and (2.10).

Employing the notation $\boldsymbol{r}_{ij} = (x_{ij}, y_{ij}, z_{ij})$, $\boldsymbol{e}_i = (e_{xi}, e_{yi}, e_{zi})$ (similarly $\boldsymbol{e}_j$), $\boldsymbol{F}_i = (F_{xi}, F_{yi}, F_{zi})$, $\boldsymbol{T}_i = (T_{xi}, T_{yi}, T_{zi})$, $\partial/\partial\boldsymbol{r} = \boldsymbol{i}\partial/\partial x + \boldsymbol{j}\partial/\partial y + \boldsymbol{k}\partial/\partial z$ and $\partial/\partial\boldsymbol{e} = \boldsymbol{i}\partial/\partial e_x + \boldsymbol{j}\partial/\partial e_y + \boldsymbol{k}\partial/\partial e_z$, the $x$-component of the force, $F_{xij}$, is evaluated from Eqs. (2.5) and (2.27) as

$$F_{xij}=-\frac{\partial U_{ij}}{\partial x_i}=-\frac{\partial U_{ij}}{\partial x_{ij}}=-\frac{\mu_0}{4\pi}\cdot\frac{-3}{r_{ij}^4}\cdot\frac{x_{ij}}{r_{ij}}\left\{\boldsymbol{m}_i\cdot\boldsymbol{m}_j-\frac{3}{r_{ij}^2}(\boldsymbol{r}_{ij}\cdot\boldsymbol{m}_i)(\boldsymbol{r}_{ij}\cdot\boldsymbol{m}_j)\right\}$$

$$-\frac{\mu_0}{4\pi}\cdot\frac{1}{r_{ij}^3}\left\{\frac{6}{r_{ij}^3}\cdot\frac{x_{ij}}{r_{ij}}(\boldsymbol{r}_{ij}\cdot\boldsymbol{m}_i)(\boldsymbol{r}_{ij}\cdot\boldsymbol{m}_j)-\frac{3}{r_{ij}^2}m_{xi}(\boldsymbol{r}_{ij}\cdot\boldsymbol{m}_j)-\frac{3}{r_{ij}^2}m_{xj}(\boldsymbol{r}_{ij}\cdot\boldsymbol{m}_i)\right\}$$

$$=-\frac{3\mu_0}{4\pi r_{ij}^4}\cdot\frac{x_{ij}}{r_{ij}}\left\{-\boldsymbol{m}_i\cdot\boldsymbol{m}_j+\frac{5}{r_{ij}^2}(\boldsymbol{r}_{ij}\cdot\boldsymbol{m}_i)(\boldsymbol{r}_{ij}\cdot\boldsymbol{m}_j)\right\}$$

$$-\frac{3\mu_0}{4\pi r_{ij}^4}\left\{-\frac{1}{r_{ij}}m_{xi}(\boldsymbol{r}_{ij}\cdot\boldsymbol{m}_j)-\frac{1}{r_{ij}}m_{xj}(\boldsymbol{r}_{ij}\cdot\boldsymbol{m}_i)\right\} \quad (2.28)$$

In this derivation, the relationship of $\partial r_{ij}/\partial x_{ij}=x_{ij}/r_{ij}$ has been used. Hence, the force $\boldsymbol{F}_{ij}$ acting on particle *i* by particle *j* is obtained by combining *x*-, *y*- and *z*-components as

$$\boldsymbol{F}_{ij}=\boldsymbol{i}F_{xij}+\boldsymbol{j}F_{yij}+\boldsymbol{k}F_{zij}=-\frac{3\mu_0}{4\pi r_{ij}^4}\left[\left\{-\boldsymbol{m}_i\cdot\boldsymbol{m}_j+\frac{5}{r_{ij}^2}(\boldsymbol{r}_{ij}\cdot\boldsymbol{m}_i)(\boldsymbol{r}_{ij}\cdot\boldsymbol{m}_j)\right\}\frac{\boldsymbol{r}_{ij}}{r_{ij}}\right.$$

$$\left.-\frac{1}{r_{ij}}(\boldsymbol{r}_{ij}\cdot\boldsymbol{m}_j)\boldsymbol{m}_i-\frac{1}{r_{ij}}(\boldsymbol{r}_{ij}\cdot\boldsymbol{m}_i)\boldsymbol{m}_j\right] \quad (2.29)$$

The torque $\boldsymbol{T}_{ij}$ exerted by particle *j* is derived from the definition in Eq. (2.10). That is,

$$\boldsymbol{T}_i=-\left(\boldsymbol{e}_i\times\frac{\partial}{\partial\boldsymbol{e}_i}\right)U_{ij}=-\boldsymbol{i}\left(e_{yi}\frac{\partial U_{ij}}{\partial e_{zi}}-e_{zi}\frac{\partial U_{ij}}{\partial e_{yi}}\right)$$

$$-\boldsymbol{j}\left(e_{zi}\frac{\partial U_{ij}}{\partial e_{xi}}-e_{xi}\frac{\partial U_{ij}}{\partial e_{zi}}\right)-\boldsymbol{k}\left(e_{xi}\frac{\partial U_{ij}}{\partial e_{yi}}-e_{yi}\frac{\partial U_{ij}}{\partial e_{xi}}\right) \quad (2.30)$$

The substitution of the expression of Eq. (2.27) into Eq. (2.30) will, after accomplishment of the mathematical manipulation, give rise to the expression for the torque. For simplicity, the *x*-component $T_{xi}$ is first derived in the following. That is,

$$T_{xi} = -\left( e_{yi}\frac{\partial U_{ij}}{\partial e_{zi}} - e_{zi}\frac{\partial U_{ij}}{\partial e_{yi}} \right) = -\frac{\mu_0 m^2}{4\pi r_{ij}^3}\left[ e_{yi}\left\{ e_{zj} - \frac{3}{r_{ij}^2}(\boldsymbol{r}_{ij}\cdot\boldsymbol{e}_j)z_{ij} \right\} - e_{zi}\left\{ e_{yj} - \frac{3}{r_{ij}^2}(\boldsymbol{r}_{ij}\cdot\boldsymbol{e}_j)y_{ij} \right\} \right]$$

$$= -\frac{\mu_0 m^2}{4\pi r_{ij}^3}\left\{ (e_{yi}e_{zj} - e_{zi}e_{yj}) - \frac{3}{r_{ij}^2}(\boldsymbol{r}_{ij}\cdot\boldsymbol{e}_j)(e_{yi}z_{ij} - e_{zi}y_{ij}) \right\} \tag{2.31}$$

Combining $x$-, $y$- and $z$-components gives rise to the final expression for the torque $\boldsymbol{T}_{ij}$ as

$$\boldsymbol{T}_{ij} = -\frac{\mu_0 m^2}{4\pi r_{ij}^3}\left\{ \boldsymbol{e}_i \times \boldsymbol{e}_j - \frac{3}{r_{ij}^2}(\boldsymbol{r}_{ij}\cdot\boldsymbol{e}_j)\, \boldsymbol{e}_i \times \boldsymbol{r}_{ij} \right\} = -\frac{\mu_0}{4\pi r_{ij}^3}\left\{ \boldsymbol{m}_i \times \boldsymbol{m}_j - \frac{3}{r_{ij}^2}(\boldsymbol{r}_{ij}\cdot\boldsymbol{m}_j)\, \boldsymbol{m}_i \times \boldsymbol{r}_{ij} \right\} \tag{2.32}$$

## 2.3 Repulsive interaction due to overlap of the steric layers

A surfactant (steric) layer attached at the surface of each magnetic particle in a suspension may be a key factor to obtain a stable dispersion by preventing the magnetic particles from aggregating and sedimenting in the gravity field. We consider here the repulsive interaction arising from the overlap of the steric layers of two magnetic particles. However, for this interaction, it is significantly difficult to derive a mathematical expression in the case of non-spherical particles, and therefore we initially concentrate on spherical particles coated with a uniform steric layer.

An expression for a steric repulsion may be derived by considering the motion of surfactant molecules in terms of the theory of entropy. Since surfactant molecules are regarded as linear chain-like polymer with a large aspect ratio, here we idealize the surfactant molecules as a long rigid rod. For analysis, it is assumed that these rods are attached to the magnetic spherical particles by a hinge and they can rotate freely in the region of a hemisphere centered about the attachment point of each rod when there is no overlap with the surfactant layer of another particle. In the following, we derive the well known expression for the interaction between the two equal-sized spheres covered by a uniform steric layer according to the study by Rosensweig et al. [2].

As shown in Fig. 2.3, it is assumed that a plane wall is coated with surfactant molecules modeled as a rigid rod anchored at the corresponding free hinge at the material surface and these two walls are located in a face to face and parallel situation. Employing the notation $\delta$ for the length of the rod, $n_s$ for the number of surfactant molecules (or rods) per unit area and $s$ for the distance between the two wall surfaces, then the interaction energy of the two walls in the situation of overlap of the surfactant layers is evaluated from the decrease in the number of available microscopic (orientational) states. If the compression rate $\kappa$ is defined as

$$\kappa = (\delta - s/2)/\delta \tag{2.33}$$

then the interaction energy $E^{(wall)}$ due to the overlap of the steric rods per unit area per one wall surface is derived as [3]

$$E^{(wall)} = n_s k_B T\kappa = n_s k_B T(\delta - s/2)/\delta \tag{2.34}$$

in which $k_B$ is Boltzmann's constant and $T$ is the liquid temperature. It is noted that the compression rate $\kappa$ is equal to the surface area ratio of the head portion of the hemisphere cut by the center plane of the two walls, $S_{cap}$, to the whole hemisphere, $S_0$ (= $2\pi\delta^2$): that is, $\kappa = S_{cap}/S_0$. In the evaluation of the steric interaction between the two spherical particles, the key quantity is the compression rate defined by Eq. (2.33). In the following, we derive an expression for the steric interaction energy due

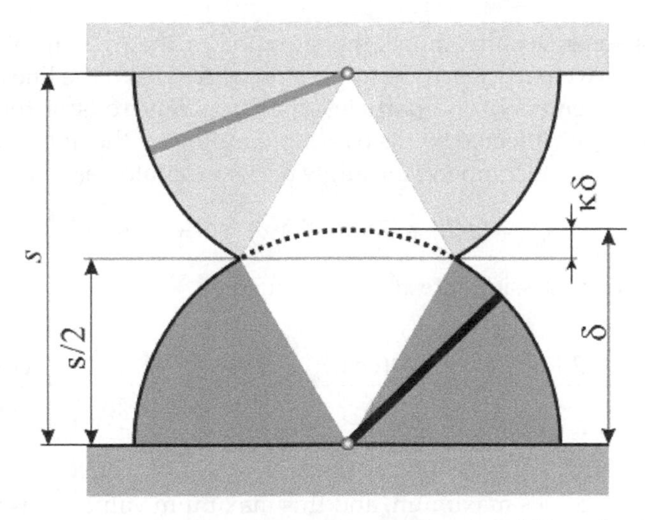


**Figure 2.3.** Decrease in entropy due to the overlap of surfactant rod-like molecules anchored at the plane material surfaces in their parallel situation.

to the overlap of the steric layers of the equal-sized spherical particles. In this derivation procedure, we use the same notation $\delta$ for the length of the rod, $n_s$ for the number of surfactant molecules (or rods) per unit area, $s$ (= $r - d$) for the surface-to-surface distance, and also $r$ for the center-to-center distance of the two particles, as shown in Fig. 2.4.

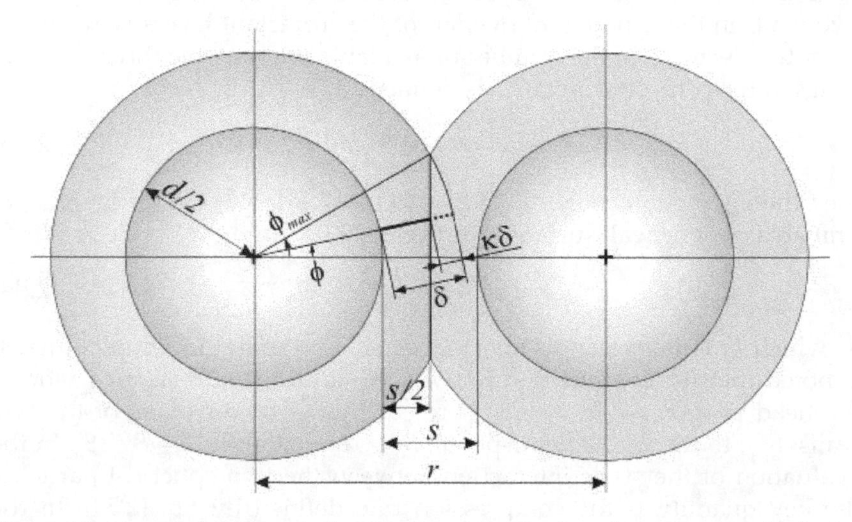


**Figure 2.4.** Compression rate $\kappa$ in the situation of the overlap of two same-sized spherical particles coated with surfactant rod-like molecules with length $\delta$.

Figure 2.4 schematically shows the situation of the overlap of the steric layers of the two particles. If the angle $\phi$ is defined from the line connected between the centers of the particles, the surfactant rods in the range of $-\phi_{max} \leq \phi \leq \phi_{max}$ are affected by the overlap. Employing the notation $d$ for the diameter of the solid sphere, the angle $\phi_{max}$ is evaluated as

$$\cos\phi_{max} = (d/2 + s/2)/(d/2 + \delta) = (d + s)/(d + 2\delta) \tag{2.35}$$

Hence, the compression rate $\kappa(\phi)$ is obtained as

$$\kappa(\phi) = \frac{(d/2+\delta)-(d/2+\delta)(\cos\phi_{max}/\cos\phi)}{\delta} = \left(1+\frac{d}{2\delta}\right)\left(1-\frac{\cos\phi_{max}}{\cos\phi}\right) \tag{2.36}$$

At $\phi = 0°$, $\kappa$ becomes maximum, and this maximum value $\kappa_{max}$ is written as

$$\kappa_{max} = (2\delta - s)/(2\delta) \tag{2.37}$$

The surface area $A_\phi$ on the solid sphere with radius $d/2$ within the angle range of $-\phi$ and $\phi$ is expressed as

$$A_\phi = 2\pi(d/2)^2(1-\cos\phi) \tag{2.38}$$

From Eqs. (2.35) and (2.36), $\cos\phi$ is expressed as a function of $\kappa$ as

$$\cos\phi = \frac{d+s}{d+2\delta(1-\kappa)} \tag{2.39}$$

Substitution of Eq. (2.39) into Eq. (2.38) leads to the final expression for $A_\phi$ as

$$A_\phi(\kappa) = 2\pi(d/2)^2\left(1-\frac{d+s}{d+2\delta(1-\kappa)}\right) = 2\pi(d/2)^2\left(1-\frac{r}{d+2\delta(1-\kappa)}\right) \tag{2.40}$$

In this expression, for a given value of the center-to-center distance $r$, the surface area $A_\phi$ on the solid sphere becomes a function of the compression rate $\kappa$.

Since the number of surfactant rods in the infinitesimal angle range of $\phi \sim (\phi + d\phi)$ is $n_s dA_\phi$ $(= 2\pi n_s(d/2)^2 \sin\phi\, d\phi)$, the interaction energy $E^S$ between the two spherical particles coated with a uniform steric layer is expressed by applying Eq. (2.34) to the present case as

$$E^S = 2\times k_B T \int_0^{A_\phi^{(max)}} \kappa(\phi)(n_s dA_\phi) = 2n_s k_B T \int_{\kappa_{max}}^{0} \kappa \frac{dA_\phi}{d\kappa} d\kappa = -2n_s k_B T \int_0^{\kappa_{max}} \kappa \frac{dA_\phi}{d\kappa} d\kappa \tag{2.41}$$

Now it is necessary to express $dA_\phi/d\kappa$ as a function of $\kappa$ rather than $\phi$. Using Eq. (2.40), this expression is obtained as

$$\frac{dA_\phi(\kappa)}{d\kappa} = 2\pi(d/2)^2\left[\frac{(d+s)(-2\delta)}{\{d+2\delta(1-\kappa)\}^2}\right] = -4\pi(d/2)^2\left[\frac{(d+s)\delta}{\{d+2\delta(1-\kappa)\}^2}\right]$$
$$= -4\pi(d/2)^2\left[\frac{r\delta}{\{d+2\delta(1-\kappa)\}^2}\right] \tag{2.42}$$

Substitution of Eq. (2.42) into Eq. (2.41) leads to

$$E^S = -2n_s k_B T \int_0^{\kappa_{max}} \kappa \frac{dA_\phi}{d\kappa} d\kappa = 8\pi n_s k_B T (d/2)^2 \int_0^{\kappa_{max}} \kappa \left[ \frac{(d+s)\delta}{\{d+2\delta(1-\kappa)\}^2} \right] d\kappa$$

$$= 8\pi n_s k_B T (d/2)^2 (d+s)\delta \int_0^{\kappa_{max}} \left[ \frac{\kappa}{\{(2\delta)\kappa - d - 2\delta)\}^2} \right] d\kappa \tag{2.43}$$

In order to calculate the above integral, the following formula is used.

$$\int \frac{x}{(ax+b)^2} dx = \frac{1}{a^2} \ln |ax+b| + \frac{b}{a^2} \cdot \frac{1}{ax+b} \tag{2.44}$$

Application of this formula to evaluation of integral in Eq. (2.43) leads to the following equation:

$$E^S = 8\pi n_s k_B T (d/2)^2 (d+s)\delta \int_0^{\kappa_{max}} \left[ \frac{\kappa}{\{(2\delta)\kappa - d - 2\delta)\}^2} \right] d\kappa$$

$$= 8\pi n_s k_B T (d/2)^2 (d+s)\delta \left[ \frac{1}{4\delta^2} \ln |2\delta\kappa - d - 2\delta| - \frac{d+2\delta}{4\delta^2} \cdot \frac{1}{2\delta\kappa - d - 2\delta} \right]_0^{\kappa_{max}}$$

$$= 8\pi n_s k_B T (d/2)^2 (d+s)\delta \left[ -\frac{1}{4\delta^2} \ln \frac{d+2\delta}{d+2\delta(1-\kappa_{max})} + \frac{d+2\delta}{4\delta^2 (d+2\delta(1-\kappa_{max}))} - \frac{1}{4\delta^2} \right] \tag{2.45}$$

Taking into account the relationship in Eq. (2.37) and $s = r - d$, the above equation finally leads to

$$E^S = \frac{n_s k_B T \pi d^2}{2} \left[ 2 - \frac{r}{\delta} \ln \frac{d+2\delta}{r} - \frac{r-d}{\delta} \right] \tag{2.46}$$

This is the expression previously derived by Rosensweig et al. [2] and is widely recognized as being applicable for molecular simulations on ferrofluids that are composed of magnetic spherical particles covered by a steric layer.

From an educational point of view, we derive a different expression for a steric repulsive interaction using the theory of statistical mechanics where the number of possible microscopic states is treated for determining thermodynamic quantities such as the entropy and interaction energy. This alternative procedure may provide an idea for developing an expression suitable for the interaction potential for the more complex non-spherical particles such as the spheroidal, rod-like, disk-like and cube-like particles. Nevertheless, in current practice, the above equation derived by Rosensweig et al. [2] is generally admitted and widely used.

Statistical quantities are related to thermodynamic quantities by the following famous Boltzmann's principle [4]:

$$S = k_B \log W \tag{2.47}$$

in which the thermodynamic quantity $S$ is the entropy, the statistical quantity $W$ is the number of microscopic states of a system, and $k_B$ is Boltzmann's constant. Hence, the evaluation of $W$ leads to specification of the entropy. If the entropy is known, a change in the free energy, $\Delta F_{free}$, is obtained by evaluating the difference in the entropy between a no-overlap and an overlap situation of the steric layers of the two particles. This equation is written as [4]

$$\Delta F_{free} = -T\Delta S \tag{2.48}$$

It is noted that the entropy change $\Delta S$ is as a function of the center-to-center distance of the two particles, and also that Eq. (2.48) may be valid with the assumption that the internal potential or the enthalpy of the system does not change as a function of the particle-particle distance.

Now we consider the number of microscopic states of the rod (or surfactant molecule) to evaluate the entropy. It is reasonable to regard microscopic states as being in proportion to the surface area where the end of the rod can orient about the hinge point. Figure 2.5 schematically shows the restriction in possible orientations of the rod due to the other plane wall. In this situation, the rod cannot penetrate the wall and this leads to a restriction in the number of possible orientations. Employing the notation $\delta$ for the length of the rod, the surface area of the hemisphere is $2\pi\delta^2$, so that the number of microscopic states $W_0$ is expressed as

$$W_0 = 2\pi C_0 \delta^2 \tag{2.49}$$

in which $C_0$ is a constant that is regarded as the number of microscopic states per unit area. If the angle of the two planes is denoted by $\alpha$, as shown in Fig. 2.5, then a portion of the hemisphere is cut by the opposite wall and the thickness of the cut portion is taken as $\kappa\delta$, measured from the tip of the head. In this situation, the number of possible microscopic states (orientations), $W_\kappa$, is evaluated as

$$W_\kappa = W_0 - 2\pi C_0 \kappa \delta^2 = 2\pi C_0 \delta^2 (1 - \kappa) \tag{2.50}$$

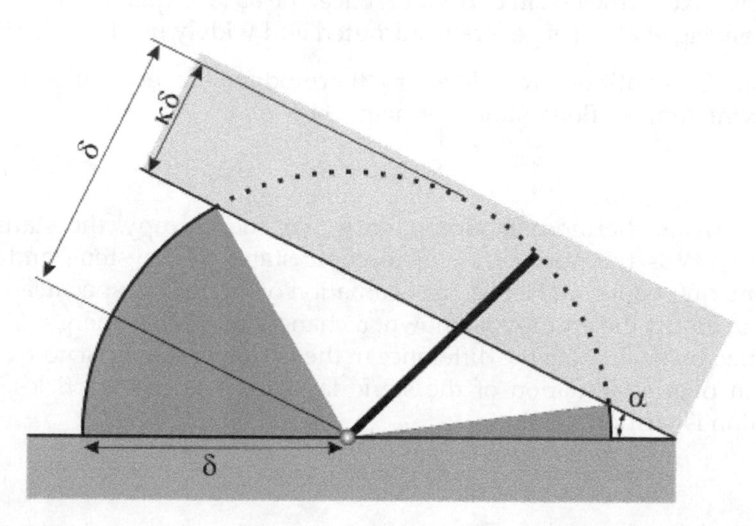


**Figure 2.5.** Restriction in possible microscopic states of a surfactant molecule due to an obstacle of the other plane wall.

We next proceed to the evaluation of the number of microscopic states in the case of two same-sized spherical particles covered by a uniform surfactant layer with thickness $\delta$, by applying the above expression to the overlap of their steric layers. In the situation of the overlapping steric layers shown in Fig. 2.6, and with reference to the line of particle centers (the $x$-axis), the symbol $s$ is used for the surface-to-surface distance between the (inside) solid spheres and the angle $\alpha$ is used to describe the position of a hinge point on the particle surface. For larger values of angle $\alpha$ the hinge point is farther from the $x$-axis and the surface area of the head portion cut by the cross-sectional interface decreases. At a certain maximum angle $\alpha_{max}$, the hemisphere is no longer cut by the overlap interface and therefore the rod is free to orient anywhere within the hemisphere. From a simple geometric analysis, the maximum angle $\alpha_{max}$ is solved as

$$\cos\alpha_{max} = \frac{d/2 + s/2 - \delta}{d/2} = 1 + \frac{s}{d} - \frac{2\delta}{d} \tag{2.51}$$

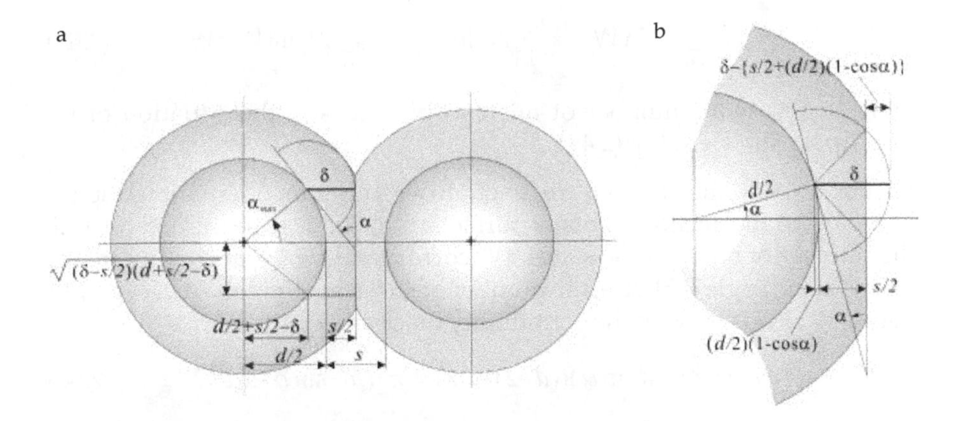


**Figure 2.6.** Calculation of possible microscopic states of a surfactant molecule in the situation of the overlap of two spherical particles coated with surfactant molecules.

As shown in Fig. 2.6(b), in respect to the rod at the hinge point $\alpha$, the quantity (or compression rate) $\kappa$ is evaluated as

$$\kappa = 1 - \left\{ \frac{s}{2\delta} + \frac{d}{2\delta}(1 - \cos\alpha) \right\} \tag{2.52}$$

We have now completed the preparation for calculating the number of possible microscopic states for the two spherical particles as a function of the center-to-center distance of the particles.

For straightforward understanding, we first treat the microscopic states in a discrete manner. If $N_i$ ($i = 1,2,\ldots$) number of rods have $W_i$ number of microscopic states, the entropy $S$ is expressed from Eq. (2.47) as

$$S = k_B \ln W = k_B \ln\left[ (W_1)^{N_1} (W_2)^{N_2} (W_3)^{N_3} \cdots \right] = k_B \sum_{i=1} N_i \ln W_i \tag{2.53}$$

The total number of surfactant molecules per magnetic particle is the sum of $N_i$ ($i = 1, 2, \ldots$) as

$$N = N_1 + N_2 + N_3 + \ldots \tag{2.54}$$

Hence, the change in the entropy, $\Delta S$, from no-overlapping to an overlapping situation is

$$\Delta S = k_B \sum_{i=1} N_i \ln W_i - k_B \sum_{i=1} N_i \ln W_0 = k_B \sum_{i=1} N_i \ln(W_i / W_0) \tag{2.55}$$

in which $W_0$ is the number of microscopic states in the situation of no overlap, as shown in Eq. (2.49).

We next expand the previous treatment to a continuum orientational space in regard to the anchor points of surfactant molecules. In this expansion, the number of microscopic states, $W_i$, corresponds to $W_\alpha$ at circumference points at the angle $\alpha$. Hence, the number of surfactant molecules, $N_i$, giving rise to $W_i$, leads to $N_\alpha$ in the continuum space as

$$N_\alpha = n_s(\pi d \sin\alpha)((d/2)\cdot d\alpha) = \pi n_s d^2 \sin\alpha \cdot d\alpha / 2 \tag{2.56}$$

where $N_\alpha$ implies the number of surfactant molecules in the infinitesimal angle range from $\alpha$ to $(\alpha + d\alpha)$. Considering this transformation, the entropy change in Eq. (2.55) is now to be expressed as

$$\Delta S = \frac{\pi n_s d^2 k_B}{2} \int_0^{\alpha_{max}} \ln\frac{W(\alpha)}{W_0} \sin\alpha \cdot d\alpha \tag{2.57}$$

in which $n_s$ is the number of surfactant molecules per unit area on the particle surface. Using Eqs. (2.50) and (2.52), the integral in Eq. (2.57) may be expressed as

$$\begin{aligned} \int_0^{\alpha_{max}} \ln\frac{W(\alpha)}{W_0} \sin\alpha \cdot d\alpha &= \int_0^{\alpha_{max}} \ln(1-\kappa(\alpha))\cdot \sin\alpha \cdot d\alpha \\ &= \int_0^{\alpha_{max}} \ln\left\{ \frac{s}{2\delta} + \frac{d}{2\delta}(1-\cos\alpha) \right\} \sin\alpha \cdot d\alpha \end{aligned} \tag{2.58}$$

If the integral variable is transformed from $\alpha$ into $z$ as $z = 1 - \cos\alpha$, the above integral is straightforwardly evaluated as

$$\begin{aligned} \int_0^{\alpha_{max}} \ln\frac{W(\alpha)}{W_0} \sin\alpha \cdot d\alpha &= \int_{z_1}^{z_2} \ln\left( \frac{s}{2\delta} + \frac{d}{2\delta} z \right) dz = \left[ (s/d+z)\ln\left( \frac{s}{2\delta} + \frac{d}{2\delta} z \right) - z \right]_{z_1}^{z_2} \\ &= \left[ (s/d+1-\cos\alpha)\ln\left( \frac{s}{2\delta} + \frac{d}{2\delta}(1-\cos\alpha) \right) - (1-\cos\alpha) \right]_0^{\alpha_{max}} \\ &= \frac{s}{d}\ln\left(\frac{2\delta}{s}\right) - \frac{2\delta - s}{d} = \frac{r-d}{d}\ln\left(\frac{2\delta}{r-d}\right) + \frac{r-2\delta}{d} - 1 \end{aligned} \tag{2.59}$$

In this derivation, the following formula regarding the partial integral has been used:

$$\int \ln(az+b)dz = \frac{az+b}{a}\ln(az+b) - z \tag{2.60}$$

Hence, we obtain the final expression for the entropy change in Eq. (2.57) as

$$\Delta S = \frac{\pi n_s d^2 k_B}{2}\left\{\frac{s}{d}\ln\left(\frac{2\delta}{s}\right) - \frac{2\delta - s}{d}\right\} = \frac{\pi n_s d^2 k_B}{2}\left\{\frac{r-d}{d}\ln\left(\frac{2\delta}{r-d}\right) + \frac{r-2\delta}{d} - 1\right\} \tag{2.61}$$

The entropy change in this equation is per one particle, and therefore $2\Delta S$ contributes the interaction energy $E^S = \Delta F_{free}$ in Eq. (2.48) as

$$\begin{aligned} E^S = -2T\Delta S &= \frac{\pi n_s d^2 k_B T}{2}\left\{2 - \frac{s}{d/2}\ln\left(\frac{2\delta}{s}\right) - \frac{d+s-2\delta}{d/2}\right\} \\ &= \frac{\pi n_s d^2 k_B T}{2}\left\{2 - \frac{r-d}{d/2}\ln\left(\frac{2\delta}{r-d}\right) - \frac{r-2\delta}{d/2}\right\} \end{aligned} \tag{2.62}$$

This is a potential energy of the two magnetic particles covered by a uniform steric layer with thickness $\delta$ in the situation of an overlap, and therefore is valid for $0 \le s \le 2\delta$ or $d \le r \le d + 2\delta$. It is noted that $E^S$ is zero in the range of $r \ge d + 2\delta$ or in the situation of no overlap.

Figure 2.7 shows the steric repulsive potential curves for the above two approaches, that is, the expression as previously derived by Rosensweig et al. [2], Eq. (2.46), and the alternative expression derived above from the theory of statistical mechanics, Eq. (2.62). Two different cases are shown for the thickness of the steric layer, $\delta^*(= \delta/d) = 0.1$ and 0.15, where the abscissa is the non-dimensional center-to-center distance $r^*(= r/d)$ between the two particles and the ordinate is the energy $\tilde{E}^s(= E^s/(\pi n_s d^2 k_B T/2))$. It is seen from Fig. 2.7 that the repulsive interaction gradually decreases to zero in the steric interaction range. Moreover, it is seen that the potential curves from Eq. (2.62) are in good agreement with Rosensweig's potential from Eq. (2.46) in the region of a shallow overlap situation, whilst the discrepancy between the two expressions becomes relatively significant with decreasing particle-particle distance.

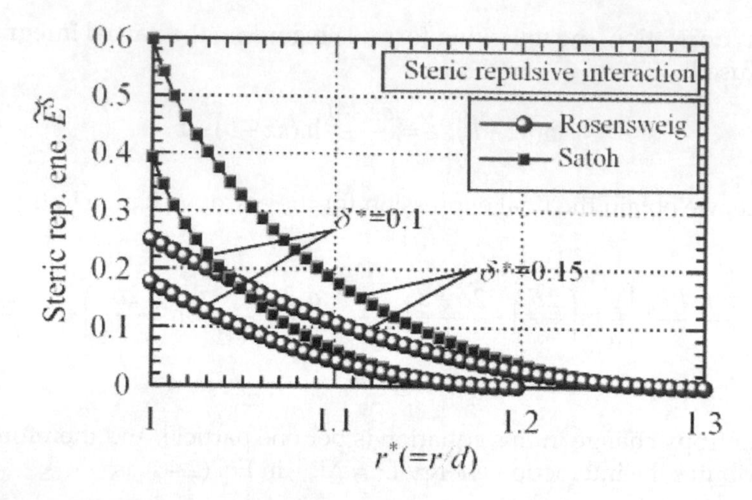


**Figure 2.7.** Steric repulsive interaction potential for the two same-sized spherical particles covered by a uniform steric layer with thickness $\delta^*$.

## 2.4 Repulsive interaction due to overlap of the electric double layers

In a suspension of magnetic particles dispersed in a base liquid, an electrical double layer is generally formed around magnetic particles. The interaction arising from the overlap of the electric layers induces a repulsive force acting between the two magnetic particles of interest. An expression for the interaction can be derived in terms of the DLVO (Derjaguin-Landau-Verwey-Overbeek) theory [5–8]. In the present section, we briefly describe this theory to show final expressions that may be useful for the simulation of certain types of magnetic suspensions in a certain situation. Readers are recommended to refer to relevant textbooks if they are interested in the background of the theory or the mathematical derivation procedure.

The DLVO theory is based on the diffuse double layer model that was presented by Gouy-Chapman. In this model, ions are assumed to be a point charge and the density of ions around each dispersed particle is assumed to follow the Boltzmann distribution and solvent is treated as a continuum medium with a fixed dielectric constant. This model enables one to express an electric potential distribution near a plate as a solution of the Poisson-Boltzmann equation. However, since the basic equation is nonlinear, a solution is obtained by introducing the Debye-Hückel approximation which yields the linearized equation. In addition, the Derjaguin approximation gives rise to an interaction energy for the two spherical particles. We now show these expressions in the following.

We first consider a suspension of identical spherical particles dispersed in a base liquid. Employing the notation of $a$ for the radius of the particle ($d$ for diameter) and $r$ for the center-to-center distance between the two particles, an expression for the interaction energy due to the overlap of the electrical double layers, $V^E$, is expressed as [5–7]

$$V^E = 2\pi\varepsilon a\psi^2 \ln[1+\exp\{-\kappa(r-2a)\}] \tag{2.63}$$

in which

$$\kappa = \sqrt{\frac{2000\, e^2 z^2 N_A C}{\varepsilon k_B T}} \tag{2.64}$$

In these equations, $\varepsilon$ is the dielectric constant of the solvent, $\psi$ is the surface potential of the spherical particles, $\kappa$ is the Debye parameter, $e$ is the elementary electric charge, $z$ is the ionic valency, $N_A$ is the Avogadro number, $C$ is the concentration of the electrolyte, $k_B$ is Boltzmann's constant and $T$ is the temperature of the suspension. It is to be noted that the reciprocal number $1/\kappa$ corresponds to the thickness of the electrical double layer.

Using the interaction energy from Eq. (2.63), the force acting on one arbitrary particle by another, $\boldsymbol{F}^E$, is derived according to the definition from Eq. (2.5) as

$$\boldsymbol{F}^E = -\nabla V^E = 2\pi\varepsilon a\psi^2\kappa \boldsymbol{t} \frac{\exp[-\kappa(r-2a)]}{1+\exp[-\kappa(r-2a)]} \tag{2.65}$$

in which $\boldsymbol{t} = \boldsymbol{r}/r$ ($r = |\boldsymbol{r}|$).

We next consider a suspension of different spherical particles and the interaction between two particles, $i$ and $j$, that have the radius $a_i$ and $a_j$ (or $d_i$ and $d_j$ for diameter) and the surface potential $\psi_i$ and $\psi_j$, respectively. From the hetero-aggregation theory, an interaction energy $V_{ij}^E$ due to the overlap of the electrical double layers of particles $i$ and $j$ is expressed as [9]

$$V_{ij}^E = \frac{\pi\varepsilon a_i a_j(\psi_i^2+\psi_j^2)}{a_i+a_j}\left\{ \frac{2\psi_i\psi_j}{\psi_i^2+\psi_j^2}\ln\left(\frac{1+\exp\{-\kappa(r_{ij}-(a_i+a_j))\}}{1-\exp\{-\kappa(r_{ij}-(a_i+a_j))\}}\right)\right.$$

$$+\ln[1-\exp\{-2\kappa(r_{ij}-(a_i+a_j))\}]\Big\} \tag{2.66}$$

The force acting on particle $i$ by particle $j$, $\boldsymbol{F}_{ij}^{E}$, is obtained as

$$\boldsymbol{F}_{ij}^{E} = -\frac{\partial V_{ij}^{E}}{\partial \boldsymbol{r}_i} = -\frac{\partial V_{ij}^{E}}{\partial r_{ij}} = 2\kappa \boldsymbol{t}_{ij} \frac{\pi\varepsilon a_i a_j (\psi_i^2+\psi_j^2)}{a_i+a_j}$$
$$\times \left\{ \frac{\psi_i\psi_j}{\psi_i^2+\psi_j^2} \cdot \frac{2\exp[-\kappa(r_{ij}-(a_i+a_j))]}{1-\exp[-2\kappa(r_{ij}-(a_i+a_j))]} - \frac{\exp[-2\kappa(r_{ij}-(a_i+a_j))]}{1-\exp[-2\kappa(r_{ij}-(a_i+a_j))]} \right\} \tag{2.67}$$

in which $\boldsymbol{t}_{ij} = \boldsymbol{r}_{ij}/r_{ij}$ $(r_{ij} = |\boldsymbol{r}_{ij}|)$.

Figure 2.8 shows the electric repulsive potential curves of the expression in Eq. (2.63), for the three different cases of the thickness of the electric double layer, $1/\kappa^*(=1/(\kappa/d)) = 0.1$, 0.15 and 0.2. In this figure, the abscissa is the non-dimensional center-to-center distance $r^*(=r/d)$ between the two particles and the ordinate is the energy $\tilde{V}^E$ $(= V^E/(2\pi\varepsilon a\psi^2))$. It is seen from Fig. 2.8 that as expected, the thickness of the electric double layer is approximately equal to the value of $1/\kappa^*$, and the repulsive interaction gradually decreases and tends to approximately zero within this range.

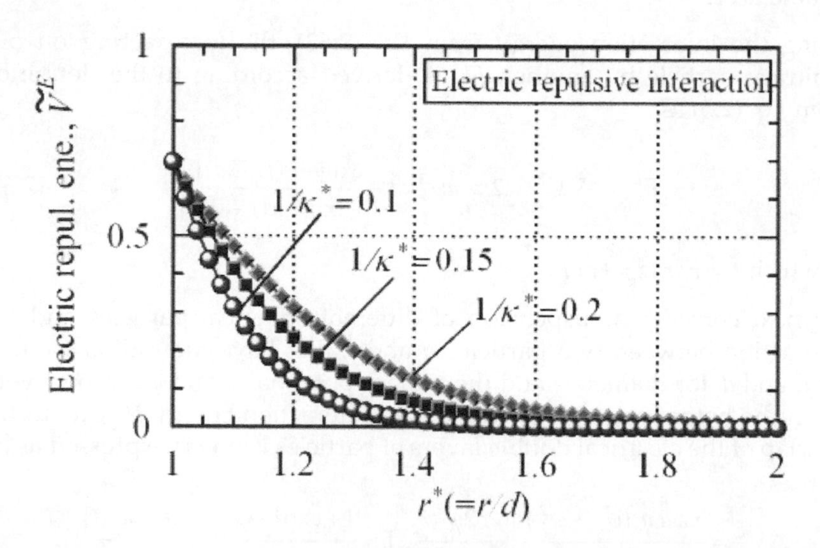


**Figure 2.8.** Electric repulsive interaction potential for the two same-sized spherical particles due to the overlap of electric double layers around dispersed particles.

Moreover, it is noted that this potential does not exhibit the same short-range order as compared with the following van der Waals attractive interaction.

## 2.5 Interaction due to van der Waals attraction

Next we show expressions for van der Waals interaction energies. This interaction potential is of relatively short range order with a potential energy that is inversely proportional to the 6th power of the inter-atomic distance. However, the summation of all the interactions between the constituent atoms of the two dispersed particles leads to a relatively long-range order interaction between the two dispersed particles. Hamaker showed an expression of the interaction energy $V^A$ between the same kind of two particles as [5–7]

$$V^A = -\frac{A}{6}\left[\frac{2a^2}{r^2-4a^2}+\frac{2a^2}{r^2}+\ln\left(\frac{r^2-4a^2}{r^2}\right)\right] \tag{2.68}$$

Using the interaction energy in Eq. (2.68), the force acting on an arbitrary particle by the other, $\boldsymbol{F}^A$, is derived as

$$\boldsymbol{F}^A = \frac{A}{6}\boldsymbol{t}\left[-\frac{4a^2r}{(r^2-4a^2)^2}-\frac{4a^2}{r^3}+\frac{2r}{r^2-4a^2}-\frac{2}{r}\right] \tag{2.69}$$

We next consider two different spherical particles, $i$ and $j$, that have the radius $a_i$ and $a_j$ ($d_i$ and $d_j$ for diameter) and the surface potential $\psi_i$ and $\psi_j$ for particles $i$ and $j$, respectively. From the hetero-aggregation theory, an interaction energy $V_{ij}^A$ due to the van der Waals interaction of particles $i$ and $j$ is expressed as [10]

$$V_{ij}^A = -\frac{A_{ij}}{6}\left[\frac{2a_ia_j}{r_{ij}^2-(a_i+a_j)^2}+\frac{2a_ia_j}{r_{ij}^2-(a_i-a_j)^2}+\ln\left(\frac{r_{ij}^2-(a_i+a_j)^2}{r_{ij}^2-(a_i-a_j)^2}\right)\right] \tag{2.70}$$

The force acting on particle $i$ by particle $j$, $\boldsymbol{F}_{ij}^A$, is obtained as

$$\boldsymbol{F}_{ij}^A = \frac{A_{ij}}{6}\boldsymbol{t}_{ij}\left[-\frac{4a_ia_jr_{ij}}{\{r_{ij}^2-(a_i+a_j)^2\}^2}-\frac{4a_ia_jr_{ij}}{\{r_{ij}^2-(a_i-a_j)^2\}^2}+\frac{2r_{ij}}{r_{ij}^2-(a_i+a_j)^2}-\frac{2r_{ij}}{r_{ij}^2-(a_i-a_j)^2}\right] \tag{2.71}$$

In the above equations, $A$ and $A_{ij}$ are the Hamaker constants. Since the magnetic particles are dispersed in a base liquid, the effective Hamaker constants need to be evaluated by taking into account the effect of the solvent molecules. For explanation of Hamaker constants, we consider the two situations, one where only one kind of particles are dispersed in a base liquid and the other where there are two kinds of particles. We now use the notation $A_1$ for the Hamaker constant for the interaction between the first kind of particles and $A_2$ for the interaction between the second kind of particles. The Hamaker constant $A_{12}$ is for the interaction between the first and the second kind of particles, and $A^{(0)}$ for the characteristic of a base liquid. It should be noted that $A_1$ and $A_2$ are for the case of no base liquid. Employing this notation, under the circumstance of solvent medium, the Hamaker constants $A_1^{(0)}$ for the interaction between the first kind of particles, $A_2^{(0)}$ between the second kind of particles, and $A_{12}^{(0)}$ between the first and the second kind of particles, respectively, are expressed as

$$A_1^{(0)} = (\sqrt{A_1} - \sqrt{A^{(0)}})^2, A_2^{(0)} = (\sqrt{A_2} - \sqrt{A^{(0)}})^2, A_{12}^{(0)} = (\sqrt{A_1} - \sqrt{A^{(0)}})(\sqrt{A_2} - \sqrt{A^{(0)}}) \tag{2.72}$$

It is noted that a stable particle suspension may be obtained by using a repulsive layer to cover the dispersed particles such as a steric (surfactant) layer or an electric double layer.

Figure 2.9 shows the potential curve for the interaction energy $V^A$ between a pair of the same kind of spherical particles, shown in Eq. (2.68): the abscissa is the non-dimensional center-to-center distance $r^*$ $(=r/d)$ between the two particles and the ordinate is the energy $\tilde{V}^A(= V^A/(A/6)$. It is seen from Fig. 2.9 that the van der Waals interaction potential is of significantly short-range order, where the potential energy is approximately zero in the range longer than $r^* \simeq 1.2$. Hence, if a steric or electric repulsive layer is effectively employed in a functional manner, the van der Waals attraction does not substantially contribute to the stability of a particle suspension.

## 2.6 Maxwell stress tensor

In this section, we briefly consider the Maxwell stress tensor. The derivation procedure for the Maxwell stress tensor makes it clear how a microscopic quantity, such as the magnetic moment in a magnetic particle suspension, is connected to a macroscopic quantity, such as the stress tensor that is usually addressed in the macroscopic analysis of flow problems. In a macroscopic point of view, for instance, a magnetic suspension is

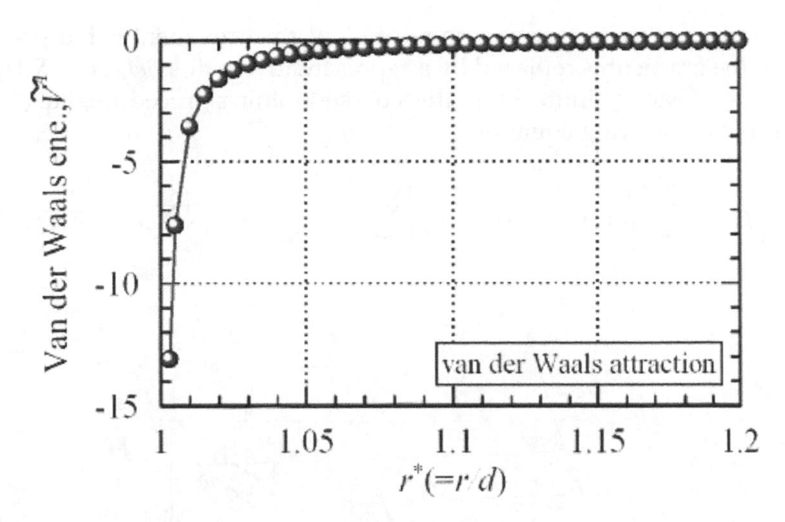


**Figure 2.9.** Van der Waals attractive interaction potential for the two same-sized spherical particles.

regarded as a continuum medium with a magnetic property as a whole so that it is unnecessary to treat each individual magnetic particle in order to obtain the flow characteristics. Therefore a macroscopic magnetic particle suspension is regarded as an apparent pure liquid with its own magnetic properties.

As shown in Eq. (2.14), a force acts on a magnetic particle with a constant magnetic dipole moment if the external magnetic field is not uniform, that is, the magnitude of the field is a function of position in the system. In a macroscopic treatment of these forces in a magnetic particle suspension, the force acting on a certain volume of suspension is evaluated by summing over the forces acting on each magnetic particle that is within this volume. As shown in Fig. 2.10, if an infinitesimal volume, taken to be large with respect to the magnetic particle diameter, is denoted by $\Delta V$, and an arbitrary magnetic particle in this volume is denoted by subscript $i$, where $i = 1, 2, \ldots, N_{\Delta V}$, then a force $\boldsymbol{F}_{\Delta V}$ exerted by the interactions between the magnetic particles and a non-uniform applied magnetic field $\boldsymbol{H}$ is expressed as

$$\boldsymbol{F}_{\Delta V} = \sum_{i=1}^{N_{\Delta V}} \{\mu_0(\boldsymbol{m}_i \cdot \nabla_i)\boldsymbol{H}\} = \sum_{i=1}^{N_{\Delta V}} \left\{ \mu_0(\boldsymbol{m}_i \cdot \frac{\partial}{\partial \boldsymbol{r}_i})\boldsymbol{H} \right\} \tag{2.73}$$

In this expression, the gradient of the magnetic field, $\partial \boldsymbol{H}/\partial \boldsymbol{r}_i$, where $i = 1,2, \ldots, N_{\Delta V}$, is to be evaluated at the center of each magnetic particle $i$.

However, from a macroscopic point of view, this treatment is too precise and so the gradient is replaced by a representative value $\partial \boldsymbol{H}/\partial \boldsymbol{r}$ (= $\nabla \boldsymbol{H}$) at the center of each volume. From this coarse-graining procedure, Eq. (2.73) leads to the following equation:

$$\boldsymbol{F}_{\Delta V} = \sum_{i=1}^{N_{\Delta V}} \left\{ \mu_0 (\boldsymbol{m}_i \cdot \frac{\partial}{\partial \boldsymbol{r}_i}) \boldsymbol{H} \right\} = \mu_0 \left\{ \sum_{i=1}^{N_{\Delta V}} \boldsymbol{m}_i \right\} \cdot \frac{\partial}{\partial \boldsymbol{r}} \boldsymbol{H} = \mu_0 \left\{ \sum_{i=1}^{N_{\Delta V}} \boldsymbol{m}_i \right\} \cdot \nabla \boldsymbol{H} \quad (2.74)$$

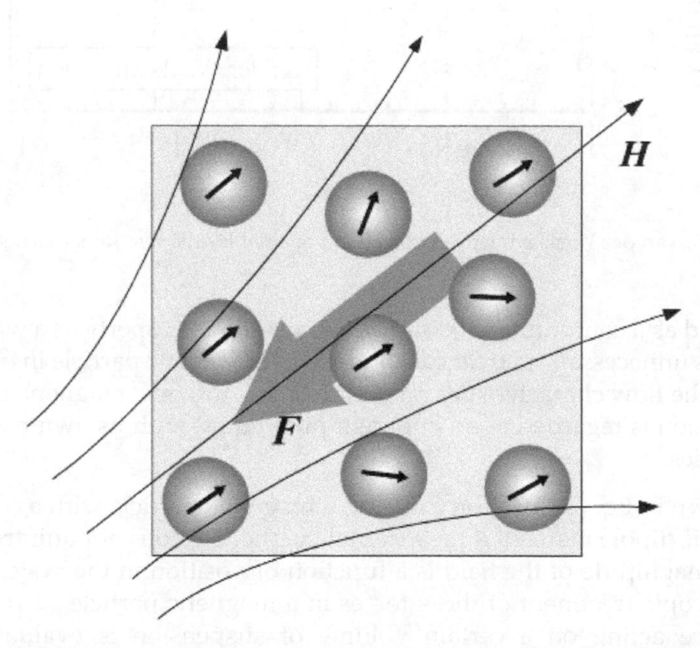


**Figure 2.10.** Body force acting on an infinitesimal volume arising from the interactions between the magnetic moments of particles and a non-uniform external magnetic field.

Hence, the force $\boldsymbol{F}$ acting on the volume $\Delta V$ per unit volume is obtained as

$$\boldsymbol{F} = \mu_0 \boldsymbol{M} \cdot \nabla \boldsymbol{H} = \mu_0 (\boldsymbol{M} \cdot \nabla) \boldsymbol{H} \quad (2.75)$$

in which $\boldsymbol{M}$ is the magnetization of the volume $\Delta V$ at position $\boldsymbol{r}$, expressed as

$$\boldsymbol{M} = \frac{1}{\Delta V} \sum_{i=1}^{N_{\Delta V}} \boldsymbol{m}_i \quad (2.76)$$

In the following, we derive an expression for the Maxwell stress tensor from Eq. (2.75). By considering the relationship of $\boldsymbol{B} = \mu_0(\boldsymbol{H}+\boldsymbol{M})$, Eq. (2.75) is rewritten as

$$\boldsymbol{F} = \mu_0 \{(\boldsymbol{B}/\mu_0 - \boldsymbol{H})\cdot\nabla\}\boldsymbol{H} = (\boldsymbol{B}\cdot\nabla)\boldsymbol{H} - \mu_0(\boldsymbol{H}\cdot\nabla)\boldsymbol{H} \tag{2.77}$$

Then we use the following formulae regarding vector analysis about arbitrary vectors $\boldsymbol{a}$ and $\boldsymbol{b}$, and constant $c$:

$$(\boldsymbol{a}\cdot\nabla)\boldsymbol{b} = \nabla\cdot(\boldsymbol{ab}) - \boldsymbol{b}(\nabla\cdot\boldsymbol{a}),\quad (\boldsymbol{a}\cdot\nabla)\boldsymbol{a} = \frac{1}{2}\nabla(\boldsymbol{a}\cdot\boldsymbol{a}) - \boldsymbol{a}\times(\nabla\times\boldsymbol{a}),\quad \nabla c = \nabla\cdot(c\boldsymbol{I}) \tag{2.78}$$

in which $\boldsymbol{I}$ is the unit tensor. If a magnetic suspension is assumed to be dielectric and the relationship of $\nabla\cdot\boldsymbol{B} = 0$ is taken into account, Eq. (2.77) is reformed as

$$\boldsymbol{F} = \nabla\cdot(\boldsymbol{BH}) - \frac{1}{2}\mu_0\nabla(\boldsymbol{H}\cdot\boldsymbol{H}) = \nabla\cdot(\boldsymbol{BH}) - \frac{1}{2}\mu_0\nabla H^2 = \nabla\cdot\left\{\boldsymbol{BH} - \frac{1}{2}\mu_0 H^2\boldsymbol{I}\right\} \tag{2.79}$$

in which $H$ is the magnitude of $\boldsymbol{H}$. If the second rank tensor $\boldsymbol{\tau}$ is expressed as

$$\boldsymbol{\tau} = \boldsymbol{BH} - \frac{1}{2}\mu_0 H^2\boldsymbol{I} \tag{2.80}$$

then this tensor $\boldsymbol{\tau}$ is called the Maxwell stress tensor and acts in the magnetic material medium. If the Maxwell stress tensor $\boldsymbol{\tau}$ is known, the force $\boldsymbol{F}$ acting on a magnetic suspension per unit volume may be evaluated as

$$\boldsymbol{F} = \nabla\cdot\boldsymbol{\tau} \tag{2.81}$$

From a macroscopic point of view, this example shows that the Maxwell stress tensor may be useful in the analysis of the flow problems of magnetic suspensions.

## Bibliography


[1] Chikazumi, S. 1997. Physics of Ferromagnetism, 2nd ed. Oxford Science Publications, London.

[2] Rosensweig, R. E., Nestor, J. W. and Timmins, R. S. 1965. Ferrohydrodynamic fluids for direct conversion of heat energy. *In*: Mater. Assoc. Direct Energy Convers. Proc. Symp. AIChE-I. Chem. Eng. Ser. 5: 104–108.

[3] Mackor, E. L. 1951. A theoretical approach of the colloidal-chemical stability of dispersions in hydrocarbons. J. Colloid Sci. 6: 492–495.
[4] McClelland, B. J. 1973. Statistical Thermodynamics. Chapman and Hall & Science Paperbacks, London.
[5] Kitahara, F. and Watanabe, M. (eds.). 1972. Electrical Interfacial Phenomena: Fundamentals, Measurements and Applications, Kyoritsu Shuppan, Tokyo (in Japanese).
[6] Kitahara, F. 1994. Fundamentals of Interface and Colloid Chemistry, Koudansha Scientific. Tokyo (in Japanese).
[7] Tachibana, T., Meguro, K., Kitahara, F., Morimoto, T., Watanabe, M., Yoshizawa, K. and Senoo, M. 1981. Colloid Chemistry: Its New Expansion, Kyoritsu Shuppan, Tokyo (in Japanese).
[8] Van de Ven, T. G. M. 1989. Colloidal Hydrodynamics, Academic Press, London.
[9] Hogg, R., Healy, T. W. and Fuerstenau, D. W. 1966. Mutual coagulation of colloidal dispersions. Trans. Faraday Soc. 62: 1638.
[10] Russel, W. B., Saville, D. A. and Schowalter, W. R. 1989. Colloidal Dispersions. Cambridge University Press, Cambridge.

CHAPTER 3

# Modeling of Magnetic Particles for Particle-Based Simulations

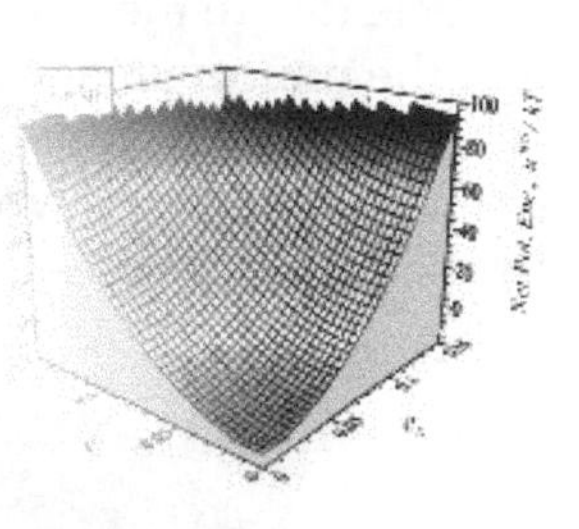

In this chapter, from a simulation point of view, we consider the modeling of magnetic particles with a variety of shapes. With an educational objective, we focus on the simpler models that are fully applicable to actual particle-based simulations, and moreover the following treatment of these models is intended to provide the reader with a valuable insight into the methodology for developing their own, more sophisticated magnetic particle models. In the following, we first show the modeling of different types of magnetic particles and then explain the expressions of the energies, forces and torques that arise from the interactions between the particles themselves and between the particles and an external magnetic field. In order to help the reader to develop an understanding of the attractive and the repulsive interactions acting between the dispersion particles, we will discuss the potential curves for various cases of a two-particle configuration. This is also a significant aid for analyzing the aggregate structures of magnetic particles in both the situations of an applied magnetic field and an external flow field.

## 3.1 Spherical particles

In this section, we first address the modeling of the magnetic spherical particle, which is the simplest model for a magnetic particle in a magnetic particle suspension.

As shown in Fig. 3.1, a magnetic particle is idealized as a spherical particle with a permanent magnetic point dipole positioned at the particle center and covered by a uniform repulsive surface layer with thickness $\delta$. We now consider two such particles, $i$ and $j$, placed in a uniform external magnetic field. In this situation, there will be no body force acting on a particle due to the interaction between the magnetic moment of a particle and the uniform magnetic field. In a suspension, however, there generally is a force acting on a particle due to the interaction with the ambient magnetic moments where they induce a non-uniform magnetic field at the particle of interest.

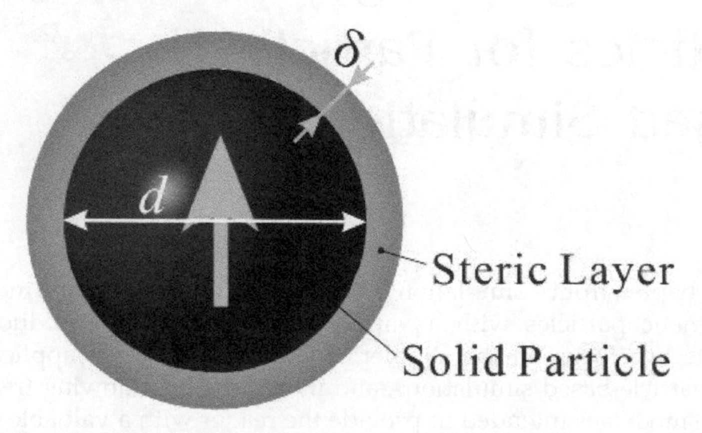


**Figure 3.1.** Spherical particle model with a constant magnetic dipole moment at the particle center covered by a uniform steric layer or a uniform electric double layer.

We employ the notation $\boldsymbol{r}_i$ for the position vector of particle $i$, $\boldsymbol{r}_{ij} = \boldsymbol{r}_i - \boldsymbol{r}_j$ (where $r_{ij} = |\boldsymbol{r}_{ij}|$) for the relative position vector to particle $j$, $\boldsymbol{m}_i$ for the magnetic moment of particle $i$, $\boldsymbol{H}$ (where $H = |\boldsymbol{H}|$) for the external magnetic field, and $\mu_0$ for the permeability of free space. Using this notation, the interaction energy of particle $i$ with the magnetic field, $u_i^{(H)}$, and the interaction energy between particles $i$ and $j$, $u_{ij}^{(m)}$, are expressed by Eqs. (2.9) and (2.18), respectively. Moreover, the magnetic force, $\boldsymbol{F}_{ij}^{(m)}$, between particles $i$ and $j$, the magnetic torque $\boldsymbol{T}_{ij}^{(m)}$, and the torque acting on particle $i$ due to the deviation of the magnetic moment from the direction of the applied magnetic field, $\boldsymbol{T}_i^{(H)}$, are expressed in Eqs. (2.29), (2.32) and (2.11), respectively.

We consider here a steric layer and show an expression for a repulsive interaction arising from the overlap of these layers for the case of two particles. As shown in Fig. 3.1, if the magnetic particle is assumed to be coated with a uniform surfactant layer with thickness $\delta$, then the repulsive interaction energy, $u_i^{(V)}$, due to the overlap of the steric layers is expressed in Eq. (2.46). Also, the force due to the overlap of the steric layers, $\boldsymbol{F}_{ij}^{(V)}$, is derived from the energy expression in Eq. (2.46) as

$$\boldsymbol{F}_{ij}^{(V)} = kT\lambda_V \frac{1}{\delta}\boldsymbol{t}_{ij} \ln(\frac{d+2\delta}{r_{ij}}) \qquad (d \le r_{ij} \le d+2\delta) \tag{3.1}$$

in which $\lambda_V = \pi d^2 n_s/2$, $n_s$ is the number of surfactant molecules per unit area on the surface of a spherical particle, $\boldsymbol{t}_{ij}$ is the unit vector given by $\boldsymbol{t}_{ij} = \boldsymbol{r}_{ij}/r_{ij}$, $k$ is Boltzmann's constant and $T$ is the system temperature.

A non-dimensional system is usually employed for performing simulations. The procedure for non-dimensionalization of the basic equations and the related physical quantities makes it clear what kinds of non-dimensional parameters (i.e., numbers) govern the physical phenomenon of interest. These non-dimensional parameters generally imply the ratio of two factors, for instance, the ratio of a representative magnetic force to a representative viscous force. An important point for the non-dimensionalization procedure is to employ a set of appropriate representative quantities that will characterize the physical phenomenon of interest in the most suitable manner. Non-dimensional expressions are obtained by expressing each quantity, say $q$, as a non-dimensional quantity, say $q^*$, multiplied by the corresponding representative quantity, say $Q$, which gives rise to $q = Qq^*$, and by substituting these quantities into dimensional equations. In the present magnetic spherical particle suspension, it may be reasonable to employ the following representative quantities: $d$ for distances, $kT$ for energies, $kT/d$ for forces, and $kT$ for torques. Employing these representative quantities, the above-mentioned energies and forces are written in non-dimensional form as

$$u_i^{(H)*} = -\xi \boldsymbol{n}_i \cdot \boldsymbol{h} \tag{3.2}$$

$$u_{ij}^{(m)*} = \frac{\lambda}{r_{ij}^{*3}}\left\{ \boldsymbol{n}_i \cdot \boldsymbol{n}_j - 3(\boldsymbol{n}_i \cdot \boldsymbol{t}_{ij})(\boldsymbol{n}_j \cdot \boldsymbol{t}_{ij}) \right\} \tag{3.3}$$

$$u_{ij}^{(V)*} = \lambda_V \left( 2 - \frac{2r_{ij}^*}{t_\delta} \ln\left(\frac{1+t_\delta}{r_{ij}^*}\right) - 2\frac{r_{ij}^* - 1}{t_\delta} \right) \qquad (1 \le r_{ij}^* \le 1 + t_\delta) \tag{3.4}$$

$$\boldsymbol{F}_{ij}^{(m)*} = -\frac{3\lambda}{r_{ij}^{*4}}\left[ -(\boldsymbol{n}_i \cdot \boldsymbol{n}_j)\boldsymbol{t}_{ij} + 5(\boldsymbol{n}_i \cdot \boldsymbol{t}_{ij})(\boldsymbol{n}_j \cdot \boldsymbol{t}_{ij})\boldsymbol{t}_{ij} - \left\{ (\boldsymbol{n}_j \cdot \boldsymbol{t}_{ij})\boldsymbol{n}_i + (\boldsymbol{n}_i \cdot \boldsymbol{t}_{ij})\boldsymbol{n}_j \right\} \right] \tag{3.5}$$

$$\boldsymbol{F}_{ij}^{(V)*} = \lambda_V \frac{1}{\delta^*} \boldsymbol{t}_{ij} \ln\left(\frac{1+t_\delta}{r_{ij}^*}\right) \quad (1 \le r_{ij}^* \le 1+t_\delta) \tag{3.6}$$

$$\boldsymbol{T}_{ij}^{(m)*} = \frac{\lambda}{r_{ij}^{3*}}\left[ 3(\boldsymbol{n}_j \cdot \boldsymbol{t}_{ij})\boldsymbol{n}_i \times \boldsymbol{t}_{ij} - \boldsymbol{n}_i \times \boldsymbol{n}_j \right] \tag{3.7}$$

$$\boldsymbol{T}_i^{(H)*} = \xi \boldsymbol{n}_i \times \boldsymbol{h} \tag{3.8}$$

in which $\boldsymbol{n}_i$ and $\boldsymbol{n}_j$ are the unit vectors denoting the direction of particles $i$ and $j$, $\boldsymbol{h}$ is the unit vector denoting the magnetic field direction, i.e., $\boldsymbol{h} = \boldsymbol{H}/H$, $\boldsymbol{t}_{ij}$ is the unit vector denoting the relative position, i.e., $\boldsymbol{t}_{ij} = \boldsymbol{r}_{ij}/r_{ij}$, and $t_\delta$ is the ratio of the thickness of the steric layer to the particle radius, i.e., $t_\delta = 2\delta/d$.

The quantities $\xi$, $\lambda$ and $\lambda_V$ that have appeared in Eqs. (3.2) to (3.8) are the non-dimensional parameters that govern the present spherical magnetic particle system. The non-dimensional parameters $\xi$ and $\lambda$ are written as

$$\xi = \mu_0 mH/kT, \; \lambda = \mu_0 m^2/4\pi d^3 kT \tag{3.9}$$

and the expression for $\lambda_V$ has already been shown in Eq. (3.1). The quantities $\xi$ and $\lambda$ represent the strengths of particle-field and particle-particle interactions relative to the thermal energy, and $\lambda_V$ represents the strength of the steric repulsive interaction relative to the thermal energy. It is important to note that if values of $\xi$ and $\lambda$ are much larger than unity, the effect of the magnetic particle-field and particle-particle interactions are both much more dominant than the thermal motion, whereas in the opposite case the effect of the thermal motion is much more dominant; a similar interpretation is also valid for the non-dimensional parameter $\lambda_V$.

In the following we discuss the characteristics of the potential energies that are characterized by the above-mentioned expressions. This manner of discussion regarding the potential curves for a set of typical configurations may enable one to analyze, at a deeper level, the internal structure of the aggregation of the magnetic particles in a suspension. In the case of a spherical magnetic particle suspension, the four typical arrangements shown in Fig. 3.2 are a key step in understanding the preferred configurations in both the aggregation situations of an applied magnetic field and no external magnetic field. The configuration shown in Fig. 3.2(a) gives rise to a minimum interaction energy and therefore magnetic particles have a strong tendency to aggregate in this formation, under the condition mentioned above where the magnetic particle-particle interaction is much stronger than the thermal energy $kT$. The configuration shown in Fig. 3.2(b) yields a second minimum energy, and in certain situations this configuration may appear. However, a strong applied magnetic field prohibits magnetic particles from aggregating with this kind of cluster formation. Since the magnetic moment of each magnetic particle strongly tends to align in the direction of a strong external magnetic field, the configuration in Fig. 3.2(a) is the most preferable for the cluster formation in the situation of a strong external magnetic field. In contrast, the configuration in Fig. 3.2(d) is the least favorable situation under the influence of a strong magnetic field. Figure 3.3 quantitatively shows the potential energy as a function of the center-to-center distance of two particles in the above four configurations. The quantity of the ordinate, $\tilde{U}^m$, is defined using Eq. (3.3) as $\tilde{U}^m = u_{ij}^{(m)*}/\lambda$. It is seen from Fig. 3.3 that the magnetic potential energy between the two magnetic particles is of long-range order, so that in simulations a sufficiently long cut-off distance, for instance longer than $r^* \simeq 6$, must be employed for calculating this interaction, or a special treatment such as the Ewald sum [1] has to be introduced specifically for taking into account the long-range interactions.

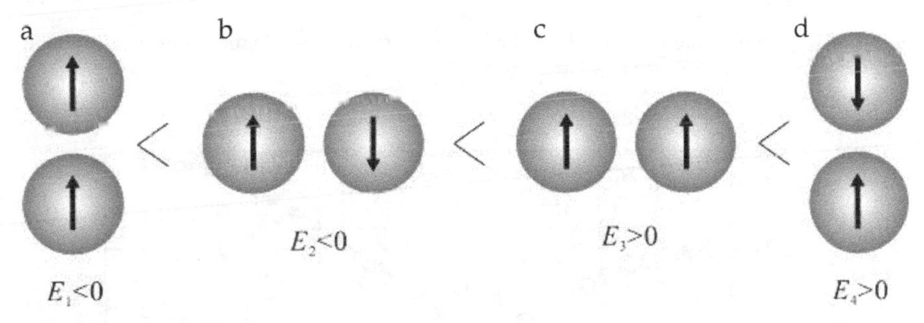


**Figure 3.2.** Interaction energy dependent on the typical four arrangements: (a) minimum energy, and (b) second minimum energy (attractive); (d) maximum energy and (c) second maximum energy (repulsive).

Figure 3.4 shows the change in the potential energy for the two typical configurations of a pair of particles in contact as the position of the second particle is described by an angle $\theta$, measured from the $x$-axis. It is seen from Fig. 3.4 that although the configuration in Fig. 3.2(c) gives rise to a positive potential energy, the approach to the more linear situation shown in Fig. 3.2(a) that occurs with increasing value of the $\theta$-angle leads to zero value at $\theta = 35.265°$ and thereafter the curve tends toward larger negative values. The understanding of this dependence of the magnetic potential energy on the angle $\theta$ is particularly valuable for understanding the aggregation of thick chain-like clusters formed along the applied magnetic field direction, in which the magnetic interactions between linear chain-like clusters play an important role.

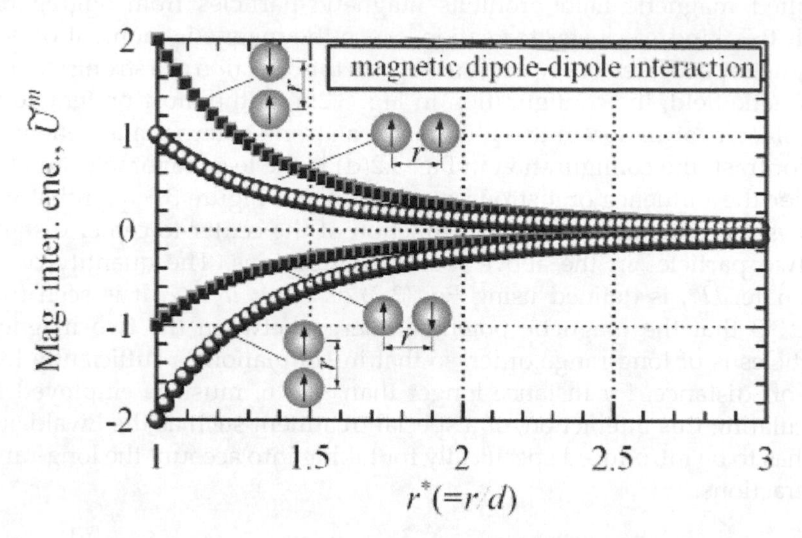


**Figure 3.3.** Magnetic interaction potential between two point dipole moments as a function of the distance for the four cases of arrangement.

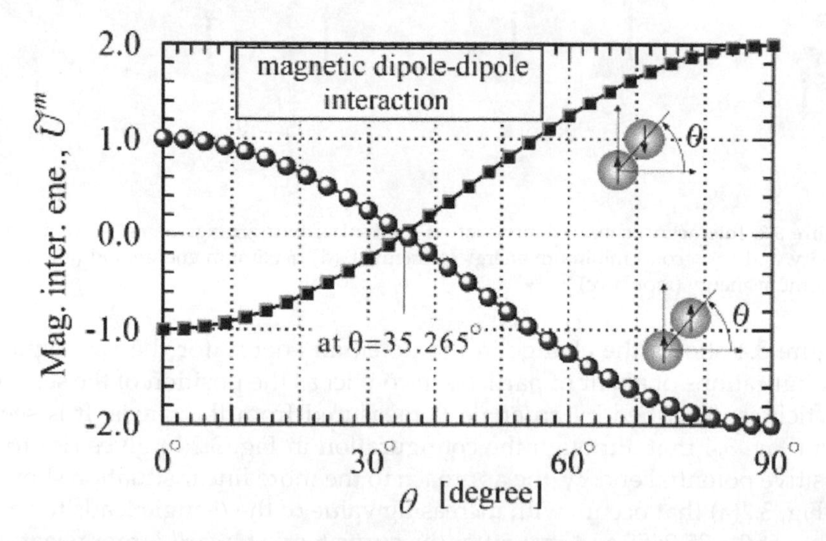


**Figure 3.4.** Magnetic interaction potential between two point dipole moments in the contact situation as a function of angle $\theta$ for the two cases of arrangement.

Finally, we discuss the change in the force and torque of a pair of particles in contact as a function of the position of the second particle described by the angle $\theta$, measured from the $x$-axis, shown in Fig. 3.5. The quantity of the abscissa, $\tilde{F}_x$ and $\tilde{F}_y$, are defined using Eq. (3.5) as $F_x^{(m)*}/\lambda$ and $F_y^{(m)*}/\lambda$, respectively, where $F_x^{(m)*}$ and $F_y^{(m)*}$ are the $x$- and $y$-component of $\boldsymbol{F}^{(m)*}$. Also, $\tilde{T}_z$ is defined using Eq. (3.7) as $T_z^{(m)*}/\lambda$, where $T_z^{(m)*}$ is the $z$-component of $\boldsymbol{T}^{(m)*}$.

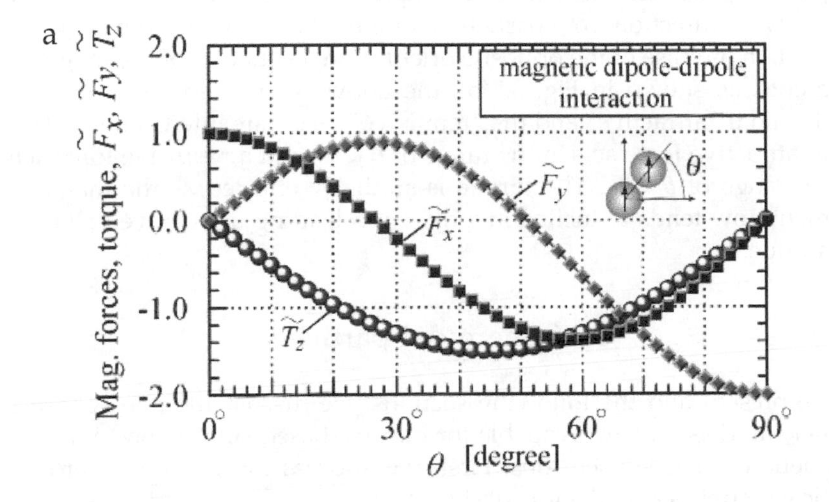


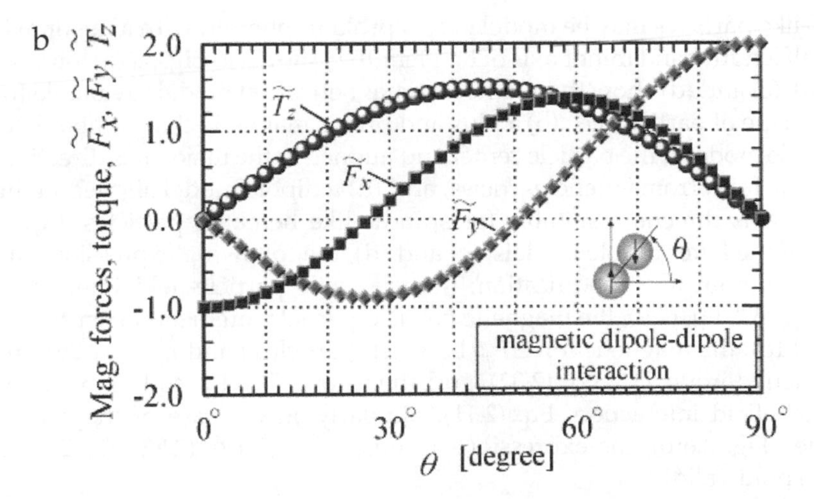


**Figure 3.5.** Magnetic force components and torque acting on the neighboring particle $j$ in the contact situation as a function of angle $\theta$ for the two cases of magnetic moment directions: (a) the same direction and (b) the opposite directions.

In the orientational position where $\theta = 0°$, a repulsive force acts on particle $j$, shown in Fig. 3.5(a), which decreases with increasing values of theta and finally vanishes at a critical angle $\theta_0$. As the angle increases from this critical angle $\theta_0$, the force becomes increasingly attractive and attains to a maximum at $\theta = 90°$. The critical angle $\theta_0$ is evaluated from $\boldsymbol{F}_{ij}^{(m)*} \cdot \boldsymbol{t}_{ij} = 0$ using Eq. (3.5) to obtain $\theta_0 = \sin^{-1}(1/3^{1/2}) = 35.265°$, which is the same angle that gives rise to zero potential energy in Eq. (3.3). The torque acts on particle $j$ in such a way that the magnetic moment tends to incline in the outward direction from particle $i$ along the line by which the angle $\theta$ is defined. In the case of the opposite orientational situation of the magnetic moments, as shown in Fig. 3.5(b), the above-mentioned characteristics hold valid if "attractive" and the "repulsive" terms are interchanged. That is, an attractive force acts in the range of $\theta < \theta_0$, and a repulsive force acts in the range of $\theta > \theta_0$. The torque is similarly effected as the magnetic moment now tends to incline in a direction leading to positive values of the torque.

## 3.2 Spheroidal particles

In the present and the following sections, we discuss the non-spherical particle models that are available for particle-based simulations of actual magnetic particle suspensions. First, we address the prolate spheroidal particle model as a model for rod-like particles.

Rod-like particles may be modeled as a prolate spheroid with a major axis of half length $a$ and minor axis of half length $b$, shown in Fig. 3.6(a). Figures 3.6(b), (c) and (d) show three different magnetization models available for this shape of particle, i.e., (b) a plus and minus magnetic charge model, (c) a dipole model at the particle center and aligned in the major axis direction, suitable for ferromagnetic particles, and (d) a dipole model aligned in the minor axis direction, suitable for spindle-like hematite particles. In the case of the latter dipole models, (c) and (d), the following expressions are valid without any modification; the magnetic particle-field interaction energy $u_i^{(H)}$, Eq. (2.9), the magnetic particle-particle interaction energy $u_{ij}^{(m)}$, Eq. (2.18), the magnetic force $\boldsymbol{F}_{ij}^{(m)}$ between particles $i$ and $j$, Eq. (2.29), the magnetic torque $\boldsymbol{T}_{ij}^{(m)}$, Eq. (2.32), and the torque $\boldsymbol{T}_i^{(H)}$ due to the magnetic particle-field interaction, Eq. (2.11). Similarly, in the case of the charge model, Fig. 3.6(b), the expressions in Eqs. (2.9), (2.19), (2.22), (2.32) and (2.11) hold valid.

One of the difficult problems related to the use of the spheroidal particle model arises in treating the repulsive layer covering each dispersed particle which is indispensable for obtaining a stable particle suspension. A mathematical expression for the overlap of spheroidal steric layers is not currently available and therefore a possible technique for treating this repulsive interaction is to apply an expression for two spherical particles previously shown in Eq. (2.46) to the present spheroidal particle model. As shown in Fig. 3.7, a spheroid may be modeled as a linear sphere-connected model where spheres with different diameters are linearly located along the major axis direction in an inscribed and contact situation where each sphere is assumed to be coated by a uniform steric layer. In this model, the interaction energy for the overlap of the steric layers of two spheroidal particles is evaluated by a summation of the interaction energy over all the possible overlapping pairs of their constituent spherical particles.

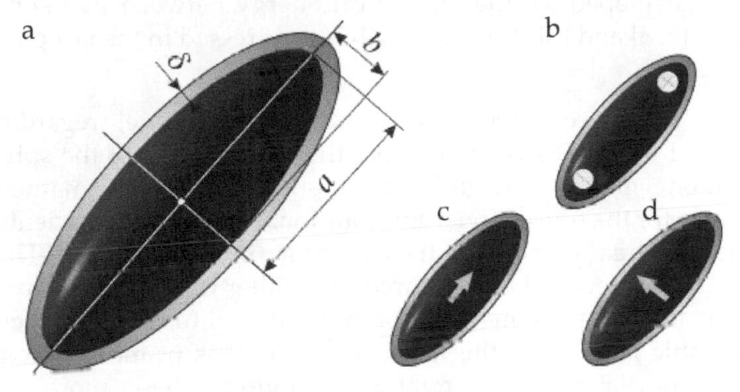


**Figure 3.6.** Spheroidal particle and several magnetization models: (a) spheroid covered by a soft repulsive layer, (b) magnetic charges model, (c) ferromagnetic particle model and (d) hematite particle model.

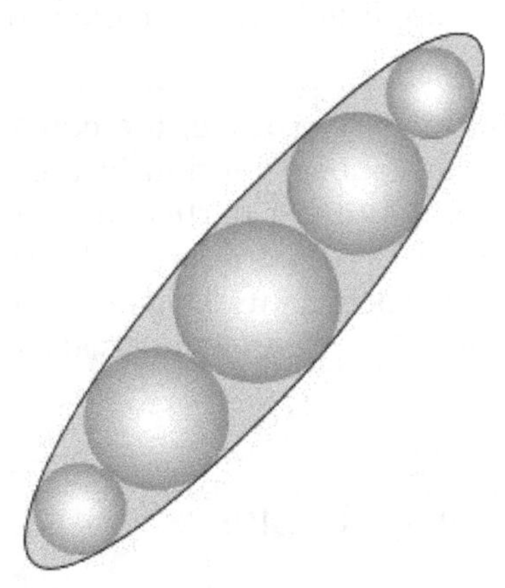

**Figure 3.7.** Linear sphere-connected model for evaluation of the repulsive interaction between two spheroidal particles.

The force and torque acting on a spheroidal particle of interest are also evaluated by a summation over all possible pairs of their constituent spherical particles. Although the above-mentioned sphere-connected model is composed of spheres with a fixed linear position in the spheroid, a more sophisticated linear sphere-connected particle model may be obtained by devising the position of the spheres. That is, the location of the first spheres on the center lines is determined by the minimum surface-to-surface distance between the spheroids and then the other spheres are located on the major axis around the first sphere. This method of treating a steric layer will be explained in more detail in the following chapter that addresses the criterion of the particle overlap. Another quite different approach to the modeling may be feasible by modifying the potential that has been developed for the interaction energy between two spheroidal molecules [2, 3] and this treatment will be addressed in the last part of the present section.

Although there are difficulties as mentioned above, regarding the treatment of the repulsive interaction, the introduction of the spheroidal particle model may have a significant advantage because the mathematical expressions for the translational and rotational diffusion coefficients of the spheroid are already known in the case of a dilute suspension. This is an important consideration when employing either the molecular dynamics or Brownian dynamics methods because the diffusion coefficients are indispensable for formulating the basic equations of motion to include both the translational and rotational motion. These more complex methods are required in order to investigate the dynamic characteristics of a suspension, such as magneto-rheological properties, whereas the Monte Carlo method is more applicable to a suspension in thermodynamic equilibrium.

In the case of an axisymmetric particle, the equation of motion can be decomposed into the translational motion in the particle axis direction and in the direction normal to the particle axis. If components normal and parallel to the particle axis are denoted by superscripts $\perp$ and $||$, respectively, the diffusion coefficients $D_{||}^{T}$ and $D_{\perp}^{T}$ for a slender spheroid with particle aspect ratio $r_p$ $(= a/b)$ are expressed as [4]

$$D_{||}^{T} = \frac{kT}{2\pi l \eta_s}\left[\frac{2}{2\ln 2r_p - 1} - \frac{4\ln 2r_p - 3}{\left\{4(\ln 2r_p)^2 - 4\ln 2r_p + 1\right\}\, r_p^2}\right]^{-1},$$

$$D_{\perp}^{T}=\frac{kT}{4\pi l\eta_s}\left[\frac{2}{2\ln 2r_p+1}-\frac{1}{\left\{4(\ln 2r_p)^2+4\ln 2r_p+1\right\}r_p^2}\right]^{-1} \tag{3.10}$$

These expressions are valid for $r_p$>>1.

Similarly, the rotational motion can be decomposed into a spin motion about the major axis direction and an ordinary rotational motion about a line normal to the particle axis. The rotational diffusion coefficients $D_{\parallel}^{R}$ and $D_{\perp}^{R}$ in these two cases are written as [4]

$$D_{\parallel}^{R}=\frac{3kT}{2\pi l^3\eta_s}r_p^2,\quad D_{\perp}^{R}=\frac{3kT}{\pi l^3\eta_s}\left[\frac{2}{2\ln 2r_p-1}+\frac{1}{\left\{4(\ln 2r_p)^2-4\ln 2r_p+1\right\}r_p^2}\right]^{-1} \tag{3.11}$$

In the above equations, $\eta_s$ is the viscosity of a base liquid, and $l$ is the particle length where $l$ = 2$a$. These expressions are valid for $r_p$>>1.

As already pointed out, these diffusion coefficients are required for the use of molecular dynamics or Brownian dynamics simulations.

Finally, we discuss another modeling technique for the repulsive interaction potential between two spheroidal particles covered by a soft repulsive layer. If an analytic expression for the interaction energy between two spheroidal particles covered by a soft repulsive layer were available, then modeling methods such as the sphere-connected particle model would not be necessary. Unfortunately, such a theoretical expression has not currently been proposed and it is desirable, from the viewpoint of accuracy, to develop a model more sophisticated than the sphere-connected particle model. In a non-spherical molecular system, the Gaussian and the modified Gaussian overlap potential model have been successfully proposed to describe the interaction energy between two molecules with the shape of spheroid [2, 3]. These models are based on the Lennard-Jones 12-6 potential which exhibits a significantly steep repulsive characteristic in the immediate vicinity of the overlap region, with a more gentle attractive energy outside this region. It is not possible to apply these potential models directly to a magnetic particle suspension because an effective repulsive interaction is required to act over the whole range of the thin soft repulsive layer (such as a steric layer) without a significant component of attractive interaction. However, it seems that these molecular potential models hold a key in order to develop an expression for the repulsive interaction that results from the overlap of the steric or electric layers covering the magnetic spheroidal particles. Thus, in the following, we briefly describe the Gaussian and the modified Gaussian overlap potential model for a spheroidal molecular system [2, 3, 5].

A spheroidal molecule $i$ is described by the center position vector $\boldsymbol{r}_i$ and the unit vector $\boldsymbol{e}_i$ denoting the particle direction. Then, for two particles, $i$ and $j$, the relative position vector $\boldsymbol{r}_{ji}$ is expressed as $\boldsymbol{r}_{ji} = \boldsymbol{r}_j - \boldsymbol{r}_i$, and the symbols $r_{ji}$ and $\boldsymbol{t}_{ji}$ are described as $r_{ji} = |\boldsymbol{r}_{ji}|$ and $\boldsymbol{t}_{ji} = \boldsymbol{r}_{ji}/r_{ji}$. Employing this notation, the simple Gaussian overlap potential model, $U_{ij}$, proposed for the interaction of two spheroidal molecules [5], is expressed as

$$U_{ij}(\boldsymbol{e}_i, \boldsymbol{e}_j, \boldsymbol{r}_{ji}) = \varepsilon(\boldsymbol{e}_i, \boldsymbol{e}_j) \exp\left\{ -\left( \frac{r_{ji}}{\sigma(\boldsymbol{e}_i, \boldsymbol{e}_j, \boldsymbol{t}_{ji})} \right)^2 \right\} \tag{3.12}$$

in which $\varepsilon(\boldsymbol{e}_i, \boldsymbol{e}_j)$ is the orientation-dependent strength and $\sigma(\boldsymbol{e}_i, \boldsymbol{e}_j, \boldsymbol{t}_{ji})$ is the orientation-dependent range of interaction energy. These parameters are expressed as

$$\varepsilon(\boldsymbol{e}_i, \boldsymbol{e}_j) = \varepsilon_0 \left\{ 1 - \chi^2 (\boldsymbol{e}_i \cdot \boldsymbol{e}_j)^2 \right\}^{-1/2} \tag{3.13}$$

$$\sigma(\boldsymbol{e}_i, \boldsymbol{e}_j, \boldsymbol{t}_{ji}) = \sigma_0 \left[ 1 - \frac{1}{2} \chi \left\{ \frac{(\boldsymbol{t}_{ji} \cdot \boldsymbol{e}_i + \boldsymbol{t}_{ji} \cdot \boldsymbol{e}_j)^2}{1 + \chi\, \boldsymbol{e}_i \cdot \boldsymbol{e}_j} + \frac{(\boldsymbol{t}_{ji} \cdot \boldsymbol{e}_i - \boldsymbol{t}_{ji} \cdot \boldsymbol{e}_j)^2}{1 - \chi\, \boldsymbol{e}_i \cdot \boldsymbol{e}_j} \right\} \right]^{-1/2} \tag{3.14}$$

In this equation, $\varepsilon_0$ and $\sigma_0$ are the scaling constant parameters denoting the strength and the range of the interaction whilst $\chi$ characterizes the anisotropy of the spheroid and is expressed using the particle aspect ratio $r_p = a/b$ as

$$\chi = (a^2 - b^2)/(a^2 + b^2) = (r_p{}^2 - 1)/(r_p{}^2 + 1) \tag{3.15}$$

The interaction between the two spheroidal molecules usually has a steep repulsive potential and a gentle attractive potential away from the vicinity of the near contact region. Hence, the simple Gaussian overlap potential $U_{ij}$, shown in Eq. (3.12), cannot describe these characteristics for the case of two spheroidal molecules, so that this overlap model potential is not suitable for application.

In order to improve the simple Gaussian overlap model, the following Gaussian overlap potential model based on the Lennard-Jones 12-6 potential, $U_{ij}$, was proposed for the interaction of two spheriodal molecules as [2]

$$U_{ij}(\boldsymbol{e}_i,\boldsymbol{e}_j,\boldsymbol{r}_{ji}) = 4\varepsilon(\boldsymbol{e}_i,\boldsymbol{e}_j)\left\{\left(\frac{\sigma(\boldsymbol{e}_i,\boldsymbol{e}_j,\boldsymbol{t}_{ji})}{r_{ji}}\right)^{12} - \left(\frac{\sigma(\boldsymbol{e}_i,\boldsymbol{e}_j,\boldsymbol{t}_{ji})}{r_{ji}}\right)^{6}\right\} \tag{3.16}$$

in which $\varepsilon(\boldsymbol{e}_i, \boldsymbol{e}_j)$ and $\sigma(\boldsymbol{e}_i, \boldsymbol{e}_j, \boldsymbol{t}_{ji})$ have already been shown in Eqs. (3.13) and (3.14). This modified potential clearly improves the original simple model in that there are positive and negative energy areas with zero interaction energy at $r_{ji} = \sigma(\boldsymbol{e}_i, \boldsymbol{e}_j, \boldsymbol{t}_{ji})$. The zero interaction energy may actually imply that the spheroidal molecules are in a surface-to-surface contact situation. However, this modified potential clearly has a weak point because the energy strength $\varepsilon(\boldsymbol{e}_i, \boldsymbol{e}_j)$ depends only on the orientational situation but not on the distance of the two spheroidal molecules. That is, in a more realistic model, the energy strength should be influenced not only by the orientational but also by the translational configuration of the two spheroidal molecules.

In order to remove this drawback and modify the above potential model, the modified potential model has been expressed as [3]

$$U_{ij}(\boldsymbol{e}_i,\boldsymbol{e}_j,\boldsymbol{r}_{ji}) = 4\varepsilon(\boldsymbol{e}_i,\boldsymbol{e}_j,\boldsymbol{t}_{ji})\left\{\left(\frac{\sigma_0}{r_{ji} - \sigma(\boldsymbol{e}_i,\boldsymbol{e}_j,\boldsymbol{t}_{ji}) + \sigma_0}\right)^{12} - \left(\frac{\sigma_0}{r_{ji} - \sigma(\boldsymbol{e}_i,\boldsymbol{e}_j,\boldsymbol{t}_{ji}) + \sigma_0}\right)^{6}\right\} \tag{3.17}$$

In this equation, the energy strength $\varepsilon$ is a function of the particle orientation $\boldsymbol{e}_i$ and $\boldsymbol{e}_j$ and the particle relative position (unit) vector $\boldsymbol{t}_{ji}$ as

$$\varepsilon(\boldsymbol{e}_i,\boldsymbol{e}_j,\boldsymbol{t}_{ji}) = \varepsilon_0\,(\varepsilon_1(\boldsymbol{e}_i,\boldsymbol{e}_j))^{\nu}\,(\varepsilon_2(\boldsymbol{e}_i,\boldsymbol{e}_j,\boldsymbol{t}_{ji}))^{\mu} \tag{3.18}$$

in which $\varepsilon_1$ and $\varepsilon_2$ are written as

$$\varepsilon_1(\boldsymbol{e}_i,\boldsymbol{e}_j) = \left\{1 - \chi_1^{\,2}(\boldsymbol{e}_i \cdot \boldsymbol{e}_j)^2\right\}^{-1/2} \tag{3.19}$$

$$\varepsilon_2(\boldsymbol{e}_i,\boldsymbol{e}_j,\boldsymbol{t}_{ji}) = 1 - \frac{1}{2}\chi_2\left\{\frac{(\boldsymbol{t}_{ji}\cdot\boldsymbol{e}_i + \boldsymbol{t}_{ji}\cdot\boldsymbol{e}_j)^2}{1 + \chi_2\,\boldsymbol{e}_i\cdot\boldsymbol{e}_j} + \frac{(\boldsymbol{t}_{ji}\cdot\boldsymbol{e}_i - \boldsymbol{t}_{ji}\cdot\boldsymbol{e}_j)^2}{1 - \chi_2\,\boldsymbol{e}_i\cdot\boldsymbol{e}_j}\right\} \tag{3.20}$$

The constants $\chi_1$ and $\chi_2$ have been introduced in order to adjust the ratio of side-by-side and end-to-end energy well depths, expressed as

$$\chi_1 = \frac{r_p^2 - 1}{r_p^2 + 1}, \quad \chi_2 = \frac{1 - (\varepsilon_e / \varepsilon_s)^{1/\mu}}{1 + (\varepsilon_e / \varepsilon_s)^{1/\mu}} \tag{3.21}$$

in which $\varepsilon_s$ signifies the strength parameter for a side-by-side configuration and $\varepsilon_e$ is the value for an end-to-end configuration.

The present modified potential shown in Eq. (3.17) has the form of exhibiting a gentle attractive interaction apart from zero energy at $r_{ji} = \sigma(\boldsymbol{e}_i, \boldsymbol{e}_j, \boldsymbol{t}_{ji})$ with a significantly steep repulsive core in the vicinity range of the surface-to-surface contact region. Moreover, the strength of the interaction energy is now dependent on the translational and rotational configuration of the two particles, so that this modified Gaussian overlap potential seems to be significantly more suitable as a model for the interaction of spheroidal particles than the previous Gaussian overlap model expressed by Eq. (3.16). In the application of the potential expressed by Eq. (3.17) to molecular simulations of spheroidal molecules, it is necessary to specify the values of constants denoted by $\sigma_0$, $\varepsilon_0$, $\nu$, $\mu$ and $\varepsilon_e/\varepsilon_s$. On the other hand, the particle aspect ratio $r_p$ is known if the spheroidal molecules of interest are specified by the investigation. It had been shown that the following specification leads to reasonable results for the aspect ratio $r_p = 3$: $\nu = 1$, $\mu = 2$ and $\varepsilon_e/\varepsilon_s = 0.2$ [3, 6]. In the following, we briefly discuss the characteristics of the potential curves for the above Gaussian and modified Gaussian interaction potentials for the situation of these specified parameters and with the particle aspect ratio $r_p = 3$.

Figure 3.8 shows the potential curves as a function of the center-to-center distance between the two spheroidal molecules with the same size: Fig. 3.8(a) is for the Gaussian and Fig. 3.8(b) is for the modified Gaussian interaction model. Each figure shown has results for both the side-by-side configuration and the end-to-end configuration of the particles. It is noted that the energy and the distance are non-dimensionalized by the strength constant $\varepsilon_0$ and the range constant $\sigma_0$, respectively, and the abscissa $r^*$ in the figures implies the quantities $r_y/\sigma_0$ or $r_x/\sigma_0$.

It is seen from Fig. 3.8(a) that for the case of the side-by-side configuration, the potential curve has a significantly steep repulsive core in the overlap range, becomes zero energy at $r_x = 1$, attains to a minimum with a potential of $-1.666$ at the center-to-center distance $r_x = 1.12$, and then increases and finally tends to converge to zero at $r_x \simeq 2.5$, which is smaller than the long axis length $r_p = 3$. These short-range characteristics are certainly reflected by the Lennard-Jones potential shown in Eq. (3.16). For the case of the end-to-end configuration, the above-mentioned characteristics are similarly described and have the same minimum value, but the attractive interaction range is significantly longer than the previous configuration as the interaction energy tends to zero at $r_y \simeq 7.5$, which corresponds to almost 2.5 times the long axis length $r_p = 3$.

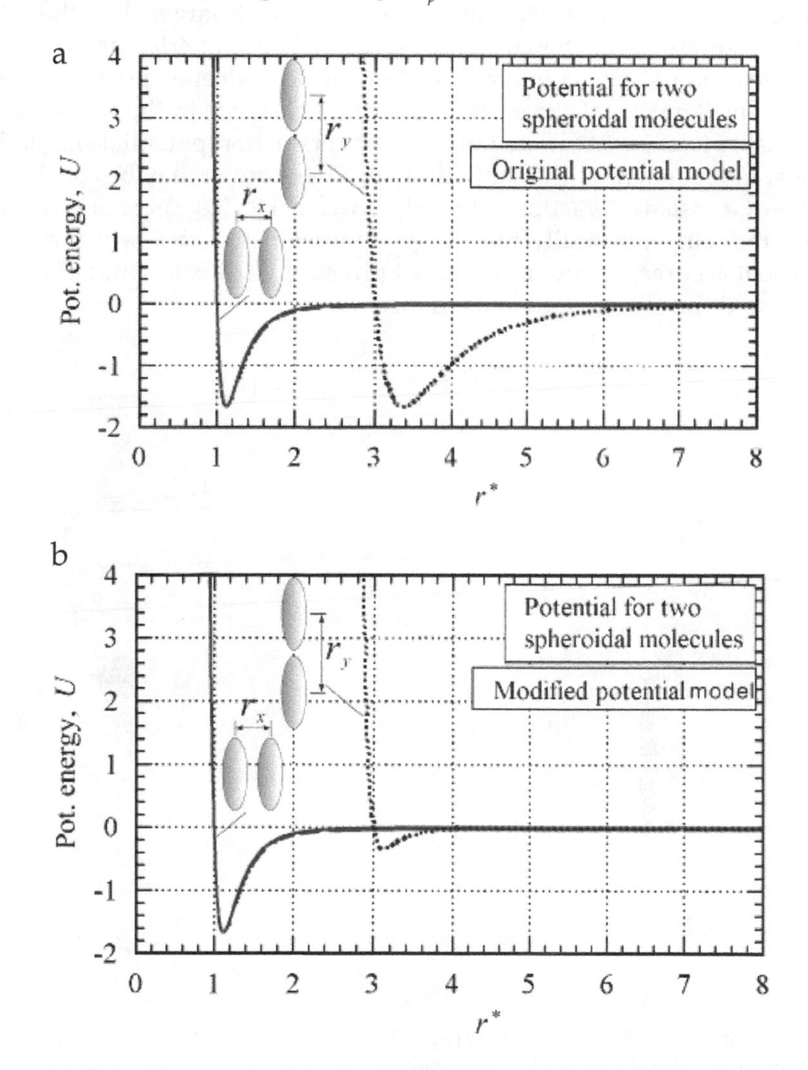


**Figure 3.8.** Potential energy for the spheroidal particle model: (a) original potential model and (b) modified potential model.

On physical grounds, it seems unreasonable that the Gaussian interaction potential has the same minimum energy depth for the end-to-end and side-by-side configurations and the former configuration has a longer range for the influence of the potential. The end-to-end configuration should be expected to yield a minimum well depth shallower than the side-by-side configuration from the following considerations demonstrated using Fig. 3.9. For the same overlap depth, the overlap area for the side-by-side configuration, shown in Fig. 3.9(a), is clearly much larger than the end-to-end configuration, shown in Fig. 3.9(b). A larger overlap area implies that there is a stronger interaction, leading to a deeper potential well. From consideration of the surface-to-surface distance in the figures, it is reasonable to expect that the range of the interaction potential should be almost independent of the configuration. Therefore, it may be concluded that the Gaussian interaction potential, shown in Fig. 3.8(a), cannot succeed in reproducing physically reasonable characteristics and will not have sufficient accuracy in order to conduct physically realistic simulations for a system composed of spheroidal molecules.

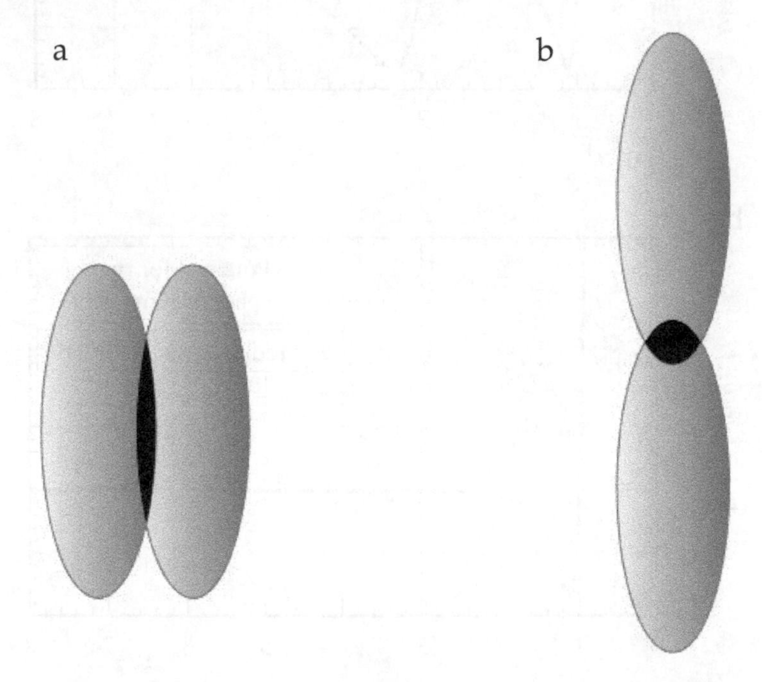


**Figure 3.9.** Strength of interaction energy dependent on the configuration of the two particles: (a) side-by-side interaction and (b) end-to-end interaction.

The modified Gaussian interaction potential, Fig. 3.8(b), exhibits the same characteristic for the side-by-side configuration as the Gaussian interaction discussed above. On the other hand, the modified Gaussian potential curve for the end-to-end configuration, certainly implies an improvement to the drawbacks of the original Gaussian potential and shows an improvement toward physically reasonable characteristics. That is, the depth of the potential well is much shallower, a 5 times reduction of the side-by-side configuration and also the interaction potential for the end-to-end configuration has a much shorter range of influence than the original. With these considerations, we conclude that the modified Gaussian interaction potential is a realistic potential model for the molecular simulation of a system of spheroidal molecules with short-range interactions between molecules.

The modified Gaussian potential shown in Eq. (3.17) is not directly applicable to a suspension composed of magnetic particles with spheroidal shape, because the interaction potential exhibits only a relatively short range repulsive interaction for the model of the overlap of a steric or electric repulsive layer covering the magnetic particles. Hence, in order to apply the modified Gaussian potential to the magnetic particle suspension, an interaction potential is required to have a significant positive energy barrier for preventing magnetic particles from aggregating within a short influence range of the repulsive overlap situation. Development of the interaction potential between magnetic spheroidal particles covered by a repulsive layer will be advanced in future.

## 3.3 Spherocylinder particles

The spherocylinder particle model has similarities with the previous spheroidal particle model in shape, but the geometry of this model has a significant advantage in that it is straightforward to access the criteria for the overlap of the repulsive layers, such as steric or electrical layers, of the two particles. On the other hand, the translational and rotational diffusion coefficients for spherocylinders are not known as mathematical expressions, which is seen to be a weak point in employing this particle model in molecular dynamics or Brownian dynamics simulations. However, this is not a serious problem because analytical expressions for the diffusion coefficient are well known for the cylindrical particle and may be satisfactorily applied in the manner of an approximation to the present particle model. Figure 3.10 shows a spherocylinder particle with diameter $d$ and length $l$, covered by a uniform steric layer with thickness $\delta$.

Similar to the previous spheroidal particle, Figs. 3.10(b), (c) and (d) show three magnetization models, i.e., (b) a plus and minus magnetic charge model, (c) a dipole model at the particle center, aligned in the major axis direction, that is suitable for ferromagnetic particles, and (d) a dipole model at the particle center, aligned in the minor axis direction, that are suitable for spindle-like hematite particles.

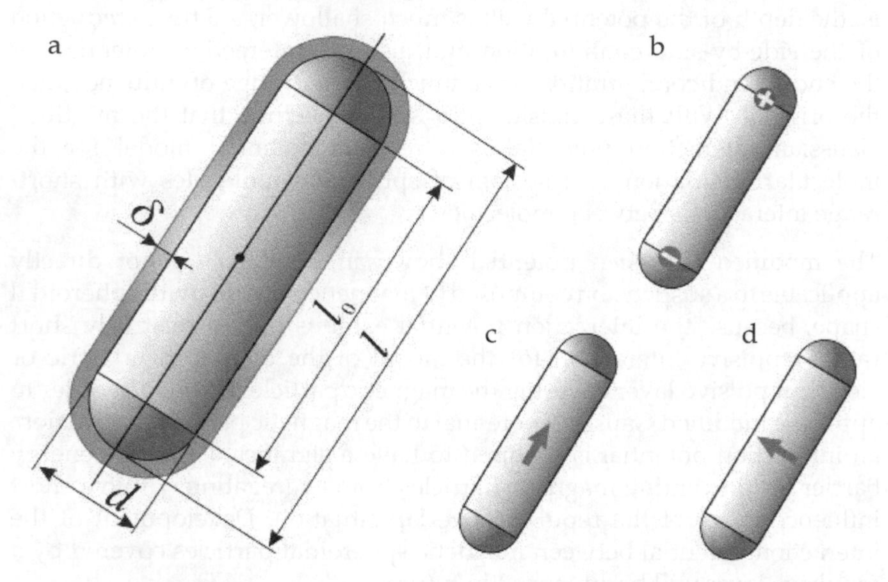


**Figure 3.10.** Spherocylinder particle model with several magnetization models: (a) spherocylinder covered by a soft repulsive layer, (b) magnetic charges model, (c) ferromagnetic particle model and (d) hematite particle model.

Forces, energies and torques acting on the spherocylinder particles due to the interactions both between the particles and between the particles and an external magnetic field are expressed by the same equations corresponding to the previous case for spheroidal particles. Currently the mathematical expressions for repulsive interactions due to the overlap of the soft repulsive layers have not been established for the spherocylinder particle, and therefore, as in the spheroidal particle case, the linear sphere-connected model also may be applicable. Since treatment of the spherocylinder model is slightly different from the spheroidal model explained in Section 3.2, we explain the present sphere-connected particle model in detail in the following.

The simplest linear sphere-connected model is obtained by having the same-sized spheres located in linear contact along the particle axis direction where the position of each constituent sphere is fixed. However, in this simple model, a physically unreasonable torque may act on the spherocylinder particle, simply arising in certain overlapping situations from a geometric anomaly regarding the constituent spheres. Therefore a more sophisticated linear sphere-connected model is explained in the following, in order to avoid this kind of unreasonable physical anomaly.

In a more sophisticated model, the constituent spheres are not fixed at constant positions for the analysis of overlapping situations, but rather, a constituent sphere is initially located on each particle axis line at the position of nearest overlap. The neighboring spheres are then added in contact along each particle axis to complete the construction of two sphere-connected rod-like particles. If the positions of spherocylinder particles $i$ and $j$ are denoted by $\boldsymbol{r}_i$ and $\boldsymbol{r}_j$, respectively, then the total force acting on particle $i$, $\boldsymbol{F}_{ij}^{(V)}$, can be obtained by summing the interaction force $\boldsymbol{F}_{ab}^{(V)}$ between sphere $a$ belonging to particle $i$ and sphere $b$ belonging to $j$. Hence, the torque $\boldsymbol{T}_{ij}^{(V)}$ is evaluated as

$$\boldsymbol{T}_{ij}^{(V)} = \sum_{a}\sum_{b}(\boldsymbol{r}_a - \boldsymbol{r}_i)\times \boldsymbol{F}_{ab}^{(V)} \tag{3.22}$$

in which $\boldsymbol{r}_a$ is the position vector of constituent sphere $a$. The detailed explanation regarding the criterion of the particle overlap will be provided in the following chapter.

As already pointed out, the expressions of the diffusion coefficients for the cylinder with diameter $d$ and length $l$ may be applicable to the current case of a spherocylinder. These diffusion coefficients are written as $D_{\parallel}^{T}$, $D_{\perp}^{T}$, $D_{\parallel}^{R}$ and $D_{\perp}^{R}$, expressed as [7–10]

$$D_{\parallel}^{T} = \frac{kT}{2\pi l\eta_s}\left\{\ln 2r_p + \ln 2 - 3/2\right\}, \quad D_{\perp}^{T} = \frac{kT}{4\pi l\eta_s}\left\{\ln 2r_p + \ln 2 - 1/2\right\} \tag{3.23}$$

$$D_{\parallel}^{R} = \frac{kT}{\pi l^3\eta_s} r_p^2, \quad D_{\perp}^{R} = \frac{3kT}{\pi l^3\eta_s}\left\{\frac{1}{\ln r_p}\left(1+\frac{\ln 2 - 1}{\ln r_p}\right) + \frac{3\times 5.45}{8\pi}\cdot\frac{1}{r_p^2}\right\}^{-1} \tag{3.24}$$

In the above equations, $\eta_s$ is the viscosity of a base liquid and $r_p$ is the particle aspect ratio, expressed as $r_p = l/d$ or $a/b$. These expressions are valid for the cylinder particle with $r_p$>>1.

Finally, we discuss characteristics of the potential curves for typical translational and orientational configurations. For the case of the dipole model of magnetization, the dependence of the potential energy on the particle center-to-center distance is exactly the same as for the spherical particle system in the case of the parallel configurations, shown in Fig. 3.3. Hence we focus here on characteristics of the potential energy as a function of the center-to-center distance in the $y$-axis direction for the two parallel configurations, i.e., with the same and the opposite orientation of the magnetic moments, which are shown in Fig. 3.11. For the case of the opposite-orientation of the magnetic moments, the potential energy is a minimum when in the side-by-side contact situation and increases through a positive maximum value at $r_y^* \simeq 1.2$ before converging to zero. When the configuration of the moments has the same-orientational direction, the configuration shows the opposite characteristics, that is, there is a maximum energy barrier in the contact situation. These characteristics clearly imply that in the case of no external magnetic field, a suspension of magnetic spherocylinder particles exhibits a raft-like cluster formation where the magnetic moments of the neighboring particles tend to incline in opposite directions to each other, which therefore leads to a lower system energy and a more stable system. On the other hand, in the situation of a strong applied magnetic field, a linear chain formation along the magnetic field direction is certainly more preferable as in the case of a magnetic spherical particle suspension shown in Fig. 3.3.

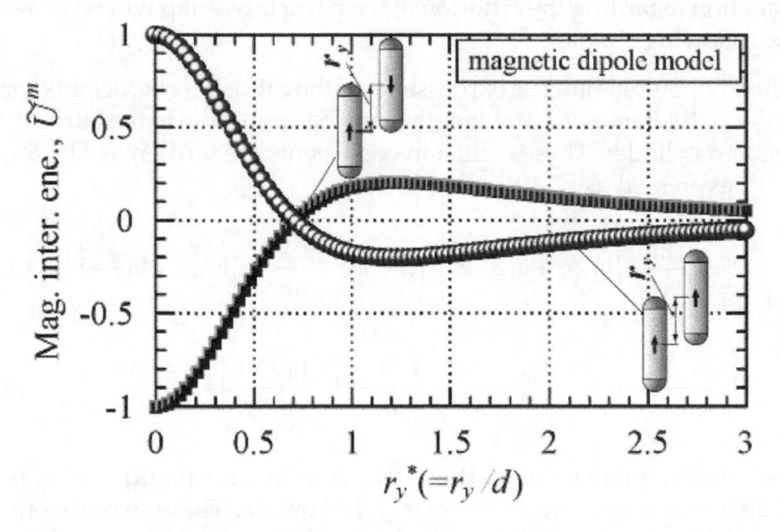


**Figure 3.11.** Potential energy for the dipole model as a function of the center-to-center distance along the major axis direction.

Next we discuss characteristics of the potential curves for the case of the magnetic charge model, shown in Figs. 3.12(a) and (b): it is noted that $\tilde{U}^m = U_{ij}/(\mu_0 q^2/4\pi d)$ where $U_{ij}$ is given in Eq. (2.19). We first discuss the dependence of the potential energy on the center-to-center distance in the $x$-direction, shown in Fig. 3.12(a). In the case of the opposite-charge configuration, the energy exhibits a minimum in the contact situation and monotonically increases toward zero with increasing distance.

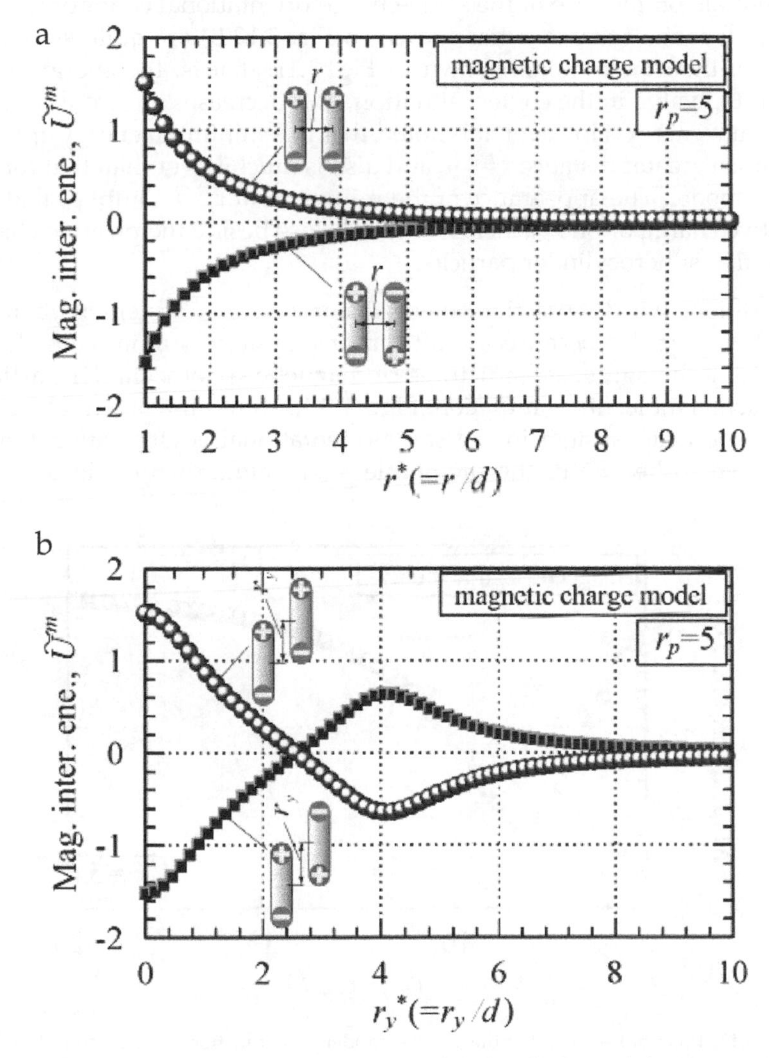


**Figure 3.12.** Potential energy for the charge model as a function of the center-to-center distance (a) along the minor axis direction and (b) along the major axis direction.

The same-charge orientational configuration of magnetic moments shows the completely opposite characteristics to those of the previous configuration. Figure 3.12(b) shows results for the dependence of the potential energy on the center-to-center distance $r_y^*$ in the $y$-direction. As in the previous cases, the opposite-charge orientational configuration of the magnetic moments exhibits the completely opposite characteristics to those of the same-charge orientation configuration, and therefore we concentrate on the case of the same-charge orientational configuration. It is seen that the characteristics shown in Fig. 3.12(b) are quite similar to those of the dipole model shown in Fig. 3.11, that is, the energy has a maximum value in the contact situation, and decreases to zero through a minimum energy. However, it is noted that this minimum energy appears at a much greater distance $r_y^* \simeq 4$, and also is much lower than that for the dipole model. The appearance of the minimum at $r_y^* \simeq 4$ implies that the negative charge of one particle is located close beside the positive charge of the first spherocylinder particle.

In addition, we discuss the potential curves for a system with more than two particles because a multi-particle system is more suitable for discussing the aggregate structures of magnetic spherocylinder particles in an actual molecular simulation. Figure 3.13 shows the potential energy for a 4-particles system in the same-orientational configuration that is presumed to be one of the preferable aggregate structures in a strong

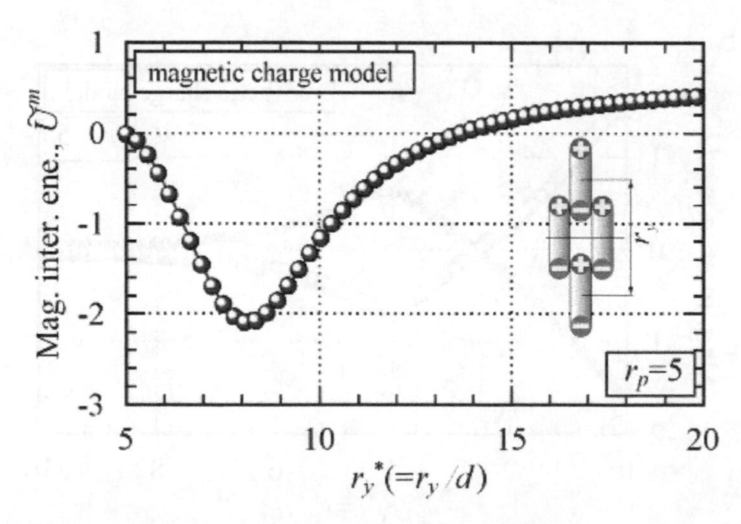


**Figure 3.13.** Potential energy for the charge model as a function of the center-to-center distance along the major axis direction for 4-particles system in a strong external magnetic field.

applied magnetic field. The abscissa implies the center-to-center distance between the two particles existing on the center line ($y$-axis): the value of $r_y^* = 5$ corresponds to the end-to-end contact configuration. It is seen from Fig. 3.13 that even for this 4-particles system, a minimum potential energy appears in the configuration of two linear center particles with charges being located near the opposite charges of the two adjacent particles. The minimum energy is (–2.099) and the position giving rise to the minimum energy is $r_y^* = 8.14$, which is slightly longer than the minimum alignment of positive and negative charges of $r_y^* = 8$. This is due to the influence of the interactions between the center-line particles and outside particles with the same (positive or negative) charge.

From these considerations, we may expect the formation of larger clusters in a multi-particle system in a strong applied magnetic field and these may be constructed by connecting together the basic structure of the above-mentioned 4-particles system.

Finally, we briefly describe the commonest aggregate formation of the hematite spherocylinder particles that are illustrated in Fig. 3.14. In a suspension of magnetic particles magnetized in the minor axis direction,

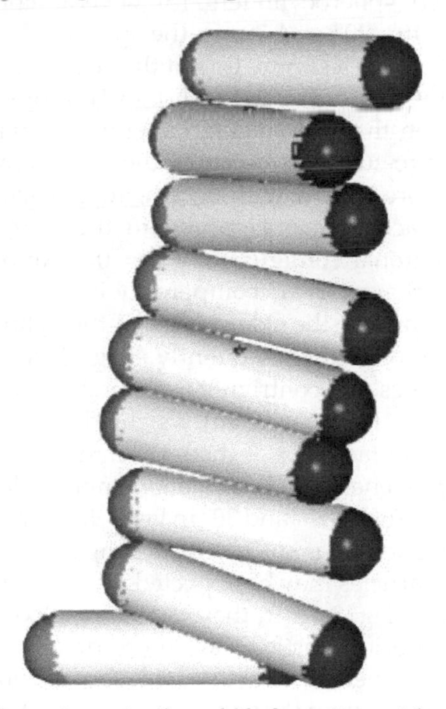

**Figure 3.14.** Raft-like cluster formation for rod-like hematite particles.

the most preferable configuration in a strong external magnetic field is that all magnetic moments align along the magnetic field direction or in the case of no applied magnetic field they will align along a non-specific direction which leads to a lower system energy. Hence, in a suspension of rod-like hematite particles in which the magnetic particle-particle interactions are much stronger than the thermal energy, the raft-like cluster formation, shown in Fig. 3.14, is significantly preferable.

Consideration of repulsive interactions due to the overlap of the steric layers enables one to expect possible cluster formation of magnetic rod-like particles in a strongly-interacting system, as shown in Fig. 3.14. We focus here on the sphere-connected particle model previously mentioned for calculating the interaction energy arising from the overlap of steric layers. For a given translational arrangement and orientational configuration of two spherocylinders that have no magnetic properties, the net potential energy of the repulsive interaction is evaluated by computing the interaction energy between all pairs of spheres constituting each spherocylinder particle. Figure 3.15 shows the potential curves as a function of the separation distance, along the $z$-axis direction, between the centers of two spherocylinders for several configurations, with orientation $\theta = 0°$, $10°$, $20°$ and $30°$ in the $xy$-plane. The quantity of the ordinate, $\tilde{U}^s$, is defined as a summation of the energy $u_{ij}^{(V)*}/\lambda_V$, for each pair of spheres, where $u_{ij}^{(V)*}$ is expressed in Eq. (3.4). From the results shown in Fig. 3.15, it is seen that the potential energy gradually increases from zero at the surface-to-surface contact of the steric layers ($r_z^* = 1.3$) and tends to increase more steeply with decreasing particle-particle distance. Although this characteristic is common to the four orientation cases, the parallel orientational configuration, $\theta = 0°$, exhibits a considerably stronger repulsive characteristic compared with the other cases. This is because the contact area of the side surface of the cylinders is lengthened in the parallel situation and consequently a significantly greater number of surfactant molecules will contribute to the repulsive interaction energy between the particles. Hence, a slightly slanted configuration leads to a significantly lower contact area of the surfactant layers and therefore to much lower potential energies. This phenomenon is clearly recognized in the curves for the case of $\theta = 20°$ and $30°$ in Fig. 3.15. If the magnetic particle-particle interaction strength is sufficiently strong for cluster formation, the magnetic rod-like particles may be expected to aggregate to form raft-like clusters, as shown in Fig. 3.14. In these stable raft-like clusters, the steric layers of the neighboring particles are more deeply compressed in the case of a stronger magnetic interaction, thereby causing larger repulsive interactions. In order to avoid the instability of the raft-like clusters arising from these steric interactions, the neighboring particles tend to incline in a slightly slanted orientational configuration. This is expected from the previous consideration regarding the potential characteristics of the sphere-connected particle model shown in Fig. 3.15.

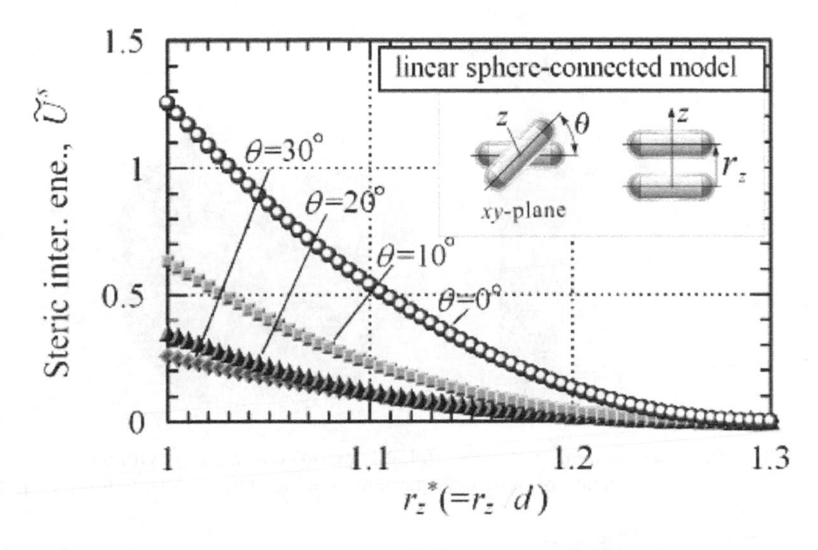


**Figure 3.15.** Potential energy as a function of the center-to center distance along the minor axis direction for the two spherocylinder particles coated by a uniform steric layer for several cases of the slanted orientational configuration in the *xy*-plane.

Hence, we may conclude that if a strongly-interacting system of magnetic rod-like particles is treated, the particles aggregate to form raft-like clusters where the magnetic moments of the neighboring particles in a cluster incline in a slightly slanted orientational configuration. The cluster formation, of course, depends on the magnetic properties of the rod-like particles and also the strength of an applied magnetic field. Consequently, it is the direction of the magnetic moment in the particle body that overwhelmingly determines the cluster formation.

## 3.4 Disk-like particles

A disk-like hematite particle [11] is modeled as a short cylinder or disk, surrounded by a torus with cross section of semicircle shape, as shown in Fig. 3.16. The disk-like hematite particle is modeled with a magnetic moment in the plane of the disk at the particle center. Since the model of the dipole moment is employed, the magnetic particle-field interaction energy $u_i^{(H)}$, the magnetic particle-particle interaction energy $u_{ij}^{(m)}$, the magnetic force $\boldsymbol{F}_{ij}^{(m)}$ between particles $i$ and $j$, the magnetic torque $\boldsymbol{T}_{ij}^{(m)}$, and the torque due to the magnetic particle-field interaction, $\boldsymbol{T}_i^{(H)}$, are all the same as in the magnetic spherical particle system, explained in Section 3.1.

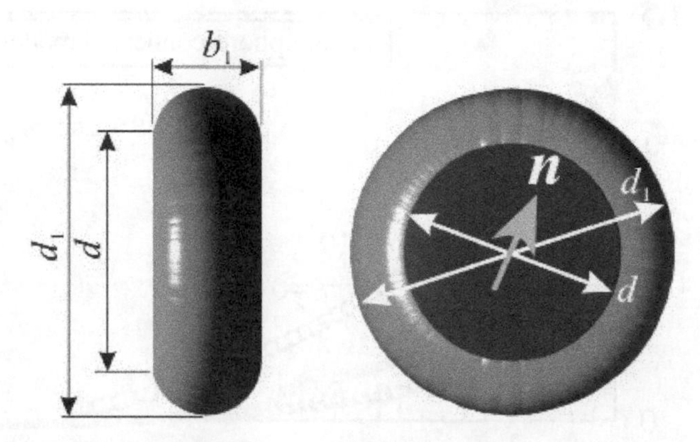


**Figure 3.16.** Disk-like hematite particle model: it is modeled as a short cylinder covered by a torus part with cross section of semicircle shape with a magnetic moment along the plane at the particle center.

Similar to a suspension of particles with spheroidal or spherocylinder geometry, from a simulation point of view, it is a difficult problem to treat the steric or electric layer covering these disk-like particles, and for the same reason, a possible approach is to employ the expression used for spherical particles. To do so, the disk-like particle is modeled as being composed of spherical particles. However, a simple modeling with constituent spherical particles located at specific positions in the disk-like particle may give rise to an unfavorable torque that induces unreasonable rotational motion. Hence, we employ the following sophisticated modeling in order to remove this kind of unreasonable behavior arising from the simpler model. Instead of locating spherical particles at predefined or fixed positions in the disk, the spherical particles are relocated at appropriate positions whenever particles overlap. Firstly, we evaluate the nearest or deepest overlapping positions of the two disk-like particles of interest. For instance, with reference to the configuration on the left-hand side in Fig. 3.17, the torus part of the upper particle is assumed to be in a configuration of nearest contact with the disk plane, i.e., not with the torus section on the lower particle. In this configuration, a spherical particle with diameter $b_1$ is placed in each disk-like particle in the following manner. A spherical particle on the upper disk-like particle is moved around the torus part and a spherical particle on the lower disk-like particle is moved in the disk plane in order to obtain the positions of these spherical particles that give rise to the shortest distance. Then, starting with a spherical particle placed at each of these positions, the other spherical particles are placed in a close-packed configuration to complete the formation of each disk-like particle. The concrete method of locating the other spherical particles is explained below in detail. With the configuration of the two disk-like particles as viewed in Fig. 3.17, the lower particle is seen as a circle. Then, the other particles are located on a line that is determined by the centers of the starting particles. Finally, the other particles are located on lines parallel to this line in a compact and close-packed manner in order to complete the form of the disk-like particles. Each constituent spherical particle is assumed to be coated with a uniform steric layer. The net repulsive interaction between the two disk-like particles due to the overlap of the steric layers may now be evaluated by summing the interaction between all the pairs of the two constituent spherical particles (belonging to the two different disk-like particles). This approach in terms of the multi-sphere model (sphere-constituting model) for evaluating the steric repulsion between two disk-like particles does not give rise to a torque that induces unreasonable particle rotations.

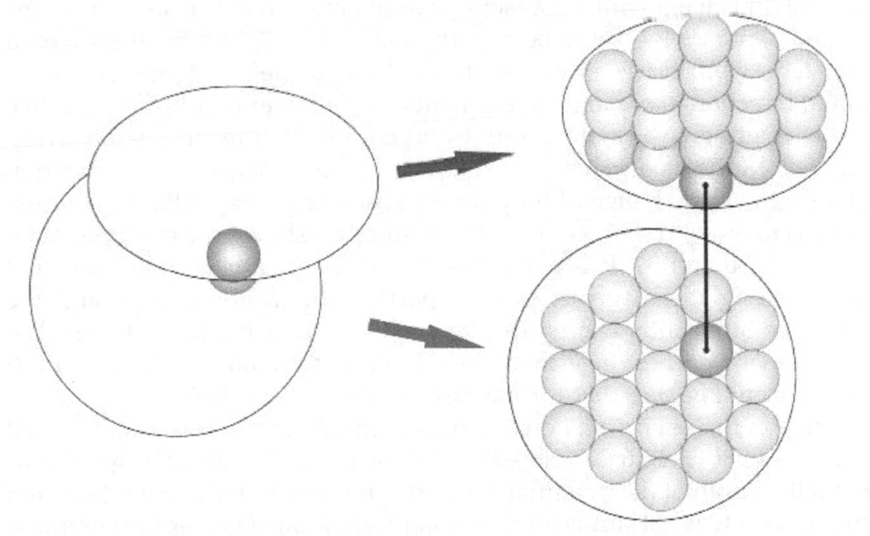

**Figure 3.17.** Evaluation of steric interactions between two disk-like particles based on the sphere-constituting model: a disk-like particle is assumed to be composed of spherical particles in order to apply the steric interaction between two spherical particles to the present interaction of two disk-like particles.

Here we address the potential interaction energy between the two magnetic disk-like particles with aspect ratio $r_p (= d_1/b_1) = 5$, coated with a steric layer with thickness $\delta = 0.15b_1$ and modeled with a dipole moment in a direction within the disk-plane at the particle center. The above steric interaction model is used for calculating the repulsive interaction energy, and the notation shown in Fig. 3.16 is used. Moreover, as shown in Section 3.1, the non-dimensional parameter $\lambda$ is used for the magnetic dipole-dipole interaction strength and a value of $\lambda_V = 150$ has been adopted for the parameter $\lambda_V$ implying the steric interaction strength. The following discussion for understanding why a strong particle-particle interaction, such as implied by the value of $\lambda \simeq 40$, is necessary in order for the magnetic disk-like particles to aggregate to form column-like clusters, shown in Fig. 3.20, whereas in contrast, a value of only $\lambda \simeq 4$ is a sufficient strength for cluster formation in the case of a magnetic spherical particle suspension.

We discuss here how the potential energy varies as a function of a slight angle change between the particle directions from an initial configuration where the two disk-like particles are located in parallel with the magnetic moments aligning completely in opposite directions. It is noted that the angle between the two magnetic moments varies for a change in the particle direction. In detail, the two disk-like particles are located on the $z$-axis with alignment in the $z$-direction, and the center-to-center separation is fixed with a contact situation between the surfactant layers, i.e., the separation is set to be $r^* (= (b_1 + 2\delta)/b_1) = 1.3$, and the thickness of the steric repulsive layer $\delta$ is set to be $0.15b_1$ as before. Moreover, as an initial direction, the magnetic moments of the lower and upper disk-like particles are set to be in the positive and negative $x$-direction, respectively. Figure 3.18 shows the change in the net potential energy $u^{net}$ as a function of a slight angle change of the upper particle from the initial orientation (0, 0, 1) to $(e_x, e_y, (1 - e_x^2 - e_y^2)^{1/2})$. The results are shown for the three cases of $\lambda = 10$, 30 and 50. It is noted that the interaction energy is evaluated by summing the magnetic particle-particle interaction energy and the repulsive interaction energy due to the overlap of the steric layers, but the magnetic particle-field interaction energy is not taken into account. It is seen from Fig. 3.18 that for the case of $e_x = e_y = 0$ with the two particles aligning in the same direction, the net potential curve has a potential well of $u^{net} = -4.55kT$, $-13.65kT$ and $-22.76\ kT$ for $\lambda = 10$, 30 and 50, respectively. From the depth of the potential wells, it is understood that the interaction strength $\lambda = 10$ is not sufficient for cluster formation, whereas the latter two strengths are expected to be sufficiently strong for the formation of stable aggregates in the parallel configuration. On the other hand, for the case of a slightly slanted configuration, $e_x = 0.1$ and $e_y = 0$, high energy barriers of

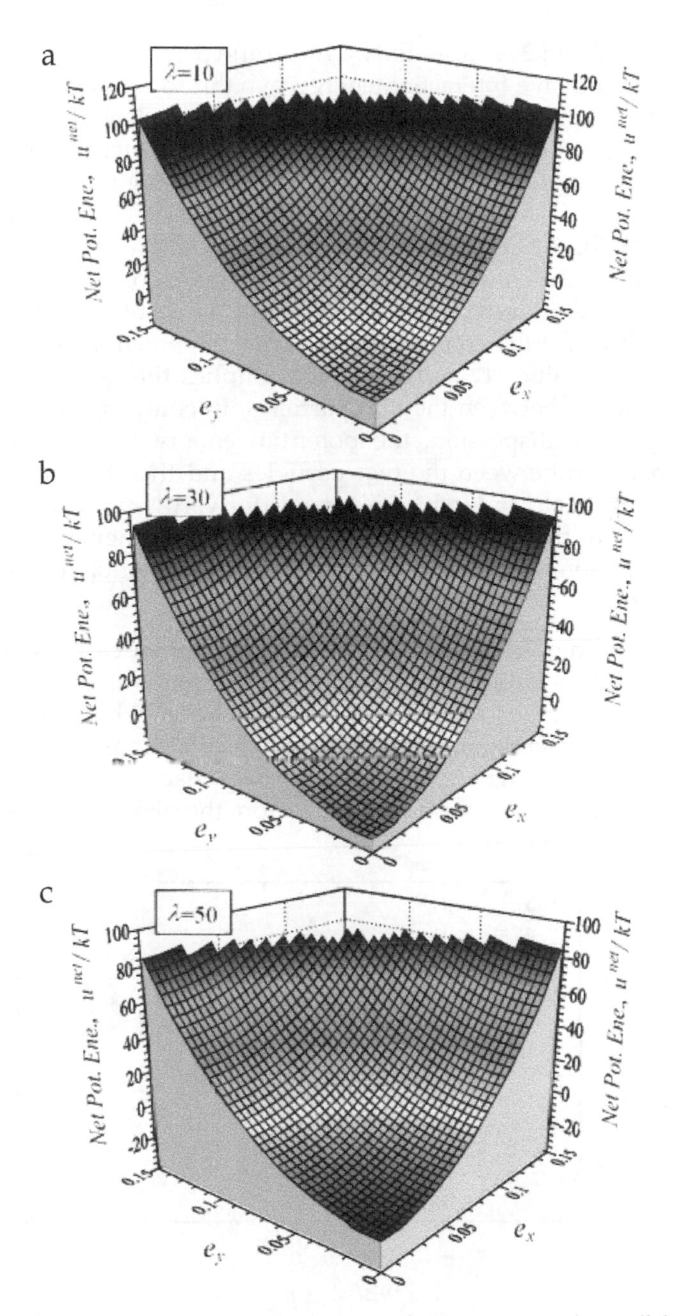


**Figure 3.18.** Steep increase in the interaction energy between two nearly parallel particles by change in the particle axis directions deviating from each other for (a) $\lambda = 10$, (b) $\lambda = 30$ and (c) $\lambda = 50$: 3D image.

$u^{net} \simeq 50kT$, $40kT$ and $20kT$ arise for $\lambda = 10$, 30 and 50, respectively, implying a significant repulsive interaction energy even in the situation of a small slant angle configuration. In order to discuss these characteristics of the potential energy more clearly, Fig. 3.19 shows the potential curves as a function of a slanted angle $\theta$ of the upper particle, changing from the $z$-axis toward the $x$-axis, for the various cases of the particle-particle interaction strength $\lambda$. We focus here on the curve for $\lambda = 20$ in Fig. 3.19. It is seen from Fig. 3.19 that for the case of the completely parallel configuration, $\theta = 0°$, the potential well exhibits a negative depth, $u^{net} = -9.10kT$, but for the case of just a slightly slanted configuration, $\theta = 4°$, the potential energy yields a large positive value, $u^{net} \simeq 10kT$, which implies that a large repulsive force ought to act between the two particles. In contrast, for a magnetic spherical particle dispersion, the potential energy is dependent only on the separation between the two particles and therefore the particles start to aggregate at $\lambda \simeq 4$, which is straightforwardly predicted from the characteristics of the potential curve; for $\lambda \gtrsim 4$, the magnetic interaction becomes more dominant than the particle Brownian motion for a spherical particle system.

From these considerations, we may conclude that for the case of the present disk-like particle, a large repulsive force is sufficiently likely to arise due to even a slight change in the configuration which may be induced by the rotational and translational Brownian motion of the particles. Hence, in comparison with a magnetic spherical particle suspension, a sufficiently strong magnetic interaction is necessary before the disk-like particles tend

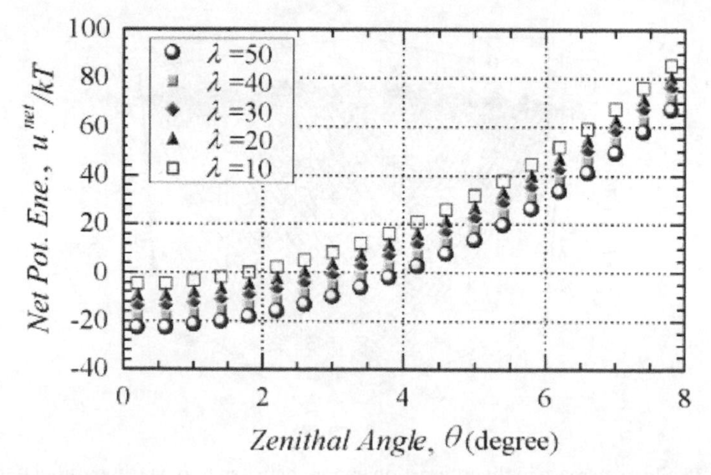


**Figure 3.19.** Steep increase in the interaction energy between two nearly parallel particles by change in the particle axis directions deviating from each other: 2D potential profile.

to aggregate to form clusters. In the above representative example of a disk-like particle with aspect ratio $r_p = 5$, the column-like clusters start to be significantly formed from a magnetic interaction strength of $\lambda \simeq 40$.

The diffusion coefficients for the present disk-like particle are not known as analytical expressions, and therefore as a first approximation the approach is to employ the expressions for the oblate spheroidal particle with diameter $d_1$ and thickness $b_1$. The diffusion coefficients in the particle direction, $D_{||}^T$, and in the perpendicular direction, $D_{\perp}^T$, for the oblate spheroid are expressed as [4]

$$D_{||}^T \Big/ \frac{kT}{6\pi(d_1/2)\eta} = \frac{3}{4} \cdot \frac{(2\hat{s}^2-1)Q + \hat{s}(1-\hat{s}^2)^{1/2}}{\hat{s}^3},$$

$$D_{\perp}^T \Big/ \frac{kT}{6\pi(d_1/2)\eta} = \frac{3}{8} \cdot \frac{(2\hat{s}^2+1)Q - \hat{s}(1-\hat{s}^2)^{1/2}}{\hat{s}^3} \tag{3.25}$$

Similarly, the diffusion coefficient $D_{||}^R$ for the spin rotational motion about major axis direction and the diffusion coefficient $D_{\perp}^R$ for the ordinary rotational motion about a line normal to the particle axis line are written as [4]

$$D_{||}^R \Big/ \frac{kT}{\pi d_1^3 \eta} = \frac{3}{2} \cdot \frac{Q - \hat{s}(1-\hat{s}^2)^{1/2}}{\hat{s}^3}, \quad D_{\perp}^R \Big/ \frac{kT}{\pi d_1^3 \eta} = \frac{3}{2} \cdot \frac{\hat{s}(1-\hat{s}^2)^{1/2} - (1-2\hat{s}^2)Q}{\hat{s}^3(2-\hat{s}^2)} \tag{3.26}$$

in which $\hat{s}$ and $Q$ are defined as

$$\hat{s} = \sqrt{1 - b_1^2/d_1^2}, \quad Q = Q(\hat{s}) = \cot^{-1}\left(\sqrt{1-\hat{s}^2}/\hat{s}\right) \tag{3.27}$$

Finally, we consider the phase of the aggregate structures that may be presumed in a suspension composed of the disk-like hematite particles in thermodynamic equilibrium. This consideration may certainly be possible because the simple dipole model is employed in the present particle model. For discussion regarding the aggregate structures, it is very useful to take into account the characteristics of the potential curves for the two magnetic spherical and spherocylinder particles, shown in Fig. 3.3 and Fig. 3.11. For the case with no external magnetic field, the disk-like particles aggregate to form column-like clusters shown in Fig. 3.20(a). In these column-like clusters, the magnetic moments of the neighboring particles align in directions that are opposite to each other and in consequence leading to a smaller system energy and a more stable system. In the case of a strong external magnetic field, each magnetic moment of the constituent

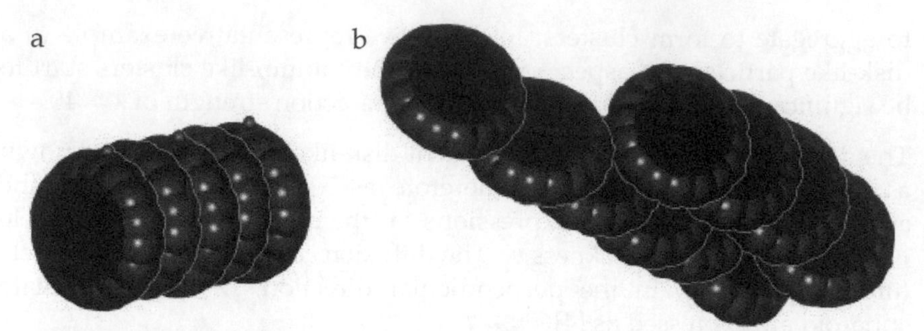


**Figure 3.20.** Typical aggregate regimes of hematite particles: (a) for no magnetic field and (b) for a strong external magnetic field.

particles strongly tends to incline in the magnetic field direction, and as a consequence, the upper disk-like cluster may slide in the field direction along the disk plane by a certain distance, as shown in Fig. 3.20(b). This is supported by the potential curve shown in Fig. 3.11 for the case of the same-direction of the magnetic moments and in the magnetic disk-like particle system, the sliding distance of approximately $1.2b_1$ gives rise to a minimum interaction energy.

## 3.5 Cube-like particles

Modern development of particle synthesis technologies enables us to develop cube-like hematite particles [12], which are quite unlike the previous axisymmetric particles such as spheroids, spherocylinders and disk-like particles. As shown in Fig. 3.21, the cubic hematite particle is

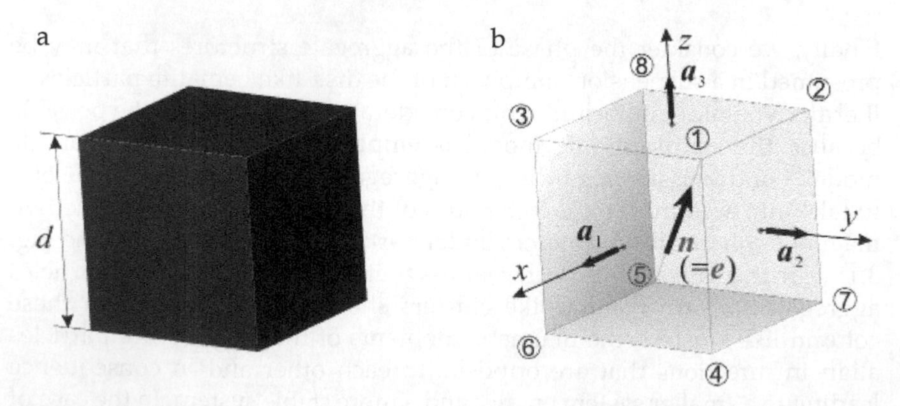


**Figure 3.21.** Notation describing a cube: (a) cubic particle model: (b) magnetization model with a magnetic moment inclining in the diagonal direction at the particle center.

regarded as being approximately magnetized in the diagonal direction of the cube. Since the dipole magnetization model is employed, as in the disk-like hematite particles, the magnetic particle-field interaction energy $u_i^{(H)}$, the magnetic particle-particle interaction energy $u_{ij}^{(m)}$, the magnetic force $\boldsymbol{F}_{ij}^{(m)}$ between particles $i$ and $j$, the magnetic torque $\boldsymbol{T}_{ij}^{(m)}$, and the torque $\boldsymbol{T}_i^{(H)}$ due to the magnetic particle-field interaction are applicable without any modifications, as in the magnetic spherical particle system, explained in Section 3.1.

In order to specify the state of the particle orientation, one unit vector is sufficient for an axisymmetric particle such as the spheroidal, spherocylinder and disk-like particles that have a magnetic moment in the particle axis direction. Two unit vectors are necessary for the case of an axisymmetric particle magnetized in a direction different from the particle axis direction. The present cubic particle is not axisymmetric and therefore the state of the orientation is specified in terms of the vectors $\boldsymbol{a}_1$ and $\boldsymbol{a}_2$, and $\boldsymbol{a}_3$ defined as $\boldsymbol{a}_3 = \boldsymbol{a}_1 \times \boldsymbol{a}_2$, normal to the $yz$-, $zx$- and $xy$-plane, respectively, in the particle coordinate system, as shown in Fig. 3.21(b). These vectors have the magnitude of $d/2$ in the present definition. With the three vectors known, the direction $\boldsymbol{e}$ of the magnetic moment, which is in the diagonal direction of the cube for hematite particles, is evaluated as

$$\boldsymbol{e} = \frac{\boldsymbol{a}_1 + \boldsymbol{a}_2 + \boldsymbol{a}_3}{|\boldsymbol{a}_1 + \boldsymbol{a}_2 + \boldsymbol{a}_3|} = \frac{\boldsymbol{a}_1 + \boldsymbol{a}_2 + \boldsymbol{a}_3}{\sqrt{3}d/2} \tag{3.28}$$

In addition, the eight vertices of the cube must be specified for assessing the overlap of two cubic particles, which will be discussed in a later chapter. If the numbering of the vertices is specified as shown in Fig. 3.21(b), an arbitrary vertex vector $\boldsymbol{\alpha}_i$ ($i$ = 1,2, ..., 8) is written as

$$\boldsymbol{\alpha}_1 = \boldsymbol{a}_1 + \boldsymbol{a}_2 + \boldsymbol{a}_3,\ \boldsymbol{\alpha}_2 = -\boldsymbol{a}_1 + \boldsymbol{a}_2 + \boldsymbol{a}_3,\ \boldsymbol{\alpha}_3 = \boldsymbol{a}_1 - \boldsymbol{a}_2 + \boldsymbol{a}_3,\ \boldsymbol{\alpha}_4 = \boldsymbol{a}_1 + \boldsymbol{a}_2 - \boldsymbol{a}_3,$$

$$\boldsymbol{\alpha}_5 = -\boldsymbol{\alpha}_1,\ \boldsymbol{\alpha}_6 = -\boldsymbol{\alpha}_2,\ \boldsymbol{\alpha}_7 = -\boldsymbol{\alpha}_3,\ \boldsymbol{\alpha}_8 = -\boldsymbol{\alpha}_4 \tag{3.29}$$

Moreover, the following mathematical results regarding these vectors are useful in the analysis of the motion of the cubic particles:

$$\boldsymbol{\alpha}_1 \cdot \boldsymbol{\alpha}_1 = 3d^2/4,\ \boldsymbol{\alpha}_1 \cdot \boldsymbol{\alpha}_2 = \boldsymbol{\alpha}_1 \cdot \boldsymbol{\alpha}_3 = \boldsymbol{\alpha}_1 \cdot \boldsymbol{\alpha}_4 = d^2/4,$$

$$\boldsymbol{\alpha}_1 \cdot \boldsymbol{\alpha}_6 = \boldsymbol{\alpha}_1 \cdot \boldsymbol{\alpha}_7 = \boldsymbol{\alpha}_1 \cdot \boldsymbol{\alpha}_8 = -d^2/4,\ \boldsymbol{\alpha}_1 \cdot \boldsymbol{\alpha}_5 = -3d^2/4 \tag{3.30}$$

$$\boldsymbol{\alpha}_1 \cdot (\boldsymbol{\alpha}_2 - \boldsymbol{\alpha}_1) = \boldsymbol{\alpha}_1 \cdot (\boldsymbol{\alpha}_3 - \boldsymbol{\alpha}_1) = \boldsymbol{\alpha}_1 \cdot (\boldsymbol{\alpha}_4 - \boldsymbol{\alpha}_1) = -\,d^2/2 \tag{3.31}$$

$$\boldsymbol{a}_1 \cdot \boldsymbol{\alpha}_1 = \boldsymbol{a}_1 \cdot \boldsymbol{\alpha}_3 = \boldsymbol{a}_1 \cdot \boldsymbol{\alpha}_4 = \boldsymbol{a}_1 \cdot \boldsymbol{\alpha}_6 = d^2/4,$$
$$\boldsymbol{a}_1 \cdot \boldsymbol{\alpha}_2 = \boldsymbol{a}_1 \cdot \boldsymbol{\alpha}_5 = \boldsymbol{a}_1 \cdot \boldsymbol{\alpha}_7 = \boldsymbol{a}_1 \cdot \boldsymbol{\alpha}_8 = -d^2/4 \tag{3.32}$$

For the case of the present cubic particle suspension, mathematical expressions for the overlap of the soft repulsive layers of the two particles are not available. Hence, in a manner similar to the treatment applied to the previous non-spherical particles, a modeling method such as the sphere-constituted particle may be necessary in order to treat the soft repulsive interactions between cubic particles coated with a soft repulsive layer. This type of simulation approach does not seem to have been currently undertaken by researchers, and therefore a study regarding this interaction will be a subject to be clarified in future work.

In order to develop a simulation technique based on Brownian dynamics for a cube-like particle suspension, useful analytical expressions for translational and rotational diffusion coefficients are required. In the case of axisymmetric particles, two translational and two rotational diffusion coefficients are sufficient for conducting molecular dynamics or Brownian dynamics situations. In contrast, the present cubic particle suspension requires a larger number of translational and rotational diffusion coefficients for these types of simulations. However, it is noted that these diffusion coefficients are not necessary for conducting a non-dynamic microscopic approach such as Monte Carlo. Moreover, if a simulation of a suspension composed of solid cubic particles can be regarded as a first approximation, then a Monte Carlo simulation is surely possible even in the situation where a mathematical expression for the overlap of the soft repulsive layers is not available.

Finally, we discuss characteristics of the magnetic interaction potential between the two cubic particles. The cubic particle has a characteristic geometry in that it is composed of six plane surfaces and this feature makes two cubes most likely to be in a face-to-face contact situation in the aggregate structures if the magnetic interaction is sufficiently strong for cluster formation. Hence, it seems important to discuss characteristics of the magnetic interaction potential with consideration to this preferred face-to-face contact configuration.

Since the interaction between the two cubic particles for a general three-dimensional configuration is quite difficult to analyze, we address a quasi two-dimensional case where the cubic particles are placed face down on a plane surface in a face-to-face contact situation, as shown in Fig. 9.1. In this face-to-face contact situation, there are only two primary orientations of the magnetic moment: that is, the magnetic moment may point toward the plane bottom surface (i.e., in the lower diagonal direction of the cube) or point away from the plane bottom surface (i.e. in the upper diagonal direction), as shown in Fig. 9.1. In the following, therefore, we discuss two types of orientational configurations: that is, (1) the magnetic moment of both the first and second particle is pointing away from the plane bottom surface and (2) the magnetic moment of only the second particle is pointing toward the plane surface. We first discuss characteristics of the potential curves for the former case and then for the latter case.

Figure 3.22(a) shows typical configurations of location and orientation for two cubic particles: the first (criterion) cube with magnetic moment

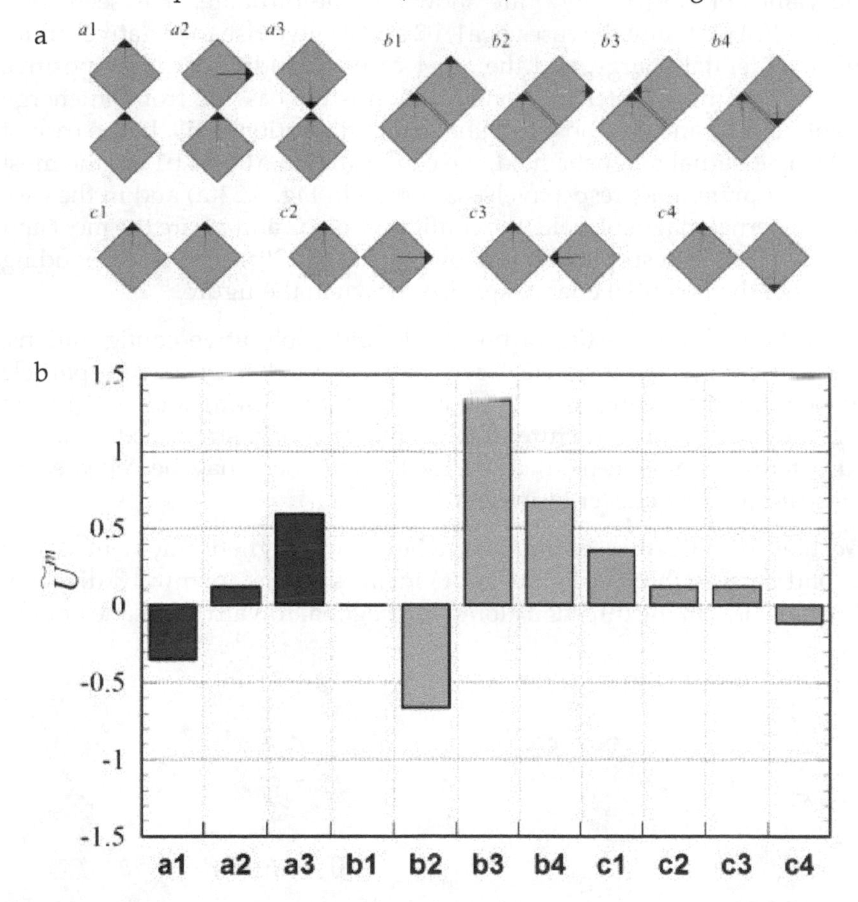


**Figure 3.22.** Interaction energies for different configurations: (a) typical configurations of location and orientation and (b) a potential energy for each case: both magnetic moments of the first (criterion) and the second cubes incline in the upper diagonal direction (or they point away from the page).

pointing along the diagonal away from the plane bottom surface is seen at either the lower or the left position, and the second cube with its magnetic moment also pointing away from the plane bottom surface, seen at either the upper or the right position, is placed with several typical positions and orientations of the magnetic moment. The second cube is placed at the upper vertex-vertex position for the cases of a1, a2 and a3, at the right diagonal face-to-face position for the cases of b1, b2, b3 and b4, and at the right vertex-vertex position for the cases of c1, c2, c3 and c4. Figure 3.22(b) shows the potential energy for each case of the above-mentioned eleven situations of the configuration, where as before the values of $\tilde{U}^m$ (= $u_{ij}^{(m)*}/\lambda$) are shown as the ordinate. It is seen from Fig. 3.22(b) that only the cases of a1, b2 and c4 give rise to negative values of the potential energy, and the other cases excluding b1 yield positive values. The most preferable configuration is the case b2 from an energy point of view and the most prohibitive configuration is b3. In the case of a strong external magnetic field, the configuration a1 and b1 are the most and next preferable, respectively, as shown in Fig. 3.23(a) and in the case of no external magnetic field, the configuration b2 and a1 are the most and next preferable, respectively, as shown in Fig. 3.23(b). The corresponding values of the potential energy are also shown in the figure.

From these characteristics of preferable and prohibitive configurations, from an energy point of view, we may expect that in a multi-particle suspension the configurations shown in Figs. 3.24(a) and 3.24(b) are regarded as a basic structure unit for the case of a strong external and no external magnetic field, respectively, and they may be repeated to construct a larger cluster formation.

We now discuss the second case where the magnetic moment of the second particle (blue or lighter cube) inclines in the downward diagonal direction, i.e., in the direction along the diagonal toward the plane bottom

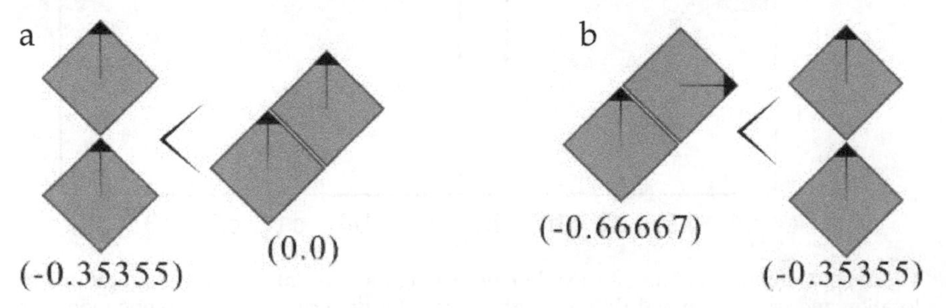


**Figure 3.23.** The most and next preferable configurations for the cases of (a) a strong external magnetic field and (b) no magnetic field.

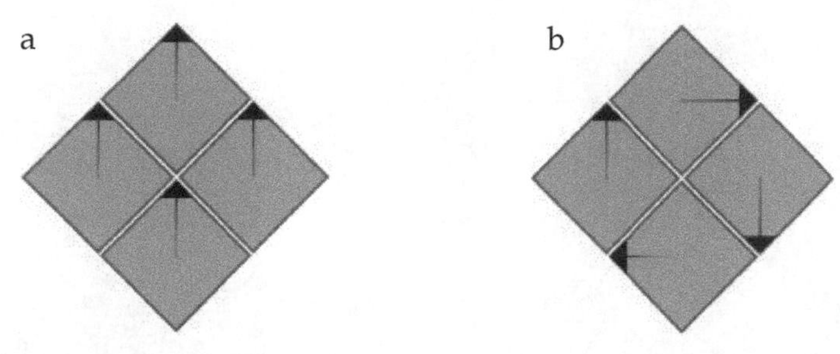


**Figure 3.24.** The most preferable configuration for a 4-particles system: (a) a strong external magnetic field case and (b) no magnetic field case.

surface. Using the conventions of the previous case, Fig. 3.25(a) shows typical configurations for two cubic particles where the second cube is placed at the upper vertex-vertex position for the cases of d1, d2 and d3, at the right diagonal face-to-face position for the cases of e1, e2, e3 and e4, and at the right vertex-vertex position for the cases of f1, f2, f3 and f4. Figure 3.25(b) shows the potential energy for each case of the above-mentioned eleven configurations. From the results of interaction energies shown in Fig. 3.25(b), it is seen that the case e2 gives rise to the minimum energy value –1.33333 and the cases e1 and d1 yield the second and third minimum energy values. Attention is drawn to the fact that the minimum energy value –1.33333 for the configuration e2 in the present case is much smaller than the minimum energy value –0.66667 for the configuration b2 in the previous case.

It is seen from Fig. 3.25 that in the case of a strong external magnetic field, the configuration e1 and d1 are expected to be the most and next preferable, respectively, as shown in Fig. 3.26(a), and in the case of no external magnetic field, the configuration e2, e1 and d1 are the most, next and third preferable situation, respectively, as shown in Fig. 3.26(b). These characteristics imply that the configurations e1 and d1 are preferable in a strong applied magnetic field, and the configurations e2 is most likely preferable and the configurations e1 and d1 are next and third preferable in the no external magnetic field situation.

For a multi-particle suspension, we may presume from the above results regarding the interaction energies that the configurations shown in Figs. 3.27(a) and (b) are the most preferable in aggregate structures for the cases of a strong external and no external magnetic field, respectively. Large clusters in a multi-particle suspension may be formed by repeating these structure units shown in Figs. 3.27(a) and (b) in each respective situation of an external magnetic field.

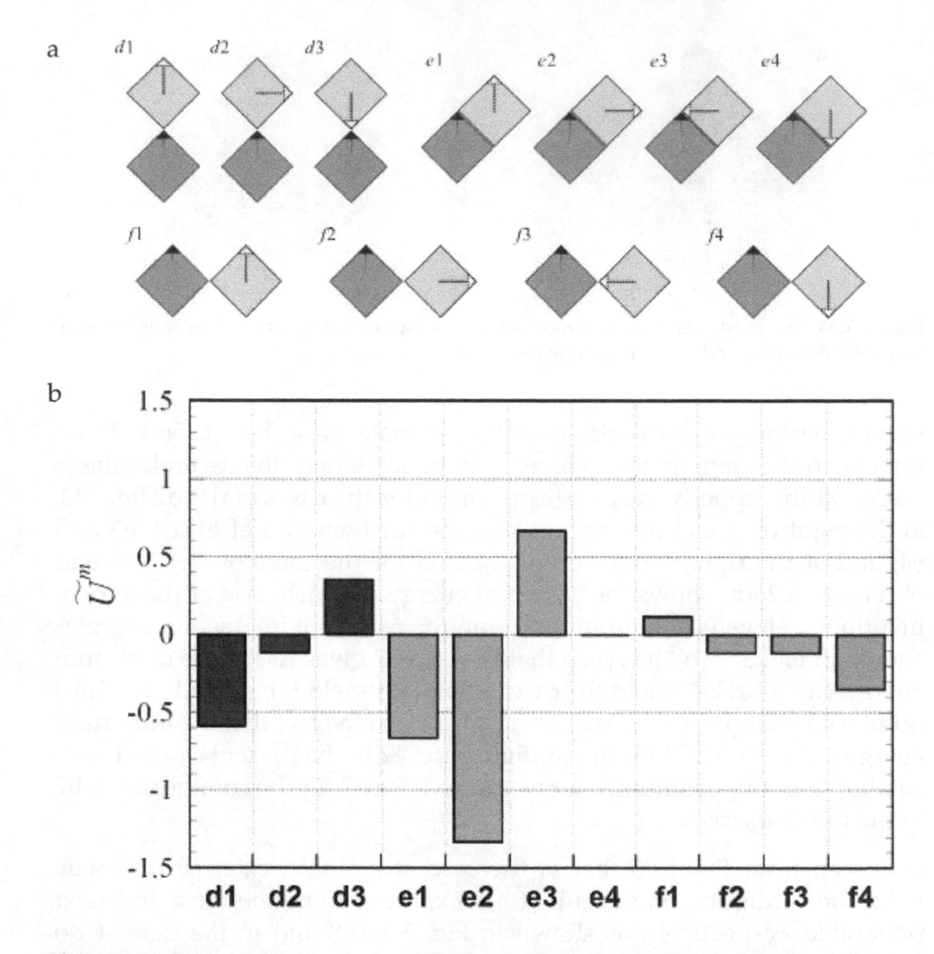


**Figure 3.25.** Interaction energies for different configurations: (a) typical configurations of location and orientation and (b) a potential energy for each case: the magnetic moment of the first (criterion) cube inclines in the upper diagonal direction and the magnetic moment of the second cube inclines in the lower diagonal direction (or only the magnetic moment of the second cube points toward the page).

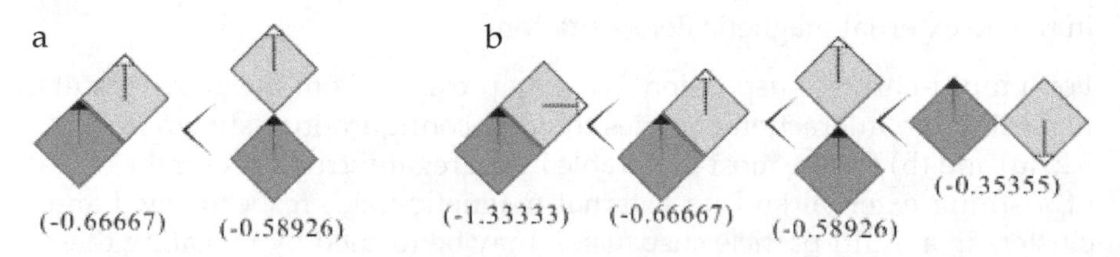


**Figure 3.26.** The most and next preferable configurations for the cases of (a) a strong external magnetic field and (b) no magnetic field.

Finally, we discuss the potential curves as a function of the distance between the two cubic particles. Since the face-to-face configuration of the cubes is most likely to be preferable even for the three-dimensional system, we discuss characteristics of the potential curves by moving the second cube on a surface plane of the first (base) cube. In the following, we focus only on the situation of a strong external magnetic field because the situation with no external magnetic field is quite difficult to treat in regard to expanding the suggested structure unit, shown in Fig. 3.27(b), to the aggregate formation for a three-dimensional multi-particle system.

Figure 3.28 shows results of the potential curves as a function of the distance between the two cubic particles measured in the surface plane of the first cube in the $x$-direction and in the $y$-direction shown on the figure.

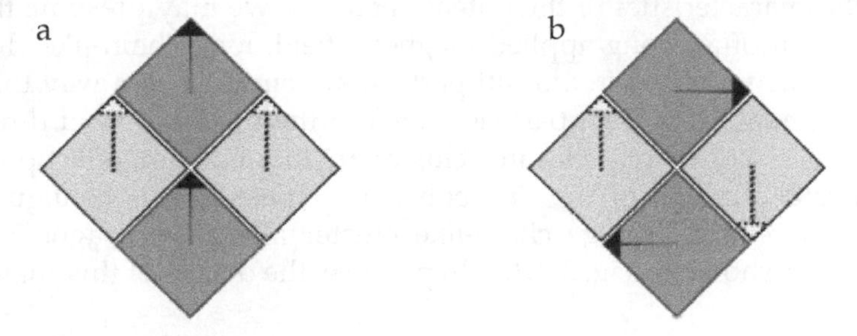


**Figure 3.27.** The most preferable configuration for a 4-particles system: (a) a strong external magnetic field case and (b) no magnetic field case.

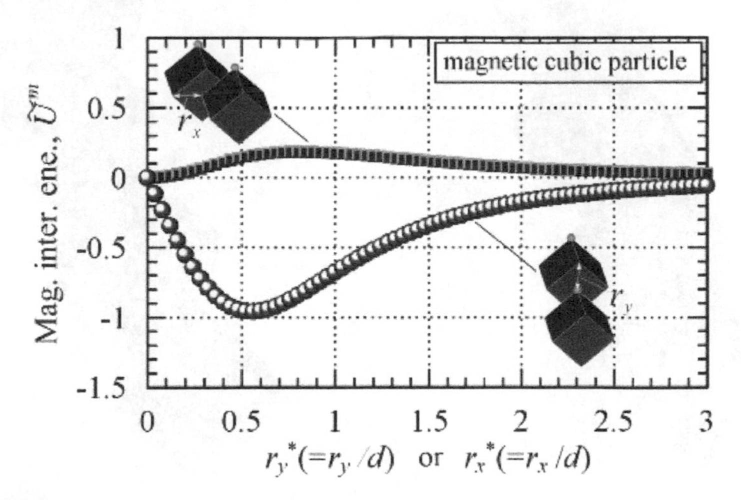


**Figure 3.28.** Potential energy curves as a function of the distance of the second cube in the $x$- and $y$-direction.

In these configurations of the two cubic particles, the magnetic moment of each cube is aligned in the same direction on the diagonal; in the figure a small sphere attached at each cube implies the magnetic moment direction form the particle center. First, it is noted from Fig. 3.28 that the complete face-to-face configuration at $r_x = 0$ or $r_y = 0$ yields zero interaction energy for both cases of the separating direction as is expected from case b1 in Fig. 3.22(b). For the case of movement of the second cube in the *x*-direction, the interaction energy shows relatively small positive values but does not become negative at any separation distance. In contrast, for the case of the *y*-direction movement, the potential energy decreases in the negative region, attains to the minimum value of approximately −1 at $r_y^* \simeq 0.5$ and finally increases to approach zero with increasing separation distance.

From the characteristics of the potential curves, we may presume that in the situation of a strong applied magnetic field, long chain-like clusters are most likely formed in a multi-particle system in such a way that the magnetic moment of each particle aligns in the magnetic field direction and the neighboring particles in a cluster are located at an offset position on opposite surfaces of the first cube; the repeat of this configuration leads to a relatively large chain-like cluster along the magnetic field direction, as shown in Fig. 3.29(a). In practice, the details of this chain-like

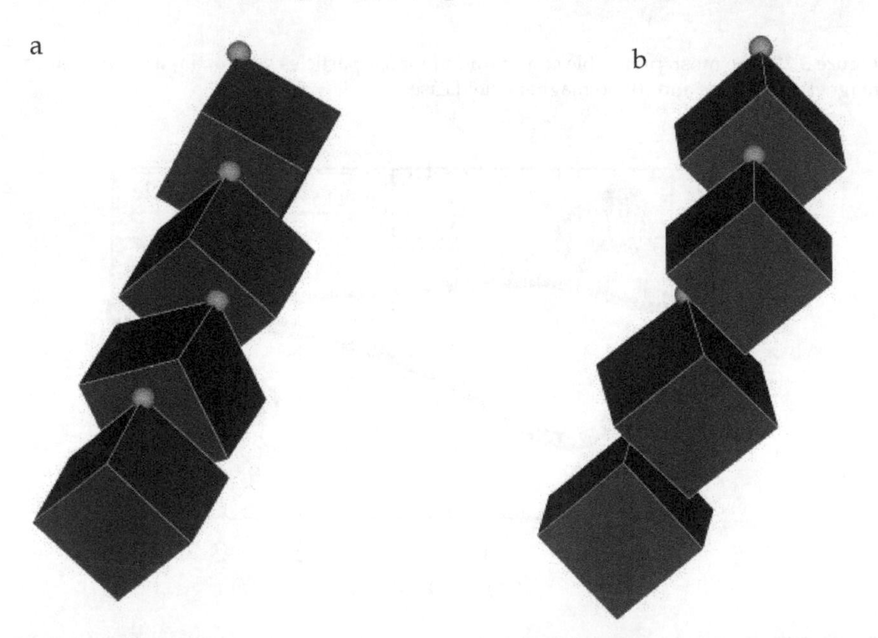


**Figure 3.29.** Cluster formation expected in a strong applied magnetic field: (a) simple chain-like formation and (b) twisted chain-like formation.

cluster formation is more complex because the second cube has an equal probability for any of the several possibilities available for being placed in the surface-to-surface configuration. In other words, if the vertex vector (implying magnetic moment direction) of the first cube aligns along the magnetic field direction, then there are three symmetrical equal surfaces of the first cube that are available for the second cube to be located in the surface-to-surface contact situation. This may lead to the linear-twisted cluster formation that is shown in Fig. 3.29(b) in addition to the simple linear cluster formation that is shown in Fig. 3.29(a).

## Bibliography


[1] Ewald, P. P. 1921. Die Berechnung optischer und elektrostatischer gitterpotentiale. Ann. Phys. 64: 253–287.

[2] Kushick, J. and Berne, B. J. 1976. Computer simulation of anisotropic molecular fluids. J. Chem. Phys. 64: 1362–1367.

[3] Gay, J. G. and Berne, B. J. 1981. Modification of the overlap potential to mimic a linear site-site potential. J. Chem. Phys. 74: 3316–3319.

[4] Kim, S. and Karrila, S. J. 1991. Microhydrodynamics: Principles and Selected Applications, Butterworth-Heinemann, Stoneham.

[5] Berne, B. J. and Pechukas, P. 1972. Gaussian model potentials for molecular interactions. J. Chem. Phys. 56: 4213–4216.

[6] Antypov, D. and Cleaver, D. J. 2004. The role of attractive interactions in rod-sphere mixtures. J. Chem. Phys. 120: 10307–10316.

[7] Brenner, H. 1974. Rheology of a dilute suspension of axisymmetric Brownian particles. Int. J. Multiphase Flow. 1: 195–341.

[8] Tirado, M. M. and de la Torre, J. G. 1979. Translational friction coefficients of rigid, symmetric top macromolecules: application to circular cylinders. J. Chem. Phys. 71: 2581–2587.

[9] Tirado, M. M. and de la Torre, J. G. 1980. Rotational dynamics of rigid, symmetric top macromolecules: application to circular cylinders. J. Chem. Phys. 73: 1986–1993.

[10] Tirado, M. M., Martinez, C. L. and de la Torre, J. G. 1984. Comparison of theories for the translational and rotational diffusion coefficients of rod-like macromolecules: application to short DNA fragments. J. Chem. Phys. 81: 2047–2052.

[11] Ozaki, M., Ookoshi, N. and Matijević, E. 1990. Preparation and magnetic properties of uniform hematite platelets. J. Colloid Interface Sci. 137: 546–549.

[12] Aoshima, M., Ozaki, M. and Satoh, A. 2012. Structural analysis of self-assembled lattice structures composed of cubic hematite particles. J. Phys. Chem. 116: 17862–17871.

CHAPTER 4

# Two Coordinate Systems for Description of Particle Orientation

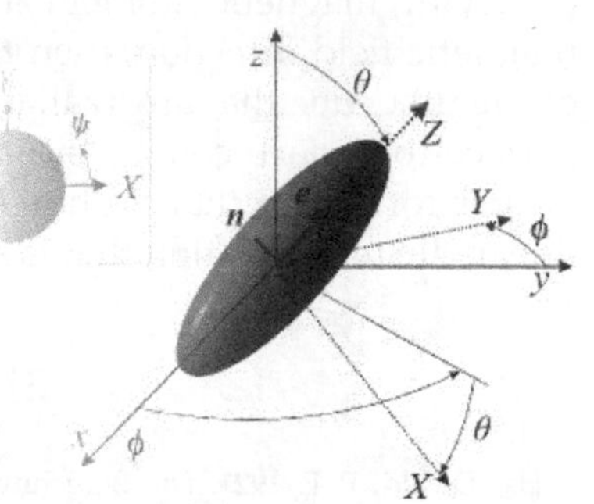


## 4.1 Rotation matrix

For a non-magnetic spherical particle system, for symmetry reasons, it is not necessary to specify the orientational situation of the particles. However, for non-spherical particles and for magnetic spherical particles, it is necessary to evaluate the orientational state of the system. In order to describe the orientational state of non-spherical particles and the magnetic spherical particles, it is frequently convenient in simulations to use two different coordinate systems. These are the absolute coordinate system and the local or particle coordinate system that are fixed at each particle body. In the mathematical treatment of the rotational motion, the absolute coordinate system is transformed in parallel so that the origin coincides with that of the particle coordinate system. In order to perform particle-based simulations, quantities in the particle coordinate system must be expressed in terms of the absolute coordinate system. In the following, therefore, we show the relationship between the quantities that are expressed in these two coordinate systems.

Figure 4.1 schematically shows how the particle coordinate system $XYZ$ is obtained from rotation of the absolute coordinate system $xyz$; as already mentioned, the latter system has been transformed in parallel to the origin of the former coordinate system. It is quite common for the direction of the $Z$-axis to be taken in the particle direction $\boldsymbol{e}$, as shown in Fig. 4.1, and then the particle coordinate system $XYZ$ is obtained in the following manner. In the first step, the $xyz$-coordinate system is rotated about the $z$-axis by an angle $\phi$ to obtain the rotated coordinate system $x'y'z'$-system and this rotation is expressed using the following rotation matrix $\boldsymbol{R}_1$:

$$\boldsymbol{R}_1 = \begin{bmatrix} \cos\phi & \sin\phi & 0 \\ -\sin\phi & \cos\phi & 0 \\ 0 & 0 & 1 \end{bmatrix} \tag{4.1}$$

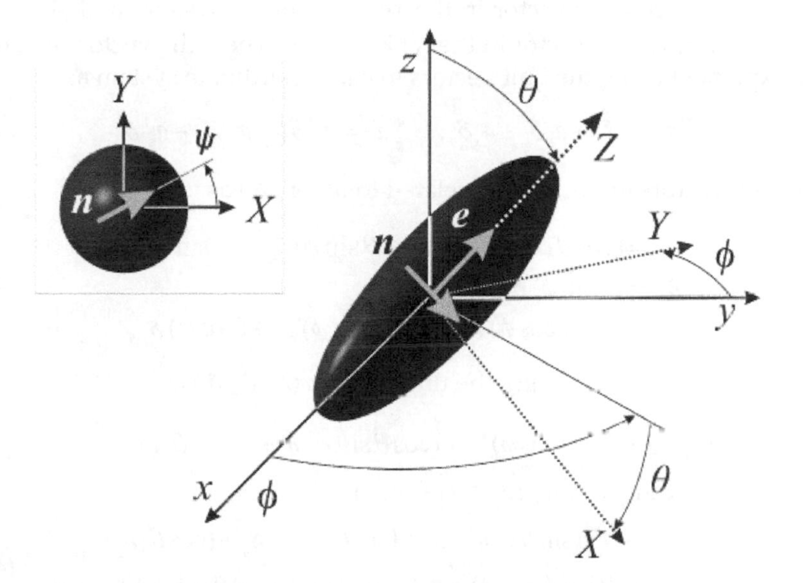


**Figure 4.1.** Particle coordinate system $XYZ$ and the absolute coordinate system $xyz$.

It is noted that this rotation matrix is related to the rotation of the $xyz$-coordinate system and not the points in the original coordinate system. In the second step, the $x'y'z'$-coordinate system is further rotated about the $y'$-axis by an angle $\theta$ to generate the $XYZ$-coordinate system, and the matrix $\boldsymbol{R}_2$ for this rotation is expressed as

$$\boldsymbol{R}_2 = \begin{bmatrix} \cos\theta & 0 & -\sin\theta \\ 0 & 1 & 0 \\ \sin\theta & 0 & \cos\theta \end{bmatrix} \tag{4.2}$$

Hence, the total rotation matrix $\boldsymbol{R}$ for transformation from the absolute coordinate system to the particle coordinate is obtained as

$$\boldsymbol{R} = \boldsymbol{R}_2 \cdot \boldsymbol{R}_1 = \begin{bmatrix} \cos\theta\cos\phi & \cos\theta\sin\phi & -\sin\theta \\ -\sin\phi & \cos\phi & 0 \\ \sin\theta\cos\phi & \sin\theta\sin\phi & \cos\theta \end{bmatrix} \tag{4.3}$$

Next we address the relationship between quantities in the absolute and in the particle coordinate systems. We employ the notation ($\boldsymbol{\delta}_x$, $\boldsymbol{\delta}_y$, $\boldsymbol{\delta}_z$) for the unit vectors in each axis-direction for the absolute coordinate system $xyz$, and similarly ($\boldsymbol{\delta}_X$, $\boldsymbol{\delta}_Y$, $\boldsymbol{\delta}_Z$) for the particle coordinate system $XYZ$, $\boldsymbol{a} = (a_x, a_y, a_z)$ for an arbitrary vector in the $xyz$-coordinate system, and similarly $\boldsymbol{a}^b = (a_x^b, a_y^b, a_z^b)$ for the vector in the $XYZ$-system. Then, the vectors $\boldsymbol{a}$ and $\boldsymbol{a}^b$ are expressed using the unit vectors in each coordinate system as

$$\boldsymbol{a} = a_x\boldsymbol{\delta}_x + a_y\boldsymbol{\delta}_y + a_x\boldsymbol{\delta}_x, \qquad \boldsymbol{a}^b = a_x{}^b\boldsymbol{\delta}_X + a_y{}^b\boldsymbol{\delta}_Y + a_z{}^b\boldsymbol{\delta}_Z \tag{4.4}$$

The unit vectors ($\boldsymbol{\delta}_X$, $\boldsymbol{\delta}_Y$, $\boldsymbol{\delta}_Z$) are related to ($\boldsymbol{\delta}_x$, $\boldsymbol{\delta}_y$, $\boldsymbol{\delta}_z$) as

$$\begin{aligned} \boldsymbol{\delta}_X &= (\cos\theta\cos\phi)\,\boldsymbol{\delta}_x + (\cos\theta\sin\phi)\,\boldsymbol{\delta}_y + (-\sin\theta)\boldsymbol{\delta}_z, \\ \boldsymbol{\delta}_Y &= (-\sin\phi)\,\boldsymbol{\delta}_x + (\cos\phi)\boldsymbol{\delta}_y, \\ \boldsymbol{\delta}_Z &= (\sin\theta\cos\phi)\,\boldsymbol{\delta}_x + (\sin\theta\sin\phi)\,\boldsymbol{\delta}_y + (\cos\theta)\,\boldsymbol{\delta}_z \end{aligned} \tag{4.5}$$

Hence, $\boldsymbol{a}^b$ is expressed using the unit vectors ($\boldsymbol{\delta}_x$, $\boldsymbol{\delta}_y$, $\boldsymbol{\delta}_z$) as

$$\begin{aligned} \boldsymbol{a}^b &= a_x^b((\cos\theta\cos\phi)\,\boldsymbol{\delta}_x + (\cos\theta\sin\phi)\,\boldsymbol{\delta}_y + (-\sin\theta)\boldsymbol{\delta}_z) \\ &\quad + a_y^b((-\sin\phi)\,\boldsymbol{\delta}_x + (\cos\phi)\boldsymbol{\delta}_y) \\ &\qquad + a_z^b((\sin\theta\cos\phi)\,\boldsymbol{\delta}_x + (\sin\theta\sin\phi)\,\boldsymbol{\delta}_y + (\cos\theta)\boldsymbol{\delta}_z) \\ &= (a_x^b(\cos\theta\cos\phi) + a_y^b(-\sin\phi) + a_z^b(\sin\theta\cos\phi))\,\boldsymbol{\delta}_x \\ &\qquad + (a_x^b(\cos\theta\sin\phi) + a_y^b(\cos\phi) + a_z^b(\sin\theta\sin\phi))\,\boldsymbol{\delta}_y \\ &\qquad + (a_x^b(-\sin\theta) + a_z^b(\cos\theta))\,\boldsymbol{\delta}_z \end{aligned} \tag{4.6}$$

This equation can be expressed in the form using the vector components as

$$\begin{bmatrix} a_x \\ a_y \\ a_z \end{bmatrix} = \begin{bmatrix} \cos\theta\cos\phi & -\sin\phi & \sin\theta\cos\phi \\ \cos\theta\sin\phi & \cos\phi & \sin\theta\sin\phi \\ -\sin\theta & 0 & \cos\theta \end{bmatrix} \begin{bmatrix} a_x^b \\ a_y^b \\ a_z^b \end{bmatrix} \tag{4.7}$$

From Eqs. (4.3) and (4.7), it is seen that the vector components $\boldsymbol{a} = (a_x, a_y, a_z)$ and $\boldsymbol{a}^b = (a_x^b, a_y^b, a_z^b)$ are related using the rotation matrix $\boldsymbol{R}$ as

$$\boldsymbol{a} = \boldsymbol{R}^{t} \cdot \boldsymbol{a}^{b}, \ \boldsymbol{a}^{b} = \boldsymbol{R} \cdot \boldsymbol{a} \tag{4.8}$$

in which it should be noted that the inverse rotation matrix $\boldsymbol{R}^{-1}$ is equal to the transposed matrix $\boldsymbol{R}^{t}$.

If the magnetic moment is magnetized in a direction normal to the particle axis such as in the case of spindle-like hematite particles, the magnetic moment $\boldsymbol{n}^{b}$ is described by specifying an angle $\psi$ in the counter-clockwise direction from the X-axis in the *XYZ*-coordinate system, where $\boldsymbol{n}^{b} = (\cos\psi, \sin\psi, 0)$, as shown in Fig. 4.1. The vector $\boldsymbol{n}$ in the *xyz*-coordinate system is obtained from Eq. (4.8) as $\boldsymbol{n} = \boldsymbol{R}^{t} \cdot \boldsymbol{n}^{b}$.

## 4.2 Rotation of axisymmetric particles

In the case of the axisymmetric particle shown in Fig. 4.1, the particle axis direction $\boldsymbol{e}$ is always taken in the Z-axis direction, and therefore $\boldsymbol{e}$ is equal to $\boldsymbol{\delta}_{Z}$. From a simulation point of view, it is important to derive the rotation matrix $\boldsymbol{R}$ from the particle direction $\boldsymbol{e}$, the components of which will be solved by the particle-based simulation methods such as Monte Carlo and Brownian dynamics. Since the particle direction is $\boldsymbol{e}^{b} = (0,0,1)$ in the particle coordinate system, the relationship expressed in Eq. (4.8) gives rise to the particle direction $\boldsymbol{e} = (e_x, e_y, e_z)$ in the absolute coordinate system as

$$e_x = \sin\theta\cos\phi, \ e_y = \sin\theta\sin\phi, \ e_z = \cos\theta \tag{4.9}$$

Since the ranges of $\theta$ and $\phi$ are defined for $0 \le \theta < \pi$ and $0 \le \phi < 2\pi$, respectively, the following expressions are obtained:

$$\sin\theta = \sqrt{1-e_z^2}, \quad \cos\phi = e_x \Big/ \sqrt{1-e_z^2}, \quad \sin\phi = e_y \Big/ \sqrt{1-e_z^2} \tag{4.10}$$

in which it has been assumed that $\sin\theta$ is not equal to zero. A special treatment is, therefore, necessary for the situation where the particle direction coincides with the z-axis direction.

We may employ the method of changing the particle direction by a small angle, which is necessary in Monte Carlo simulations, but in contrast, is not necessary in molecular dynamics or Brownian dynamics simulations because the particle direction there is evaluated in a deterministic manner. We will consider this rotational procedure in detail in Section 7.1.

In the case of the rod-like hematite particles that are assumed to have a magnetic moment at the particle center with a direction normal to the particle axis direction, the magnetic moment direction $\boldsymbol{n}$ is also changed with a change in the particle direction. In the following, we derive the new magnetic moment direction $\boldsymbol{n}'$ from the original direction $\boldsymbol{n}$, resulting from the change in the particle orientation from $\boldsymbol{e}$ to $\boldsymbol{e}'$. The unit vector $\boldsymbol{\delta}_\perp$ along the line about which the vector $\boldsymbol{e}$ is rotated is written as

$$\boldsymbol{\delta}_\perp = \boldsymbol{e} \times \Delta\boldsymbol{e} / |\boldsymbol{e} \times \Delta\boldsymbol{e}| \tag{4.11}$$

in which $\Delta\boldsymbol{e} = \boldsymbol{e}' - \boldsymbol{e}$. The vector $\boldsymbol{n}$ is decomposed into the components $\boldsymbol{n}_p$ and $\boldsymbol{n}_n$ parallel and normal to the vector $\boldsymbol{\delta}_\perp$, respectively, expressed as $\boldsymbol{n} = \boldsymbol{n}_p + \boldsymbol{n}_n$. The vector component $\boldsymbol{n}_p$ does not change but $\boldsymbol{n}_n$ undergoes a small change with a change in the particle orientation. The vector components are expressed using the vector $\boldsymbol{\delta}_\perp$ as

$$\boldsymbol{n}_p = (\boldsymbol{n} \cdot \boldsymbol{\delta}_\perp)\boldsymbol{\delta}_\perp, \; \boldsymbol{n}_n = \boldsymbol{n} - \boldsymbol{n}_p = \boldsymbol{n} - (\boldsymbol{n} \cdot \boldsymbol{\delta}_\perp)\boldsymbol{\delta}_\perp \tag{4.12}$$

Then, the change $\Delta\boldsymbol{n}_n (= \boldsymbol{n}_n' - \boldsymbol{n}_n)$ is expressed as

$$\Delta\boldsymbol{n}_n = |\Delta\boldsymbol{e}| (\boldsymbol{\delta}_\perp \times \boldsymbol{n}_n) \tag{4.13}$$

Finally, the new direction $\boldsymbol{n}'$ from the original direction $\boldsymbol{n}$, resulting from the change in the particle orientation from $\boldsymbol{e}$ to $\boldsymbol{e}'$, is obtained as

$$\boldsymbol{n}' = \boldsymbol{n}_p + \boldsymbol{n}_n + \Delta\boldsymbol{n}_n = \boldsymbol{n} + \Delta\boldsymbol{n}_n \tag{4.14}$$

Since $\boldsymbol{n}'$ in Eq. (4.14) is not a unit vector, it must be normalized to obtain the final unit vector for the new magnetic moment direction, i.e., $\boldsymbol{n}'/|\boldsymbol{n}'| \rightarrow \boldsymbol{n}'$. It is noted that Eq. (4.14) provides the new magnetic moment direction without using the rotation matrix or without using the angles defining the rotation matrix.

## 4.3 Rotation of cubic particles

As explained in Section 3.5, the orientational situation of the cubic particle is described in terms of the three unit vectors $\boldsymbol{a}_1$, $\boldsymbol{a}_2$ and $\boldsymbol{a}_3$ ($= \boldsymbol{a}_1 \times \boldsymbol{a}_2$) that are normal to the *yz*-, *zx*- and *xy*-plane of the cube, respectively. Moreover, it may also be convenient to use the vertex vector $\boldsymbol{\alpha}_i$ ($i = 1, 2, \ldots, 8$) that are defined in Eq. (3.29), as shown in Fig. 3.21. In the case of the cubic particle that has a magnetic moment in the diagonal direction, the particle direction is taken in the *z*-axis direction in the particle coordinate system. Hence, the following two procedures are necessitated to change the rotational states, these being, (1) the first procedure for changing the particle axis orientation in a general direction by a small angle and (2) the second procedure for changing the particle about the magnetic moment direction by a small angle. These procedures concerning the particle orientation are indispensable in performing Monte Carlo simulations, as already mentioned.

Referring to the definition of the symbols shown in Fig. 3.21 and the rotation matrix $\boldsymbol{R}$ explained in Section 4.1, the rotation matrix $\boldsymbol{Q}_z$ for rotating an arbitrary point in the $xyz$-coordinate system about the $z$-axis by a small angle $\theta_z$ is written as

$$\boldsymbol{Q}_z = \begin{bmatrix} \cos\theta_z & -\sin\theta_z & 0 \\ \sin\theta_z & \cos\theta_z & 0 \\ 0 & 0 & 1 \end{bmatrix} \tag{4.15}$$

Similarly, the rotation matrix $\boldsymbol{Q}_y$ for rotation about the $y$-axis by a small angle $\theta_y$ and $\boldsymbol{Q}_x$ for rotation about the $x$-axis by a small angle $\theta_x$ are written as

$$\boldsymbol{Q}_y = \begin{bmatrix} \cos\theta_y & 0 & \sin\theta_y \\ 0 & 1 & 0 \\ -\sin\theta_y & 0 & \cos\theta_y \end{bmatrix}, \quad \boldsymbol{Q}_x = \begin{bmatrix} 1 & 0 & 0 \\ 0 & \cos\theta_x & -\sin\theta_x \\ 0 & \sin\theta_x & \cos\theta_x \end{bmatrix} \tag{4.16}$$

Employing the following abbreviations, $C_x = \cos\theta_x$, $S_x = \sin\theta_x$, $C_y = \cos\theta_y$, $S_y = \sin\theta_y$, $C_z = \cos\theta_z$ and $S_z = \sin\theta_z$, for the small angle changes $\theta_x$, $\theta_y$ and $\theta_z$, the cubic particle can be rotated as a whole in an arbitrary direction from the original orientation by using the following rotation matrix $\boldsymbol{Q}$:

$$\begin{aligned} \boldsymbol{Q} = \boldsymbol{Q}_z \cdot \boldsymbol{Q}_y \cdot \boldsymbol{Q}_x &= \begin{bmatrix} C_z & S_z & 0 \\ S_z & C_z & 0 \\ 0 & 0 & 1 \end{bmatrix} \begin{bmatrix} C_y & 0 & S_y \\ 0 & 1 & 0 \\ S_y & 0 & C_y \end{bmatrix} \begin{bmatrix} 1 & 0 & 0 \\ 0 & C_x & -S_x \\ 0 & S_x & C_x \end{bmatrix} \\ &= \begin{bmatrix} C_y C_z & S_x S_y C_z - C_x S_z & C_x S_y C_z + S_x S_z \\ C_y S_z & S_x S_y S_z \; C_x C_z & C_x S_y S_z - S_x C_z \\ S_y & S_x C_y & C_x C_y \end{bmatrix} \end{aligned} \tag{4.17}$$

We introduce the notation $\hat{\boldsymbol{a}}_i^b (i = 1,2,3)$ defined by $\hat{\boldsymbol{a}}_i^b = \boldsymbol{a}_i^b / |\boldsymbol{a}_i^b|$, and similarly $\hat{\boldsymbol{a}}_i = \boldsymbol{a}_i / |\boldsymbol{a}_i|$. Using this rotation matrix, the three unit vectors $\hat{\boldsymbol{a}}_1^b$, $\hat{\boldsymbol{a}}_2^b$ and $\hat{\boldsymbol{a}}_3^b$ in the particle coordinate system are slightly displaced in orientation to give a new position with unit vectors $\hat{\boldsymbol{a}}_1^{b'}$, $\hat{\boldsymbol{a}}_2^{b'}$ and $\hat{\boldsymbol{a}}_3^{b'}$, which are expressed as

$$\hat{a}_1^b{}' = Q \cdot \hat{a}_1^b, \quad \hat{a}_2^b{}' = Q \cdot \hat{a}_2^b, \quad \hat{a}_3^b{}' = Q \cdot \hat{a}_3^b, \quad \hat{a}_4^b{}' = -\hat{a}_1^b{}', \quad \hat{a}_5^b{}' = -\hat{a}_2^b{}' \quad \hat{a}_6^b{}' = -\hat{a}_3^b{}' \tag{4.18}$$

The present rotation matrix $\boldsymbol{R}$, which relates an arbitrary vector $\boldsymbol{b}$ in the absolute coordinate system to the corresponding vector $\boldsymbol{b}^b$ in the particle coordinate system, gives rise to the corresponding unit vectors $\hat{\boldsymbol{a}}_1'$, $\hat{\boldsymbol{a}}_2'$ and $\hat{\boldsymbol{a}}_3'$ in the absolute coordinate system as

$$\hat{a}_1{}' = \boldsymbol{R}^t \cdot \hat{a}_1^b{}', \quad \hat{a}_2{}' = \boldsymbol{R}^t \cdot \hat{a}_2^b{}', \quad \hat{a}_3{}' = \boldsymbol{R}^t \cdot \hat{a}_3^b{}', \quad \hat{a}_4{}' = -\hat{a}_1{}', \quad \hat{a}_5{}' = -\hat{a}_2{}' \quad \hat{a}_6{}' = -\hat{a}_3{}' \tag{4.19}$$

We now derive the rotation matrix $\boldsymbol{R}'$ for the new rotational state of the cube that is described by the unit vectors $\hat{\boldsymbol{a}}_1'$, $\hat{\boldsymbol{a}}_2'$ and $\hat{\boldsymbol{a}}_3'$ in the absolute coordinate system. In any rotational state of the cube, the unit vector in each direction in the particle coordinate system, $\boldsymbol{\delta}_x^b$, $\boldsymbol{\delta}_y^b$ and $\boldsymbol{\delta}_z^b$ are always $\boldsymbol{\delta}_x^b = (1,0,0)$, $\boldsymbol{\delta}_y^b = (0,1,0)$ and $\boldsymbol{\delta}_z^b = (0,0,1)$, respectively. Hence, the rotation matrix $\boldsymbol{R}'$ is related to these unit vectors by the following expressions:

$$\hat{a}_1{}' = \boldsymbol{R}'^t \cdot \boldsymbol{\delta}_x^b, \quad \hat{a}_2{}' = \boldsymbol{R}'^t \cdot \boldsymbol{\delta}_y^b, \quad \hat{a}_3{}' = \boldsymbol{R}'^t \cdot \boldsymbol{\delta}_z^b \tag{4.20}$$

If the rotation matrix $\boldsymbol{R}'$ is expressed in the component form as

$$\boldsymbol{R}' = \begin{bmatrix} R_{xx}{}' & R_{xy}{}' & R_{xz}{}' \\ R_{yx}{}' & R_{yy}{}' & R_{yz}{}' \\ R_{zx}{}' & R_{zy}{}' & R_{zz}{}' \end{bmatrix} \tag{4.21}$$

then Eqs. (4.20) and (4.21) yield the expressions as

$$\begin{bmatrix} R_{xx}{}' \\ R_{xy}{}' \\ R_{xz}{}' \end{bmatrix} = \hat{a}_1{}', \quad \begin{bmatrix} R_{yx}{}' \\ R_{yy}{}' \\ R_{yz}{}' \end{bmatrix} = \hat{a}_2{}', \quad \begin{bmatrix} R_{zx}{}' \\ R_{zy}{}' \\ R_{zz}{}' \end{bmatrix} = \hat{a}_3{}' \tag{4.22}$$

We summarize the above procedure regarding a small rotation of the cubic particle, which is available for Monte Carlo simulations, in the following methodology. The small angles $\theta_x$, $\theta_y$ and $\theta_z$ are stochastically sampled from a uniform random range from $-\theta_{max}$ to $\theta_{max}$ where $\theta_{max}$ is a small constant angle that is specified as a simulation parameter. The rotation matrix $\boldsymbol{Q}$ is evaluated from Eq. (4.17), and the three unit vectors $\hat{\boldsymbol{a}}_1^{b\prime}$, $\hat{\boldsymbol{a}}_2^{b\prime}$ and $\hat{\boldsymbol{a}}_3^{b\prime}$, describing the orientation of the cube in the particle coordinate system are determined from Eq. (4.18). The corresponding unit vectors $\hat{\boldsymbol{a}}_1'$, $\hat{\boldsymbol{a}}_2'$ and $\hat{\boldsymbol{a}}_3'$ in the absolute coordinate system are then obtained from Eq.

(4.19). Finally, the rotation matrix $\boldsymbol{R}'$, which relates the orientational state of the rotated cube to the absolute coordinate system, is evaluated from Eq. (4.22).

Next, we discuss the procedure regarding the orientation of the cube about the magnetic moment direction. In the case of a significantly strong applied magnetic field, the particle is expected to rotate about the magnetic moment direction which will coincide with the applied magnetic field direction under this circumstance. In the case of the cube having a magnetic moment in the diagonal direction, as shown in Fig. 4.2, it is necessitated to rotate the cube about this diagonal line for generating a new orientational state for a step in the Monte Carlo procedure. If the small rotational angle changes $\theta_x$, $\theta_y$, and $\theta_z$ are all equal to a small angle $\theta_0$, then the rotation matrix $\boldsymbol{Q}$ in Eq. (4.17) gives rise to the rotation about the diagonal line of the cube, as shown in Fig. 4.2. This is straightforwardly verified since the small angle $\theta_0$ can be assumed to be much smaller than unity and the rotation matrix reduces to the following expression:

$$\boldsymbol{Q} = \begin{bmatrix} 1 & -\sin\theta_0 & \sin\theta_0 \\ \sin\theta_0 & 1 & -\sin\theta_0 \\ -\sin\theta_0 & \sin\theta_0 & 1 \end{bmatrix} \tag{4.23}$$

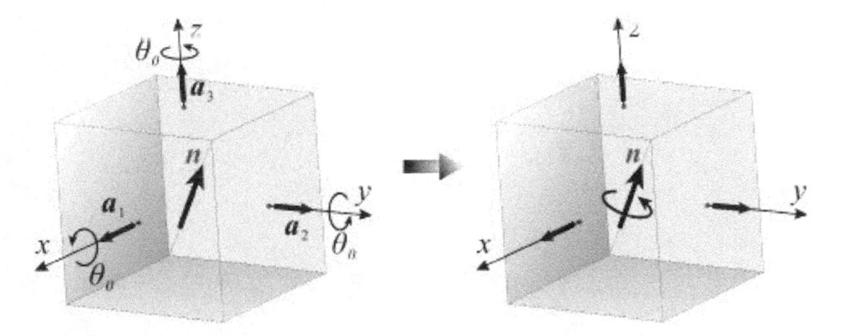


**Figure 4.2.** Orientation about the magnetic moment direction.

Hence, an arbitrary point $\boldsymbol{p}^b = (p_x^b, p_y^b, p_z^b)$ is transformed into the new position $\boldsymbol{p}^{b'} = (p_x^{b'}, p_y^{b'}, p_z^{b'})$ by the following relations:

$$\begin{bmatrix} p_x^{b}{}' \\ p_y^{b}{}' \\ p_z^{b}{}' \end{bmatrix} = \begin{bmatrix} 1 & -\sin\theta_0 & \sin\theta_0 \\ \sin\theta_0 & 1 & -\sin\theta_0 \\ -\sin\theta_0 & \sin\theta_0 & 1 \end{bmatrix} \begin{bmatrix} p_x^b \\ p_y^b \\ p_z^b \end{bmatrix} = \begin{bmatrix} p_x^b - p_y^b\sin\theta_0 + p_z^b\sin\theta_0 \\ p_x^b\sin\theta_0 + p_y^b - p_z^b\sin\theta_0 \\ -p_x^b\sin\theta_0 + p_y^b\sin\theta_0 + p_z^b \end{bmatrix} \tag{4.24}$$

If the point $\boldsymbol{p}^b$ is on the diagonal line along the magnetic moment direction $\boldsymbol{n}$, expressed as $\boldsymbol{p}^b = (p_0^b, p_0^b, p_0^b)$, Eq. (4.24) yields the results of $\boldsymbol{p}^{b'} = (p_0^b, p_0^b, p_0^b)$. This clearly shows that any points on the diagonal line remain at the original position for the rotational procedure, which verifies that the rotation by the matrix $\boldsymbol{Q}$ in Eq. (4.17) with the same small angle $\theta_0$ for $\theta_x$, $\theta_y$, and $\theta_z$ certainly implies the rotation of the cube about both the magnetic moment direction and the diagonal line of the cube.

CHAPTER 5

# Criterion of Particle Overlap

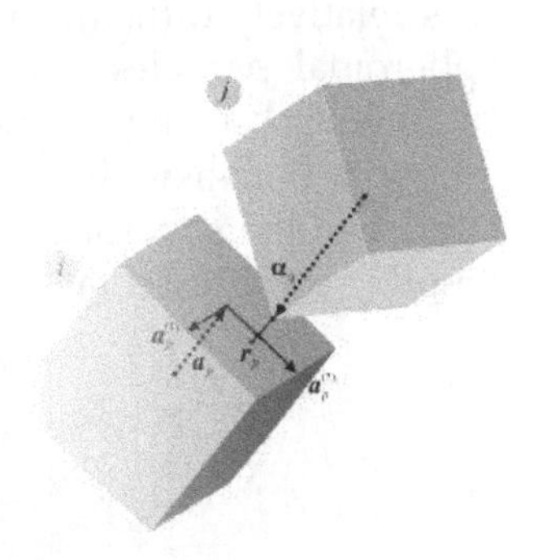

## 5.1 Spheroidal particles

In particle-based simulation methods, an appropriate treatment to prevent the overlap of solid particles or to evaluate the repulsive forces or energies between the soft layers covering particles, is very important, in order to avoid an instability of a system resulting from a physically unreasonable particle overlap. In development of a particle based simulation code for non-spherical particle suspensions, one of the more difficult tasks is to treat particle overlap and evaluate the minimum distance between two dispersed particles. Hence, before tackling development of a simulation code, it is required to conduct a systematic analysis in regard to the particle overlap or the minimum distance between particles. This analytical assessment of the particle overlap leads to a successful implementation in computer language. Moreover, knowledge of the minimum distance between two particles that is obtained from the assessment procedure, is indispensable for evaluating the cluster size distribution of the aggregate structures of dispersed particles. In the present and following sections, therefore, we analyze the criterion of the particle overlap for two non-spherical particles such as spherocylinder, disk-like and cube-like particles. The overlap criterion for two spherical particles is straightforwardly expressed using the center-to-center distance, hence, in the present section we begin with a discussion concerning spheroidal particles [1, 2].

If a solid spheroid is modeled as the sphere-connected model, as shown in Fig. 3.7, the assessment of the overlap between these two spheroids is made quite straightforward by applying the criterion for spheres to all the pairs of constituent spheres belonging to the different spheroids. Hence, in the following, we focus on the criterion for particle overlap for the case of the solid spheroidal particle [1, 2].

It is relatively difficult to assess the criterion of the overlap of the solid spheroidal particles, and therefore we first discuss the mathematical treatment of a single spheroid. As shown in Fig. 5.1(a), if the spheroid aligns in the $\hat{z}$-axis direction in the (particle) $\hat{x}\hat{y}\hat{z}$-coordinate system, the surface of the spheroid with half minor axis length $a$ and half major axis length $c$ is expressed as

$$\left(\frac{\hat{x}}{a}\right)^2 + \left(\frac{\hat{y}}{a}\right)^2 + \left(\frac{\hat{z}}{c}\right)^2 = 1 \tag{5.1}$$

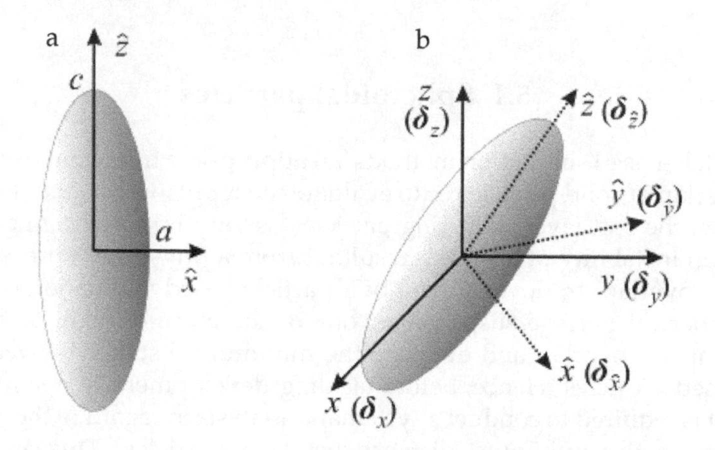


**Figure 5.1.** Coordinate systems for the spheroidal particle: (a) the particle coordinate system $\hat{x}\hat{y}\hat{z}$ and (b) its relationship with the absolute coordinate system $xyz$.

If a positive function $F_0(\hat{\boldsymbol{r}})$ is defined by the following expression:

$$F_0(\hat{\boldsymbol{r}}) = \left(\frac{\hat{x}}{a}\right)^2 + \left(\frac{\hat{y}}{a}\right)^2 + \left(\frac{\hat{z}}{c}\right)^2 \tag{5.2}$$

then any general point $\hat{\boldsymbol{r}} = (\hat{x}, \hat{y}, \hat{z})$ in the particle coordinate system may be assessed with the following criteria:

$$F_0(\hat{\boldsymbol{r}}) = \begin{cases} <1: \text{implies the inside of the spheroid} \\ =1: \text{implies the surface of the spheroid} \\ >1: \text{implies the outside of the spheroid} \end{cases} \tag{5.3}$$

For simplicity, we employ the following matrix $\boldsymbol{\alpha}_0$:

$$\boldsymbol{\alpha}_0 = \begin{bmatrix} 1/a & 0 & 0 \\ 0 & 1/a & 0 \\ 0 & 0 & 1/c \end{bmatrix} \tag{5.4}$$

Using this matrix, the function $F_0(\hat{\boldsymbol{r}})$ is expressed in simple form as

$$F_0(\hat{\boldsymbol{r}}) = (\boldsymbol{\alpha}_0 \cdot \hat{\boldsymbol{r}}) \cdot (\boldsymbol{\alpha}_0 \cdot \hat{\boldsymbol{r}}) \tag{5.5}$$

It is noted that the matrix $\boldsymbol{\alpha}_0$ is determined only by the shape of the spheroid.

Next, by applying the above function to a general situation of the spheroid, we discuss whether or not any point in the absolute coordinate system is inside the spheroid that inclines with orientation described by $(\theta, \phi)$, as shown in Fig. 4.1. The rotation matrix $\boldsymbol{R}$ has already been expressed in Eq. (4.3). If the unit vectors in each axis direction in the particle coordinate system are denoted by $\hat{\boldsymbol{\delta}}_{\hat{x}} = (1,0,0)$, $\hat{\boldsymbol{\delta}}_{\hat{y}} = (0,1,0)$ and $\hat{\boldsymbol{\delta}}_{\hat{z}} = (0,0,1)$, they are expressed as the unit vectors $\boldsymbol{\delta}_{\hat{x}}$, $\boldsymbol{\delta}_{\hat{y}}$ and $\boldsymbol{\delta}_{\hat{z}}$ in the absolute coordinate system, using the rotation matrix $\boldsymbol{R}$, as

$$\boldsymbol{\delta}_{\hat{x}} = \boldsymbol{R}^{-1} \cdot \hat{\boldsymbol{\delta}}_{\hat{x}} = \boldsymbol{R}^{t} \cdot \hat{\boldsymbol{\delta}}_{\hat{x}} = \begin{bmatrix} R_{11} \\ R_{12} \\ R_{13} \end{bmatrix}, \quad \boldsymbol{\delta}_{\hat{y}} = \boldsymbol{R}^{t} \cdot \hat{\boldsymbol{\delta}}_{\hat{y}} = \begin{bmatrix} R_{21} \\ R_{22} \\ R_{23} \end{bmatrix}, \quad \boldsymbol{\delta}_{\hat{z}} = \boldsymbol{R}^{t} \cdot \hat{\boldsymbol{\delta}}_{\hat{z}} = \begin{bmatrix} R_{31} \\ R_{32} \\ R_{33} \end{bmatrix} \tag{5.6}$$

Hence, if the unit vectors $\boldsymbol{\delta}_{\hat{x}}$, $\boldsymbol{\delta}_{\hat{y}}$ and $\boldsymbol{\delta}_{\hat{z}}$ in the absolute coordinate are known, the rotation matrix is straightforwardly obtained from Eq. (5.6).

In order to assess whether or not any point in the absolute coordinate system is inside the spheroid, it is more straightforward to transform this point to the particle coordinate system and to assess the criteria according to a modified function of the expression in Eq. (5.5). Employing the notation $\boldsymbol{x} = (x, y, z)$ for an arbitrary point in the absolute coordinate system, this point is expressed as $\hat{\boldsymbol{x}} = (\hat{x}, \hat{y}, \hat{z}) = \boldsymbol{R} \cdot \boldsymbol{x}$ in the particle coordinate system using the rotation matrix $\boldsymbol{R}$, as shown in Fig. 5.1(b). Hence, whether or not the point $x$ is inside an arbitrary spheroid $a$ with the center position $r_a$ may now be assessed with the following criteria:

$$F_0(R_a \cdot (x - r_a)) = \begin{cases} <1: \text{implies the inside of the spheroid} \\ =1: \text{implies the surface of the spheroid} \\ >1: \text{implies the outside of the spheroid} \end{cases} \tag{5.7}$$

in which subscript $a$ is attached to the rotation matrix because each particle has its own rotation matrix with its center position vector. For simplicity, the notation $F_1(x - r_a, e_a)$ is employed rather than $F_0(R_a \cdot (x - r_a))$, where $e_a$ is the unit vector denoting the particle direction.

The function $F_1(x - r_a, e_a)$ is explicitly written as

$$\begin{aligned} &F_1(x - r_a, e_a) = \alpha_0 \cdot (R_a \cdot (x - r_a)) \cdot \alpha_0 \cdot (R_a \cdot (x - r_a)) \\ &= (x - r_a) \cdot (\alpha_0 \cdot R_a)^t \cdot \alpha_0 \cdot (R_a \cdot (x - r_a)) \\ &= (x - r_a) \cdot (R_a^t \cdot \alpha_0^t) \cdot \alpha_0 \cdot (R_a \cdot (x - r_a)) = (x - r_a) \cdot (R_a^{-1} \cdot \alpha_0) \cdot \alpha_0 \cdot (R_a \cdot (x - r_a)) \\ &= (x - r_a) \cdot (R_a^{-1} \cdot \alpha_0 \cdot \alpha_0 \cdot R_a) \cdot (x - r_a) \end{aligned} \tag{5.8}$$

in which the following relationships have been taken into account: the inverse of $R_a$, $R_a^{-1}$, is equal to $R_a^t$, and the matrix $\alpha_0$ has only diagonal comments.

Employing new matrices $\alpha$ and $A_a$ expressed as

$$\alpha = \alpha_0 \cdot \alpha_0 = \begin{bmatrix} 1/a^2 & 0 & 0 \\ 0 & 1/a^2 & 0 \\ 0 & 0 & 1/c^2 \end{bmatrix}, \quad A_a = R_a^{-1} \cdot \alpha^{-1} \cdot R_a \tag{5.9}$$

then Eq. (5.8) is expressed in simple form as

$$F_1(x - r_a, e_a) = (x - r_a) \cdot A_a^{-1} \cdot (x - r_a) \tag{5.10}$$

The matrices $\alpha$ and $A_a$ have the following characteristics:

$$\alpha^{-1} = \begin{bmatrix} a^2 & 0 & 0 \\ 0 & a^2 & 0 \\ 0 & 0 & c^2 \end{bmatrix}, \quad A_a^{-1} = R_a^{-1} \cdot \alpha \cdot R_a, \quad (A_a^{-1})^t = A_a^{-1}, \quad (A_a)^t = A_a \tag{5.11}$$

We have now completed the preparation for conducting analysis of the assessment of the overlap criterion for the two solid spheroidal particles and will now consider the overlap problem regarding spheroidal particles $a$ and $b$ that are located at $\boldsymbol{r}_a$ and $\boldsymbol{r}_b$ with directions $\boldsymbol{e}_a$ and $\boldsymbol{e}_b$, respectively, as shown in Fig. 5.2.

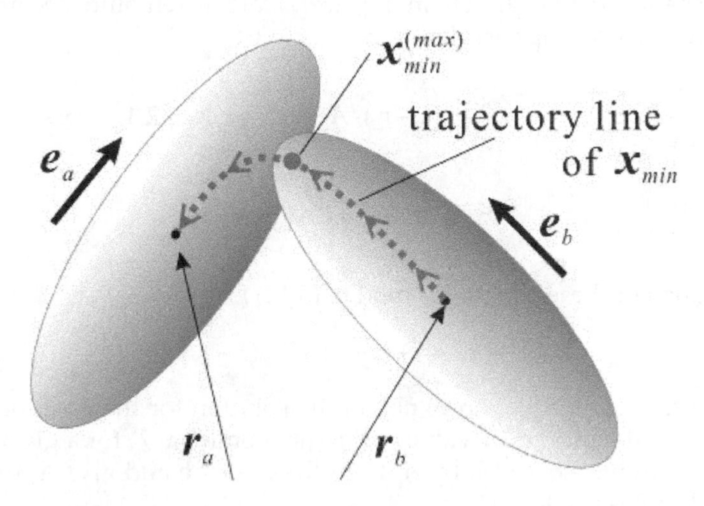


**Figure 5.2.** Trajectory line of the $\boldsymbol{x}_{min}$ for determining the position of the overlap point of the two spheroidal particles.

Equation (5.10) holds for particle $a$ and a similar equation will apply to particle $b$. In order to assess the overlap of these two spheroids, the following function $F_{ab}^{(0)}(\boldsymbol{x}, \lambda)$ is introduced:

$$F_{ab}^{(0)}(\boldsymbol{x},\lambda) = \lambda F_1(\boldsymbol{x}-\boldsymbol{r}_a, \boldsymbol{e}_a) + (1-\lambda)F_1(\boldsymbol{x}-\boldsymbol{r}_b, \boldsymbol{e}_b) \tag{5.12}$$

This function is also always positive and has the following characteristics:

$$\begin{aligned} &(1)\ \text{At}\ \lambda = 0,\ F_{ab}^{(0)}(\boldsymbol{x},\lambda)\ \text{becomes minimum at}\ \boldsymbol{x} = \boldsymbol{r}_b\ \text{to yield}\ F_{ab}^{(0)}(\boldsymbol{x},\lambda) = 0. \\ &(2)\ \text{At}\ \lambda = 1,\ F_{ab}^{(0)}(\boldsymbol{x},\lambda)\ \text{becomes minimum at}\ \boldsymbol{x} = \boldsymbol{r}_a\ \text{to yield}\ F_{ab}^{(0)}(\boldsymbol{x},\lambda) = 0. \end{aligned} \tag{5.13}$$

From these results, it may be expected that the trajectory of $\boldsymbol{x}$, which gives rise to a minimum value of $F_{ab}^{(0)}(\boldsymbol{x}, \lambda)$, with a change in $\lambda$ passes through the overlap area of the two spheroidal particles. This expectation will be validated by the following mathematical procedure.

We now find the position $\boldsymbol{x}$ where the function $F_{ab}^{(0)}(\boldsymbol{x}, \lambda)$ becomes minimum for a certain given value of $\lambda$ in the situation of the two spheroidal particles given by ($\boldsymbol{r}_a$, $\boldsymbol{e}_a$) and ($\boldsymbol{r}_b$, $\boldsymbol{e}_b$), as shown Fig. 5.2. For a given value of $\lambda$, this position is obtained if the following equation is satisfied:

$$\frac{\partial}{\partial \boldsymbol{x}} F_{ab}^{(0)}(\boldsymbol{x},\lambda)=0 \tag{5.14}$$

If the characteristics shown in Eq. (5.11) are taken into account, the following relationships are derived:

$$\frac{\partial}{\partial \boldsymbol{x}} F_1(\boldsymbol{x}-\boldsymbol{r}_a,\boldsymbol{e}_a)=\frac{\partial}{\partial \boldsymbol{x}}\Big[(\boldsymbol{x}-\boldsymbol{r}_a)\cdot \boldsymbol{A}_a^{-1}\cdot(\boldsymbol{x}-\boldsymbol{r}_a)\Big]=2\boldsymbol{A}_a^{-1}\cdot(\boldsymbol{x}-\boldsymbol{r}_a) \tag{5.15}$$

$$\frac{\partial}{\partial \boldsymbol{x}} F_1(\boldsymbol{x}-\boldsymbol{r}_b,\boldsymbol{e}_b)=2\boldsymbol{A}_b^{-1}\cdot(\boldsymbol{x}-\boldsymbol{r}_b) \tag{5.16}$$

Substitution of these equations into Eq. (5.14) leads to

$$\lambda \boldsymbol{A}_a^{-1}\cdot(\boldsymbol{x}-\boldsymbol{r}_a)+(1-\lambda)\boldsymbol{A}_b^{-1}\cdot(\boldsymbol{x}-\boldsymbol{r}_b)=0 \tag{5.17}$$

We now solve this equation to obtain the solution for the position $\boldsymbol{x}$ that gives rise to the minimum value of the function $F_{ab}^{(0)}(\boldsymbol{x}, \lambda)$ for a given value of $\lambda$. Multiplying Eq. (5.17) by $\boldsymbol{A}_a$ from the left side and also $\boldsymbol{A}_b$ from the right side gives rise to

$$(\lambda \boldsymbol{A}_b+(1-\lambda)\boldsymbol{A}_a)\cdot \boldsymbol{x}=\lambda \boldsymbol{A}_b\cdot \boldsymbol{r}_a+(1-\lambda)\boldsymbol{A}_a\cdot \boldsymbol{r}_b \tag{5.18}$$

If the matrix $\boldsymbol{C}^{-1}$ is defined by the following equation:

$$\boldsymbol{C}^{-1}=\lambda \boldsymbol{A}_b+(1-\lambda)\boldsymbol{A}_a \tag{5.19}$$

then Eq. (5.18) reduces to

$$\boldsymbol{C}^{-1}\cdot \boldsymbol{x}=\boldsymbol{C}^{-1}\cdot \boldsymbol{r}_b-\lambda \boldsymbol{A}_b\cdot(\boldsymbol{r}_b-\boldsymbol{r}_a) \tag{5.20}$$

It is noted that $\boldsymbol{C}$ and $\boldsymbol{C}^{-1}$ are both symmetric tensors.

Finally, we obtain the solution, denoted by $\boldsymbol{x}_{min}$, as

$$\boldsymbol{x}_{min}=\boldsymbol{r}_b-\lambda \boldsymbol{C}\cdot \boldsymbol{A}_b\cdot \boldsymbol{r}_{ba} \quad \text{or} \quad \boldsymbol{x}_{min}=\boldsymbol{r}_a+(1-\lambda)\boldsymbol{C}\cdot \boldsymbol{A}_a\cdot \boldsymbol{r}_{ba} \tag{5.21}$$

in which $\boldsymbol{r}_{ba}=\boldsymbol{r}_b-\boldsymbol{r}_a$ and Eq. (5.19) has been used for obtaining the second equation. Hence, the minimum value $F_{ab}^{(0)}(\boldsymbol{x}_{min}, \lambda)$ is obtained as

$$
\begin{aligned}
F_{ab}^{(0)}(\boldsymbol{x}_{min},\lambda) &= \lambda(\boldsymbol{x}_{min}-\boldsymbol{r}_a)\cdot \boldsymbol{A}_a^{-1}\cdot(\boldsymbol{x}_{min}-\boldsymbol{r}_a)+(1-\lambda)(\boldsymbol{x}_{min}-\boldsymbol{r}_b)\cdot \boldsymbol{A}_b^{-1}\cdot(\boldsymbol{x}_{min}-\boldsymbol{r}_b)\\
&= \lambda(1-\lambda)^2\,\boldsymbol{C}\cdot \boldsymbol{A}_a\cdot \boldsymbol{r}_{ba}\cdot \boldsymbol{A}_a^{-1}\cdot \boldsymbol{C}\cdot \boldsymbol{A}_a\cdot \boldsymbol{r}_{ba}+\lambda^2(1-\lambda)\boldsymbol{C}\cdot \boldsymbol{A}_b\cdot \boldsymbol{r}_{ba}\cdot \boldsymbol{A}_b^{-1}\cdot \boldsymbol{C}\cdot \boldsymbol{A}_b\cdot \boldsymbol{r}_{ba}\\
&= \lambda(1-\lambda)^2\,\boldsymbol{r}_{ba}\cdot \boldsymbol{C}\cdot \boldsymbol{C}\cdot \boldsymbol{A}_a\cdot \boldsymbol{r}_{ba}+\lambda^2(1-\lambda)\boldsymbol{r}_{ba}\cdot \boldsymbol{C}\cdot \boldsymbol{C}\cdot \boldsymbol{A}_b\cdot \boldsymbol{r}_{ba}\\
&= \lambda(1-\lambda)\boldsymbol{r}_{ba}\cdot \boldsymbol{C}\cdot \boldsymbol{C}\cdot\{(1-\lambda)\boldsymbol{A}_a+\lambda\,\boldsymbol{A}_b\}\cdot \boldsymbol{r}_{ba}\\
&= \lambda(1-\lambda)\boldsymbol{r}_{ba}\cdot \boldsymbol{C}\cdot \boldsymbol{C}\cdot \boldsymbol{C}^{-1}\cdot \boldsymbol{r}_{ba}\\
&= \lambda(1-\lambda)\boldsymbol{r}_{ba}\cdot \boldsymbol{C}\cdot \boldsymbol{r}_{ba}
\end{aligned}
\tag{5.22}
$$

If we draw the trajectory line of the position $\boldsymbol{x}_{min}$, which gives rise to the minimum value $F_{ab}^{(0)}(\boldsymbol{x}_{min}, \lambda)$, for change in the value of $\lambda$ from zero to unity, then this line should start from the center of particle *b*, through the overlap or minimum distance area of the two spheroidal particles, and finally ends at the center of particle *a*, as shown in Fig. 5.2. This curve of $F_{ab}^{(0)}(\boldsymbol{x}_{min}, \lambda)$ as a function of $\lambda$ has a maximum value at a certain value of $\lambda$, and this point should be in the overlap or minimum distance area of the two particles. We employ the notation $F_{ab}^{(0)\,max}$ for the maximum value, $\boldsymbol{x}_{min}^{(max)}$ for the position yielding the maximum value, and $\lambda^{(max)}$ for the value of $\lambda$ providing $\boldsymbol{x}_{min}^{(max)}$. Employing these symbols, $F_{ab}^{(0)max}$ is expressed as

$$
F_{ab}^{(0)max} = F_{ab}^{(0)}(\boldsymbol{x}_{min}^{(max)}, \lambda^{(max)}) \tag{5.23}
$$

Hence, the assessment regarding the overlap between the two spheroidal particles is conducted according to the following manner:

$$
F_{ab}^{(0)max} = \begin{cases} <1: \text{implies overlap of the two spheroidal particles} \\ =1: \text{implies surface–to–surface contact} \\ >1: \text{implies no overlap} \end{cases} \tag{5.24}
$$

We summarize the assessment procedure regarding the overlap between the two solid spheroidal particles *a* and *b* that are located at $\boldsymbol{r}_a$ and $\boldsymbol{r}_b$ with directions $\boldsymbol{e}_a$ and $\boldsymbol{e}_b$, respectively.

(1) Calculate $\cos\theta_a$, $\sin\theta_a$, $\cos\phi_a$, $\sin\phi_a$, $\cos\theta_b$, $\sin\theta_b$, $\cos\phi_b$ and $\sin\phi_b$ from the known values of $\boldsymbol{e}_a$ and $\boldsymbol{e}_b$,
(2) Calculate the rotation matrices $\boldsymbol{R}_a$ and $\boldsymbol{R}_b$ from Eq. (5.6),
(3) Calculate the matrices $\boldsymbol{A}_a$ and $\boldsymbol{A}_b$ from Eq. (5.9),
(4) Set a value of $\lambda$ ($\lambda = 0.5$ is appropriate for an initial value),
(5) Calculate $\boldsymbol{C}^{-1}$ from Eq. (5.19), and evaluate $\boldsymbol{C}$,
(6) Calculate $F_{ab}^{(0)}(\boldsymbol{x}_{min}, \lambda)$ from Eq. (5.22),

(7) If $F_{ab}^{(0)}(\boldsymbol{x}_{min}, \lambda)$ cannot be recognized as a maximum value $F_{ab}^{(0)max}$ for the change of $\lambda$, then return to step (4),

(8) Using the value of $F_{ab}^{(0)\,max}$, the following assessment procedure is conducted regarding the particle overlap:

 (8)-1 If $F_{ab}^{(0)max} < 1$, there is an overlap between the two spheroidal particles,

 (8)-2 If $F_{ab}^{(0)max} = 1$, there is a surface-to-surface contact between the two spheroidal particles,

 (8)-3 If $F_{ab}^{(0)max} > 1$, there is no overlap between the two spheroidal particles.

Using $\lambda^{(max)}$ giving rise to $F_{ab}^{(0)\,max}$, the position yielding the maximum value, $\boldsymbol{x}_{min}^{(max)}$, is then evaluated from Eq. (5.21).

## 5.2 Spherocylinder particles

As already mentioned in the previous section, the spheroid model is quite difficult to be assessed regarding the particle overlap, so a spherocylinder is frequently employed as a model particle for spindle-like particles that are experimentally synthesized. A spherocylinder is a cylindrical body caped by hemispheres at both the ends of the cylinder, as shown in Fig. 3.10, so that overlaps between two spherocylinders are classified into several typical overlap configurations. Figure 5.3 schematically shows these configurations, that is, configuration (a) is an overlap on a direct line, configuration (b) is an overlap between parallel particles, configuration (c) is an overlap in the perpendicular situation, and configurations (d) and (e) are an overlap in a general arrangement of the particles. In all cases, the key factor is a point $P_i$ on the major axis line of particle $i$ from which a perpendicular line is drawn to the center of the nearest hemisphere of particle $j$. As shown in Fig. 3.10, we use symbol $d$ for the particle diameter, $l_0$ for the length of the cylindrical body and $l$ for the length of the spherocylinder.

Configurations (a) to (c) are quite straightforward for assessing the overlap of the two spherocylinders. In the case of configuration (a), if the center-to-center distance between two particles, $r_{ij}$, is smaller than the length $l$, i.e., $r_{ij} < l$, then the two particles overlap. In the case of configuration (b), point $P_i$ is a key factor for assessment of the overlap. In the case of the point $P_i$ being inside the cylindrical body, the distance between the longer axis lines of the particles, $r_{ij}^{\perp}$, is smaller than $d$, i.e., $r_{ij}^{\perp} < d$, then the particles overlap. In the case of the point $P_i$ being outside the cylindrical body, the distance between the centers of the hemispheres belonging to particles $i$ and $j$, $r_{ij}^{hscap}$, is smaller than $d$, i.e., $r_{ij}^{hscap} < d$, then the particles overlap. In configurations (c) to (e), the following similar assessment procedure is applicable and the position $P_i$ is also a key factor for assessment of the overlap.

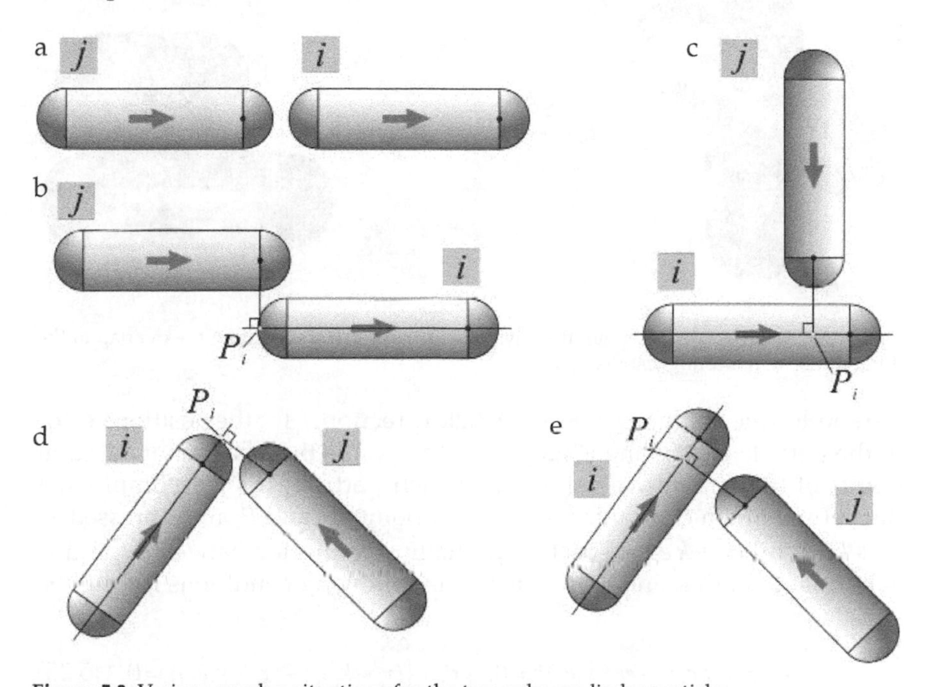


**Figure 5.3.** Various overlap situations for the two spherocylinder particles.

In Figs. 5.3(d) and (e), if the position of the particle center and the point $P_i$ are denoted by $\boldsymbol{r}_i$ and $\boldsymbol{r}_i^{(P)}$, respectively, then the regime of the particle overlap depends on the distance $l_i^{(P)} = |\boldsymbol{r}_i^{(P)} - \boldsymbol{r}_i|$. If $l_i^{(P)} < l_0/2$, as shown in Fig. 5.3(e), there is a possibility of the overlap between the cylindrical body of particle $i$ and the hemisphere of particle $j$. We use the notation $r_{ij}^{(P)}$ for the length of the line between the center of the hemisphere of particle $j$ and the point $P_i$. In the above regime, if $r_{ij}^{(P)} < d$, then the particles overlap. In the regime of $l_i^{(P)} \geq l_0/2$, if $r_{ij}^{hscap} < d$, there is an overlap between the hemispheres of particles $i$ and $j$.

In the case of a three-dimensional system, the above assessment procedure is applicable, if the point $P_i$ is obtained from the straightforward vector analysis regarding the orthogonality condition of the two vectors. The overlap between the cylindrical bodies of the two spherocylinders is a specific event in the three-dimensional system, which is schematically shown in Fig. 5.4 and in the following, we concentrate on developing the criteria for this type of overlap.

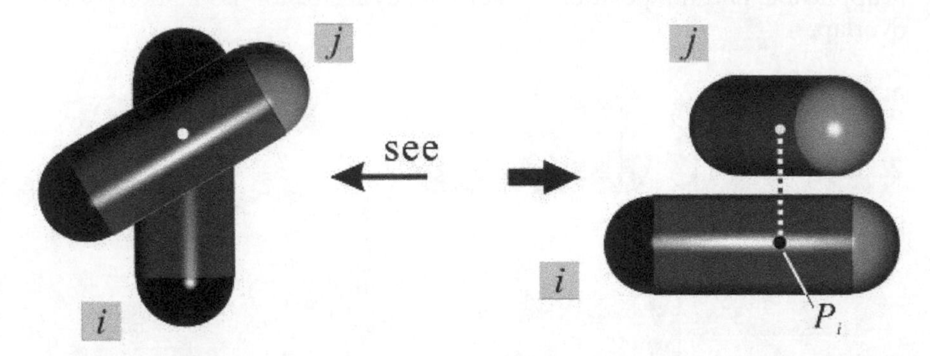


**Figure 5.4.** Appropriate viewpoint for a systematic assessment regarding the overlap in the three-dimensional configuration.

We employ the notation $\boldsymbol{e}_i$ for the particle direction, $\boldsymbol{r}_i$ for the position vector of the particle center, and $P_i$ and $P_j$ for the points that are the intersection points of the line drawn vertically to each particle axis line. Employing the yet unknown constants $k_i$ and $k_j$, the points $P_i$ and $P_j$ are expressed as $(\boldsymbol{r}_i + k_i\boldsymbol{e}_i)$ and $(\boldsymbol{r}_j + k_j\boldsymbol{e}_j)$, respectively. The line connected between $P_i$ and $P_j$ is normal to both $\boldsymbol{e}_i$ and $\boldsymbol{e}_j$, therefore, the following conditions have to be satisfied:

$$\boldsymbol{e}_i \cdot \{(\boldsymbol{r}_i + k_i\boldsymbol{e}_i) - (\boldsymbol{r}_j + k_j\boldsymbol{e}_j)\} = 0, \quad \boldsymbol{e}_j \cdot \{(\boldsymbol{r}_i + k_i\boldsymbol{e}_i) - (\boldsymbol{r}_j + k_j\boldsymbol{e}_j)\} = 0 \quad (5.25)$$

These equations are straightforwardly solved by using a vector analysis procedure to obtain the solutions of $k_i$ and $k_j$ as

$$\begin{bmatrix} k_i \\ k_j \end{bmatrix} = \frac{1}{1-(\boldsymbol{e}_i \cdot \boldsymbol{e}_j)^2} \begin{bmatrix} -1 & \boldsymbol{e}_i \cdot \boldsymbol{e}_j \\ -\boldsymbol{e}_i \cdot \boldsymbol{e}_j & 1 \end{bmatrix} \begin{bmatrix} \boldsymbol{e}_i \cdot \boldsymbol{r}_{ij} \\ \boldsymbol{e}_j \cdot \boldsymbol{r}_{ij} \end{bmatrix} \quad (5.26)$$

Using these values of $k_i$ and $k_j$, the overlap between the cylindrical bodies is assessed in the following manner. In the case of $|k_i| < l_0/2$ and $|k_j| < l_0/2$, if $|(\boldsymbol{r}_i + k_i\boldsymbol{e}_i) - (\boldsymbol{r}_j + k_j\boldsymbol{e}_j)| < d$, there is an overlap between the two particles. If $|k_i| \geq l_0/2$ and $|k_j| < l_0/2$ or if $|k_i| < l_0/2$ and $|k_j| \geq l_0/2$, there is a possibility of the overlap between the cylindrical body and the hemisphere. If $|k_i| \geq l_0/2$ and $|k_j| \geq l_0/2$, there is a possibility of the overlap between the hemispheres of the two spherocylinders. Assessment of the overlap for the latter two regimes is conducted according to a procedure similar to that for the previous two-dimensional system.

From the assessment of the particle overlap according to the above conditions, we are able to find the nearest points on the axis lines of each spherocylinder particle. If we use the notation $\boldsymbol{r}_i^{(min)}$ and $\boldsymbol{r}_j^{(min)}$ for points on the axis lines of particles *i* and *j*, respectively, which give rise to the nearest distance, the interaction energy or force due to the overlap of the steric layers may be evaluated in the following manner. If the sphere-connected particle model is adopted for evaluation of the interaction energy or force between the steric layers, a spherical particle is located at the position $\boldsymbol{r}_i^{(min)}$ and the neighboring spheres are placed in close-contact to finally complete the sphere-connected particle *i*. The sphere-connected particle *j* is then formed in the same manner around the initial sphere located at the position $\boldsymbol{r}_j^{(min)}$. The force or energy acting between the two spherocylinder particles due to the overlap of the steric layers is evaluated by summing the interactions between all the pairs of constituent spheres belonging to the two different spherocylinder particles. Moreover, the torque acting on a particle can be evaluated by conducting the cross product of these forces and their position vector relative to the center of the particle of interest.

The main points of the assessment procedure concerning the particle overlap of the spherocylinder particles are summarized below in an algorithmic form:

(1) If the conditions of $|\boldsymbol{e}_i \cdot \boldsymbol{e}_j| = |\boldsymbol{e}_i \cdot (\boldsymbol{r}_i - \boldsymbol{r}_j)/|(\boldsymbol{r}_i - \boldsymbol{r}_j)|| = |\boldsymbol{e}_j \cdot (\boldsymbol{r}_i - \boldsymbol{r}_j)/|(\boldsymbol{r}_i - \boldsymbol{r}_j)|| = 1$ are satisfied, there is a possibility of the overlap in the situation of the two spherocylinders located on the center-to-center line, aligning along this line, as shown in Fig. 5.3(a).

(1)-1 If the separation distance between the nearest pair of hemispheres is smaller than the diameter *d* of the cylindrical body, there is an overlap between the hemispheres.

(1)-2 If not, there is no overlap.

(2) If the conditions of $|\boldsymbol{e}_i \cdot \boldsymbol{e}_j| = 1$, $|\boldsymbol{e}_i \cdot (\boldsymbol{r}_i - \boldsymbol{r}_j)/|(\boldsymbol{r}_i - \boldsymbol{r}_j)|| \neq 1$ and $|\boldsymbol{e}_j \cdot (\boldsymbol{r}_i - \boldsymbol{r}_j)/|(\boldsymbol{r}_i - \boldsymbol{r}_j)|| \neq 1$ are satisfied, there is a possibility of the overlap in a parallel configuration, as shown in Fig. 5.3(b).

(2)-1 If the separation distance between the two parallel lines is larger than or equal to the diameter *d*, there is no overlap between the particles.

(2)-2 If not, the location of point $P_i$, shown in Fig. 5.3(b), is evaluated.

(2)-2-1 If point $P_i$ is inside the range of the cylindrical body, there is an overlap.

(2)-2-2 If the separation distance between the nearest pair of hemispheres is smaller than diameter $d$, there is an overlap.

(3) If the condition of $|e_i \cdot e_j| = 0$ is satisfied, there is a possibility of the overlap in the perpendicular configuration, as shown in Fig. 5.3(c). Then the location of point $P_i$ is evaluated.

(3)-1 If point $P_i$ is inside the range of the cylindrical body, and the shortest distance between the nearest hemisphere and point $P_i$ is smaller than diameter $d$, then there is an overlap.

(3)-2 If point $P_i$ is outside the range of the cylindrical body, and the separation distance between the nearest pair of hemispheres is smaller than diameter $d$, there is an overlap.

(4) At this final stage, a general configuration is treated, as shown in Figs. 5.3 (d) and (e). We assume that the configuration satisfies the condition $|k_j| \geq |k_i|$, as shown in Figs. 5.3(d) and (e).

(4)-1 If point $P_i$ is inside the range of the cylindrical body, as shown in Fig. 5.3(e), and the distance between the nearest hemisphere and point $P_i$ is smaller than diameter $d$, then there is an overlap.

(4)-2 If point $P_i$ is outside the range of the cylindrical body, as shown in Fig. 5.3(d), and the separation distance between the nearest pair of hemispheres is smaller than diameter $d$, there is an overlap.

## 5.3 Disk-like particles

Although it may be reasonable to employ oblate spheroidal particles for flat-like particles that are frequently synthesized experimentally, they are significantly difficult to treat in simulations in terms of the particle overlap, similar to the case of prolate spheroidal particles. From this background, a disk-like particle model may be employed for simulations because the assessment of particle overlap and evaluation of the interactions between repulsive layers covering particles are relatively tractable compared with the oblate spheroidal particle model. In the following, therefore, we focus on the disk-like particle model to analyze the criterion of the overlap for two disk-like particles. The criterion for overlap between the disk-like particles shown in Fig. 3.16 is relatively difficult, and therefore we initially treat a particle with negligible thickness, as shown in Fig. 5.5, which is referred to in the following as a plate-like particle.

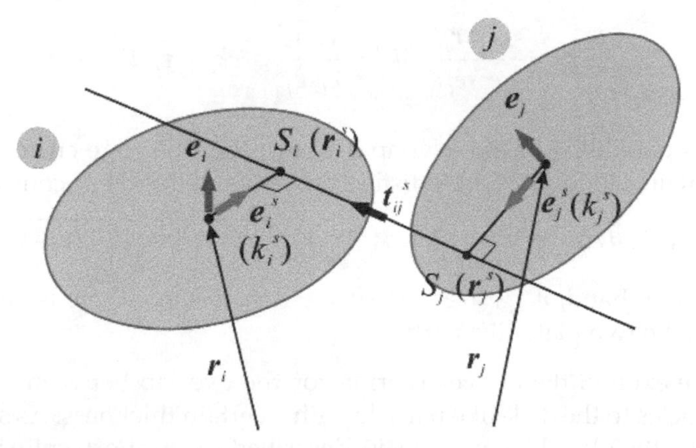


**Figure 5.5.** Analysis of circles with radius $r_0$ (= $d/2$).

The key point for the systematic analysis of particle overlap is the line of intersection generated by the two plates. We employ symbol $d$ for the particle diameter, $\boldsymbol{r}_i$ of particle center, $\boldsymbol{e}_i$ for the particle orientation, normal to the plate surface, $S_i$ for the intersection point of the perpendicular line drawn from $\boldsymbol{r}_i$ to the intersection line, $\boldsymbol{r}_i^s$ for the position vector of $S_i$, $\boldsymbol{e}_i^s$ for the unit vector denoting the direction of $(\boldsymbol{r}_i^s - \boldsymbol{r}_i)$, and $k_i^s$ for a constant defined by $k_i^s \boldsymbol{e}_i^s = (\boldsymbol{r}_i^s - \boldsymbol{r}_i)$. Similar symbols are employed for particle $j$. Moreover, the notation $t_{ij}^s$ is adopted as the unit vector denoting the direction of the line drawn from points $S_j$ to $S_i$, and $k_{ij}^s$ for a constant defined by $k_{ij}^s\,\boldsymbol{t}_{ij}^s = (\boldsymbol{r}_i^s - \boldsymbol{r}_j^s)$, as shown in Fig. 5.5.

From the orthogonality conditions concerning $\boldsymbol{t}_{ij}^s$ and $\boldsymbol{e}_i$, $\boldsymbol{t}_{ij}^s$ and $\boldsymbol{e}_j$, $\boldsymbol{e}_i$ and $\boldsymbol{e}_i^s$, $\boldsymbol{e}_j$ and $\boldsymbol{e}_j^s$, $\boldsymbol{e}_i^s$ and $\boldsymbol{t}_{ij}^s$, and $\boldsymbol{e}_j^s$ and $\boldsymbol{t}_{ij}^s$, the vectors $\boldsymbol{t}_{ij}^s$, $\boldsymbol{e}_i^s$ and $\boldsymbol{e}_j^s$ are obtained from elementary vector analysis as

$$\boldsymbol{t}_{ij}^{s} = \boldsymbol{e}_j \times \boldsymbol{e}_i \Big/ \left|\boldsymbol{e}_j \times \boldsymbol{e}_i\right|, \quad \boldsymbol{e}_i^{s} = -\boldsymbol{e}_i \times \boldsymbol{t}_{ij}^{s}, \quad \boldsymbol{e}_j^{s} = \boldsymbol{e}_j \times \boldsymbol{t}_{ij}^{s} \tag{5.27}$$

These relationships are valid for the configuration shown in Fig. 5.5 where particle $j$ is on the right-hand side of particle $i$. The expression of point $S_i$ in the two different forms yields the following equation:

$$\boldsymbol{r}_i + k_i^{s}\boldsymbol{e}_i^{s} = \boldsymbol{r}_j + k_j^{s}\boldsymbol{e}_j^{s} + k_{ij}^{s}\boldsymbol{t}_{ij}^{s} \tag{5.28}$$

The left and right-hand sides in this equation are related to the same position vector $\boldsymbol{r}_i^s$ which is traced from the center of particles $i$ and $j$, respectively. From the orthogonality condition of the unit vectors and Eq. (5.28), the constants $k_i^s$, $k_j^s$ and $k_{ij}^s$ are solved as

$$k_i^s = -\frac{\boldsymbol{e}_j \cdot \boldsymbol{r}_{ij}}{\boldsymbol{e}_j \cdot \boldsymbol{e}_i^s}, \quad k_j^s = \frac{\boldsymbol{e}_i \cdot \boldsymbol{r}_{ij}}{\boldsymbol{e}_i \cdot \boldsymbol{e}_j^s}, \quad k_{ij}^s = \boldsymbol{r}_{ij} \cdot \boldsymbol{t}_{ij}^s \tag{5.29}$$

Hence, a possibility of the overlap between the two edge circles arises if the following conditions are satisfied by the values of these constants:

$$|k_i^s| < d/2, \quad |k_j^s| < d/2, \quad |k_{ij}^s| < \sqrt{(d/2)^2 - k_i^s} + \sqrt{(d/2)^2 - k_j^s} \tag{5.30}$$

On the other hand, if $|k_i^s| \geq 1$ or $|k_j^s| \geq 1$ is satisfied, there is no overlap between the two plate-like particles.

Next, we expand the above criterion for the overlap between the plate-like particles to the disk-like particle with a certain thickness. As shown in Fig. 3.16, the disk-like particle is described as a short cylinder with diameter $d$ and thickness $b_1$ surrounded by a torus part with a cross section of semicircle shape, and therefore, it has the thickness $b_1$ and the diameter $d_1$ of the maximum outline circle.

The two simple configurations of two particles being in the same plane and in the two parallel planes can be straightforwardly treated, and therefore we do not address these cases here.

Figure 5.6 shows the typical regimes for the particle overlap of the disk-like particles. Although the particles have a thickness $b_1$, only the circle of the cross-section at the center of the cylindrical part is displayed in order to clarify the essence of the configuration. In Fig. 5.6(a), the intersection line passes through both particles; in Figs. 5.6(b) and (c), it passes only through the inside of particle $i$; in Fig. 5.6(d), it passes outside of both particles. For all the configurations, the positions of the points $Q_i$ and $Q_j$ are the main factor for assessing particle overlap, that is, the points in the vicinity of $Q_i$ and $Q_j$ give rise to the minimum distance between the two disk-like particles. In the cases of Figs. 5.6(a), (b) and (d), the assessment procedure is approximately the same, and therefore we first address the configuration in Fig. 5.6(c), because here point $Q_i$ has characteristics different from those for the other three configurations.

In Fig. 5.6(c), point $Q_i$ on the plane of particle $i$ is the intersection point of the line, on the plane surface of particle $i$, that is perpendicularly drawn from the point $Q_j$ of the circle of particle $j$. The position vector $\boldsymbol{r}_{i(j)}^{Q}$ of point $Q_i$ is obtained from vector analysis using the geometric relationship of the particle configurations as

$$\boldsymbol{r}_{i(j)}^{Q} = \boldsymbol{r}_j + (d/2)\,\boldsymbol{e}_j^s - k_{i(j)}^{Q}\,\boldsymbol{e}_i \tag{5.31}$$

in which $k_{i(j)}^{Q}$ is a constant and expressed as

$$k_{i(j)}^{Q} = (k_j^s - d/2)\left|\boldsymbol{e}_j^s \cdot \boldsymbol{e}_i\right| \tag{5.32}$$

Hence, if point $Q_i$ is on the plane or inside the circle of particle $i$ in the configuration shown in Fig. 5.6(c), then the satisfaction of $|k_{i(j)}^{Q}| < b_1$ ensures an overlap between the plane of particle $i$ and the torus part of particle $j$.

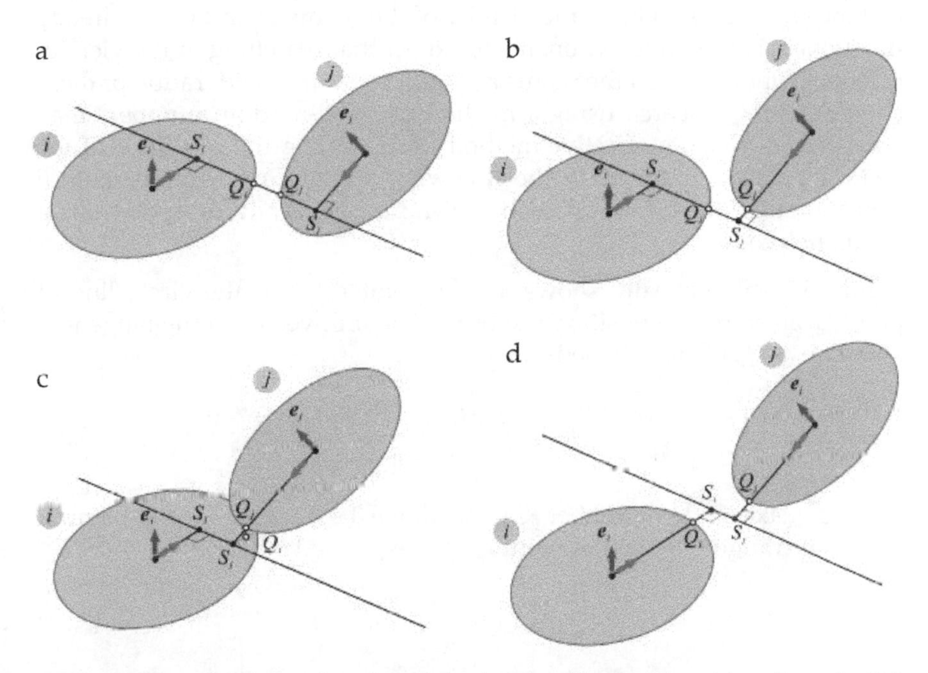


**Figure 5.6.** Typical regimes of overlap of circular plate-like particles with radius $r_0$ (= $d/2$) and infinitesimal thin thickness.

Next, we consider the overlap criterion for the other configurations shown in Figs. 5.6(a), (b) and (d). The nearest position between two points on each circle should be in the vicinity of $Q_i$ and $Q_j$, and we denote these positions by $\boldsymbol{r}_i^{(min)}$ and $\boldsymbol{r}_j^{(min)}$ on particles $i$ and $j$, respectively. If these positions are obtained by a suitable method, which is discussed in the following, the shortest distance given by $r_{ij}^{(min)} = |\boldsymbol{r}_i^{(min)} - \boldsymbol{r}_j^{(min)}|$ is straightforwardly used for assessing the overlap criterion. For instance, if the shortest distance is greater than the particle thickness, that is, $r_{ij}^{(min)} > b_1$, there will be no overlap between the two disk-like particles.

In the overlap assessment procedure, it is a key factor to evaluate the nearest positions $\boldsymbol{r}_i^{(min)}$ and $\boldsymbol{r}_j^{(min)}$, and there may be several methods for making these evaluations in terms of the different analytical and numerical approaches. Here we show a numerical approach, which from a simulation point of view, is relatively straightforward to treat and is also applicable to the development of a simulation code. Since the circle is a simple shape and the configuration of particles $i$ and $j$ is relatively straightforward, it is clear that the nearest points, $\boldsymbol{r}_i^{(min)}$ and $\boldsymbol{r}_j^{(min)}$, giving rise to the minimum distance $r_{ij}^{(min)}$, should be in the vicinity of the points $Q_i$ and $Q_j$, as already mentioned. Hence, it is reasonable to expect that searching in the vicinity of these point will yield the points $\boldsymbol{r}_i^{(min)}$ and $\boldsymbol{r}_j^{(min)}$. In consideration of these characteristics, a search using a method of quasi-random numbers may be expected to be an effective method for obtaining the positions of the nearest points. Therefore in the following we discuss a mathematical aspect for this stochastic approach, which may be an aid for developing a simulation code.

Figure 5.7 schematically shows a trial point $\boldsymbol{r}_{i(try)}^{(min)}$ on the circle line of point $Q_i$ given by the position vector $\boldsymbol{r}_i^Q$. The unit vector in the tangential direction at $Q_i$, $\boldsymbol{n}_i^Q$, is obtained as

$$\boldsymbol{n}_i^Q = \frac{\boldsymbol{e}_i \times (\boldsymbol{r}_i^Q - \boldsymbol{r}_i)}{\left|\boldsymbol{e}_i \times (\boldsymbol{r}_i^Q - \boldsymbol{r}_i)\right|} \tag{5.33}$$

Using a quasi-random number $R$, in the range $0 \le R \le 1$, and the maximum trial angle range $\Delta\beta$, the trial point $\boldsymbol{r}_{i(try)}^{(min)}$ is expressed as

$$\boldsymbol{r}_{i(try)}^{(min)} = \boldsymbol{r}_i^Q + (2R-1)(r_0 \Delta\beta\, \boldsymbol{n}_i^Q) \tag{5.34}$$

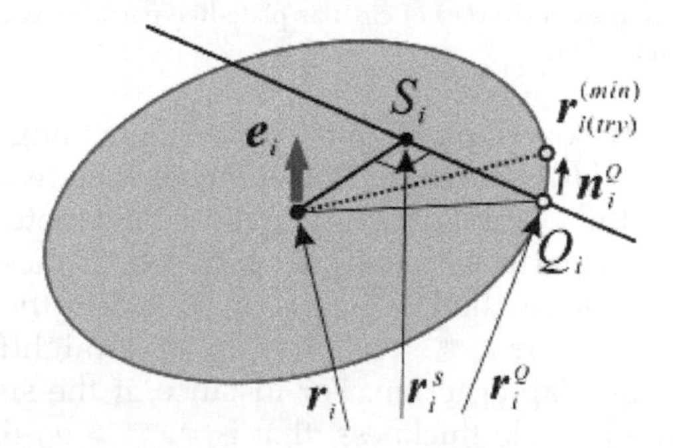


**Figure 5.7.** Attempt at setting a trial position for searching for the position of $\boldsymbol{r}_i^{(min)}$ giving rise to the minimum distance $r_{ij}^{(min)}$ : the radius of the circle is $r_0$ (= $d/2$).

Since $|\boldsymbol{r}_{i(try)}^{(min)} - \boldsymbol{r}_i|$ is required to be equal to $r_0$ (= $d/2$), the following $\boldsymbol{r}_{i(try)}^{(min)}$ is used instead of Eq. (5.34), unless $\Delta\beta$ is sufficiently small:

$$\boldsymbol{r}_{i(try)}^{(min)} = \boldsymbol{r}_i + r_0 \frac{\boldsymbol{r}_i^{Q} + (2R-1)(r_0 \Delta\beta\, \boldsymbol{n}_i^{Q}) - \boldsymbol{r}_i}{\left| \boldsymbol{r}_i^{Q} + (2R-1)(r_0 \Delta\beta\, \boldsymbol{n}_i^{Q}) - \boldsymbol{r}_i \right|} \tag{5.35}$$

The maximum trial angle range $\Delta\beta$ may be chosen, for instance, as $\Delta\beta = 10(\pi/180)$.

Figure 5.8 demonstrates the trial points $p_1, p_2, \ldots, p_7$ that are generated in the above procedure using a random number taken from a sequence of quasi-random numbers. Similar trial points are generated for particle $j$. If point $p_3$ gives rise to the minimum distance between the two circles, the trial range is reassigned between points $p_2$ and $p_4$, and then the above trail procedure is repeated. For a new range between $q_2$ and $q_4$, another trial procedure is conducted to obtain the final solution. If this solution is assumed to be the point $s_3$, then this point is taken to be $\boldsymbol{r}_i^{(min)}$ and similarly $\boldsymbol{r}_j^{(min)}$ is obtained for particle $j$.

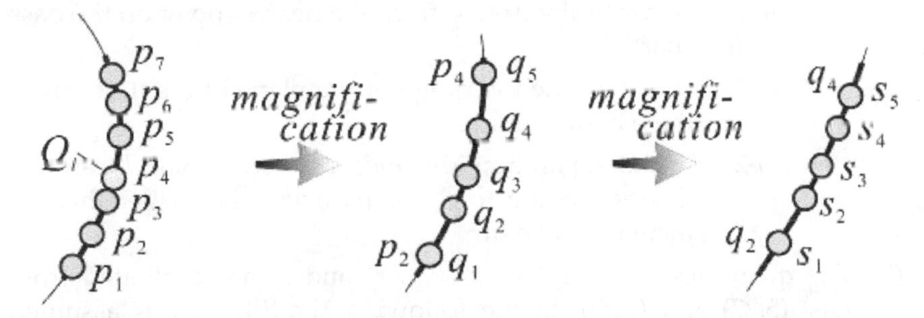


**Figure 5.8.** Searching method for obtaining the point $\boldsymbol{r}_i^{(min)}$ giving rise to the minimum distance $r_{ij}^{(min)}$ by narrowing the searching area (length) in several searching steps.

If the sphere-constituting particle model is used for the disk-like particle, locating a sphere with radius $b_1$ at the points $\boldsymbol{r}_i^{(min)}$ and $\boldsymbol{r}_j^{(min)}$ enables one to evaluate a repulsive force or energy resulting from the overlap of soft repulsive layers covering each disk-like particle by evaluating these quantities between the relevant pairs of spheres on the torus part of each disk-like particle.

The assessment procedure for particle overlap between two disk-like particles of a certain thickness is summarized below. Referring to the shape shown in Fig. 3.16, for simplicity, we name the circle with diameter $d_1$ the "outer circle", and the circle of the cylinder cross-section with diameter $d$

the "base circle": it is noted that these circles are on the cross-section plane at the center of the cylindrical body.

(1) If the conditions of $|e_i \cdot e_j| = 1$ and $|e_i \cdot (r_i - r_j)/|(r_i - r_j)|| = |e_j \cdot (r_i - r_j)/|(r_i - r_j)|| = 0$ are satisfied, there is a possibility of overlap in the situation where the two disk-like planes are located in the same plane and each particle direction is normal to this plane. If the separation distance $r_{ij}$ $(=|r_i - r_j| = |r_{ij}|)$ between the centers of the particles is smaller than the diameter $d_1$ (shown in Fig. 3.16), then there is an overlap between the torus bodies of the disk-like particles.

(2) If the conditions of $|e_i \cdot e_j| = 1$, $|e_i \cdot r_{ij}| \neq 1$ and $|e_j \cdot r_{ij}| \neq 1$ are satisfied, there is a possibility of the overlap in the parallel configuration, where the two disk planes are parallel.

(2)-1 If the separation distance between the two parallel planes is larger than (or equal to) $b_1$, there is no overlap between the particles.

(2)-2 If not, the location of point $P_i$ on the plane of the outer circle of particle $i$ is evaluated where point $P_i$ is the intersection point of the line vertically drawn from the nearest point on the base circle of particle $j$.

(2)-2-1 If $P_i$ is inside the range of the plane of the base circle, there is an overlap.

(2)-2-2 If the separation distance between point $P_i$ and $P_j$ (corresponding to $P_i$ for particle $j$) is smaller than $b_1$, there is an overlap.

(3) The quantities $e_i^s$, $e_j^s$, $t_{ij}^s$, $k_i^s$, $k_j^s$, $k_{ij}^s$, $r_i^s$ and $r_j^s$ are evaluated from Eqs. (5.27) and (5.29). In the following, $|k_j^s| \geq |k_i^s|$ is assumed for simplicity. If the condition of $|e_i \cdot e_j| = 0$ is satisfied, there is a possibility of the overlap in the perpendicular configuration. As in (2)-2, the location of point $P_i$ is evaluated.

(3)-1 If point $P_i$ is inside the base circle range, and the distance between point $P_i$ and the nearest point $P_j^{(0)}$ on the base circle of particle $j$ is smaller than $b_1$, then there is an overlap.

(3)-2 If point $P_i$ is outside the range of the cylindrical body, and the separation distance between the nearest points $P_i^{(0)}$ and $P_j^{(0)}$ on each base circle is smaller than $b_1$, there is an overlap.

(4) Next, a general configuration is treated, as shown in Fig. 5.6. As in procedure (3), we assume that the configuration satisfies the condition $|k_j^s| \geq |k_i^s|$, for which typical configurations are schematically shown in Fig. 5.6.

(4)-1 If the conditions of $|k_i^s| \le d/2$, $|k_j^s| \ge d/2$ and $|k_{ij}^s| < \sqrt{(d/2)^2 - k_i^s}$ are satisfied, point $S_j$ is inside the range of the base circle, as shown in Fig. 5.6(c). In this configurational situation, if $k_{i(j)}^Q$ in Eq. (5.32) is smaller than $b_1$, there is an overlap between the disk surface-plane of particle $i$ and the torus-part of particle $j$.

(4)-2 If the conditions of $|k_i^s| \le d/2$, $|k_j^s| \ge d/2$ and $|k_{ij}^s| < \sqrt{(d/2)^2 - k_i^s}$ are not satisfied, the configurations schematically shown in Figs. 5.6(a), (b) and (d) are considered. In these cases, the positions $\boldsymbol{r}_i^{(min)}$ and $\boldsymbol{r}_j^{(min)}$ on each base circle, which give rise to the shortest distance between the particles, are solved by the trial-and-error method using quasi-random numbers, which has been explained above, Fig. 5.8. These points should be in the vicinity of positions $Q_i$ and $Q_j$. If the shortest distance $r_{ij}^{(min)} = |\boldsymbol{r}_i^{(min)} - \boldsymbol{r}_j^{(min)}|$ is smaller than $b_1$, there is an overlap between the two disk-like particles.

## 5.4 Cube-like particles

Certain kinds of hematite particles have the shape of cube, and these cubic particles are quite difficult to treat in simulations because the assessment for particle overlap is significantly complex. However, the cube is the next simplest geometric shape after axisymmetric particles, and therefore, it seems to be important to address analysis procedures for the particle overlap between two cubic particles with the same size. Through the analysis of a particle overlap, we can obtain the location points on the surface of the two cubes that give rise to the minimum distance between the particles. These points may be indispensable to evaluate the interaction energy or force due to the overlap of the repulsive layers covering each cube-like particle. In Monte Carlo simulations, the assessment regarding whether or not there is a particle overlap is sufficient for a system composed of hard cubic particles. On the other hand, the dynamic simulation methods such as molecular and Brownian dynamics require the specific positions of the points giving rise to the minimum distance in order to evaluate repulsive forces between particles. Before we start to analyze the particle overlap of cubes, we first introduce the notation describing the orientational situation of a single cube, which enables one to conduct a systematic analysis regarding the assessment of particle overlap.

As shown in Fig. 3.21, the faces are numbered 1, 2, ..., 6 and the corners 1, 2, ..., 8; the position of an arbitrary face $m$ ($m = 1, 2, \ldots, 6$) from the particle center is denoted by $\boldsymbol{a}_m$ and the position of an arbitrary corner $n$ ($n = 1, 2,$

…, 8) is described by $\boldsymbol{a}_n$. The characteristics of these vectors have already been shown in Eqs. (3.29) to (3.32). Table 5.1 shows the relationships of each corner with its neighboring corners and related faces. Similarly, Table 5.2 shows the relationships of each face with its related corners. These relationships will be an important aid for analyzing the particle overlap.

**Table 5.1.** Relationship between corners and faces of the cube.

| Corners | 1 | 2 | 3 | 4 | 5 | 6 | 7 | 8 |
|---|---|---|---|---|---|---|---|---|
| Neighboring Corners | 2, 3, 4 | 1, 7, 8 | 1, 6, 8 | 1, 6, 7 | 6, 7, 8 | 3, 4, 5 | 2, 4, 5 | 2, 3, 5 |
| Related Faces | 1, 2, 3 | 2, 3, 4 | 1, 3, 5 | 1, 2, 6 | 4, 5, 6 | 1, 5, 6 | 2, 4, 6 | 3, 4, 5 |

**Table 5.2.** Relationship between faces and corners of the cube.

| Faces | 1 | 2 | 3 | 4 | 5 | 6 |
|---|---|---|---|---|---|---|
| Related Corners | 1, 4, 3, 6 | 2, 1, 7, 4 | 8, 2, 3, 1 | 5, 8, 7, 2 | 6, 5, 3, 8 | 4, 6, 7, 5 |

In order to choose corners that are possibly related to the particle overlap, the following quantity $\beta_n^{\,i}$ and $\beta_n^{\,j}$ ($n = 1, 2, \ldots, 8$) are introduced:

$$\begin{aligned} \beta_n^i &= \boldsymbol{a}_n^i \cdot (\boldsymbol{r}_j - \boldsymbol{r}_i) = \boldsymbol{a}_n^i \cdot \boldsymbol{r}_{ji} && \text{(for particle } i\text{)}, \\ \beta_n^j &= \boldsymbol{a}_n^j \cdot (\boldsymbol{r}_i - \boldsymbol{r}_j) = \boldsymbol{a}_n^j \cdot \boldsymbol{r}_{ij} && \text{(for particle } j\text{)} \end{aligned} \tag{5.36}$$

Similarly, the quantity $\gamma_m^{\,i}$ and $\gamma_m^{\,j}$ ($m = 1, 2, \ldots, 6$) are used to determine which face is possibly related to the particle overlap:

$$\begin{aligned} \gamma_m^i &= \boldsymbol{a}_m^i \cdot \boldsymbol{r}_{ji} && \text{(for particle } i\text{)}, \\ \gamma_m^j &= \boldsymbol{a}_m^j \cdot \boldsymbol{r}_{ij} && \text{(for particle } j\text{)} \end{aligned} \tag{5.37}$$

In these equations, $\boldsymbol{r}_i$ is the position vector of the center of particle $i$, and $\boldsymbol{r}_{ij}$ is the relative position from particle $j$. It is noted that a larger value of these quantities leads to a higher possibility of relating to the particle overlap. We have now finished preparation for starting the analysis of the assessment of the particle overlap between two same-sized cubic particles.

Figure 5.9 shows typical regimes of particle overlap, i.e., (a) face-to-face overlap in the parallel configuration, (b) face-to-line overlap in the parallel situation of a face and a line, (c1) face-to-corner overlap, (c2) line-to-line overlap, and (c3) line-to-line or corner-to-face overlap. The last three regimes are distinguished according to the location of the intersection point, on the plane of the face, of the line perpendicularly drawn from the corner. This point is inside the area of the face for regime (c1), outside in the one direction for regime (c2) and outside in the two directions (outside in the corner direction) for regime (c3). In the following, we discuss the assessment of the particle overlap in this order.

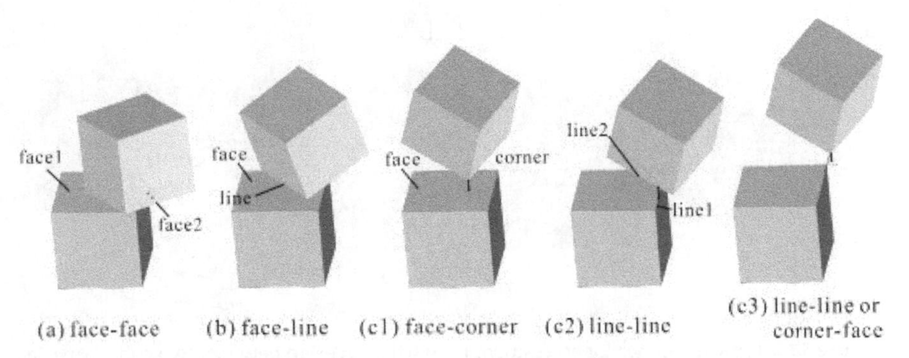


**Figure 5.9.** Typical regimes for assessment for particle overlap between two cubic particles.

## 5.4.1 *Face-to-face overlap*

We first consider the parallel situation of two faces being parallel, as shown in case (a) in Fig. 5.9. According to the values $\gamma_m^{\ i}$ and $\gamma_m^{\ j}$ ($m$ = 1, 2, … , 6), we are able to determine the faces that are possibly related to the overlap. The faces giving the first, second and third largest values of $\gamma_m^{\ i}$ are chosen as the faces of interest, which are denoted by faces $m_1^i$, $m_2^i$ and $m_3^i$, respectively. Similarly, the notation $m_1^j$, $m_2^j$ and $m_3^j$ are used for particle $j$. If face $m_1^i$ and face $m_1^j$ are in the parallel configuration, the following expression must be satisfied:

$$\boldsymbol{a}_{m_1^i} \cdot \boldsymbol{a}_{m_1^j} = -d^2/4 \tag{5.38}$$

For simplicity of the following analysis, face $p$ of particle $i$ is assumed to be parallel to face $q$ of particle $j$.

Figure 5.10 shows the configuration of the two cubes viewed from the direction normal to the parallel faces. Since we assume that particle $j$ is located at a certain positive distance above the paper, the front face of particle $i$ and the rear face of particle $j$ will be addressed for assessing the overlap. The four corners of face $p$ are named as corners $p_1$, $p_2$, $p_3$ and $p_4$, which are straightforwardly recognized from Table 5.2 for the known face number. Similar corner numbers $q_1$, $q_2$, $q_3$ and $q_4$ are employed for face $q$ of particle $j$. The position vectors $\boldsymbol{b}_{p_k}$ ($k$ = 1, 2, 3 and 4) and $\boldsymbol{b}_{q_k}$ ($k$ = 1, 2, 3 and 4) for each corner relative to the center of the corresponding face are expressed as

$$\boldsymbol{b}_{p_k} = \boldsymbol{a}_{p_k} - \boldsymbol{a}_p \text{ (for particle } i\text{)}, \boldsymbol{b}_{q_k} = \boldsymbol{a}_{q_k} - \boldsymbol{a}_q \text{ (for particle } j\text{)} \tag{5.39}$$

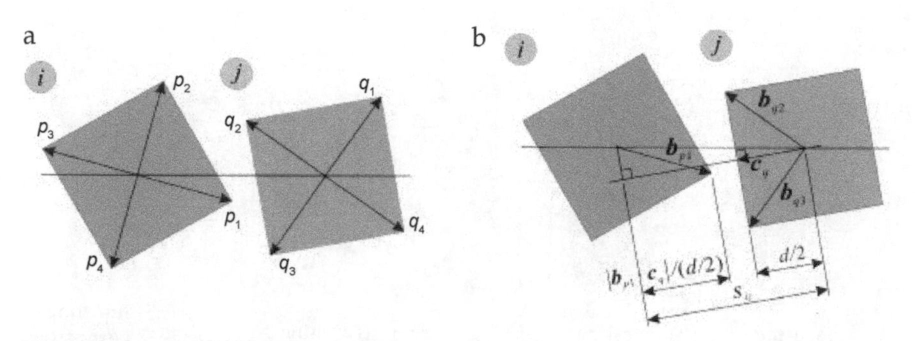


**Figure 5.10.** Typical regimes for assessment for particle overlap between two cubic particles.

We first find a corner that has the largest value of $\beta_n^{\,i}$ or $\beta_n^{\,j}$, and this corner is assumed to be corner $p_1$ of particle $i$. Then we find a face of the other particle (particle $j$) that has the largest value of $\gamma_m^{\,j}$, and this face is assumed to be the face described by the vectors $\boldsymbol{b}_{q_2}$ and $\boldsymbol{b}_{q_3}$. This configuration is shown in Fig. 5.10(b) with introduction to a new notation $\boldsymbol{c}_q$ expressed as

$$\boldsymbol{c}_q = \frac{\boldsymbol{b}_{q_2} + \boldsymbol{b}_{q_3}}{2} \tag{5.40}$$

Refer to Fig. 5.10(b), where the lengths projected on a line along the vector $\boldsymbol{c}_q$ enables one to describe the criterion for the particle overlap. We use the notation $s_{ij}$ for the projected length of the particle center-to-center separation along the surface of the parallel faces. Employing these symbols, if the following condition is satisfied, there is an overlap between the two cubic particles:

$$s_{ij} < \frac{\left|\boldsymbol{b}_{p_1} \cdot \boldsymbol{c}_q\right|}{d/2} + \frac{d}{2} \quad \text{and} \quad \frac{\left|\boldsymbol{r}_{ij} \cdot \boldsymbol{a}_p\right|}{d/2} < d \tag{5.41}$$

The second condition implies that the two particles are required to also overlap in the direction along the vector $\boldsymbol{a}_p$.

## 5.4.2 Face-to-line overlap in the parallel situation

Next, we consider the configuration (b), shown in Fig. 5.9, where a face of particle $i$ possibly overlaps with a line of particle $j$ in the parallel situation. As before, the faces giving the first, second and third largest values of $\gamma_m^{\ i}$ are chosen as the faces of interest, which are denoted by faces $m_1^i$, $m_2^i$ and $m_3^i$, respectively. Similarly, the notation $m_1^j$, $m_2^j$ and $m_3^j$ are used for particle $j$. If the value of $\gamma_{m_1}^i$ is larger than that of $\gamma_{m_1}^j$, the face $m_1^i$ of particle $i$ is chosen as a criterion surface for the following assessment procedure, and for simplicity, this face is named "face $p$". Then, according to the values $\beta_n^{\ j}(n = 1, 2, \ldots, 8)$ of particle $j$, the corners giving the first, second and third largest values of $\beta_n^{\ j}$ are chosen as the objects for analysis, which are denoted by corners $n_1^j$, $n_2^j$ and $n_3^j$, respectively. For all these corners, we evaluate the following quantities:

$$\boldsymbol{b}_{n_1^j}^{(1)} = \boldsymbol{\alpha}_{n_1^j}^{(1)} - \boldsymbol{\alpha}_{n_1^j}, \quad \boldsymbol{b}_{n_1^j}^{(2)} = \boldsymbol{\alpha}_{n_1^j}^{(2)} - \boldsymbol{\alpha}_{n_1^j}, \quad \boldsymbol{b}_{n_1^j}^{(3)} = \boldsymbol{\alpha}_{n_1^j}^{(3)} - \boldsymbol{\alpha}_{n_1^j} \ \text{(for corner } n_1^j) \tag{5.42}$$

Similar quantities are evaluated for corners $n_2^j$ and $n_3^j$. In Eq. (5.42), $\boldsymbol{\alpha}_{n_1^j}^{(1)}, \boldsymbol{\alpha}_{n_1^j}^{(2)}$ and $\boldsymbol{\alpha}_{n_1^j}^{(3)}$ are the vectors of the three corners related to the criterion corner $n_1^j$. Then, we evaluate the following vector products with the face vector $\boldsymbol{a}_p$ of particle $i$:

$$\begin{aligned}
&\zeta_1^{(1)} = \boldsymbol{a}_p \cdot \boldsymbol{b}_{n_1^j}^{(1)},\ \zeta_1^{(2)} = \boldsymbol{a}_p \cdot \boldsymbol{b}_{n_1^j}^{(2)},\ \zeta_1^{(3)} = \boldsymbol{a}_p \cdot \boldsymbol{b}_{n_1^j}^{(3)} \ \text{(for corner } n_1^j), \\
&\zeta_2^{(1)} = \boldsymbol{a}_p \cdot \boldsymbol{b}_{n_2^j}^{(1)},\ \zeta_2^{(2)} = \boldsymbol{a}_p \cdot \boldsymbol{b}_{n_2^j}^{(2)},\ \zeta_2^{(3)} = \boldsymbol{a}_p \cdot \boldsymbol{b}_{n_2^j}^{(3)} \ \text{(for corner } n_2^j), \\
&\zeta_3^{(1)} = \boldsymbol{a}_p \cdot \boldsymbol{b}_{n_3^j}^{(1)},\ \zeta_3^{(2)} = \boldsymbol{a}_p \cdot \boldsymbol{b}_{n_3^j}^{(2)},\ \zeta_3^{(3)} = \boldsymbol{a}_p \cdot \boldsymbol{b}_{n_3^j}^{(3)} \ \text{(for corner } n_3^j)
\end{aligned} \tag{5.43}$$

Using these values, we are able to determine the corners of particle $j$ required for analysis of the particle overlap. If all these three values are positive or equal to zero, then this corner is chosen as an object for analysis. Since we are considering the situation where there are no pairs of parallel faces between particles $i$ and $j$, there should be at most two corners that are related to the overlap with face $p$ or the corner lines of particle $i$. If there is an overlap between face $p$ and a certain corner line in the parallel situation, one of zeta values in Eq. (5.43) should be zero, in fact two corners should give rise to a zero zeta value regarding the two neighboring corners. For advancing the analysis, we will assume that the line connected with the first neighboring corner of corner $n_1^j$ is parallel to face $p$ of particle $i$, i.e., $\zeta_1^{(1)} = 0$; for simplicity, this corner is named "corner $q$". Hence, if the following condition is satisfied, then there is an overlap between the face of particle $i$ and the corner line of particle $j$:

$$s_{ij} < \frac{|\boldsymbol{\alpha}_q \cdot \boldsymbol{a}_p|}{d/2} + \frac{d}{2} \tag{5.44}$$

in which $s_{ij}$ is the length of the particle center-to-center separation along the vector $\boldsymbol{a}_p$.

### 5.4.3 *Face-to-corner overlap*

In this section, we consider a general case where the overlap criterion is assessed by treating a face of particle $i$ and a corner of particle $j$, which may result in the face-to-corner or line-to-line overlap. For analysis of this situation, the values in Eq. (5.43) are also used and in this case only one corner of particle $j$ should satisfy the condition of all the three values of zeta being positive. We assume here that the corner $n_1^j$ satisfies this condition (i.e., $\zeta_1^{(1)} > 0$, $\zeta_1^{(2)} > 0$ and $\zeta_1^{(3)} > 0$) and this corner is renamed "corner $q$" for simplicity of the following analysis. Therefore, we will treat face $p$ of particle $i$ and corner $q$ of particle $j$ in the analysis regarding the overlap between particles $i$ and $j$.

As shown in Fig. 5.11, the criterion for analysis of the particle overlap is the intersection point $\boldsymbol{r}_p$ on face $p$ of the line vertically drawn from corner $q$. In the following we evaluate the location of this point using the vectors $\boldsymbol{a}_p$, $\boldsymbol{\alpha}_q$, etc. From Fig. 3.21(b), it is seen that the vector $\boldsymbol{a}_p$ inevitably has the two (face) vectors, which are normal to each other and also normal to $\boldsymbol{a}_p$, and these vectors are parallel to face $a$; we employ the notation $\boldsymbol{a}_p^{(1)}$ and $\boldsymbol{a}_p^{(2)}$ for these face vectors, as shown in Fig. 5.11. An arbitrary point on the face, $\boldsymbol{r}_{c'}$, is expressed using the vectors $\boldsymbol{a}_p^{(1)}$ and $\boldsymbol{a}_p^{(2)}$ with the introduction of unknown constants $k^{(1)}$ and $k^{(2)}$ as

$$\boldsymbol{r}_c \quad k^{(1)}\boldsymbol{a}^{(1)}_p + k^{(2)}\boldsymbol{a}^{(2)}_p + {}_i + \boldsymbol{a}_p \tag{5.45}$$

With employment of appropriate values for $k^{(1)}$ and $k^{(2)}$, this point will be the intersection point $\boldsymbol{r}_p$ of the line vertically drawn from the corner.

On the other hand, we consider the path to attain the intersection point $\boldsymbol{r}_p$ through the center of particle $j$. Using an appropriate value of an unknown constant $k^{(q)}$, the point $\boldsymbol{r}_p$ is expressed as

$$\boldsymbol{r}_p = k^{(q)}(-\boldsymbol{a}_p) + \boldsymbol{r}_j + \boldsymbol{\alpha}_q \tag{5.46}$$

Hence, Eqs. (5.45) and (5.46) give rise to the following relationship:

$$k^{(1)}\boldsymbol{a}_p^{(1)}+k^{(2)}\boldsymbol{a}_p^{(2)}+k^{(q)}\boldsymbol{a}_p=\boldsymbol{r}_{ji}+\boldsymbol{\alpha}_q-\boldsymbol{a}_p \tag{5.47}$$

By taking into account the following relationships:

$$\boldsymbol{a}_p\cdot\boldsymbol{a}_p^{(1)}=\boldsymbol{a}_p\cdot\boldsymbol{a}_p^{(2)}=\boldsymbol{a}_p^{(1)}\cdot\boldsymbol{a}_p^{(2)}=0 \tag{5.48}$$

the solutions of the unknown constants are obtained as

$$\begin{aligned}&k^{(1)}=\frac{4}{d^2}\left(\boldsymbol{a}_p^{(1)}\cdot\boldsymbol{r}_{ji}+\boldsymbol{a}_p^{(1)}\cdot\boldsymbol{\alpha}_q\right),\quad k^{(2)}=\frac{4}{d^2}\left(\boldsymbol{a}_p^{(2)}\cdot\boldsymbol{r}_{ji}+\boldsymbol{a}_p^{(2)}\cdot\boldsymbol{\alpha}_q\right),\\&k^{(q)}=\frac{4}{d^2}\left(\boldsymbol{a}_p\cdot\boldsymbol{r}_{ji}+\boldsymbol{a}_p\cdot\boldsymbol{\alpha}_q\right)-1\end{aligned} \tag{5.49}$$

As shown in Fig. 5.11, if the intersection point is inside the surface of face $p$, it is straightforward to assess whether or not there is an overlap between particles $i$ and $j$. The following condition is required for the particle overlap:

$$k^{(q)}<0 \quad \text{for} \quad -1\le k^{(1)}\le 1 \quad \text{and} \quad -1\le k^{(2)}\le 1 \tag{5.50}$$

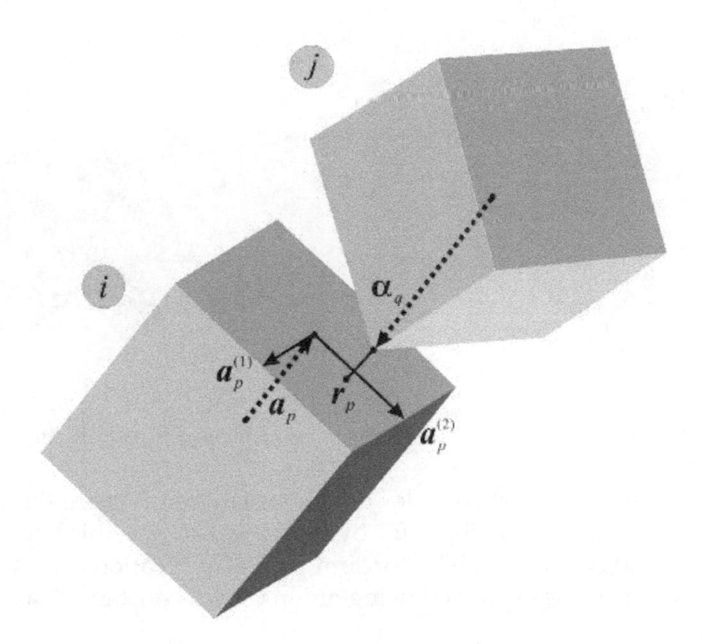


**Figure 5.11.** Assessment for the overlap between the face of particle $i$ and the corner of particle $j$.

### 5.4.4 Line-to-line overlap outside a face in a certain face direction

In this section we consider the configuration shown in Fig. 5.12, where the intersection point $\boldsymbol{r}_p$ is outside face $p$ in the direction of $\boldsymbol{a}_p^{(2)}$ alone and in this case, $k^{(1)}$ and $k^{(2)}$ satisfy the condition of $|k^{(1)}| < 1$ and $|k^{(2)}| > 1$. In the configuration shown in Fig. 5.12, there is a possibility of overlap between the corner lines of the particles. For the case of particle $i$, the corner line for the object is the line normal to the vector $\boldsymbol{a}_p^{(2)}$ through the head of this vector and for simplicity, we call this corner line "line $p$". The objects for particle $j$ are the corner lines $\boldsymbol{b}_q^{(1)}$, $\boldsymbol{b}_p^{(2)}$ and $\boldsymbol{a}_p^{(3)}$ related to the corner $q$ of particle $j$, and these vectors are expressed as

$$\boldsymbol{b}_q^{(1)} = \boldsymbol{a}_q^{(1)} - \boldsymbol{a}_q,\ \boldsymbol{b}_q^{(2)} = \boldsymbol{a}_q^{(2)} - \boldsymbol{a}_q,\ \boldsymbol{b}_q^{(3)} = \boldsymbol{a}_q^{(3)} - \boldsymbol{a}_q \text{ (for corner } q \text{ of particle } j) \quad (5.51)$$

in which $\boldsymbol{a}_q^{(1)}$, $\boldsymbol{a}_q^{(2)}$ and $\boldsymbol{a}_q^{(3)}$ are the vectors of the three corners related to corner $q$. For instance, if $\boldsymbol{a}_p^{(2)} \cdot \boldsymbol{b}_q^{(3)} \geq 0$, the line $\boldsymbol{b}_q^{(3)}$ never overlaps with line $p$, and therefore, this line is removed from the objects treated in the following overlap analysis. There are two lines, at most, that satisfy $\boldsymbol{a}_p^{(2)} \cdot \boldsymbol{b}_q^{(1)} < 0$ and/or $\boldsymbol{a}_p^{(2)} \cdot \boldsymbol{b}_q^{(2)} < 0$ if $\boldsymbol{a}_p^{(2)} \cdot \boldsymbol{b}_q^{(3)} \geq 0$.

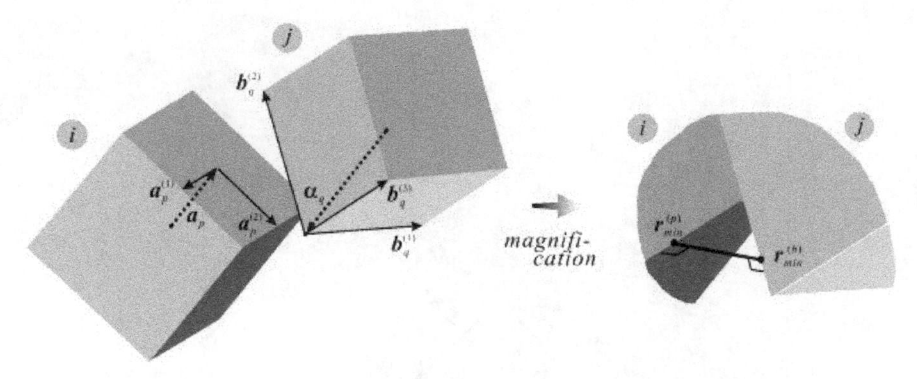


**Figure 5.12.** Assessment for the overlap between the corner lines of particles $i$ and $j$.

In the following, we first consider a simpler situation where the line $\boldsymbol{b}_q^{(2)}$ alone is assumed to be an object for overlap analysis, i.e., this line satisfies $\boldsymbol{a}_p^{(2)} \cdot \boldsymbol{b}_q^{(2)} < 0$ and is called "line $b$," for simplicity. The location of an arbitrary point on line $p$, $\boldsymbol{r}^{(p)}$, is expressed using an unknown constant $k^{(p)}$ as

$$\boldsymbol{r}^{(p)} = k^{(p)}\boldsymbol{a}_p^{(1)} + \boldsymbol{a}_p^{(2)} + \boldsymbol{r}_i + \boldsymbol{a}_p \quad (5.52)$$

Similarly, the location of an arbitrary point on line $b$, $\boldsymbol{r}^{(b)}$, is expressed using an unknown constant $k^{(b)}$ as

$$\boldsymbol{r}^{(b)} = k^{(b)}\boldsymbol{b}_q^{(2)} + \boldsymbol{r}_j + \boldsymbol{\alpha}_q \tag{5.53}$$

For mathematical manipulation, we take into consideration the relationship of $\boldsymbol{a}_p^{(1)} \cdot \boldsymbol{a}_p^{(2)} = 0$.

As shown in Fig. 5.12, the minimum distance between lines $p$ and $q$ requires the following condition:

$$(\boldsymbol{r}^{(p)} - \boldsymbol{r}^{(b)}) \cdot \boldsymbol{a}_p^{(1)} = (\boldsymbol{r}^{(p)} - \boldsymbol{r}^{(b)}) \cdot \boldsymbol{b}_q^{(2)} = 0 \tag{5.54}$$

From Eqs. (5.52) to (5.54), the unknown constants $k^{(p)}$ and $k^{(b)}$ are obtained to specify the positions $\boldsymbol{r}_{min}^{(p)}$ and $\boldsymbol{r}_{min}^{(b)}$ that give rise to the minimum distance. That is,

$$\begin{bmatrix} k^{(p)} \\ k^{(b)} \end{bmatrix} = \frac{1}{|\boldsymbol{a}_p^{(1)} \cdot \boldsymbol{b}_q^{(2)}|^2 - d^4/4} \begin{bmatrix} d^2 & -\boldsymbol{a}_p^{(1)} \cdot \boldsymbol{b}_q^{(2)} \\ \boldsymbol{a}_p^{(1)} \cdot \boldsymbol{b}_q^{(2)} & -d^2/4 \end{bmatrix} \begin{bmatrix} \boldsymbol{a}_p^{(1)} \cdot (\boldsymbol{r}_{ij} - \boldsymbol{\alpha}_q) \\ \boldsymbol{b}_q^{(2)} \cdot (\boldsymbol{r}_{ij} + \boldsymbol{a}_p^{(2)} + \boldsymbol{a}_p - \boldsymbol{\alpha}_q) \end{bmatrix} \tag{5.55}$$

Since line $p$ is not parallel to line $b$ in the present configuration, the denominator in Eq. (5.55) never becomes zero. Using these values for $k^{(p)}$ and $k^{(b)}$, the following conditions are required for the overlap between lines $p$ and $b$:

$$\frac{|(\boldsymbol{r}_{min}^{(b)} - \boldsymbol{r}_{min}^{(p)}) \cdot \boldsymbol{r}_{ij}|}{|(\boldsymbol{r}_{min}^{(b)} - \boldsymbol{r}_{min}^{(p)})|} < \frac{|(\boldsymbol{r}_{min}^{(b)} - \boldsymbol{r}_{min}^{(p)}) \cdot (\boldsymbol{a}_p + \boldsymbol{a}_p^{(2)} + k^{(p)}\boldsymbol{a}_p^{(1)})| + |(\boldsymbol{r}_{min}^{(b)} - \boldsymbol{r}_{min}^{(p)}) \cdot (\boldsymbol{\alpha}_q + k^{(b)}\boldsymbol{b}_q^{(2)})|}{|(\boldsymbol{r}_{min}^{(b)} - \boldsymbol{r}_{min}^{(p)})|} \tag{5.56}$$

This condition is valid under the condition of $|k^{(p)}| < 1$; the configuration where this condition is not satisfied will be addressed later.

We now consider a more complex situation where both the lines $\boldsymbol{b}_q^{(1)}$ and $\boldsymbol{b}_q^{(2)}$ are assumed to be objects for the overlap analysis, i.e., the conditions of $\boldsymbol{a}_p^{(2)} \cdot \boldsymbol{b}_q^{(1)} < 0$ and $\boldsymbol{a}_p^{(2)} \cdot \boldsymbol{b}_q^{(2)} < 0$ are satisfied (it is noted that $\boldsymbol{a}_p^{(2)} \cdot \boldsymbol{b}_q^{(3)} \geq 0$); these lines are called "line $b_1$," and "line $b_2$," for simplicity. As in the above configuration, the quantities $k_1^{(p)}$ and $k_1^{(b)}$, corresponding to $k^{(p)}$ and $k^{(b)}$, are expressed as equations similar to Eqs. (5.55), and the positions $\boldsymbol{r}_{(min1)}^{(p)}$ and $\boldsymbol{r}_{(min1)}^{(b)}$, corresponding to $\boldsymbol{r}_{min}^{(p)}$ and $\boldsymbol{r}_{min}^{(b)}$, give rise to the minimum distance between lines $p$ and $b_1$.

Following a similar procedure, $k_2^{(p)}$, $k_2^{(b)}$, $\boldsymbol{r}^{(p)}_{(min2)}$ and $\boldsymbol{r}^{(p)}_{(min2)}$ are obtained that give rise to the minimum distance between lines $p$ and $b_2$. Hence, if either of the following two conditions is satisfied, there is an overlap between particles $i$ and $j$:

$$\frac{|(\boldsymbol{r}^{(b)}_{(min1)}-\boldsymbol{r}^{(p)}_{(min1)})\cdot\boldsymbol{r}_{ij}|}{|(\boldsymbol{r}^{(b)}_{(min1)}-\boldsymbol{r}^{(p)}_{(min1)})|} < \frac{|(\boldsymbol{r}^{(b)}_{(min1)}-\boldsymbol{r}^{(p)}_{(min1)})\cdot(\boldsymbol{a}_p+\boldsymbol{a}^{(2)}_p+k_1^{(p)}\boldsymbol{a}^{(1)}_p)|+|(\boldsymbol{r}^{(b)}_{(min1)}-\boldsymbol{r}^{(p)}_{(min1)})\cdot(\boldsymbol{a}_q+k_1^{(b)}\boldsymbol{b}^{(1)}_q)|}{|(\boldsymbol{r}^{(b)}_{(min1)}-\boldsymbol{r}^{(p)}_{(min1)})|} \tag{5.57}$$

$$\frac{|(\boldsymbol{r}^{(b)}_{(min2)}-\boldsymbol{r}^{(p)}_{(min2)})\cdot\boldsymbol{r}_{ij}|}{|(\boldsymbol{r}^{(b)}_{(min2)}-\boldsymbol{r}^{(p)}_{(min2)})|} < \frac{|(\boldsymbol{r}^{(b)}_{(min2)}-\boldsymbol{r}^{(p)}_{(min2)})\cdot(\boldsymbol{a}_p+\boldsymbol{a}^{(2)}_p+k_2^{(p)}\boldsymbol{a}^{(1)}_p)|+|(\boldsymbol{r}^{(b)}_{(min2)}-\boldsymbol{r}^{(p)}_{(min2)})\cdot(\boldsymbol{\alpha}_q+k_2^{(b)}\boldsymbol{b}^{(2)}_q)|}{|(\boldsymbol{r}^{(b)}_{(min2)}-\boldsymbol{r}^{(p)}_{(min2)})|} \tag{5.58}$$

It is noted that this criterion for the particle overlap is applicable in the situation of $|k_1^{(p)}|<1$ and $|k_2^{(p)}|<1$.

We expect that the above situation for a line-to-line overlap will be the one most frequently encountered, but the situation of $|k_1^{(p)}|<1$ and $|k_2^{(p)}| > 1$ may arise occasionally. In this exceptional configuration, the overlap assessment is conducted according to the following procedures. That is, (1) an assessment is made for an overlap for line $p$ of particle $i$ with line $b_1$ of particle $j$, and (2) an assessment is made for an overlap for a corner of particle $i$ with the face of particle $j$ defined by the vectors $\boldsymbol{b}_q^{(1)}$ and $\boldsymbol{b}_q^{(2)}$. The first and second assessments are conducted using conditions that are similar to Eqs. (5.56) and (5.50), respectively.

Although they have a significantly low occurrence, there are two other possible overlap configurations which are schematically shown in Fig. 5.13. That is, the line $\boldsymbol{b}_q^{(2)}$ alone is possibly an object for an overlap analysis, i.e., this line satisfies $\boldsymbol{a}_p^{(2)}\cdot\boldsymbol{b}_q^{(2)}<0$, but the nearest point of particle $i$ is the corner or on line $p_3$, as shown in Fig. 5.13(a). In this situation, the corner line parallel to the vector $\boldsymbol{a}_p$, named "line $p_3$," in Fig. 5.13(a), may possibly overlap with line $q_2$: the vector $\boldsymbol{b}_p^{(3)}$ is defined in Fig. 5.13(a). Moreover, the corner defined by lines $p_1$ and $p_2$ possibly overlaps with the face defined by lines $q_2$ and $q_3$ in the configuration in Fig. 5.13(b). As before, the latter configuration is expected to appear with a significantly small possibility, because the face of particle $i$ has been chosen as a criterion surface under the assumption that the value of $\gamma^i_{m_1}$ is larger than $\gamma^i_{m_2}$ and

$\gamma^i_{m3}$. The position of the point on the face that gives rise to the minimum distance between particles $i$ and $j$ can be evaluated in a similar manner to the previous case for the corner-to-face configuration in Section 5.4.3. Hence, in the following, we concentrate on the former situation of the overlap between line $p_3$ and line $q_2$.

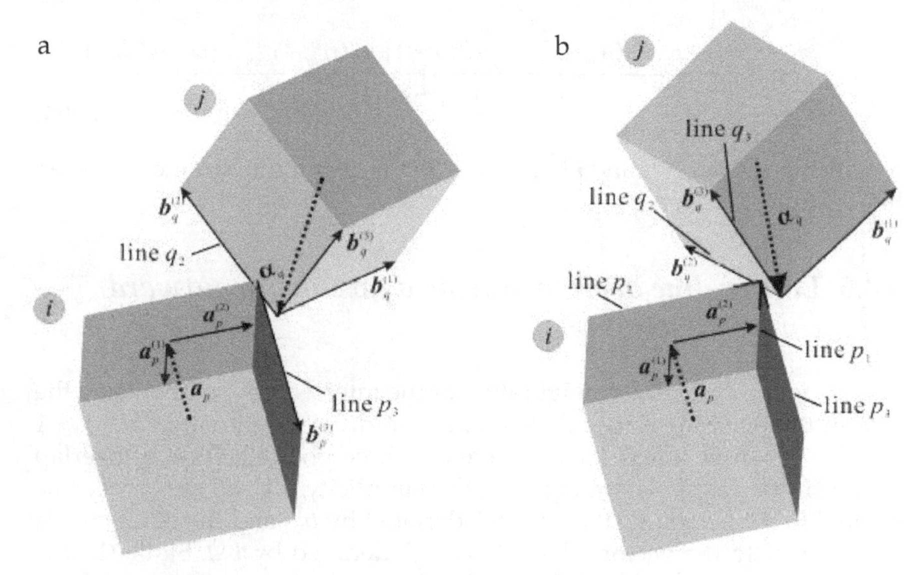


**Figure 5.13.** Assessment for the other cases of overlap between the corner lines of particles $i$ and $j$ or between the corner of particle $i$ and the face of particle $j$.

The location of an arbitrary point on line $p_3$, $\boldsymbol{r}^{(p)}$, is expressed using an unknown constant $k^{(p)}$ as

$$\boldsymbol{r}^{(p)} = k^{(p)}\boldsymbol{b}_p^{(3)} + \boldsymbol{r}_i + \boldsymbol{a}_p + \boldsymbol{a}_p^{(2)} - \boldsymbol{a}_p^{(1)} = -2k^{(p)}\boldsymbol{a}_p + \boldsymbol{r}_i + \boldsymbol{a}_p + \boldsymbol{a}_p^{(2)} - \boldsymbol{a}_p^{(1)} \tag{5.59}$$

Similarly, the location of an arbitrary point on line $b$ (i.e., line $q_2$), $\boldsymbol{r}^{(b)}$, is expressed using an unknown constant $k^{(b)}$ as

$$\boldsymbol{r}^{(b)} = k^{(b)}\boldsymbol{b}_q^{(2)} + \boldsymbol{r}_j + \boldsymbol{\alpha}_q \tag{5.60}$$

From similar mathematical procedures by which Eq. (5.55) has been obtained, we obtain the solutions of $k^{(p)}$ and $k^{(b)}$ as

$$\begin{bmatrix} k^{(p)} \\ k^{(b)} \end{bmatrix} = \frac{1}{|\boldsymbol{b}_p^{(3)}\cdot\boldsymbol{b}_q^{(2)}|^2 - d^4} \begin{bmatrix} -d^2 & \boldsymbol{b}_p^{(3)}\cdot\boldsymbol{b}_q^{(2)} \\ -\boldsymbol{b}_p^{(3)}\cdot\boldsymbol{b}_q^{(2)} & d^2 \end{bmatrix} \begin{bmatrix} \boldsymbol{b}_p^{(3)}\cdot(-\boldsymbol{r}_{ij} + \boldsymbol{\alpha}_q) + d^2/2 \\ -\boldsymbol{b}_q^{(2)}\cdot(\boldsymbol{r}_{ij} + \boldsymbol{a}_p + \boldsymbol{a}_p^{(2)} - \boldsymbol{a}_p^{(1)} - \boldsymbol{\alpha}_q) \end{bmatrix} \tag{5.61}$$

Using these values for $k^{(p)}$ and $k^{(b)}$, the following conditions are required for the overlap between lines $p_3$ and $q_2$ for the configuration shown in Fig. 5.13(a):

$$\frac{|(\boldsymbol{r}_{min}^{(b)}-\boldsymbol{r}_{min}^{(p)})\cdot\boldsymbol{r}_{ij}|}{|(\boldsymbol{r}_{min}^{(b)}-\boldsymbol{r}_{min}^{(p)})|} < \frac{|(\boldsymbol{r}_{min}^{(b)}-\boldsymbol{r}_{min}^{(p)})\cdot(\boldsymbol{a}_p+\boldsymbol{a}_p^{(2)}-\boldsymbol{a}_p^{(1)}+k^{(p)}\boldsymbol{b}_p^{(3)})|+|(\boldsymbol{r}_{min}^{(b)}-\boldsymbol{r}_{min}^{(p)})\cdot(\boldsymbol{\alpha}_q+k^{(b)}\boldsymbol{b}_q^{(2)})|}{|(\boldsymbol{r}_{min}^{(b)}-\boldsymbol{r}_{min}^{(p)})|} \tag{5.62}$$

in which positions $\boldsymbol{r}_{min}^{(p)}$ and $\boldsymbol{r}_{min}^{(b)}$ give rise to the minimum distance between lines $p_3$ and $q_2$.

### 5.4.5 Line-to-line overlap outside a face in the outward corner direction

Finally, we consider the configuration of the intersection point outside the face in the outward corner direction, i.e., address the situation $|k^{(1)}| > 1$ and $|k^{(2)}| > 1$. In this situation, there are three possible cases of overlap configuration, as shown in Fig. 5.14. For simplicity, we call the corner line normal to the vector $\boldsymbol{a}_p^{(2)}$ the "line $p_1$" denoted by $\boldsymbol{b}_p^{(1)}$ and the other corner line normal to the vector $\boldsymbol{a}_p^{(1)}$ the "line $p_2$" denoted by $\boldsymbol{b}_p^{(2)}$. Similarly, the corner and the lines of particle $j$ as objects for overlap criterion are called the "corner $q$" denoted by the vector $\boldsymbol{\alpha}_q$ and the "line $q_1$", "line $q_2$", and "line $q_3$", denoted by $\boldsymbol{b}_p^{(1)}$, $\boldsymbol{b}_p^{(2)}$ and $\boldsymbol{b}_p^{(3)}$, respectively.

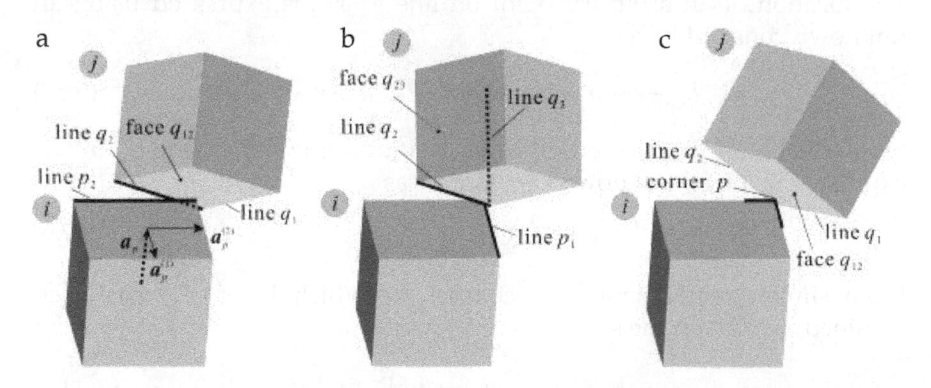


**Figure 5.14.** Assessment for the overlap for the case of the intersection point being outside the face of particle $i$ in the outward corner direction: a possibility of an overlap (a) between line $p_2$ of particle $i$ and line $q_2$ of particle $j$, (b) between line $p_1$ of particle $i$ and line $q_2$ of particle $j$ and (c) between corner $p$ of particle $i$ and face $q_{12}$ of particle $j$.

Regarding particle $j$, the position $\boldsymbol{r}_q$ of corner $q$, and the line vectors $\boldsymbol{b}_q^{(1)}$, $\boldsymbol{b}_q^{(2)}$ and $\boldsymbol{b}_q^{(3)}$ related to corner $q$ are expressed as

$$\boldsymbol{r}_q = \boldsymbol{r}_j + \boldsymbol{\alpha}_q \tag{5.63}$$

$$\boldsymbol{b}_q^{(1)} = \boldsymbol{\alpha}_q^{(1)} - \boldsymbol{\alpha}_q,\ \boldsymbol{b}_q^{(2)} = \boldsymbol{\alpha}_q^{(2)} - \boldsymbol{\alpha}_q,\ \boldsymbol{b}_q^{(3)} = \boldsymbol{\alpha}_q^{(3)} - \boldsymbol{\alpha}_q \text{ (for corner } q \text{ of particle } j) \tag{5.64}$$

Similarly, regarding particle $i$, the position $\boldsymbol{r}_p$ of corner $p$, and the line vectors $\boldsymbol{b}_p^{(1)}$ and $\boldsymbol{b}_p^{(2)}$ related to corner $p$ are written as

$$\boldsymbol{r}_p = \boldsymbol{r}_i + \boldsymbol{\alpha}_p \tag{5.65}$$

$$\boldsymbol{b}_p^{(1)} = \boldsymbol{\alpha}_p^{(1)} - \boldsymbol{\alpha}_p,\ \boldsymbol{b}_p^{(2)} = \boldsymbol{\alpha}_p^{(2)} - \boldsymbol{\alpha}_p \text{ (for corner } p \text{ of particle } i) \tag{5.66}$$

There are two types of configurations for a line-to-line overlap between particles $i$ and $j$, that is, an overlap between line $p_2$ and line $q_2$ shown in Fig. 5.14(a) and an overlap between line $p_1$ and line $q_2$ shown in Fig. 5.14(b). Which configuration arises in a certain situation is determined in the following manner. We evaluate the quantity $\eta^{(2)}$ that is defined by the following expression:

$$\eta^{(2)} = \{(\boldsymbol{\alpha}_p - \boldsymbol{\alpha}_q) \times \boldsymbol{a}_p\} \cdot \boldsymbol{b}_q^{(2)} \tag{5.67}$$

This value of $\eta^{(2)}$ enables one to choose an appropriate configuration as follows:

(1) If $\eta^{(2)} \geq 0$, there is a possibility of an overlap of line $q_2$ with line $p_2$ (no overlap with line $p_1$).

(2) If $\eta^{(2)} < 0$, there is a possibility of an overlap of line $q_2$ with line $p_1$ (no overlap with line $p_2$).

In either case, a face of particle $j$ may possibly overlap with corner $p$ of particle $i$, although the probability for this kind of overlap seems to be significantly low. In the case of the overlap between line $p_2$ and line $q_2$ shown in Fig. 5.14(a), the overlap between corner $p$ and the face defined by line $q_2$ and $q_1$, named "face $q_{12}$", must also be assessed in addition to this line-to-line overlap, as shown in Fig. 5.14(c). Similarly, in the case of the overlap between line $p_1$ and line $q_2$ shown in Fig. 5.14(b), the overlap between corner $p$ and the face defined by line $q_2$ and $q_3$, named "face $q_{23}$," must also be assessed in addition to this line-to-line overlap.

For simplicity, we focus on the former case where there is a possibility of the overlap between line $p_2$ and line $q_2$ and also between corner $p$ and face $q_{12}$, as shown in Figs. 5.14(a) and (c). For the case of the former line-to-line configuration, the assessment procedure that has been explained in Section 5.4.4 can be employed almost directly. If no overlap is recognized to have occurred in this procedure, we will then proceed with another assessment regarding the latter face-to-corner overlap. In this procedure, the assessment criterion explained in Section 5.4.3 can be adopted. If there is no overlap between corner $p$ and face $q_{12}$, then it is concluded that particle $i$ does not overlap with particle $j$.

### 5.4.6 *Summary of the assessment procedure regarding the particle overlap between two cube-like particles*

We summarize here the systematic assessment procedure regarding particle overlap by taking into account the results of the above-mentioned analysis. This may be a significant aid for the reader to develop their own particle-based simulation program for a suspension composed of cube-like particles using methods such as Monte Carlo, molecular dynamics and Stokesian dynamics.

The main algorithm regarding the assessment of the particle overlap is as follows:

(1) For particles $i$ and $j$, the quantities $\beta_n^{\,i}$ and $\beta_n^{\,j}$ ($n = 1, 2, \dots, 8$), and also $\gamma_m^{\,i}$ and $\gamma_m^{\,j}$ ($m = 1, 2, \dots, 6$) are evaluated from Eqs. (5.36) and (5.37).

(2) For particle $i$, the faces giving rise to the first, second and third largest values of $\gamma_m^{\,i}$ are chosen as faces $m_1^i$, $m_2^i$ and $m_3^i$, respectively. Similarly, faces $m_1^j$, $m_2^j$ and $m_3^j$ are determined for particle $j$.

(3) If there is a pair of parallel faces where the condition in Eq. (5.38) is satisfied, the assessment procedure in Section 5.4.1 is carried out. In the following, it is assumed that there are no pairs of parallel faces of particles $i$ and $j$.

(4) By comparing the values $\gamma_{m_1^i}^i$ with $\gamma_{m_1^j}^j$, the face yielding a larger value is chosen as an object for the overlap assessment. This face is assumed to be $m_1^i$ and renamed "face $p$" for simplicity.

(5) According to the values $\beta_n^{\,j}$ ($n = 1, 2, \dots, 8$) of particle $j$, the corners giving the first, second and third largest values of $\beta_n^{\,j}$ are chosen as the objects for analysis, which are named by corners $n_1^j$, $n_2^j$ and $n_3^j$, respectively.

(6) Then, we treat another special configuration, i.e., a possibility of face-to-line overlap is considered. The values of $(\zeta_1^{(1)}, \zeta_1^{(2)}, \zeta_1^{(3)})$ for corner $n_1^j$, $(\zeta_2^{(1)}, \zeta_2^{(2)}, \zeta_2^{(3)})$ for corner $n_2^j$, and $(\zeta_3^{(1)}, \zeta_3^{(2)}, \zeta_3^{(3)})$ for corner $n_3^j$ are evaluated. If there is one (exactly two) quantity that is zero, the corner line related to this corner is parallel to face $p$ of particle $i$. This corner is named "corner $q$" and in this face-to-line configuration, the overlap is assessed according to Eq. (5.44) in Section 5.4.2 . In the following, it is assumed that there is no corner that gives rise to zero value of the above zeta quantities. Moreover, one corner of these corners should satisfy the condition of all these three values of zeta being positive, so that it is assumed that the corner $n_1^j$ satisfies this condition (i.e., $\zeta_1^{(1)} > 0$, $\zeta_1^{(2)} > 0$ and $\zeta_1^{(3)} > 0$). This corner is named "corner $q$" for simplicity. After this, we will treat the face $p$ of particle $i$ and the corner $q$ of particle $j$ in the following.

(7) If the intersection point $\boldsymbol{r}_p$, on face $p$, of the line vertically drawn from corner $q$ is inside the plane of face $p$, the assessment of a particle overlap is conducted according to the procedure shown in Section 5.4.3.

(8) If the interaction point $\boldsymbol{r}_p$ on face $p$ is outside face $p$ in a direction normal to a corner line, there is a possibility of an overlap between each corner line (also there may be a corner-to-face overlap with a sufficiently small possibility). In this configuration, the particle overlap is assessed according to the procedure shown in Section 5.4.4.

(9) If the interaction point $\boldsymbol{r}_p$ on face $p$ is outside face $p$ in a corner direction, there is a possibility of an overlap between each corner line (also there may be a corner-to-face overlap with a sufficiently small possibility). This configuration is significantly similar to the previous situation described in step (8). In this situation, the particle overlap is assessed according to the procedure shown in Section 5.4.5.

## Bibliography

[1] Perram, J. W. and Wertheim, M. S. 1985. Statistical mechanics of hard ellipsoids. I. Overlap algorithm and the contact function. J. Comp. Phys. 58: 409–416.

[2] Donev, A., Torquato, S. and Stillinger, F. H. 2005. Neighbor list collision-driven molecular dynamics simulation for nonspherical hard particles. II. Applications to ellipses and ellipsoids. J. Comp. Phys. 202: 765–793.

CHAPTER 6

# Particle-Based Simulation Methods

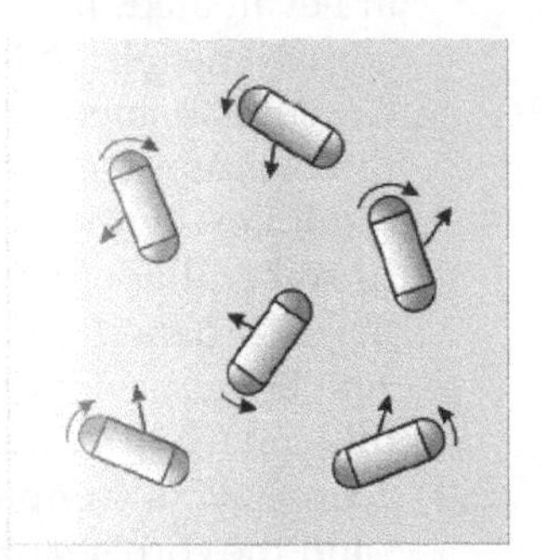

## 6.1 Monte Carlo method

There are a variety of particle-based simulation methods for investigating a particle suspension at a microscopic level. Monte Carlo method is suitable for investigating equilibrium characteristics of a suspension and other methods such as molecular dynamics and Brownian dynamics enable one to investigate its dynamics properties that are dependent on the flow situation. We first address the Monte Carlo method that is a relatively straightforward tool for investigating a particle suspension in thermodynamic equilibrium.

The Monte Carlo method is widely used for a molecular system that is in thermodynamic equilibrium [1], and furthermore is applicable to a suspension composed of fine particles. This method is based on statistical mechanics, by which microscopic states are generated in a stochastic manner. In order to grasp the methodology of the Monte Carlo method, we focus on the differences between the generation of microscopic states for the Monte Carlo method and a molecular dynamics-like method. Figure 6.1(a) schematically shows the microscopic states that are generated according to the advance of the time step in molecular dynamics-like methods, and similarly Fig. 6.1(b) shows the microscopic states generated in Monte Carlo (MC) steps. In the former method, particles (molecules) are moved according to the equation of motion, and therefore the microscopic states appear in a natural manner as a result of dynamic motion. In Monte Carlo method, the microscopic states are generated according to the probability density function that describes a thermodynamic equilibrium situation of the system. Hence, if a movie is made using a series of microscopic states generated in advance of MC steps, we cannot enjoy the natural smooth motion of the molecules or dispersed particles. However, the microscopic states appear according to physical laws expressed as stochastic probability, so that snapshots of microscopic states and ensemble averages of the physical quantities of interest are able to explain the corresponding experimental or theoretical results in quite an accurate manner. Interaction energies between particles are used for generating microscopic states in the Monte Carlo method, and therefore a physically unreasonable overlapped configuration, shown in Fig. 6.2(b), is not generated but, on the other hand, a probable configuration giving rise to a lower interaction energy, shown in Fig. 6.2(a), should be generated in MC steps. In the following, we briefly outline the equilibrium statistical mechanics necessary to understand the background of the Monte Carlo method.

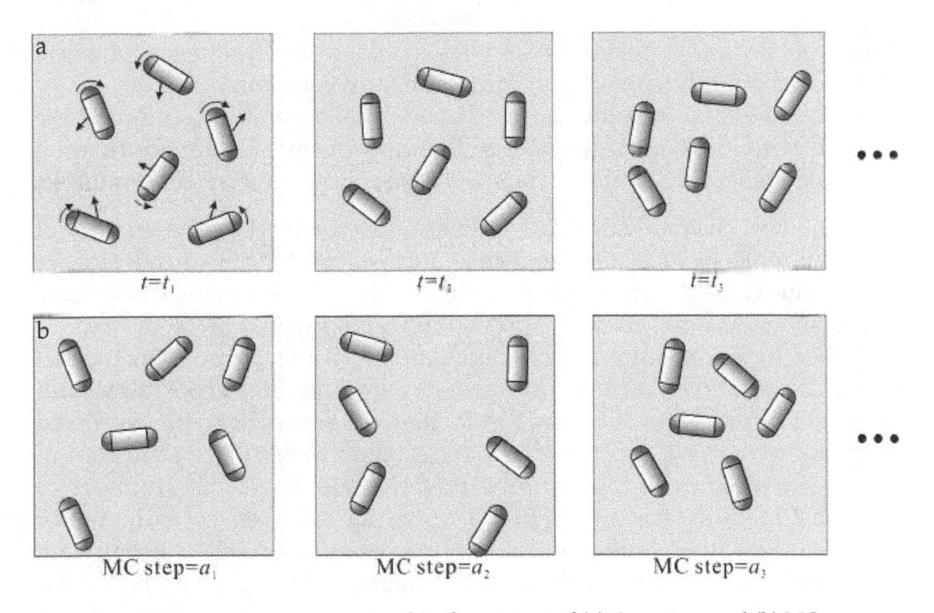


**Figure 6.1.** Microscopic states generated in the process of (a) time steps and (b) MC steps.

A statistical ensemble is composed of numerous microscopic states, and each microscopic state appears according to the probability density function that connects the actual physical quantities with statistical ones [2]. Employment of an appropriate probability density function ensures that the ensemble average of the physical quantities of interest in statistical mechanics has a physically valid meaning, that is, it can explain the corresponding experimental result with reasonable accuracy.

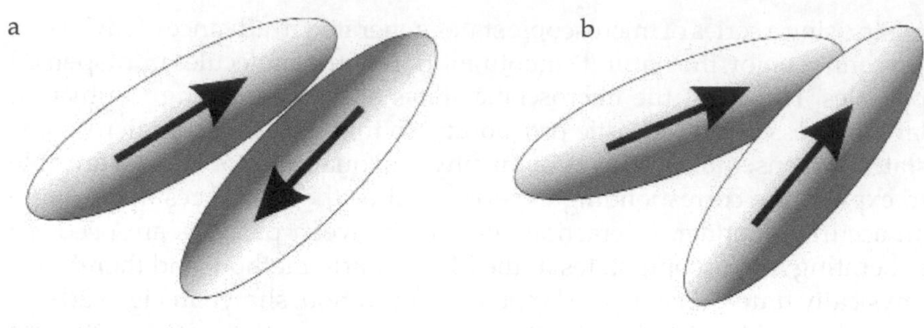


**Figure 6.2.** Configurations appearing due to the probability density function: (a) a preferable configuration and (b) a prohibitive configuration from an energy point of view.

There are several statistical ensembles available for treating a physical system in thermodynamic equilibrium, the most common being the micro-canonical, the canonical and the grand-canonical ensembles. Among these ensembles, the canonical ensemble is generally the most useful for treating a particle suspension, and therefore, here we concentrate on this ensemble.

The canonical ensemble treats a system that is specified by the number of particles (molecules), $N$, temperature $T$ and volume $V$; these quantities are given values and therefore remain constant. For the following discussion, we assume that an arbitrary non-spherical particle $i$ in a system, is described by the position $\boldsymbol{r}_i$, translational velocity $\boldsymbol{v}_i$, particle orientation $\boldsymbol{e}_i$, and angular velocity $\boldsymbol{\omega}_i$. Moreover, the generalized positions and momenta for these positions and velocities are assumed to be expressed by $q_a$ and $p_a$ ($a = 1,2, \ldots, N_0$), where $N_0$ is equal to $6N$ that is the sum of $3N$ for the number of translational motions and $3N$ for the number of rotational motions; it is noted that a restriction such as the temperature being constant reduces the number of degrees of freedom $N_0$. Employing these symbols, the probability density function $\rho_0(q_1, q_2, \ldots; p_1, p_2, \ldots)$ for the present canonical ensemble is expressed as

$$\rho_0(q_1,q_2,\ldots;p_1,p_2,\ldots) = \frac{\exp\{-H(q_1,q_2,\ldots;p_1,p_2,\ldots)/kT\}}{\int\ldots\int \exp\{-H(q_1,q_2,\ldots;p_1,p_2,\ldots)/kT\}dq_1 dq_2\ldots dq_{N_0}\, dp_1 dp_2 \ldots dp_{N_0}} \tag{6.1}$$

in which $k$ is Boltzmann's constant, and the denominator is a definite integral. The Hamiltonian function $H$ is the sum of the kinetic energy $K(p_1, p_2, \ldots)$ and the potential energy $U(q_1, q_2, \ldots)$, expressed as

$$H(q_1,q_2,\ldots;\, p_1,p_2,\ldots) = K(p_1,p_2,\ldots) + U(q_1,q_2,\ldots) \tag{6.2}$$

In general, it is possible to analytically integrate the Hamiltonian in Eq. (6.2) with respect to the generalized momenta, and consequently the integration of both sides in Eq. (6.1) leads to the simplified probability density function $\rho(q_1, q_2, \ldots)$ that is dependent only on the generalized positions:

$$\begin{aligned}\rho(q_1,q_2,\ldots) &= \int\cdots\int \rho_0(q_1,q_2,\ldots;\, p_1,p_2,\ldots)dp_1dp_2 \ldots dp_{N_0} \\ &= \frac{\exp\{-U(q_1,q_2,\ldots)/kT\}}{\int\cdots\int \exp\{-U(q_1,q_2,\ldots)/kT\}dq_1dq_2\ldots dq_{N_0}}\end{aligned} \tag{6.3}$$

This probability density function $\rho$ is called the canonical distribution function in the position and orientation space for the present ensemble, and it implies that $\rho(q_1, q_2, \ldots)dq_1dq_2 \ldots dq_{N0}$ is equal to the probability that a certain microscopic state appears within an infinitesimal position range from $(q_1, q_2, \ldots, q_{N0})$ to $(q_1+dq_1, q_2+dq_2, \ldots, q_{N0}+dq_{N0})$.

Using the canonical distribution function in Eq. (6.3), the average of an arbitrary quantity $A(q_1, q_2, \ldots)$ that is dependent on the generalized positions is evaluated as

$$\langle A\rangle = \int A(q_1,q_2,\ldots)\rho(q_1,q_2,\ldots)dq_1dq_2\ldots dq_{N_0} \tag{6.4}$$

In the non-spherical particle system, if the generalized momentum for the translational motion is taken as $\boldsymbol{p}_i = m\boldsymbol{v}_i$ for particle $i$, then the generalized position is $\boldsymbol{r}_i$, where $m$ is the mass of particles and $\boldsymbol{v}_i$ is the translational velocity, as previously defined. Similarly, if the generalized momentum for the rotational motion is expressed as $\boldsymbol{p}_i = I\boldsymbol{\omega}_i$, then the generalized position (orientation) is $\boldsymbol{\phi}_i$, where $I$ is the moment of inertia and $\boldsymbol{\phi}_i$ is related to the angular velocity $\boldsymbol{\omega}_i$ as $\boldsymbol{\omega}_i = d\boldsymbol{\phi}_i/dt$. The quantity $\boldsymbol{\phi}_i$ corresponds to the particle orientation $\boldsymbol{e}_i$, which is expressed as $(d\boldsymbol{e}_i/dt) = (d\boldsymbol{\phi}_i/dt) \times \boldsymbol{e}_i = \boldsymbol{\omega}_i \times \boldsymbol{e}_i$.

If the integrals both in the denominator in Eq. (6.3) and in the average in Eq. (6.4) are able to be analytically evaluated, a numerical simulation method such as Monte Carlo method is not necessary for evaluating the unknown physical quantities of interest. Unfortunately, these analytical evaluations are significantly restricted to several limited cases such as an ideal gas system and are impossible for almost all cases of physical phenomena. From this reason, Monte Carlo method has been developed to become a powerful simulation tool for investigating a molecular or particle system in thermodynamic equilibrium, where equilibrium statistical mechanics is basically applicable. Monte Carlo simulations enable one to evaluate the ensemble average in a relatively simple manner. We now consider the theory of Monte Carlo method in the following.

In Monte Carlo simulations, a series of microscopic states of particles are generated according to the canonical distribution function, shown in Eq. (6.3), for a system with given particle number $N$, system volume $V$ and system temperature $T$. Generation of microscopic states is performed using uniform random numbers ranging from zero to unity. Figure 6.3 schematically shows a new microscopic state $b$ generated from the present microscopic state $a$ by changing the position and the orientation of an arbitrary particle $i$. For instance, a new translational position $\boldsymbol{r}_i^{(can)} = (x_i^{(can)}, y_i^{(can)}, z_i^{(can)})$ is generated using random numbers $R_1$, $R_2$ and $R_3$ that are selected from a sequence of uniform random numbers:

$$x_i^{(can)} = x_i + \Delta r(2R_1 - 1), \quad y_i^{(can)} = y_i + \Delta r(2R_2 - 1), \quad z_i^{(can)} = z_i + \Delta r(2R_3 - 1) \tag{6.5}$$

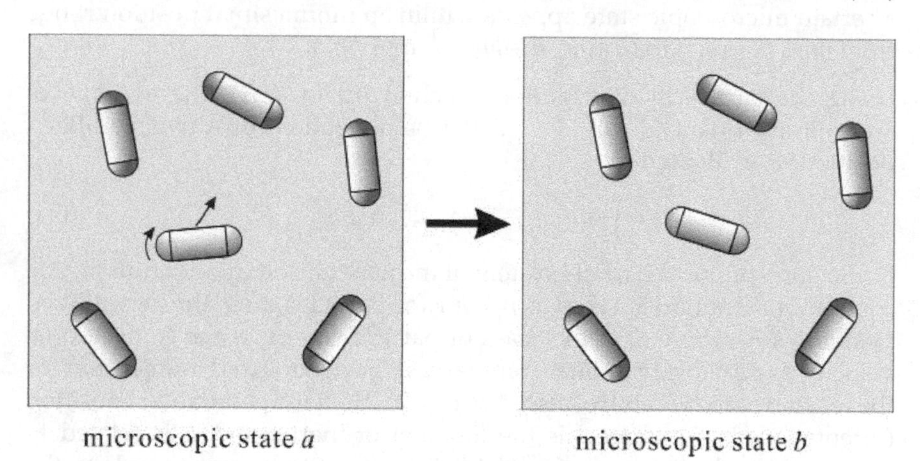


**Figure 6.3.** A new (candidate) microscopic state $b$ generated from microscopic state $a$ by moving and rotating an arbitrary particle by a small distance and a small rotation.

in which $\boldsymbol{r}_i = (x_i, y_i, z_i)$ is the present position, $\Delta r$ is the maximum displacement of the translational movement and this quantity has been specified before performing Monte Carlo simulations. This new microscopic state (i.e., a candidate state) is to be adopted according to the probability density function shown in Eq. (6.3). However, this direct assessment is not possible because the denominator in Eq. (6.3) cannot be evaluated analytically; in other words, the probability of microscopic state $b$ appearing is not known or is not able to be evaluated. Hence, the acceptance or rejection of a candidate microscopic state must be determined by a different means. The Metropolis method has overcome this difficulty by avoiding knowledge of the probability density function itself [3]; that is, the denominator is not used in the assessment procedure for this ingenious approach.

In the Metropolis method, the transition probability $p_{ba}$ of a microscopic state $b$ from state $a$ is employed as

$$p_{ba} = \begin{cases} 1 & (U_b \le U_a) \\ \rho_b / \rho_a & (U_b > U_a) \end{cases} \tag{6.6}$$

in which $U_a$ and $U_b$ are the respective interaction energies, and $\rho_a$ and $\rho_b$ are the probability density functions of the microscopic states $a$ and $b$. The ratio of the probability functions shown in Eq. (6.6) is derived from Eq. (6.3) and expressed as

$$\frac{\rho_b}{\rho_a} = \exp\left\{ -\frac{1}{kT}\Big( U(\boldsymbol{r}_1^{(can)}, \boldsymbol{r}_2^{(can)}, \cdots, \boldsymbol{r}_N^{(can)}; \boldsymbol{e}_1^{(can)}, \boldsymbol{e}_2^{(can)}, \cdots, \boldsymbol{e}_N^{(can)}) - U(\boldsymbol{r}_1, \boldsymbol{r}_2, \cdots, \boldsymbol{r}_N; \boldsymbol{e}_1, \boldsymbol{e}_2, \cdots, \boldsymbol{e}_N) \Big) \right\} \tag{6.7}$$

in which $\boldsymbol{r}_i^{(can)}$ and $\boldsymbol{e}_i^{(can)}$ ($i = 1,2,\ldots,N$) are the position and orientational vectors of an arbitrary particle, respectively. Although Eq. (6.7) is for the general case of all particles being moved and rotated to generate new microscopic state $b$, this expression is applicable to the present case where an arbitrary single particle is to be changed in position and in orientation in an attempt to transition to a new microscopic state.

According to the Metropolis method, new microscopic state $b$ is accepted if the interaction energy $U_b$ is equal to or smaller than the interaction energy $U_a$ for microscopic state $a$. If not so, microscopic state $b$ is accepted with the probability $\rho_b/\rho_a$. If microscopic state $b$ is rejected, the particle of interest is returned to the original position and orientation. In this way, a series of microscopic states are generated by repeating the attempt of changing the position and orientation of all the particles in a system. One cycle of the transition attempt, where there are $N$ number of attempts for $N$ number of particles, is called a Monte Carlo (MC) step. Acceptance of a new microscopic state $b$ with the probability $\rho_b/\rho_a$ cannot be accomplished by a single assessment process. The acceptance is performed with a statistical meaning, that is, for instance, if the acceptance probability is

10/100 then 10 attempts should be accepted from among 100 assessment procedures. Acceptance with the probability $\rho_b/\rho_a$ can be accomplished using a random number $R$ (ranging from zero to unity) in the following manner:

(1) If $\rho_b/\rho_a \geq R$, new microscopic state $b$ is accepted, and

(2) If $\rho_b/\rho_a < R$, new microscopic state $b$ is rejected.

We consider the meaning of the transition probability in Eq. (6.6) which Metropolis et al. [3] have introduced with great success in the molecular simulation field; the Metropolis method is generally and widely used in Monte Carlo simulations in a variety of fields relating to molecular or fine particle systems. If the first condition alone in Eq. (6.6) is employed, then the microscopic state converges to a minimum energy, as shown in Fig. 6.4(a) and in the case of a strongly-attractive-interacting system, the particles aggregate to form clusters that give rise to a minimum energy state. If the system temperature is significantly high so that the thermal energy is much larger than the interaction strength between particles, then the particles solely move without aggregate formation, shown in Fig. 6.4(b).

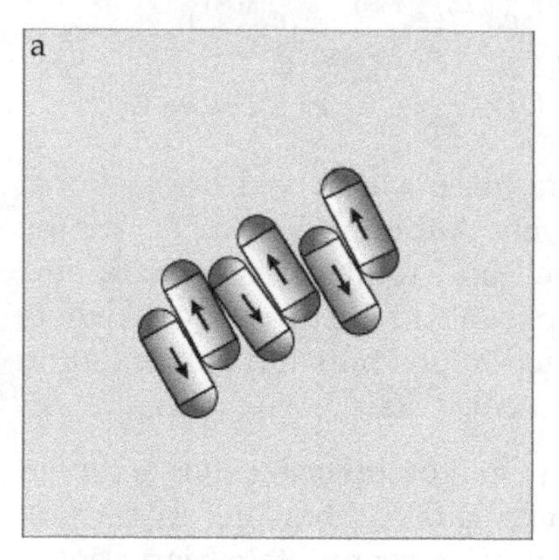


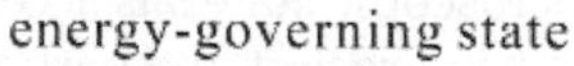


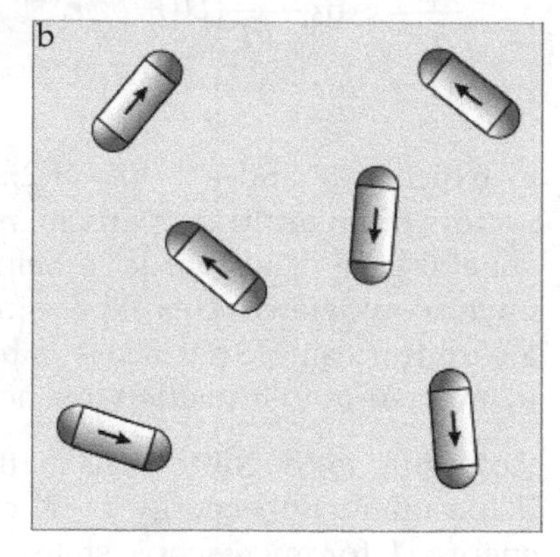


**Figure 6.4.** Dependence of appearance of microscopic states on the interaction energy and the entropy: (a) an energy-governing state and (b) an entropy-governing state.

Hence, intermediate microscopic states such as those between Figs. 6.4(a) and (b) should frequently appear in a physical system. These microscopic states are essentially generated from the second procedure of Eq. (6.6), that is, microscopic states with a higher energy are accepted with a probability that is according to the ratio of the probability density functions. If the system temperature is relatively low, microscopic states such as shown in Fig. 6.4(a) should appear frequently, and for the opposite circumstance, particles prefer to move solely as shown in Fig. 6.4(b). The dependence of the appearance of microscopic states on the system temperature may be understood in a straightforward manner by addressing the free energy and the entropy factors.

The Helmholtz free energy $F$ is defined as the sum of the potential interaction energy $E$ and the entropy $S$. That is,

$$F = E - TS \tag{6.8}$$

In the situation of thermodynamic equilibrium for a system with given $N$, $V$ and $T$, the thermodynamic state should appear so that the free energy $F$ tends to a minimum energy. The entropy will be relatively large if particles solely move without the formation of clusters, but on the other hand, the formation of clusters will tend to lower the entropy of the system. Hence, in the entropy-governing situation, the microscopic state in Fig. 6.4(b) should frequently appear and in the energy-governing situation a microscopic state in Fig. 6.4(a) should be preferred as the thermodynamic equilibrium situation. From this consideration, it is concluded that the second condition in Eq. (6.6) is quite reasonable for reflecting the entropy factor in Monte Carlo simulations while the effect of the temperature is included in the ratio of the probability density functions, as shown in Eq. (6.7).

As a result of employing the transition probability in Eq. (6.6), microscopic states will appear according to the original probability density function in Eq. (6.3). This significantly simplifies the ensemble average shown in Eq. (6.4) and allows the ensemble average to be evaluated from a simple arithmetic average of the sampling data:

$$\langle A \rangle \approx \sum_{n=1}^{M} A(\boldsymbol{r}_1^{\,n}, \boldsymbol{r}_2^{\,n}, \cdots, \boldsymbol{r}_N^{\,n}; \boldsymbol{e}_1^{\,n}, \boldsymbol{e}_2^{\,n}, \cdots, \boldsymbol{e}_N^{\,n}) / M \tag{6.9}$$

in which $M$ is the total sampling number, and ($\boldsymbol{r}_1^{\,n}$, $\boldsymbol{r}_2^{\,n}$, ..., $\boldsymbol{e}_N^{\,n}$) are the position and orientational data that are sampled at the $n$-th order.

We may now show the main part of the general algorithm of the Metropolis Monte Carlo method for a non-spherical particle system. The algorithm for determining a transition from state $a$ to state $b$ is as follows:

1. Choose a particle in a sequence or at random
2. Move this particle to a new position chosen in a random manner according to Eq. (6.5) and regard this state as microscopic state $b$

3. Compute the interaction energy $U_b$
4. Compute the ratio of the probability functions $\rho_b/\rho_a$ in Eq. (6.7)
5. If $U_b \leq U_a$, then microscopic state $b$ is accepted and this state is regarded as new microscopic state $a$ and then go to step 7
6. If $U_b > U_a$, then a random number ranging from zero to unity, $R_1$, is taken from a uniform random number sequence
   6.1 $\rho_b/\rho_a \geq R_1$, then microscopic state $b$ is accepted and this state is regarded as new microscopic state $a$ and then go to step 7
   6.2 $\rho_b/\rho_a < R_1$, then microscopic state $b$ is rejected, and go to step 7
7. Rotate the particle to a new orientation in a random manner similar to Eq. (6.5), and regard this state as microscopic state $b$
8. Compute the interaction energy $U_b$
9. Compute the ratio of the probability functions $\rho_b/\rho_a$ in Eq. (6.7)
10. If $U_b \leq U_a$, then microscopic state $b$ is accepted and this state is regarded as new microscopic state $a$ and then go to step 1
11. If $U_b > U_a$, then a random number ranging from zero to unity, $R_2$, is taken from a uniform random number sequence
    11.1 $\rho_b/\rho_a \geq R_2$, then microscopic state $b$ is accepted and this state is regarded as new microscopic state $a$ and then return to step 1
    11.2 $\rho_b/\rho_a < R_2$, then microscopic state $b$ is rejected and then return to step 1.

In the above example, the attempt of a change in the position and orientation is carried out separately, but it is possible to assess them simultaneously by combining them together and considering them as a single attempt.

In sampling the data for evaluating the physical quantities of interest, it may be reasonable to conduct the sampling procedure at intervals of every or several MC steps and this will depend on the maximum displacement $\Delta r$ that has to be specified before starting a simulation.

## 6.2 Molecular dynamics method

Molecular dynamics method has widely been used as a microscopic analysis tool for a variety of fields in both the equilibrium and non-equilibrium situations of a molecular system. In a particle suspension system, if dispersed particles are approximately larger than micron-order, Brownian motion is negligible and therefore the molecular dynamics method is also a useful simulation approach for a particle suspension in order to investigate physical phenomena from a microscopic point of view. A non-equilibrium situation is indispensable for investigating fluid characteristics such as rheology, therefore it is desirable to address molecular dynamics method in this section.

In the case of a suspension composed of dispersed particles, it is significantly difficult to simulate both the particles and the ambient flow field simultaneously, and the ambient fluid is generally regarded as a continuum medium where the effect of the interaction between the fluid and the particles appears as a friction term in the equation of motion of particles. The simultaneous simulation of the particle and the flow field may be performed by the lattice Boltzmann method or methods such as multi-particle collision dynamics and dissipative particle dynamics methods, some of which will be discussed in the following sections. Even if the ambient fluid is treated as a continuum medium in molecular dynamics simulations, it is still significantly difficult to take into account the multi-body hydrodynamic interactions among particles. Although this treatment may be performed by Stokesian dynamics method for a suspension composed of spherical particles [4, 5], it becomes too difficult for a non-spherical particle system even today. Hence, in the present section, we focus only on the case of a dilute dispersion composed of non-spherical particles where multi-body hydrodynamic interactions are negligible among dispersed particles; the effect of the ambient fluid is reflected through the friction effect. Moreover, since the particles with arbitrary shape are quite difficult to treat, we focus on the axisymmetric particles that can be represented by a spherocylinder or spheroidal particle.

In the following, we first consider the case where the inertial term plays an important role for determining the characteristics of particle motion and then address the case where the inertia factor is negligible. Here, we do not treat the ambient flow field together with particle motion and so the flow field is assumed to be linear and represented by a simple shear flow or a quiescent flow field.

### *6.2.1 For the case of the inertia being taken into account*

We consider here the motion of a single axisymmetric particle in a quiescent flow field. In this particle model, shown in Fig. 6.5, the translational motion is decomposed into a motion along the particle axis and a motion normal to the particle axis. Similarly, the rotational motion is decomposed into a spin rotational motion about the particle axis direction and a usual rotational motion about a line normal to the particle axis, as shown in Figs. 6.6 and 6.7. In this way we treat both the translational and rotational motion in order to describe the particle motion. Subscripts $\|$ and $\perp$ are employed for denoting the components along and normal to the particle axis, respectively. If we use the symbol $\boldsymbol{r}$ for the position vector of the particle center, $\boldsymbol{v}$ for the translational velocity vector and $\boldsymbol{\omega}$ for the angular velocity vector, then the translational and the rotational motion are described as

$$m\frac{d^2\boldsymbol{r}_{\|}}{dt^2} = \boldsymbol{F}_{\|} - \xi_{\|}\boldsymbol{v}_{\|}, \qquad m\frac{d^2\boldsymbol{r}_{\perp}}{dt^2} = \boldsymbol{F}_{\perp} - \xi_{\perp}\boldsymbol{v}_{\perp} \tag{6.10}$$

$$I\frac{d\omega}{dt} = \boldsymbol{T}_{\perp} - \xi_r\omega \tag{6.11}$$

in which $m$ and $I$ are the mass and the inertia moment of the particle, respectively, and $\xi_{\|}$, $\xi_{\perp}$ and $\xi_r$ are the translational friction coefficients and rotational friction coefficient, respectively. Equation (6.11) implies the expression for the rotational motion about a line passing through the

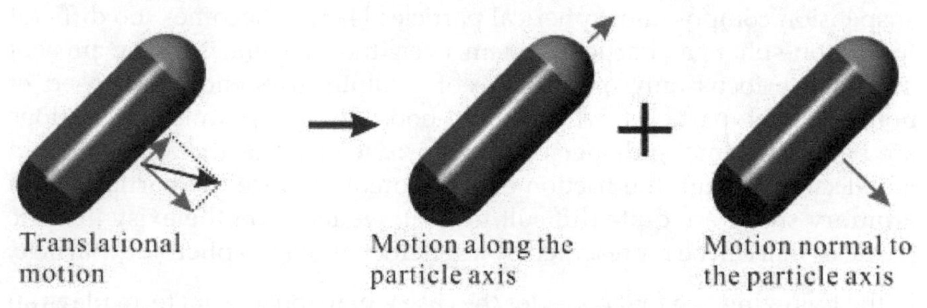


**Figure 6.5.** Translational motion for an axisymmetric particle: the motion is decomposed into the motion along and normal to the particle axis direction.

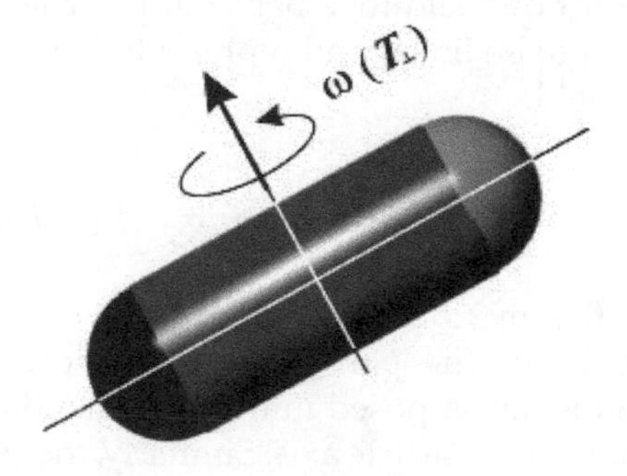


**Figure 6.6.** Ordinary rotational motion about a line through the particle center normal to the particle axis line.

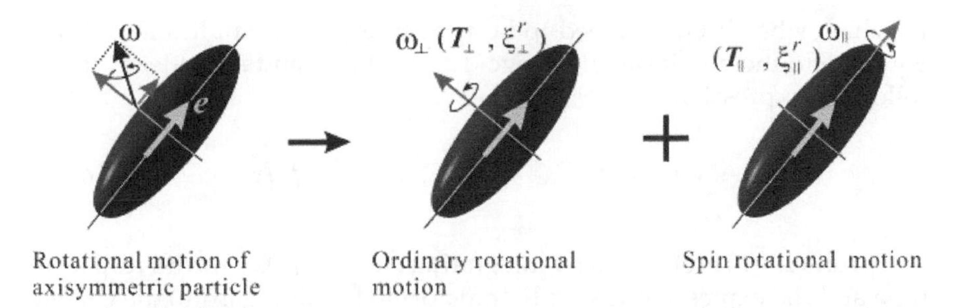


**Figure 6.7.** Rotational motion of an axisymmetric particle: the rotational motion is decomposed into the usual rotational motion about a line normal to the particle axis direction and the spin rotational motion about the particle axis.

particle center and normal to the particle axis direction. It is noted that this treatment regarding the rotational motion is generally valid for most types of particles. However, for special type of magnetic particles that are magnetized in a direction normal to the particle axis, as in the case of spindle-like hematite particles [6, 7], it is necessary to take into account the spin rotational motion about the particle axis. The translational and rotational diffusion coefficients are well known as mathematical expressions for spheres, cylinders and spheroids that are frequently employed as a particle model for simulations. The treatment of the equation of the rotational motion is quite similar to that of the translational motion and therefore we concentrate on the latter particle motion in the following. Moreover, for simplicity, the subscript denoting parallel or vertical motion is dropped in the following analysis.

Equation (6.10) is rewritten in a slightly different form to aid mathematical manipulation:

$$\frac{d\boldsymbol{v}}{dt}+(\xi/m)\,\boldsymbol{v} = \boldsymbol{F}/m \tag{6.12}$$

If subscript $\parallel$ is attached to the quantities $\boldsymbol{v}$, $\boldsymbol{F}$, and $\xi$, then this equation is implied by the first equation in Eq. (6.10). If the force $\boldsymbol{F}$ can be regarded as being constant during an infinitesimal time change, Eq. (6.12) becomes the typical linear ordinary differential equation. From formula regarding solutions of the ordinary differential equations, Eq. (6.12) is straightforwardly solved and expressed as

$$\begin{aligned}\boldsymbol{v}(t) &= \exp(-\int(\xi/m)dt)\left\{\int(\boldsymbol{F}/m)\exp(\int(\xi/m)dt)dt+\boldsymbol{c}\right\} \\ &= e^{-\frac{\xi}{m}(t-t_0)}\boldsymbol{v}_0+\frac{1}{\xi}\left\{1-e^{-\frac{\xi}{m}(t-t_0)}\right\}\boldsymbol{F}\end{aligned} \tag{6.13}$$

in which $v$ has been assumed to be $v_0$ at time $t = t_0$. Application of this equation to the infinitesimal change during time $t$ and $t + \Delta t$ leads to the following expression:

$$v(t+\Delta t) = e^{-\frac{\xi}{m}\Delta t} v(t) + \frac{1}{\xi}\{1 - e^{-\frac{\xi}{m}\Delta t)}\} F(t) \tag{6.14}$$

The position $r(t)$ is obtained by integrating $v(t)$ in Eq. (6.13) with respect to time and the expression is simply transformed with regard to the change from time $t$ to $t + \Delta t$, leading to

$$r(t+\Delta t) = r(t) + \frac{m}{\xi}\left(1 - e^{-\frac{\xi}{m}\Delta t)}\right) v(t) + \frac{1}{\xi}\left\{\Delta t - \frac{m}{\xi}\left(1 - e^{-\frac{\xi}{m}\Delta t)}\right)\right\} F(t) \tag{6.15}$$

The accuracy of the velocity expression may be improved by changing the assumption regarding the force $F$ such that the force linearly varies with time, leading to the following equation:

$$\begin{aligned} v(t+\Delta t) &= e^{-\frac{\xi}{m}\Delta t} v(t) + \frac{1}{\xi}\{1 - e^{-\frac{\xi}{m}\Delta t)}\} F(t) \\ &\quad + \frac{1}{\xi \Delta t}\{\Delta t - \frac{m}{\xi}(1 - e^{-\frac{\xi}{m}\Delta t})\}\big(F(t+\Delta t) - F(t)\big) \end{aligned} \tag{6.16}$$

It is noted that the expressions in Eqs. (6.15) and (6.16) reduce to the velocity Verlet algorithm in the limit of $(\xi \Delta t / m)$ approaching zero. If subscripts $||$ and $\perp$ are resumed in Eqs. (6.15) and (6.16), then these equations are for the translational motion parallel and normal to the particle axis.

Similarly, the angular velocity vector $\omega$ is expressed as

$$\begin{aligned} \omega(t+\Delta t) &= e^{-\frac{\xi_r}{I}\Delta t} \omega(t) + \frac{1}{\xi_r}\{1 - e^{-\frac{\xi_r}{I}\Delta t)}\} T_\perp(t) \\ &\quad + \frac{1}{\xi_r \Delta t}\{\Delta t - \frac{I}{\xi_r}(1 - e^{-\frac{\xi_r}{I}\Delta t})\}\big(T_\perp(t+\Delta t) - T_\perp(t)\big) \end{aligned} \tag{6.17}$$

The particle orientation $e(t)$ is related to the angular velocity vector as

$$\frac{de(t)}{dt} = \omega(t) \times e(t) \tag{6.18}$$

Hence, if the time interval $\Delta t$ is sufficiently small, the particle orientation $\boldsymbol{e}(t + \Delta t)$ at the next time step is obtained from Euler finite difference formula as

$$\boldsymbol{e}(t+\Delta t) = \boldsymbol{e}(t) + \Delta t \boldsymbol{\omega}(t) \times \boldsymbol{e}(t) \tag{6.19}$$

A more accurate algorithm for obtaining the particle orientation than used in the present method may be developed in a more sophisticated approach [8].

It is summarized from the above that the particle position $\boldsymbol{r}(t + \Delta t)$, translational velocity $\boldsymbol{v}(t + \Delta t)$, particle orientation $\boldsymbol{e}(t + \Delta t)$ and angular velocity $\boldsymbol{\omega}(t + \Delta t)$ at the next time step $t + \Delta t$ are evaluated from Eqs. (6.15), (6.16), (6.19) and (6.17), respectively, for the case of a suspension composed of axisymmetric particles.

## 6.2.2 *For the case of the inertia being negligible*

If the dispersed particles are in a range smaller than micron-order, the inertia term may be considered negligible and therefore the basic equations in Eqs. (6.10) and (6.11) reduce to

$$\boldsymbol{v}_{\parallel} = \boldsymbol{F}_{\parallel}/\xi_{\parallel}, \quad \boldsymbol{v}_{\perp} = \boldsymbol{F}_{\perp}/\xi_{\perp}, \quad \boldsymbol{\omega} = \boldsymbol{T}_{\perp}/\xi_{r} \tag{6.20}$$

From these expressions, it is seen that the translational velocity and angular velocity are determined if the force and torque acting on the particle are evaluated. The position $\boldsymbol{r}(t + \Delta t)$ at the next time step is evaluated using the translational velocities in Eqs. (6.20) as

$$\boldsymbol{r}(t+\Delta t) = \boldsymbol{r}_{\parallel}(t+\Delta t) + \boldsymbol{r}_{\perp}(t+\Delta t) = \boldsymbol{r}_{\parallel}(t) + \boldsymbol{r}_{\perp}(t) + \Delta t(\boldsymbol{v}_{\parallel}(t) + \boldsymbol{v}_{\perp}(t)) = \boldsymbol{r}(t) + \Delta t\, \boldsymbol{v}(t) \tag{6.21}$$

Equation (6.19) may also be used for evaluating the orientation $\boldsymbol{e}(t + \Delta t)$ of the particle at the next time step.

The quantities along and normal to the particles axis, which are necessary for evaluating the new position and orientation, are expressed using the net force $\boldsymbol{F}$ and torque $\boldsymbol{T}$ as

$$\boldsymbol{F}_{\parallel} = (\boldsymbol{F}\cdot\boldsymbol{e})\boldsymbol{F}, \quad \boldsymbol{F}_{\perp} = \boldsymbol{F} - \boldsymbol{F}_{\parallel}, \quad \boldsymbol{T}_{\parallel} = (\boldsymbol{T}\cdot\boldsymbol{e})\boldsymbol{T}, \quad \boldsymbol{T}_{\perp} = \boldsymbol{T} - (\boldsymbol{T}\cdot\boldsymbol{e})\boldsymbol{T} \tag{6.22}$$

The friction coefficients are well known for typical particle models, where for instance, $\xi_{||}$, $\xi_{\perp}$ and $\xi_r$ are written for the cylinder particle with particle aspect ratio $r_p$ and diameter $d$ as [9]

$$\xi_{\parallel} = 2\pi r_p d\eta_s \{\ln 2r_p + \ln 2 - 3/2\}^{-1}, \quad \xi_{\perp} = 4\pi r_p d\eta_s \{\ln 2r_p + \ln 2 - 1/2\}^{-1},$$
$$\xi_r = \frac{\pi d^3 r_p{}^3 \eta_s}{3}\left\{\frac{1}{\ln r_p}\left(1 + \frac{\ln 2 - 1}{\ln r_p}\right) + \frac{3\times 5.45}{8\pi}\cdot\frac{1}{r_p^2}\right\} \tag{6.23}$$

These expressions are valid for the case of $r_p >> 1$. Other particle models, for instance the spheroidal particle, have different values of the friction coefficients and they may be referenced from standard textbooks [10].

From Eq. (6.20), it is seen that the system temperature cannot be evaluated from the translational or angular velocity of the particle, in other words, the effect of the temperature appears only through the friction coefficients, i.e., the viscosity. This is reasonable because the inertia terms have been neglected in this section.

Since a linear flow field is generally employed for investigating the rheological characteristics of a particle suspension by molecular dynamics or Brownian dynamics, it is useful to apply the above theory to the simulation of a simple shear flow. Hence, in the following, we address the theory for the situation where a simple shear flow is applied to a dilute suspension composed of axisymmetric particles.

We consider a simple shear flow $\boldsymbol{U} = (\dot{\gamma}\, y, 0, 0)$ in the $x$-axis direction with shear rate $\dot{\gamma}$. This flow field is also characterized by the angular velocity $\boldsymbol{\Omega}$ and rate-of-strain tensor $\boldsymbol{E}$, which are expressed as

$$\boldsymbol{\Omega} = \frac{1}{2}\nabla\times\boldsymbol{U} = \begin{bmatrix} 0 \\ 0 \\ -\dot{\gamma}/2 \end{bmatrix}, \qquad \boldsymbol{E} = \frac{1}{2}\{\nabla\boldsymbol{U} + (\nabla\boldsymbol{U})^t\} = \begin{bmatrix} 0 & \dot{\gamma}/2 & 0 \\ \dot{\gamma}/2 & 0 & 0 \\ 0 & 0 & 0 \end{bmatrix} \tag{6.24}$$

For this case the basic equations for the particle motions are expressed as

$$\boldsymbol{v}_{\parallel} = \boldsymbol{U}_{\parallel}(\boldsymbol{r}_i) + \frac{1}{\xi_{\parallel}}\boldsymbol{F}_{\parallel}, \quad \boldsymbol{v}_{\perp} = \boldsymbol{U}_{\perp}(\boldsymbol{r}) + \frac{1}{\xi_{\perp}}\boldsymbol{F}_{\perp} \tag{6.25}$$

$$\omega = \boldsymbol{\Omega} + \frac{1}{\xi_r}\boldsymbol{T}_{\perp} - \frac{Y^H}{Y^C}(\varepsilon\cdot\boldsymbol{ee}):\boldsymbol{E} \tag{6.26}$$

in which $\varepsilon$ is a third-rank tensor (Eddington's epsilon), and the vector and tensor operator of the third term on the right-hand side in Eq. (6.26) are used with reference to the textbook [11]. The quantity of $Y^H/Y^C$ is the ratio of the resistance functions and reflects the interaction of the particle with the ambient fluid. This quantity is dependent on the particle shape and expressed for the above-mentioned cylinderical particle as [9]

$$\frac{Y^H}{Y^C}=\left\{\frac{1}{\ln r_p}\left(1+\frac{\ln 2-1}{\ln r_p}\right)-\frac{3\times 5.45}{8\pi}\cdot\frac{1}{r_p^2}\right\}\bigg/\left\{\frac{1}{\ln r_p}\left(1+\frac{\ln 2-1}{\ln r_p}\right)+\frac{3\times 5.45}{8\pi}\cdot\frac{1}{r_p^2}\right\} \tag{6.27}$$

In Eq. (6.25), $\boldsymbol{U}_{||}$ and $\boldsymbol{U}_{\perp}$ are the components of the flow velocity $\boldsymbol{U}$ in the direction parallel and normal to the particle axis direction and are evaluated from relationships similar to Eq. (6.22).

The third term on the right-hand side in Eq. (6.26) arises from the shape effect that results from the interaction between the particle and the ambient shear flow field and therefore it is expected that a longer axisymmetric particle leads to a larger influence on the rotational motion of the particle.

After the translational velocity $\boldsymbol{v}$ (= $\boldsymbol{v}_{||}$ + $\boldsymbol{v}_{\perp}$) and the angular velocity $\boldsymbol{\omega}$ are evaluated from Eqs. (6.25) and (6.26), the position $\boldsymbol{r}(t+\Delta t)$ and the orientation $\boldsymbol{e}(t+\Delta t)$ of the particle at the next time step are obtained from the following expressions:

$$\boldsymbol{r}_{||}(t+\Delta t)=\boldsymbol{r}_{||}(t)+\boldsymbol{U}_{||}\Delta t+\frac{1}{\xi_{||}}\boldsymbol{F}_{||}(t)\Delta t \tag{6.28}$$

$$\boldsymbol{r}_{\perp}(t+\Delta t)=\boldsymbol{r}_{|}(t)+\boldsymbol{U}_{\perp}\Delta t+\frac{1}{\xi_{\perp}}\boldsymbol{F}_{\perp}(t)\Delta t \tag{6.29}$$

$$\boldsymbol{e}(t+\Delta t)=\boldsymbol{e}(t)+\boldsymbol{\Omega}_{\perp}\times\boldsymbol{e}\Delta t+\frac{1}{\xi_r}\boldsymbol{T}_{\perp}(t)\times\boldsymbol{e}\,\Delta t-\frac{Y^H}{Y^C}\big((\boldsymbol{\varepsilon}\cdot\boldsymbol{ee}):\boldsymbol{E}\big)\times\boldsymbol{e}\Delta t \tag{6.30}$$

Finally we show the relevant part of the molecular dynamics algorithm for the simulation for a dilute suspension composed of axisymmetric particles that move in a simple shear flow:

(1) Assign the initial position, translational and angular velocity, and orientation of all the particles in a system
(2) Calculate the forces and torques acting on the particles
(3) Calculate the components of each related quantity parallel and normal to the particle axis direction according to Eqs. (6.22) and related relationships
(4) Calculate the translational and angular velocities in Eqs. (6.25) and (6.26) at the next time step

(5) Calculate the positions and orientations of the particles in Eqs. (6.28) to (6.30) at the next time step

(6) Advance one time step and return to step 2.

In sampling data for evaluating the physical quantities of interest, it may be reasonable to conduct the sampling procedure every time step, which is of course dependent on the employed value of the time interval; a larger time interval for the averaging procedure may be reasonable in molecular dynamics simulations when the simulation step is a relatively small time interval.

## 6.3 Brownian dynamics method

If the dispersed particles in a suspension are in a range smaller than micron-order, it is not possible to neglect the Brownian motion of the particles in a particle-based simulation and therefore the molecular dynamics is not a suitable method for simulating dispersed particles in a suspension. In this situation, the motion of the dispersed particles generally has a component of both translational and rotational Brownian motion in a flow field but in the case of spherical particles, treatment of only translational Brownian motion is sufficient for a variety of physical phenomena. On the other hand, in the case of non-spherical particles such as spheroids, it is usually important that rotational Brownian motion is also taken into account.

The dispersed particles interact with other dispersed particles and also interact with the ambient fluid, and in an actual particle suspension the motion of dispersed particles is influenced by a multi-body hydrodynamic interaction via the particle-fluid interactions. However, it is significantly difficult to treat this multi-body hydrodynamic interaction among particles and this difficulty becomes prohibitive even for the non-spherical axisymmetric particles such as cylinders and spheroids. For a suspension composed of spherical particles, the simulation method for taking into account these multi-body hydrodynamic interactions has already been developed [12], although it is necessary to treat mobility or resistance functions carefully in developing a simulation code for Brownian dynamics method.

In Brownian dynamics simulations, a pre-specified time interval is employed for performing a simulation, and therefore a difficult problem will arise for a suspension composed of hard (solid) particles without a steric layer coating the particles. This is because the addition of random displacements to particle motion may induce an overlap between the two particles after one time interval. However, the problem is not severe in the case of a colloidal dispersion because particles are generally covered by a repulsive layer in order to prevent the particles from aggregating and to create a stable dispersion. It is necessary to develop a technique for taking into account hydrodynamic interactions in the case of the non-spherical particle suspensions, and other methods such as lattice Boltzmann method seem to be desirable for this objective.

There is another problem in ordinary Brownian dynamics method. That is, only the random particle motion is simulated in the situation of a given flow field such as a quiescent or a linear flow field and the flow field itself is not solved in the simulation. In contrast, other simulation methods such as the lattice Boltzmann, dissipative dynamics and multi-particle collision dynamics methods enable one to simulate both particle motion and the ambient flow field simultaneously.

In the following, we first discuss Brownian dynamics method for a dilute suspension composed of spherical particles, and then for a suspension composed of axisymmetric particles in a quiescent or a simple shear flow field. We also address a dilute suspension composed of these particles where the hydrodynamic interaction between the dispersed particles is neglected and the ambient fluid is reflected as a friction term of the particle motion, therefore, hydrodynamic interactions are not taken into account.

### 6.3.1 For the case of a dilute suspension of spherical particles

We consider the Brownian motion of a spherical particle in a simple shear flow which is expressed as $\boldsymbol{U} = (\dot{\gamma} y, 0, 0)$ in the $x$-axis direction with shear rate $\dot{\gamma}$, angular velocity $\boldsymbol{\Omega}$ and rate-of-strain tensor $\boldsymbol{E}$, as shown in Section 6.2. In this situation, the motion of the spherical particle is described by the following Langevin equation [1]:

$$m\frac{d^2\boldsymbol{r}}{dt^2} = \boldsymbol{F} - \xi(\boldsymbol{v} - \boldsymbol{U}) + \boldsymbol{F}^B \tag{6.31}$$

in which $m$ is the mass of the particle, $\xi$ is the translational friction coefficient, $\boldsymbol{F}$ is the non-random force, and $\boldsymbol{F}^B$ is the random force inducing the translational Brownian motion. The random force $\boldsymbol{F}^B = (F_x^B, F_y^B, F_z^B)$ should be larger for a higher temperature and has the following stochastic characteristics:

$$\begin{aligned}&\langle F_x^B(t)\rangle = \langle F_y^B(t)\rangle = \langle F_z^B(t)\rangle = 0,\\&\langle F_x^B(t)F_x^B(t')\rangle = \langle F_y^B(t)F_y^B(t')\rangle = \langle F_z^B(t)F_z^B(t')\rangle = 2\xi kT\delta(t-t')\end{aligned} \tag{6.32}$$

in which $\delta(t-t')$ is the Dirac delta function, $k$ is Boltzmann's constant and $T$ is the system temperature.

Since the inertial term of the particle is generally negligible for almost all colloidal dispersions, Eq. (6.31) reduces to the algorithm of Brownian dynamics [12]:

$$\boldsymbol{r}(t+\Delta t) = \boldsymbol{r}(t) + \Delta t\boldsymbol{U} + \frac{1}{\xi}\Delta t\boldsymbol{F} + \Delta\boldsymbol{r}^B \tag{6.33}$$

in which $\Delta\boldsymbol{r}^B = (\Delta r_x^{\ B}, \Delta r_y^{\ B}, \Delta r_z^{\ B})$ is the random displacement inducing the Brownian motion. These quantities have the following stochastic characteristics:

$$\begin{aligned}&\langle \Delta r_x^B(t)\rangle = \langle \Delta r_y^B(t)\rangle = \langle \Delta r_z^B(t)\rangle = 0,\\&\left\langle\{\Delta r_x^B(t)\}^2\right\rangle = \left\langle\{\Delta r_y^B(t)\}^2\right\rangle = \left\langle\{\Delta r_z^B(t)\}^2\right\rangle = \frac{2kT}{\xi}\Delta t\end{aligned} \tag{6.34}$$

In these expressions, $\Delta t$ is the time interval, and the friction coefficient $\xi$ is expressed as $\xi = 3\pi\eta d$ from Stokes' drag formula, where $d$ is the diameter of the sphere and $\eta$ is the viscosity of the base liquid.

In conducting Brownian dynamics simulations, the random displacements are generated according to the normal distribution with the standard deviation in Eq. (6.34), and the position $\boldsymbol{r}(t + \Delta t)$ at the next time step is evaluated from Eq. (6.33) in the usual molecular dynamics procedure.

It should be noted that a theory that can take into account hydrodynamic interactions among particles has already been developed for the spherical particle [12], but a serious particle overlap may frequently appear unless careful attention is paid to this problem in the development of a simulation code.

### 6.3.2 *For the case of a dilute suspension of axisymmetric particles*

We now consider a dilute suspension composed of axisymmetric particles in a simple shear flow described previously in Section 6.2.2. The translational Brownian motion is governed by a basic equation similar to that shown in Eq. (6.31) for spherical particles. For an axisymmetric particle, the translational motion can be decomposed into the motion in the long axis direction and the motion in a direction normal to the long axis. These expressions are written as

$$\begin{aligned} m\frac{d^2\boldsymbol{r}_{\parallel}}{dt^2} &= \boldsymbol{F}_{\parallel} - \xi_{\parallel}^{t}(\boldsymbol{v}_{\parallel} - \boldsymbol{U}_{\parallel}) + \boldsymbol{F}_{\parallel}^{B}, \\ m\frac{d^2\boldsymbol{r}_{\perp}}{dt^2} &= \boldsymbol{F}_{\perp} - \xi_{\perp}^{t}(\boldsymbol{v}_{\perp} - \boldsymbol{U}_{\perp}) + \boldsymbol{F}_{\perp}^{B} \end{aligned} \tag{6.35}$$

in which the subscripts || and $\perp$ imply the quantities parallel and normal to the particle axis direction respectively, $m$ is the mass of the particle, $\boldsymbol{F}$ is the non-random force acting on the particle, $\boldsymbol{U}$ is a simple shear flow field, and $\boldsymbol{F}^B$ is the random force inducing the translational Brownian motion. The random force $\boldsymbol{F}_{\parallel}{}^{B} = F_{\parallel}{}^{B}\boldsymbol{e}$ and $\boldsymbol{F}_{\perp}{}^{B} = F_{\perp 1}{}^{B}\boldsymbol{e}_{\perp 1} + F_{\perp}{}^{B}\boldsymbol{e}_{\perp 2}$ have the following stochastic characteristics:

$$\left\langle F_{\parallel}^{B}(t)\right\rangle = 0, \qquad \left\langle F_{\parallel}^{B}(t)F_{\parallel}^{B}(t')\right\rangle = 2\xi_{\parallel}^{t}kT\delta(t-t') \tag{6.36}$$

$$\left\langle F_{\perp 1}^{B}(t)\right\rangle = \left\langle F_{\perp 2}^{B}(t)\right\rangle = 0, \qquad \left\langle F_{\perp 1}^{B}(t)F_{\perp 1}^{B}(t')\right\rangle = \left\langle F_{\perp 2}^{B}(t)F_{\perp 2}^{B}(t')\right\rangle = 2\xi_{\perp}^{t}kT\delta(t-t') \tag{6.37}$$

In these equations, $\boldsymbol{e}$ is the unit vector denoting the particle direction, the vectors $\boldsymbol{e}_{\perp 1}$ and $\boldsymbol{e}_{\perp 2}$ are the unit vectors normal to each other in a plane normal to the particle axis, and $\xi_{\parallel}^{t}$ and $\xi_{\perp}^{t}$ are the translational diffusion coefficients parallel and normal to the particle axis direction, respectively.

Since the inertia term is generally negligible in a particle colloidal dispersion, Eq. (6.35) reduces to the following basic equations:

$$\boldsymbol{r}_{\parallel}(t+\Delta t) = \boldsymbol{r}_{\parallel}(t) + \boldsymbol{U}_{\parallel}\Delta t + \frac{1}{\xi_{\parallel}^{t}}\boldsymbol{F}_{\parallel}(t)\Delta t + \Delta r_{\parallel}^{B}\boldsymbol{e}(t) \tag{6.38}$$

$$\boldsymbol{r}_{\perp}(t+\Delta t) = \boldsymbol{r}_{\perp}(t) + \boldsymbol{U}_{\perp}\Delta t + \frac{1}{\xi_{\perp}^{t}}\boldsymbol{F}_{\perp}(t)\Delta t + \Delta r_{\perp 1}^{B}\boldsymbol{e}_{\perp 1}(t) + \Delta r_{\perp 2}^{B}\boldsymbol{e}_{\perp 2}(t) \tag{6.39}$$

In these equations, the quantities $\Delta r_{\parallel}^{B}$, $\Delta r_{\perp 1}^{B}$ and $\Delta r_{\perp 2}^{B}$ are the random displacements inducing the translational Brownian motion and have the following stochastic characteristics:

$$\left.\begin{aligned} &\left\langle \Delta r_{\parallel}^{B}\right\rangle = \left\langle \Delta r_{\perp 1}^{B}\right\rangle = \left\langle \Delta r_{\perp 2}^{B}\right\rangle = 0, \\ &\left\langle (\Delta r_{\parallel}^{B})^2\right\rangle = 2kT\Delta t/\xi_{\parallel}^{t}, \\ &\left\langle (\Delta r_{\perp 1}^{B})^2\right\rangle = \left\langle (\Delta r_{\perp 2}^{B})^2\right\rangle = 2kT\Delta t/\xi_{\perp}^{t}. \end{aligned}\right\} \tag{6.40}$$

If $\boldsymbol{r}_{||}$ and $\boldsymbol{r}_{\perp}$ are evaluated, the particle position is obtained as $\boldsymbol{r} = \boldsymbol{r}_{||} + \boldsymbol{r}_{\perp}$.

As shown in Fig. 6.7, the rotational Brownian motion about a line through the particle center and normal to the particle axis direction is described by the following expression:

$$I\frac{d^2\boldsymbol{\varphi}_{\perp}}{dt^2} = \boldsymbol{T}_{\perp} + \boldsymbol{T}_{\perp}^{fl} + \boldsymbol{T}_{\perp}^{B} \tag{6.41}$$

in which $\boldsymbol{\varphi}_{\perp}$ is defined as $d\boldsymbol{\varphi}_{\perp}/dt = \boldsymbol{\omega}_{\perp}$, $I$ is the inertia moment of the particle, $\boldsymbol{\omega}_{\perp}$ is the angular velocity vector, $\boldsymbol{T}_{\perp}$ is the non-random torque acting on the particle, $\boldsymbol{T}_{\perp}^{fl}$ is the viscous torque arising from the interaction with the ambient fluid, and $\boldsymbol{T}_{\perp}^{B}$ is the random torque inducing the rotational Brownian motion. The viscous torque $\boldsymbol{T}_{\perp}^{fl}$ is expressed for a simple shear flow as

$$\boldsymbol{T}_{\perp}^{fl} = -\xi_{\perp}^{r}(\boldsymbol{\omega}_{\perp} - \boldsymbol{\Omega}_{\perp}) - \xi_{\perp}^{r}\frac{Y^H}{Y^C}(\boldsymbol{\varepsilon}\cdot\boldsymbol{ee}):\boldsymbol{E} \tag{6.42}$$

Also, $\boldsymbol{T}_{\perp}^{B} = T_{\perp 1}^{B}\boldsymbol{e}_{\perp 1} + T_{\perp}^{B}\boldsymbol{e}_{\perp 2}$ is the random torque inducing the rotational Brownian motion, and has the following stochastic characteristics:

$$\langle T_{\perp 1}^{B}(t)\rangle = \langle T_{\perp 2}^{B}(t)\rangle = 0, \qquad \langle T_{\perp 1}^{B}(t)T_{\perp 1}^{B}(t')\rangle = \langle T_{\perp 2}^{B}(t)T_{\perp 2}^{B}(t')\rangle = 2\xi_{\perp}^{r}kT\delta(t-t') \tag{6.43}$$

in which $\xi_{\perp}^{r}$ is the rotational friction coefficient for the usual rotational motion about a line normal to the particle axis direction.

In a usual colloidal dispersion, the inertia term is negligible in Eq. (6.41), and therefore the basic equation for Brownian dynamics simulations is obtained as

$$\boldsymbol{\omega}_{\perp} = \boldsymbol{\Omega}_{\perp} + \frac{1}{\xi_{\perp}^{r}}\boldsymbol{T}_{\perp} - \frac{Y^H}{Y^C}(\boldsymbol{\varepsilon}\cdot\boldsymbol{ee}):\boldsymbol{E} + \frac{1}{\xi_{\perp}^{r}}(T_{\perp 1}^{B}\boldsymbol{e}_{\perp 1} + T_{\perp 2}^{B}\boldsymbol{e}_{\perp 2}) \tag{6.44}$$

By taking into account the relationship of $d\boldsymbol{\varphi}/dt = \boldsymbol{\omega}$, Eq. (6.44) is rewritten as

$$\boldsymbol{\varphi}_{\perp}(t+\Delta t) = \boldsymbol{\varphi}_{\perp}(t) + \Delta t\boldsymbol{\Omega}_{\perp} + \Delta t\frac{1}{\xi_{\perp}^{r}}\boldsymbol{T}_{\perp} - \Delta t\frac{Y^H}{Y^C}(\boldsymbol{\varepsilon}\cdot\boldsymbol{ee}):\boldsymbol{E} + (\Delta\varphi_{\perp 1}^{B}\boldsymbol{e}_{\perp 1} + \Delta\varphi_{\perp 2}^{B}\boldsymbol{e}_{\perp 2}) \tag{6.45}$$

in which the quantities $\Delta\varphi_{\perp 1}^{B}$ and $\Delta\varphi_{\perp 2}^{B}$ are the random displacements inducing the rotational Brownian motion and have the following stochastic characteristics:

$$\left\langle \Delta\varphi_{\perp 1}^{B}(t) \right\rangle = \left\langle \Delta\varphi_{\perp 2}^{B}(t) \right\rangle = 0, \qquad \left\langle \Delta\varphi_{\perp 1}^{B}(t)\Delta\varphi_{\perp 1}^{B}(t') \right\rangle = \left\langle \Delta\varphi_{\perp 2}^{B}(t)\Delta\varphi_{\perp 2}^{B}(t') \right\rangle = 2kT\Delta t / \xi_{\perp}^{\;r} \tag{6.46}$$

Since the quantity $\boldsymbol{\varphi}$ is not an intuitive description for the particle orientation, it may be reasonable to derive the basic equation of the orientation $\boldsymbol{e}$ from Eq. (6.45). If we take into account the following relationship:

$$\boldsymbol{e}(t+\Delta t) = \boldsymbol{e}(t) + \Delta t\, \boldsymbol{\omega}_{\perp}(t) \times \boldsymbol{e}(t) \tag{6.47}$$

then the particle orientation $\boldsymbol{e}(t + \Delta t)$ at the next time step is obtained from Eqs. (6.44), (6.47) and the relationship of $d\boldsymbol{\varphi}/dt = \boldsymbol{\omega}$ as

$$\boldsymbol{e}(t+\Delta t) = \boldsymbol{e}(t) + \boldsymbol{\Omega}_{\perp} \times \boldsymbol{e}\Delta t + \frac{1}{\xi_{\perp}^{\;r}} \boldsymbol{T}_{\perp}(t) \times \boldsymbol{e}\Delta t - \frac{Y^{H}}{Y^{C}} \big( (\varepsilon \cdot \boldsymbol{ee}) : \boldsymbol{E} \big) \times \boldsymbol{e}\Delta t + \Delta\varphi_{\perp 1}^{B} \boldsymbol{e}_{\perp 1} + \Delta\varphi_{\perp 2}^{B} \boldsymbol{e}_{\perp 2} \tag{6.48}$$

In conducting Brownian dynamics simulations, if the random displacements are generated according to the normal distribution with standard deviation in Eqs. (6.40) and (6.46), then the position $\boldsymbol{r}(t + \Delta t)$ at the next time step is evaluated from Eqs. (6.38) and (6.39) and the orientation $\boldsymbol{e}(t + \Delta t)$ is determined from Eq. (6.48).

For the case of a particle with a magnetic moment normal to the particle axis, such as in the case of a rod-like hematite particle, it is necessary to specify the spin Brownian motion about the particle axis in addition to the ordinary rotational motion about a line normal to the particle axis [13]. As shown in Fig. 6.8, employing the unit vector $\boldsymbol{n}$ for denoting the direction

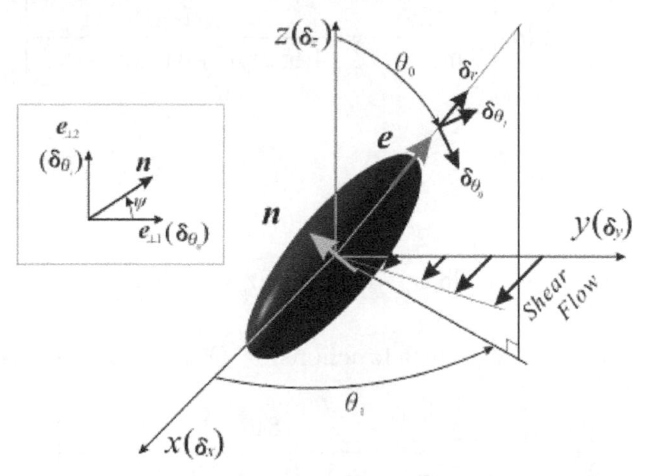


**Figure 6.8.** Magnetic spheroidal particle with a magnetic moment in a direction normal to the particle axis direction in a simple shear flow.

of the magnetic moment, the equation describing the magnetic moment direction is written as

$$\boldsymbol{n}(t+\Delta t)=\boldsymbol{n}(t)+\boldsymbol{\Omega}_{\|}\times\boldsymbol{n}(t)\Delta t+\frac{1}{\xi_{\|}^{r}}\boldsymbol{T}_{\|}(t)\times\boldsymbol{n}(t)\Delta t+\Delta\varphi_{\|}^{B}\,\boldsymbol{e}(t)\times\boldsymbol{n}(t) \quad (6.49)$$

In this equation, the quantity $\Delta\varphi_{\|}^{B}$ is the rotational random displacement inducing the spin Brownian motion and has the following stochastic characteristics:

$$\left\langle\Delta\varphi_{\|}^{B}\right\rangle=0,\quad \left\langle(\Delta\varphi_{\|}^{B})^{2}\right\rangle=2kT\Delta t/\xi_{\|}^{r} \quad (6.50)$$

The quantities in the parallel and the normal directions to the particle axis are expressed using the unit vector $\boldsymbol{e}$ denoting the particle axis direction as

$$\left.\begin{array}{lll}\boldsymbol{F}^{\|}=(\boldsymbol{F}\cdot\boldsymbol{e})\boldsymbol{e}, & \boldsymbol{F}^{\perp}=\boldsymbol{F}-\boldsymbol{F}^{\|}, & \boldsymbol{T}^{\|}=(\boldsymbol{T}\cdot\boldsymbol{e})\boldsymbol{e},\\ \boldsymbol{T}^{\perp}=\boldsymbol{T}-\boldsymbol{T}^{\|}, & \boldsymbol{\Omega}^{\|}=(\boldsymbol{\Omega}\cdot\boldsymbol{e})\boldsymbol{e}, & \boldsymbol{\Omega}^{\perp}=\boldsymbol{\Omega}-\boldsymbol{\Omega}^{\|}\end{array}\right\} \quad (6.51)$$

For reference, we show the translational and rotational friction coefficients for the prolate spheroidal particle with major (longer) length $l$, minor (shorter) length $d$ and particle aspect ratio $r_p = l/d$ [10]:

$$\left.\begin{array}{l}\xi_{\|}^{t}=2\pi l\eta_{s}\left[\dfrac{2}{2\ln 2r_{p}-1}-\dfrac{4\ln 2r_{p}-3}{\{4(\ln 2r_{p})^{2}-4\ln 2r_{p}+1\}r_{p}^{2}}\right],\\[2ex] \xi_{\perp}^{t}=4\pi l\eta_{s}\left[\dfrac{2}{2\ln 2r_{p}+1}-\dfrac{1}{\{4(\ln 2r_{p})^{2}+4\ln 2r_{p}+1\}r_{p}^{2}}\right]\end{array}\right. \quad (6.52)$$

$$\left.\begin{array}{l}\xi_{\|}^{r}=\dfrac{2\pi l^{3}\eta_{s}}{3r_{p}^{2}},\\[2ex] \xi_{\perp}^{r}=\dfrac{\pi l^{3}\eta_{s}}{3}\left[\dfrac{2}{2\ln 2r_{p}-1}+\dfrac{1}{\{4(\ln 2r_{p})^{2}-4\ln 2r_{p}+1\}r_{p}^{2}}\right]\end{array}\right. \quad (6.53)$$

Also, the ratio of the resistance functions, $Y^H/Y^C$, is expressed as

$$\frac{Y^{H}}{Y^{C}}=\left[\frac{2}{2\ln 2r_{p}-1}-\frac{8\ln 2r_{p}-5}{\{4(\ln 2r_{p})^{2}-4\ln 2r_{p}+1\}r_{p}^{2}}\right]$$

$$\times\left[\frac{2}{2\ln 2r_p - 1} + \frac{1}{\{4(\ln 2r_p)^2 - 4\ln 2r_p + 1\}r_p^2}\right]^{-1} \qquad (6.54)$$

The basic algorithm for conducting the Brownian dynamics simulation for the axisymmetric particle with a magnetic moment normal to the particle axis, based on Eqs. (6.38), (6.39), (6.48) and (6.49), is as follows:

(1) Specify the initial position, particle orientation and magnetic moment direction of all particles
(2) Calculate the forces and the torque acting on each particle
(3) Generate the random displacements shown in Eqs. (6.40), (6.46) and (6.50) from a sequence of uniform random numbers with a range of zero to unity
(4) Calculate all the particle positions at the next time step from Eqs. (6.38) and (6.39)
(5) Calculate all the particle orientations at the next time step from Eq. (6.48)
(6) Calculate the direction of the magnetic moment of all the particles at the next time step from Eq. (6.49)
(7) Return to step (2).

In order to obtain the average value of physical quantities of interest, it may be desirable to sample data every certain number of time steps, whereas a larger time interval employed in a simulation may require a more frequent sampling procedure.

## 6.4 Lattice Boltzmann method

As described previously in Section 6.1, if a particle suspension is in thermodynamic equilibrium, the Monte Carlo method is an effective and powerful tool as a microscopic analysis technique. However, this method is not applicable to non-equilibrium flow problems, which are indispensable for investigating dynamic properties such as rheological characteristics. If the main objective is to investigate rheological characteristics of a suspension composed of non-magnetic or magnetic particles, we are directed to consider the behavior of dispersed particles in a given flow field such as a simple shear flow. To do so, the Stokesian dynamics and Brownian dynamics methods may be a desirable simulation approach. The latter method is necessary to investigate the behavior of dispersed particles that perform translational and rotational Brownian motion. However, in the ordinary Brownian dynamics simulations, it is significantly difficult to take into account the multi-body hydrodynamic interactions among dispersed particles, especially for the case of non-spherical particle suspensions. In addition, the ambient liquid is regarded as a continuum medium and the effect of the base liquid on the particle motion appears only as a friction term in the basic equations of motion. This suggests that Brownian dynamics method may not be appropriate as a simulation technique to investigate both the particle motion and the flow field simultaneously.

There are several possible simulation techniques that will solve the behavior of magnetic particles and the ambient flow field simultaneously. One method along this line is dissipative particle dynamics [14–16], where groups or clusters of solvent molecules are regarded as virtual fluid particles (dissipative particles) and these fluid particles are simulated together with the dispersed particles. In this method the multi-body hydrodynamics among the dispersed particles are reflected through the interactions of the dispersed particles with the dissipative particles. This method has an additional significant merit in that the basic equation of motion of the dissipative particles has a term which can induce random motion in the dissipative particles, and consequently Brownian motion of dispersed particles is indirectly activated by this random motion.

The lattice Boltzmann method is another hopeful approach and is based on the concept of the virtual fluid particle for the flow problems of particle suspensions [17–20]. Ordinary particle-based methods such as molecular dynamics and Brownian dynamics methods simulate the particle motion by addressing the position and velocity of the dispersed particles. In contrast, the lattice Boltzmann method does not address the particle positions and velocities, but treats the particle distribution function of the fluid particles which determines the rate at which the fluid particles move between neighboring sites with advancing time. Originally, the lattice Boltzmann method was developed for numerically solving the flow field of a pure fluid, but this method is also applicable to a particle suspension for simulating both the particle motion and the ambient flow field simultaneously. However, in the case of a magnetic suspension, the magnetic particles are coated with a soft steric layer such as a surfactant or electric double layer, and therefore it is required to treat the interactions between the steric layers and the fluid particles carefully. Another point for a successful application of the lattice Boltzmann method to a particle suspension is to develop a sophisticated technique for the accurate generation of the translational and rotational Brownian motion of the dispersed particles [19, 20].

In the following, we address the theoretical background of the lattice Boltzmann method and several difficult problems that must be overcome in order to apply this method to a suspension composed of magnetic particles in a flow field, where the translational and rotational Brownian motion of dispersed particles are not negligible in a flow field.

### 6.4.1 BGK lattice Boltzmann method for a pure fluid system

In this section, we first describe the BGK lattice Boltzmann method for a two-dimensional pure liquid rather than a particle suspension and then summarize the basic equations for a three-dimensional system. In the following sections we will discuss several problems for successfully applying the BGK lattice Boltzmann method to a suspension composed of dispersed particles.

In the lattice Boltzmann method, the simulation region is regarded as a lattice system and the fluid is assumed to be composed of virtual fluid particles. These fluids particles are moved from their present site to a neighboring site in the lattice system with advancing time. The present section addresses a two-dimensional system and therefore we employ the D2Q9 lattice model, which is generally used for a two-dimensional system. In this lattice model, the fluid particles have eight possibilities for moving to a neighboring site, and on including the quiescent state, there are a total of nine possibilities to be considered, $\alpha = 0, 1, \ldots, 8$, as shown in Fig. 6.9.

a

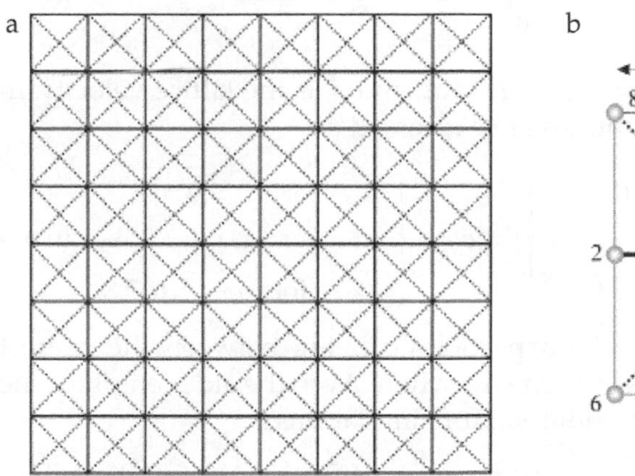

b

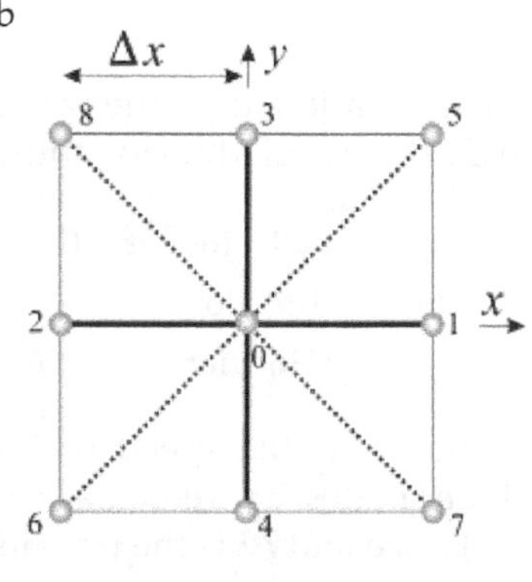


**Figure 6.9.** D2Q9 lattice model for two-dimensional lattice Boltzmann simulations: (a) the lattice model and (b) the unit cell for D2Q9 model.

The accuracy and validity of the lattice Boltzmann method are significantly dependent on the modeling of the collision dynamics among the fluid particles. There are several models available in regard to the collision dynamics, and among these models, the BGK collision model is well known and the simplest approach for the modeling of the Boltzmann equation. We will therefore focus on the BGK lattice Boltzmann method. In the BGK collision model, the particle distribution function in the $\alpha$-direction depends only on the corresponding distribution at a point on the line to the neighboring site in the opposite direction $\bar{\alpha}$, and not on any of the other sites or any other directions. This characteristic gives rise to a great merit for the BGK lattice Boltzmann method in regard to the development of a simulation program.

If an arbitrary lattice point is denoted by $\boldsymbol{r}$ and the particle distribution function in the $\alpha$-direction at time $t$ is denoted by $f_\alpha(\boldsymbol{r}, t)$, then the particle distribution $f_\alpha(\boldsymbol{r} + \boldsymbol{c}_\alpha \Delta t, t + \Delta t)$ at $\boldsymbol{r} + \boldsymbol{c}_\alpha \Delta t$ after the time interval $\Delta t$ can be obtained from the following equations [17, 18]:

$$f_\alpha(\boldsymbol{r}+\boldsymbol{c}_\alpha\Delta t, t+\Delta t) = \tilde{f}_\alpha(\boldsymbol{r},t), \quad \tilde{f}_\alpha(\boldsymbol{r},t) = f_\alpha(\boldsymbol{r},t) + \frac{1}{\tau}\{f_\alpha^{(0)}(\boldsymbol{r},t) - f_\alpha(\boldsymbol{r},t)\} \tag{6.55}$$

in which $\tau$ is the (non-dimensional) relaxation time, and $f_\alpha^{(0)}$ is the equilibrium distribution function and expressed with the macroscopic fluid velocity $\boldsymbol{u}$ ($u = |\boldsymbol{u}|$) and the density $\rho_0$ as

$$f_\alpha^{(0)} = \rho_0 w_\alpha \left\{ 1 + 3\frac{\boldsymbol{c}_\alpha \cdot \boldsymbol{u}}{c^2} - \frac{3u^2}{2c^2} + \frac{9}{2} \cdot \frac{(\boldsymbol{c}_\alpha \cdot \boldsymbol{u})^2}{c^4} \right\} \tag{6.56}$$

In this equation, $w_\alpha$ is the weight constant, and $\boldsymbol{c}_\alpha$ is the lattice velocity. In the D2Q9 model, these quantities are expressed as

$$w_\alpha = \begin{Bmatrix} 4/9 & \text{for} & \alpha = 0 \\ 1/9 & \text{for} & \alpha = 1,2,3,4 \\ 1/36 & \text{for} & \alpha = 5,6,7,8 \end{Bmatrix}, \quad |\boldsymbol{c}_\alpha| = \begin{Bmatrix} 0 & \text{for} & \alpha = 0 \\ c & \text{for} & \alpha = 1,2,3,4 \\ \sqrt{2}c & \text{for} & \alpha = 5,6,7,8 \end{Bmatrix} \tag{6.57}$$

in which $c$ is the speed of fluid particles moving between the nearest neighbor sites, expressed as $c = \Delta x / \Delta t$, where $\Delta x$ is the side length of the unit lattice and $\Delta t$ is the time interval for simulations.

Specification of the lattice velocity $\boldsymbol{c}_\alpha$ in Eq. (6.57) guarantees that fluid particles can move from site to site during the time interval. If the particle distribution $f_\alpha$ ($\alpha = 0, 1, 2, \ldots, 8$) is known at each lattice point, the macroscopic density and momentum of the fluid can be evaluated from the following equations:

$$\rho(\boldsymbol{r},t)=\sum_{\alpha=0}^{8} f_{\alpha}(\boldsymbol{r},t)\ ,\quad \rho(\boldsymbol{r},t)\boldsymbol{u}(\boldsymbol{r},t)=\sum_{\alpha=0}^{8} f_{\alpha}(\boldsymbol{r},t)\,\boldsymbol{c}_{\alpha} \tag{6.58}$$

Hence, the value of $\boldsymbol{u}$ at a certain time step, which are evaluated from Eq. (6.58) using the known particle distribution function, enables one to evaluate the new particle distribution function at the next time step from Eq. (6.55).

In the case of a three-dimensional system, the basic equation in Eq. (6.55) and the equilibrium distribution function in Eq. (6.56) may be used without any modifications, but the weight constant $w_{\alpha}$ and the lattice velocity $\boldsymbol{c}_{\alpha}$ are different from those of a two-dimensional system. Although there are a variety of lattice models that may be used for a three-dimensional system, we concentrate here on the D3Q19 lattice model, shown in Fig. 6.10, which is one of the more frequently used. In this lattice model, the fluid particles have nineteen possibilities including the quiescent state for moving to the neighboring sites, $\alpha = 0, 1, \ldots, 18$, as shown in Fig. 6.10. In this lattice model, the weight constant $w_{\alpha}$ and the lattice velocity $\boldsymbol{c}_{\alpha}$ are written as

$$w_{\alpha}=\begin{Bmatrix} 1/3 & \text{for} & \alpha=0 \\ 1/18 & \text{for} & \alpha=1,2,\ldots,6 \\ 1/36 & \text{for} & \alpha=7,8,\ldots,18 \end{Bmatrix},\quad \left|\boldsymbol{c}_{\alpha}\right|=\begin{Bmatrix} 0 & \text{for} & \alpha=0 \\ c & \text{for} & \alpha=1,2,\ldots,6 \\ \sqrt{2}c & \text{for} & \alpha=7,8,\ldots 18 \end{Bmatrix} \tag{6.59}$$

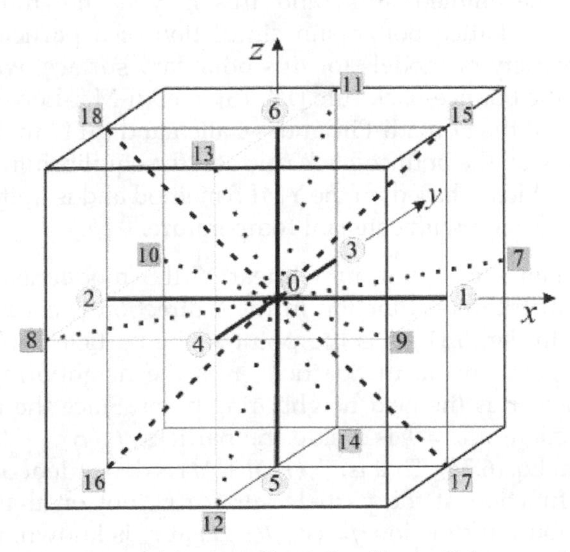


**Figure 6.10.** D3Q19 lattice model for three-dimensional lattice Boltzmann simulations.

In the case of a particle suspension, the basic equations in Eqs. (6.55) and (6.56) with the weight constant $w_\alpha$ and the lattice velocity $\boldsymbol{c}_\alpha$ in Eq. (6.57) or (6.59) are directly applicable without any modifications. However, the following treatments are necessary in regard to the boundary conditions and the interactions between the dispersed particles and the virtual fluid particles. If the dispersed particles are required to exhibit the Brownian motion, a technique for activating the Brownian motion must be combined with the basic equations of the lattice Boltzmann method. These topics will be addressed in the following sections.

## 6.4.2 Boundary conditions

The boundary surfaces that we have to treat in a lattice Boltzmann simulation are the outer boundary surfaces of the simulation region and the particle surface. If a system is in thermodynamic equilibrium, the periodic boundary condition, which is usually used in molecular dynamics simulation, is applicable to the outer boundary surfaces. For instance, in Fig. 6.9(a), the particle distribution function $f_\alpha(x,y,t)|_{left}$ ($\alpha$ = 0, 1, 2, ..., 8) at the left boundary surface is made equivalent to $f_\alpha(x,y,t)|_{right}$ at the right surface and a similar procedure is carried out for the upper and lower boundary surfaces in Fig. 6.9(a).

We now consider the modeling of the boundary surface between the particle and the ambient fluid and this may be the most important modeling in the lattice Boltzmann simulation of a particle dispersion. There are a variety of models for this boundary surface, which include, for instance, the bounce-back rule [17, 18], the Yu-Mei-Luo-Shyy (YMLS) method [21] and the Bouzidi-Firdaouss-Lallemand (BFL) method [22]. We focus here only on the bounce-back rule and the equilibrium linear YMLS method [23], which is based on the YMLS method and is quite suitable for a particle suspension with constant temperature.

We first explain the equilibrium linear YMLS model using Fig. 6.11. The particle distribution function in the $\alpha$-direction is considered ($\alpha$ = 2 in Fig. 6.11). In Fig. 6.11, $\boldsymbol{r}_w$ is the point at the particle surface, $\boldsymbol{r}_p$ is the neighboring point inside the particle, $\boldsymbol{r}_l$ is the neighboring site in the liquid area, and $\boldsymbol{r}_{l'}$ is the next neighboring point. Since the next point of $\boldsymbol{r}_l$ in the direction of $\alpha$ = 1 is inside the particle, $f_2(\boldsymbol{r}_l, t+\Delta t)$ cannot be obtained from Eq. (6.55). That is, $f_2(\boldsymbol{r}_l, t+\Delta t)$ is dependent on the particle distribution function at the particle surface $\boldsymbol{r}_w$, not on that at $\boldsymbol{r}_p$. If the particle distribution function $f_2$ $(\boldsymbol{r}_w, t+\Delta t)$ at $\boldsymbol{r}_w$ is known, then $f_2$ $(\boldsymbol{r}_l, t +$

$\Delta t$) at $\boldsymbol{r}_l$ can be evaluated from the linear interpolation method using those at $\boldsymbol{r}_{l'}$ and $\boldsymbol{r}_w$ as [21]

$$f_2\left(\boldsymbol{r}_l, t+\Delta t\right)=\frac{\Delta_w}{1+\Delta_w} f_2\left(\boldsymbol{r}_{l'}, t+\Delta t\right)+\frac{1}{1+\Delta_w} f_2\left(\boldsymbol{r}_w, t+\Delta t\right) \tag{6.60}$$

in which $\Delta_w = |\boldsymbol{r}_l - \boldsymbol{r}_w| / |\boldsymbol{r}_l - \boldsymbol{r}_p|$. Although Fig. 6.11 shows the case of $\bar{\alpha} = 2$ (i.e., the opposite direction to $\alpha = 1$), Eq. (6.60) simply holds for any direction $\alpha$ along which the connecting line crosses the particle surface. In order to evaluate $f_2(\boldsymbol{r}_l, t+\Delta t)$ from Eq. (6.60), the function $f_2(\boldsymbol{r}_w, t+\Delta t)$ at the surface is necessary, and the equilibrium linear method uses the equilibrium distribution shown in Eq. (6.56) as $f_2(\boldsymbol{r}_w, t+\Delta t)$ [23].

The bounce-back rule is the most well-known and the simplest surface model. In this model, the fluid particles start from their lattice site, next to the surface of the material, collide with the material surface, and return to the original site during a unit time interval. Hence, according to the locations of the lattice sites shown in Fig. 6.11, the particle distribution function $f_1(\boldsymbol{r}_l, t)$ is assigned to $f_2(\boldsymbol{r}_l, t+\Delta t)$ at the next time step in the model of the bounce-back rule.

It is noted that the expression in Eq. (6.60) has to be modified if the boundary surface moves with advancing time.

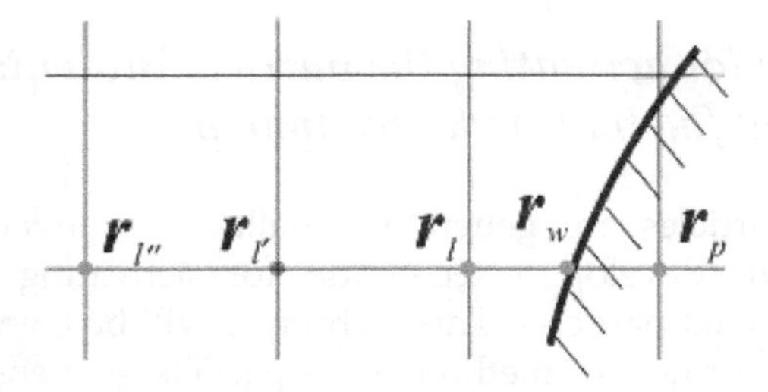


**Figure 6.11.** Boundary condition for interaction between fluid particles and the material surface.

## 6.4.3 *Forces and torques acting on particles dispersed in a suspension*

We obtain the forces and torques acting on a dispersed particle by evaluating the interactions between the dispersed and fluid particles. We focus on an arbitrary lattice point $\boldsymbol{r}_l^p$ next to the particle surface and in this case the neighboring site in the direction of $\alpha = \alpha_l^p$ is inside the particle. The momentum of the fluid particles moving toward the particle surface from $\boldsymbol{r}_l^p$ at time $t$ is $\boldsymbol{c}_{\alpha_l^p}\tilde{f}_{\alpha_l^p}(\boldsymbol{r}_l^p,t)\Delta x\Delta y$, and that after the collision at $(t+\Delta t)$ is $-\boldsymbol{c}_{\alpha_l^p} f_{\bar{\alpha}_l^p}(\boldsymbol{r}_l^p,t+\Delta t)\Delta x\Delta y$. Since the change in the momentum during the time interval $\Delta t$ is equal to the impulse $\boldsymbol{F}_{\alpha_l^p}\Delta t$, the force acting on the particle $\boldsymbol{F}_{\alpha_l^p}$ can be obtained as

$$\boldsymbol{F}_{\alpha_l^p} = \{\boldsymbol{c}_{\alpha_l^p}\tilde{f}_{\alpha_l^p}(\boldsymbol{r}_l^p,t)\Delta x\Delta y + \boldsymbol{c}_{\alpha_l^p} f_{\bar{\alpha}_l^p}(\boldsymbol{r}_l^p,t+\Delta t)\Delta x\Delta y\}/\Delta t \quad (6.61)$$

Similarly, the torque acting on the particle $\boldsymbol{T}_{\alpha_l^p}$ due to the collision with fluid particles is written as

$$\boldsymbol{T}_{\alpha_l^p} = (\boldsymbol{r}_w - \boldsymbol{r}_c)\times\{\boldsymbol{c}_{\alpha_l^p}\tilde{f}_{\alpha_l^p}(\boldsymbol{r}_l^p,t)\Delta x\Delta y + \boldsymbol{c}_{\alpha_l^p} f_{\bar{\alpha}_l^p}(\boldsymbol{r}_l^p,t+\Delta t)\Delta x\Delta y\}/\Delta t \quad (6.62)$$

in which $\boldsymbol{r}_c$ is the center of the particle, and $\boldsymbol{r}_w$ is the position vector at the particle surface on a line along the $\alpha$-direction from $\boldsymbol{r}_l^p$.

Hence, the total force $\boldsymbol{F}_i^{(lttc)}$ and torque $\boldsymbol{T}_i^{(lttc)}$ acting on particle $i$ by the ambient fluid can finally be obtained by summing the contributions from the neighboring lattice points $\boldsymbol{r}_l^p$ and the interacting directions $\alpha_l^p$:

$$\boldsymbol{F}_i^{(lttc)} = \sum_l \sum_{\alpha_l^p} \boldsymbol{F}_{\alpha_l^p}, \qquad \boldsymbol{T}_i^{(lttc)} = \sum_l \sum_{\alpha_l^p} \boldsymbol{T}_{\alpha_l^p} \quad (6.63)$$

### 6.4.4 *Method for activating the particle Brownian motion based on fluctuation hydrodynamics*

If dispersed particles are generally smaller than micron-order, it is indispensable to develop a technique for activating the Brownian motion of dispersed particles. This technique will be combined with the ordinary lattice Brownian method for a particle suspension. There are several pioneering works that have been conducted based on fluctuation hydrodynamics in order to apply the lattice Boltzmann method to particle dispersions [19, 20]. We briefly summarize this method [23].

The fluctuation $s_{ij}$ in the stress can be reproduced by introducing the corresponding fluctuation term $f_\alpha^{(fluc)}$ into the basic equation of Eq. (6.55). The fluctuation term is required to satisfy the following equations:

$$\sum_\alpha f_\alpha^{(fluc)} = 0, \qquad \sum_\alpha \boldsymbol{c}_\alpha f_\alpha^{(fluc)} = 0 \quad (6.64)$$

In order for the fluctuation term to satisfy these conditions, we assume the following form with an unknown constant $s_0$:

$$f_{\alpha}^{(fluc)} = -s_0 w_{\alpha} \sum_{k} \sum_{l} s_{kl} c_{\alpha k} c_{\alpha l} \tag{6.65}$$

From the definition of the fluctuation of the momentum flux $\Pi_{ij}^{(fluc)} (= \sum_{\alpha} c_{\alpha i} c_{\alpha j} f_{\alpha}^{(fluc)})$, this fluctuation is equal to $-s_{ij}$, so that $\Pi_{ij}^{(fluc)}$ has to satisfy the following relationship:

$$\overline{\Pi_{ij}^{(fluc)}(\boldsymbol{r}_1, t_1) \Pi_{kl}^{(fluc)}(\boldsymbol{r}_2, t_2)} = 2kT\eta(\delta_{ik}\delta_{jl} + \delta_{il}\delta_{jk} + \delta_{ij}\delta_{kl})\delta(\boldsymbol{r}_1 - \boldsymbol{r}_2)\delta(t_1 - t_2) \tag{6.66}$$

in which $\eta$ is the viscosity of a base liquid, and $\delta_{ij}$ is Kronecker's delta. Substitution of Eq. (6.65) into Eq. (6.66) with consideration of Eq. (6.64) gives rise to $s_0 = 1/(2c_s^4)$, so that the final expression for $f_{\alpha}^{(fluc)}$ is written as

$$f_{\alpha}^{(fluc)} = -\frac{1}{2c_s^4} w_{\alpha} \sum_{k} \sum_{l} s_{kl} c_{\alpha k} c_{\alpha l} \tag{6.67}$$

in which $c_s$ is the speed of sound and related to the lattice speed $c$ as $c_s = c/\sqrt{3}$.

One of the possible treatments is that the particle distribution $f_{\alpha}^{(fluc)}$ inducing the fluctuation is added to only the lattice sites which interact with dispersed particles. However, it should be noted that it is indispensable to assess whether or not the Brownian motion is activated at a physically reasonable level, and this is true when any technique that is employed for activating Brownian motion is combined with the ordinary lattice Boltzmann method.

### 6.4.5 Hybrid method of lattice Boltzmann and Brownian dynamics for activating the particle Brownian motion

In the previous technique based on fluctuation hydrodynamics, the Brownian motion of dispersed particles is induced by the motion of ambient fluid particles. If we change the perspective, it may be possible for the dispersed particles themselves to perform the translational and rotational Brownian motion in the ambient fluid particles. This idea suggests that it may be feasible to apply the Brownian dynamics method to the lattice Boltzmann method for a particle suspension [23]. This hybrid-type method does not require any changes in the basic equation (Eq. (6.55)) of the lattice Boltzmann method or does not require the addition of any new terms to Eq. (6.55).

As previously described in Section 6.3.2, the translational Brownian motion for an axisymmetric particle is expressed as

$$m\frac{d^2\boldsymbol{r}_{\|}}{dt^2}=\boldsymbol{F}_{\|}+\boldsymbol{F}_{\|}^{fl}+\boldsymbol{F}_{\|}^{B},\qquad m\frac{d^2\boldsymbol{r}_{\perp}}{dt^2}=\boldsymbol{F}_{\perp}+\boldsymbol{F}_{\perp}^{fl}+\boldsymbol{F}_{\perp}^{B} \tag{6.68}$$

in which the subscripts || and ⊥ imply the quantities parallel and normal to the particle axis direction respectively, $m$ is the mass of the particle, and $\boldsymbol{F}$ is the non-random force acting on the particle. The viscous force exerted by the ambient fluid, $\boldsymbol{F}^{fl}$, is evaluated in the present lattice Boltzmann method, from the interaction of a dispersed particle with the ambient fluid particles in a manner similar to that explained in Section 6.4.3. The random forces inducing the translational Brownian motion, $\boldsymbol{F}_{\|}^{B}=F_{\|}^{B}\boldsymbol{e}$ and $\boldsymbol{F}_{\perp}^{B}=F_{\perp 1}^{B}\boldsymbol{e}_{\perp 1}+F_{\perp}^{B}\boldsymbol{e}_{\perp 2}$, are required to satisfy the stochastic characteristics shown in Eqs. (6.36) and (6.37).

Similarly, the rotational Brownian motion about a line, through the particle center, normal to the particle axis direction, is governed by the following expression:

$$I\frac{d^2\boldsymbol{\varphi}_{\perp}}{dt^2}=\boldsymbol{T}_{\perp}+\boldsymbol{T}_{\perp}^{fl}+\boldsymbol{T}_{\perp}^{B} \tag{6.69}$$

in which $\boldsymbol{\varphi}_{\perp}$ is defined as $d\boldsymbol{\varphi}_{\perp}/dt=\boldsymbol{\omega}_{\perp}$, $I$ is the inertia moment of the particle, $\boldsymbol{\omega}_{\perp}$ is the angular velocity vector, $\boldsymbol{T}_{\perp}$ is the non-Brownian torque acting on the particle, and $\boldsymbol{T}_{\perp}^{fl}$ is the viscous torque arising from the interaction with the ambient fluid particles. Furthermore, $\boldsymbol{T}_{\perp}^{B}=T_{\perp 1}^{B}\boldsymbol{e}_{\perp 1}+T_{\perp}^{B}\boldsymbol{e}_{\perp 2}$ is the random torque inducing the rotational Brownian motion and is required to satisfy the stochastic characteristics shown in Eqs. (6.43).

The random motion of the dispersed particles is expected to be activated by introducing the random forces and torques in Eqs. (6.36), (6.37) and (6.43), which are generated in the usual way based on the normal distribution with each standard deviation [1]. The flow field around the particles is solved by means of the ordinary lattice Boltzmann method. This method for activating the Brownian motion of the dispersed particles may be called a "hybrid-type" simulation method.

Finally, as already pointed in Section 6.4.4, the above hybrid-type method does not necessarily ensure that the Brownian motion of the dispersed particles is activated at a physically reasonable level. Hence, it may be necessary to use an alternative technique in order to improve the activation level of the Brownian motion.

## 6.5 Multi-particle collision dynamics method

In addition to the lattice Boltzmann method discussed above in Section 6.4, the multi-particle collision dynamics method (or stochastic rotation dynamics method) enables one to simulate the motion of dispersed particles and the flow field simultaneously [24–26]. Similar to dissipative particle dynamics, this method simulates coarse-grained particles or virtual fluid particles that are regarded as groups or clusters of solvent molecules, as shown in Fig. 6.12. The motion of these virtual fluid particles will have a characteristic time smaller than but similar to that of the dispersed particles. This characteristic makes dissipative particle and multi-particle collision dynamics methods significantly useful for simulating a particle suspension. The dispersed particles are simulated by an ordinary molecular dynamics-like method, and the ambient flow field is solved by the above-mentioned methods based on the virtual fluid particles. Since the virtual fluid particles interact with the dispersed particles, the multi-body hydrodynamic interactions among the dispersed particles will be reflected through these interactions. Here we address only the multi-particle collision dynamics method that may be a hopeful approach in addition to the dissipative particle dynamics method as a microscopic or mesoscopic simulation technique for a particle suspension.

The lattice Boltzmann method may require an additional technique for its application to a particle suspension, since dispersed particles generally perform Brownian motion in a colloidal dispersion and the ordinary lattice Boltzmann method does not have such a term in the basic equations. In contrast, in the multi-particle collision dynamics, the thermal motion of virtual particles is induced by an inherent characteristic included in the basic equation and, as already pointed out, this random motion induces

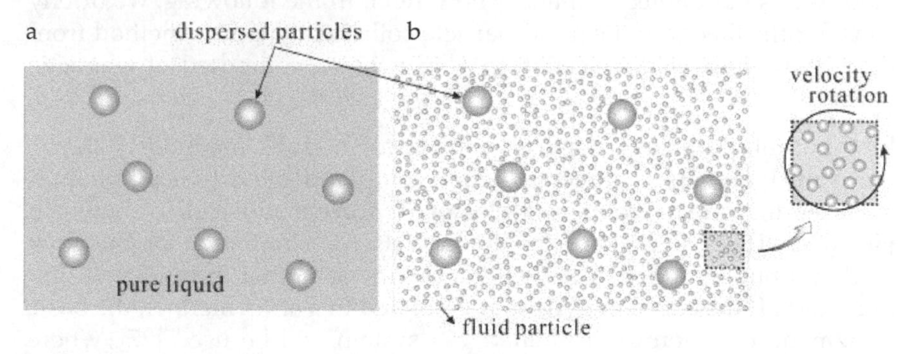


**Figure 6.12.** Different treatment of liquid in particle-based simulations: (a) continuum liquid medium and (b) coarse-grained fluid particles.

the Brownian motion of dispersed particles through the interactions between fluid and dispersed particles. Another characteristic of the multi-particle collision dynamics method is that this method treats the position and translational velocity of virtual fluid particles while the lattice Boltzmann method treats the particle distribution function and not the motion of the fluid particles themselves. In a microscopic or mesoscopic simulation of a pure liquid, the way of modeling the basic equation of the virtual fluid particles leads to a different approach such as multi-particle collision dynamics and dissipative particle dynamics.

If this method is applied to a suspension of magnetic particles, the magnetic particles are simulated by a molecular dynamics-like method rather than Brownian dynamics, and the flow field of the fluid phase is solved by the multi-particle collision dynamics method. The motion of the magnetic particles is, of course, influenced by the interactions with the ambient fluid particles, and therefore the magnetic particles are indirectly influenced by each other through the motion of the ambient fluid particles, thus leading to multi-body hydrodynamic interactions among magnetic particles. The application of this method does not require the assumption of a linear flow field such as a simple shear flow, which is generally employed in the ordinary Brownian dynamics method for analyzing the rheology of a suspension. Moreover, the interaction between fluid and magnetic particles arises through the dynamic interaction at the particle level rather than using an analytical solution regarding this interaction. Hence, it is relatively straightforward to apply this method to a suspension composed of non-spherical magnetic particles such as rod-like and disk-like particles. From these characteristics, the multi-particle collision dynamics method seems to be quite a desirable approach for analyzing the behavior of the magnetic particles in a suspension under the situation of an arbitrary flow field as it is not limited to a linear flow field. In the following, we briefly consider the theory of the multi-particle collision dynamics method from the viewpoint of its application to a suspension composed of magnetic particles.

In the multi-particle collision dynamics method, as mentioned above, the pure liquid is regarded as being composed of virtual point fluid particles that are a group or a cluster of solvent molecules, shown in Fig. 6.12. These fluid particles are objects for simulation of the flow field of a pure liquid and they interact with each other through collision dynamics. If the collision dynamics is treated in a strict manner, the basic Boltzmann equation for a rarefied gas system will be used [27], where the collision term of gas molecules (particles) is expressed in a complex integral expression. This integral term is significantly difficult to treat analytically and therefore the method of treating this integral term leads to an approximate modeling or a numerically-solving approach. The latter approach is the direct simulation Monte Carlo method where the collision term is approximately expressed by a stochastic technique [28]. The former approach is that the integral term of the collision dynamics is rendered into a simple mathematical expression; the most famous modeling is the BGK collision model [27]. In the multi-particle collision dynamics method, the treatment of the collision dynamics among virtual fluid particles is described in a stochastic manner which is very similar in concept to that of the direct simulation Monte Carlo method. In the following we show the basic equations of the multi-particle collision dynamics method.

The fluid system is assumed to be composed of a number $N$ of point fluid particles with mass $m$ and an arbitrary fluid particle $i$ is described by the position vector $\boldsymbol{r}_i$ and velocity vector $\boldsymbol{v}_i$. The multi-particle collision dynamics is performed using two procedures known by the streaming procedure and the collision procedure. In a streaming step, the particle position $\boldsymbol{r}_i(t + \Delta t)$ at time $t + \Delta t$ after a small time interval $\Delta t$ is evaluated as

$$\boldsymbol{r}_i(t+\Delta t) = \boldsymbol{r}_i(t) + \Delta t\ \boldsymbol{v}_i(t) \tag{6.70}$$

This is a finite difference scheme for the equation $\boldsymbol{v}_i(t) = d\boldsymbol{r}_i(t)/dt$.

The collision procedure treats the momentum exchange through the collision among fluid particles during the time interval $\Delta t$ and after this step the velocities of all the particles are updated. An appropriate treatment of the collision will give rise to a physically reasonable result, that is, an accurate flow field that is equivalent to the solution of the Navier-Stokes equation. The whole system is divided into numerous small cells where fluid particles are located in a cell as determined by the fluid particle position. In the multi-particle collision dynamics model (or the stochastic rotation model), the fluid particle of interest interacts with the other fluid particles in the cell to which the particles belong. The size of each cubic cell is typically chosen such that the mean number of fluid particles located in a cell is approximately 3 to 20 where the number is not constant but fluctuates with advancing time steps. If we employ the notation $N_c$ for the number of virtual fluid particles in a certain cell at a certain time step $t$ and $\boldsymbol{u}_c(t)$ for the center-of-mass velocity of the cell, then the velocity of the fluid particle $\boldsymbol{v}_i(t + \Delta t)$ at the next time step $t + \Delta t$ is expressed according to the multi-particle collision dynamics as [24–26]

$$\boldsymbol{v}_i(t+\Delta t) = \boldsymbol{u}_c(t) + \boldsymbol{R}\cdot(\boldsymbol{v}_i(t) - \boldsymbol{u}_c(t)) \tag{6.71}$$

in which $u_c(t)$ is obtained by averaging the velocities of the fluid particles belonging to the cell and expressed as

$$u_c(t) = \frac{1}{N_c} \sum_{i=1}^{N_c} v_i(t) \tag{6.72}$$

Also, $R$ is the stochastic rotation matrix about a certain axis of the normal coordinate system. This is the key quantity of the multi-particle collision dynamics and it is noted that each cell has its own rotation matrix that is stochastically determined at every time step, i.e., is not constant with advancing time steps. The axis of the cell for the rotation is randomly chosen with probability 1/3 and the velocities of the fluid particles relative to the center-of-mass velocity of the cell are all rotated about the axis by a specified constant angle $\pm\alpha$ with probability 1/2 for each direction (clockwise or counter-clockwise). In many cases, a value between $\pi/2$ and $\pi$ is employed as the angle $\alpha$.

As shown in Fig. 6.13, the axis for the rotation is chosen at random from the $x$-, $y$- and $z$-axis with equal probability 1/3. If the notation $R_x$, $R_y$ and $R_z$ is used for the rotation about the $x$-, $y$- and $z$-axis, respectively, then these expressions are written for the rotation angle $\alpha$ as

$$R_x = \begin{bmatrix} 1 & 0 & 0 \\ 0 & \cos\alpha & -\sin\alpha \\ 0 & \sin\alpha & \cos\alpha \end{bmatrix}, R_y = \begin{bmatrix} \cos\alpha & 0 & \sin\alpha \\ 0 & 1 & 0 \\ -\sin\alpha & 0 & \cos\alpha \end{bmatrix}, R_z = \begin{bmatrix} \cos\alpha & -\sin\alpha & 0 \\ \sin\alpha & \cos\alpha & 0 \\ 0 & 0 & 1 \end{bmatrix} \tag{6.73}$$

Using these rotation matrices, it is quite straightforward to evaluate the matrix manipulation in Eq. (6.71). If the unit vector of the chosen axis is denoted by $e$, Eq. (6.71) is written in different form as

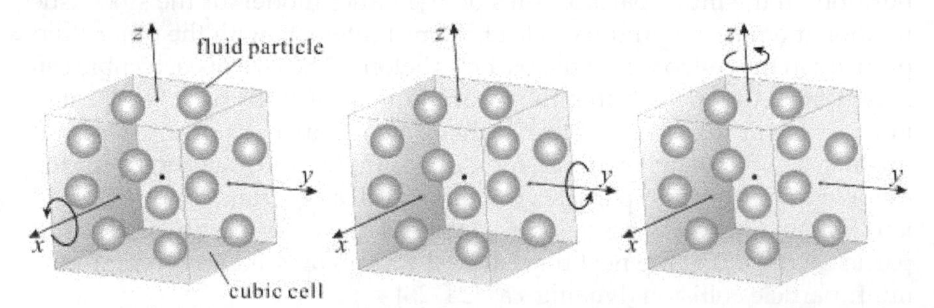


**Figure 6.13.** Rotation about each axis with equal probability 1/3 regarding the rotational treatment of the velocities of virtual fluid particles in each cell.

$$\boldsymbol{v}_i(t+\Delta t)=\boldsymbol{u}_c(t)+(\boldsymbol{v}_{i\perp}(t)-\boldsymbol{u}_{c\perp}(t))\cos\alpha+\boldsymbol{e}\times(\boldsymbol{v}_{i\perp}(t)-\boldsymbol{u}_{c\perp}(t))\sin\alpha+(\boldsymbol{v}_{i||}(t)-\boldsymbol{u}_{c||}(t)) \tag{6.74}$$

in which the subscripts || and ⊥ imply the components of the vectors that are parallel and normal to the axis direction ***e***, respectively. Equation (6.74) comes from the fact that only the component normal to the selected axis is related to the rotation procedure and the other component does not change for this rotation. An arbitrary vector ***a*** is divided into the two components parallel and normal to the unit vector ***e***, expressed as

$$\boldsymbol{a}_{||}=(\boldsymbol{e}\cdot\boldsymbol{a})\boldsymbol{e},\quad \boldsymbol{a}_{\perp}=\boldsymbol{a}-\boldsymbol{a}_{||}=\boldsymbol{a}-(\boldsymbol{e}\cdot\boldsymbol{a})\boldsymbol{e} \tag{6.75}$$

The multi-particle collision dynamics is characterized by several parameters from a simulation point of view. The first parameter is the mean free path $\lambda$ of fluid particles relative to the cell size $a_0$:

$$\lambda=\frac{\Delta t\sqrt{kT/m}}{a_0} \tag{6.76}$$

in which $k$ is Boltzmann's constant, $T$ is the system temperature and $m$ is the mass of a fluid particle. If the value of $\lambda$ is much smaller than unity, the same fluid particles repeatedly collide with each other with advancing time. This may induce a serious problem regarding the validity of the model for multi-particle collision dynamics as the molecular chaos assumption breaks down. The second parameter is the mean number $\gamma$ of fluid particles per collision cell. If the value of this parameter is too small, the collisions among fluid particles may not reflect the fluid properties in a sufficiently proper manner. By taking into account computational efficiency, the value of $\gamma \le 10$ has been recommended by several researchers [29], in order to obtain reasonable physical results. The third parameter is the rotation angle $\alpha$. Larger values of $\alpha$ imply larger Schmidt numbers. The Schmidt number *Sc* is defined as the ratio of momentum transport to mass diffusivity, i.e., $Sc = \nu/D$, where $\nu$ is the kinematic viscosity and $D$ is the diffusion coefficient. The value of $\pi/2 \le \alpha \le (170/180)\pi$ has been recommended by several researchers [29].

It is noted that during the multi-particle collision dynamics procedures, the conservation of the local mass, translational momentum and kinetic energy are satisfied at each collision step, moreover, the single particle velocity distribution evolves to a Maxwell velocity distribution. Lastly it is also noted that a random shift of the collision cell of interest must be performed at every collision procedure in order to ensure the Galilean invariance in certain situations [30–32].

# Bibliography


[1] Allen, M. P. and Tildesley, D. J. 1987. Computer Simulation of Liquids, Clarendon Press, Oxford.

[2] McClelland, B. J. 1973. Statistical Thermodynamics, Chapman and Hall, London.

[3] Metropolis, N., Rosenbluth, A. W., Rosenbluth, M. N. and Teller, A. 1953. Equation of state calculations by fast computing machines. J. Chem. Phys. 21: 1087–1092.

[4] Brady Bossis, G. and Brady, J. F. 1984. Dynamic simulation of sheared suspensions. I. General method. J. Chem. Phys. 80: 5141.

[5] Brady, J. F. and Bossis, G. 1985. The rheology of concentrated suspensions of spheres in simple shear flow by numerical simulation. J. Fluid Mech. 155: 105–129.

[6] Ozaki, M., Senna, M., Koishi, M. and Honda, H. 2000. Particles of specific functions. pp. 662–682. *In*: T. Sugimoto (ed.). Fine Particles. Marcel Dekker, New York.

[7] Van der Beek, D., Reich, H., Van der Schoot, P., Dijkstra, M., Schilling, T., Vink, R., Schmidt, M., Van Roij, R. and Lekkerkerker, H. 2006. Isotropic-nematic interface and wetting in suspensions of colloidal platelets. Phys. Rev. Lett. 97: 087801.

[8] Hockney, R. W. 1970. The potential calculation and some applications. Methods Comput. Phys. 9: 136–211.

[9] Brenner, H. 1974. Rheology of a dilute suspension of axisymmetric Brownian particles. Int. J. Multiphase Flow. 1: 195–341.

[10] Kim, S. and Karrila, S. J. 1991. Microhydrodynamics: Principles and Selected Applications, Butterworth- Heinemann, Stoneham.

[11] Bird, R. B., Armstrong, R. C. and Hassager, O. 1977. Dynamics of Polymeric Liquids, Vol. 1, Fluid Mechanics, John Wiley and Sons, New York.

[12] Ermak, D. L. and McCammon, J. A. 1978. Brownian dynamics with hydrodynamic interactions. J. Chem. Phys. 69: 1352.

[13] Satoh, A. 2015. Brownian dynamics simulations with spin Brownian motion on the negative magneto-rheological effect of a rod-like hematite particle suspension. Molec. Phys. 113: 656–670.

[14] Hoogerbrugge, P. J. and Koelman, J. M. V. A. 1992. Simulating microscopic hydrodynamic phenomena with dissipative particle dynamics. Europhys. Lett. 19: 155–160.

[15] Koelman, J. M. V. A. and Hoogerbrugge, P. J. 1993. Dynamic simulations of hard-sphere suspensions under steady shear. Europhys. Lett. 21: 263.

[16] Espanol, P. and Warren, P. 1995. Statistical mechanics of dissipative particle dynamics. Europhys. Lett. 30: 191.

[17] Succi, S. 2001. The Lattice Boltzmann Equation for Fluid Dynamics and Beyond, Clarendon Press, Oxford.

[18] Rothman, D. H. and Zaleski, S. 1997. Lattice-Gas Cellular Automata, Simple Models of Complex Hydrodynamics, Cambridge University Press, Cambridge.

[19] Ladd, A. J. C. 1993. Short-time motion of colloidal particles: numerical simulation via a fluctuating lattice-Boltzmann equation. Phys. Rev. Lett. 70: 1339–1342.

[20] Ladd, A. J. C. 1994. Numerical simulations of particulate suspensions via a discretized Boltzmann equation. Part 1. Theoretical foundation. J. Fluid Mech. 271: 285–309.

[21] Yu, D., Mei, R., Luo, L. -S. and Shyy, W. 2003. Viscous flow computations with the method of lattice Boltzmann equation. Prog. Aerospace Sci. 39: 329–367.

[22] Bouzidi, M., Firdaouss, M. and Lallemand, P. 2001. Momentum transfer of a Boltzmann-lattice fluid with boundaries. Phys. Fluids. 13: 3452–3459.

[23] Satoh, A. 2012. On the method of activating the Brownian motion for application of the lattice Boltzmann method to magnetic particle dispersions. Molec. Phys. 110: 1–15.

[24] Malevanets, A. and Kapral, R. 1999. Mesoscopic model for solvent dynamics. J. Chem. Phys. 110: 8605.

[25] Malevanets, A. and Kapral, R. 2000. Solute molecular dynamics in a mesoscale solvent. J. Chem. Phys. 112: 7260.
[26] Gompper, G., Ihle, T., Kroll, D. M. and Winkler, R. G. 2009. Multi-particle collision dynamics: a particle-based mesoscale simulation approach to the hydrodynamics of complex fluids. Adv. Polym. Sci. 221: 1–91.
[27] Vincenti, W. G. and Kruger, C. H. 1965. Introduction to Physical Gas Dynamics, John Wiley and Sons, New York.
[28] Bird, G. A. 1994. Molecular Gas Dynamics and the Direct Simulation of Gas Flows, Oxford University Press, Oxford.
[29] Tomilov, A., Videcq, A., Chartier, T., Ala-Nissila, T. and Vattulaninen, I. 2012. Tracer diffusion in colloidal suspensions under dilute and crowded conditions with hydrodynamic interactions. J. Chem. Phys. 137: 014503.
[30] Ihle, T. and Kroll, D. M. 2001. Stochastic rotation dynamics: A Galilean-invariant mesoscopic model for fluid flow. Phys. Rev. E. 63: 020201.
[31] Ihle, T. and Kroll, D. M. 2003. Stochastic rotation dynamics. I. Formalism, Galilean invariance, and Green-Kubo relations. Phys. Rev. E. 67: 066705.
[32] Ihle, T. and Kroll, D. M. 2003. Stochastic rotation dynamics. II. Transport coefficients, numerics and long-time tails. Phys. Rev. E. 67: 066706.

CHAPTER 7

# Strategy of Simulations

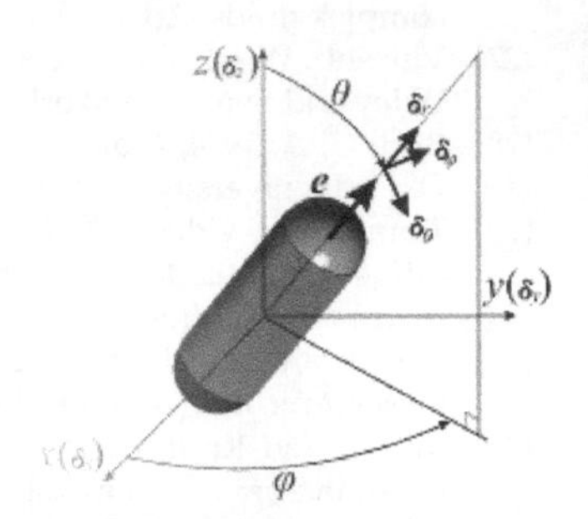


## 7.1 Generation of rotation of non-spherical particles

In this chapter, we address miscellaneous techniques for performing particle-based simulations of particle suspensions. There are a variety of techniques for developing a successful simulation program or code such as assignment of initial conditions and treatment of boundary conditions. However, we will not consider these basic techniques as they are explained sufficiently well in other textbooks, but rather consider relatively advanced strategies employed as simulation techniques. In this section, we first consider the method of randomly rotating an axisymmetric particle by a small angle, which is indispensable in developing a computational code of Monte Carlo simulations.

As shown in Fig. 7.1, the particle orientation of an axisymmetric particle is denoted by the unit vector $\boldsymbol{e}$, or specified by a polar angle $\theta$ and an azimuthal angle $\varphi$ of the orthogonal coordinate system. The unit vector $\boldsymbol{e}$ is related to the angles $\theta$ and $\varphi$ as $\boldsymbol{e} = (e_x, e_y, e_z) = (\sin\theta\cos\varphi, \sin\theta\sin\varphi, \cos\theta)$. The most straightforward way for rotation is to change the angles $(\theta, \varphi)$ to $(\theta + \delta\theta, \varphi + \delta\varphi)$ by small angle displacements $\delta\theta$ and $\delta\varphi$. However, application of this method may have significant limitations for certain situations in a magnetic particle suspension where the magnetic spherical or axisymmetric particle have a strong tendency to incline in the direction of a strong external field. This is because the constant angle displacements $\delta\theta$ and $\delta\varphi$, which are specified before starting a simulation, lead to an attempt of the rotational displacement dependent on the particle original orientation $(\theta, \varphi)$ with a non-equal probability. This shortcoming is more evident in the results of an equilibrium orientational distribution function for the situation of no applied magnetic field. Hence, we explain a more general technique for the rotation in the following, using Fig. 7.1.

As explained in Section 4.1, the particle coordinate system *XYZ* is described by the unit vectors $\boldsymbol{\delta}_r$ and ($\boldsymbol{\delta}_\theta$, $\boldsymbol{\delta}_\varphi$) that are normal to the particle axis direction $\boldsymbol{\delta}_r$ (= $\boldsymbol{e}$), as shown in Fig. 7.1. In the *XYZ* coordinate system, the particle direction is expressed as $\boldsymbol{e}^b$ = (0,0,1). If the particle direction is rotated and aligned to an arbitrary new point in the area of a small square normal to the particle direction and with side length $\delta r_\theta$, then this orientation $\boldsymbol{e}^{b\prime}$ is expressed in the particle coordinate system as

$$\boldsymbol{e}^{b\prime}=(e_x^{\ b\prime},e_y^{\ b\prime},e_z^{\ b\prime})=(\delta r_\theta(R_1-0.5),\ \delta r_\theta(R_2-0.5),\ \sqrt{1-(e_x^{\ b\prime})^2-(e_y^{\ b\prime})^2}) \tag{7.1}$$

in which $R_1$ and $R_2$ are random numbers taken from a sequence of uniform random numbers ranging from zero to unity. Employment of a small value of $\delta r_\theta$ (<< 1) ensures the attempt of rotational displacement to any point in the square area with equal probability. Hence, the new rotated direction $\boldsymbol{e}'$ in the absolute coordinate system *xyz* is obtained using the rotation matrix $\boldsymbol{R}$ in Eq. (4.3) as

$$\boldsymbol{e}'=\boldsymbol{R}^t\cdot\boldsymbol{e}^{b\prime}=\begin{bmatrix}\cos\theta\cos\varphi & -\sin\varphi & \sin\theta\cos\varphi\\ \cos\theta\sin\varphi & \cos\varphi & \sin\theta\sin\varphi\\ -\sin\theta & 0 & \cos\theta\end{bmatrix}\begin{bmatrix}e_x^{\ b\prime}\\ e_y^{\ b\prime}\\ e_z^{\ b\prime}\end{bmatrix} \tag{7.2}$$

In Monte Carlo simulations, the system energy is evaluated for the new orientation of the randomly selected particle which is generated by

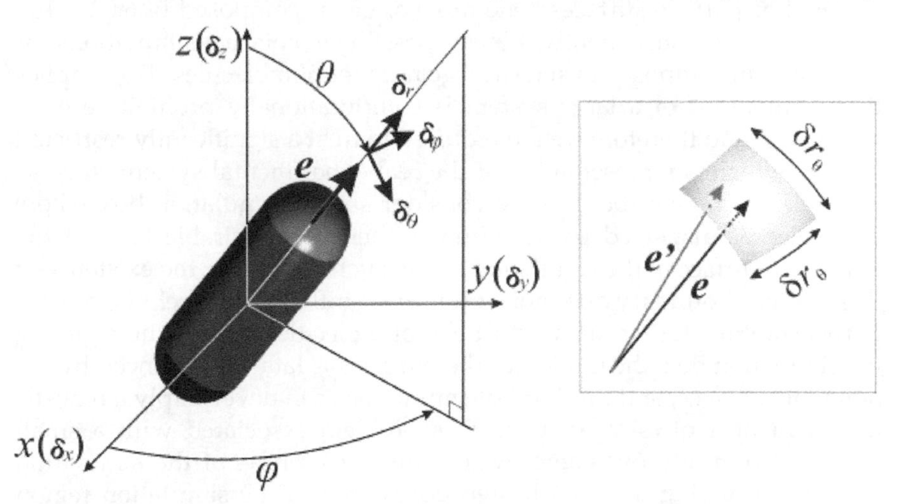


**Figure 7.1.** Rotation of an axisymmetric particle for Monte Carlo simulations.

the above-mentioned procedure and assessment regarding acceptance or rejection of the new particle orientation is conducted by the usual Metropolis method. If the new orientation is accepted, then the rotation matrix describing the relationship between the absolute and the particle coordinate systems is updated.

If the polar angle $\theta$ is defined as $0 \leq \theta < \pi$, the following treatment is necessary regarding the angles $\theta$ and $\varphi$. If a new accepted angle $\theta$ is larger than or equal to $\pi$, $(\theta, \varphi)$ are replaced with $(\theta - \pi, \varphi + \pi)$, and if $(\varphi + \pi)$ is larger than or equal to $2\pi$, then $(\varphi + \pi)$ is replaced with $(\varphi + \pi - 2\pi)$. Similar treatment is necessary if the angle $\theta$ is defined as $0 \leq \theta < \pi/2$. Moreover, if an axisymmetric particle has a magnetic moment in a direction normal to the particle axis direction, then the magnetic moment direction must be modified as a result of the particle orientation change but the attempted change of the magnetic moment direction itself is treated in a different Monte Carlo procedure. If the new particle orientation $\boldsymbol{e}'$ is determined by the above-mentioned step, then the modified direction of the magnetic moment, $\boldsymbol{n}'$, is evaluated by the method previously explained in Section 4.2.

## 7.2 Lees-Edwards boundary condition

In general simulations including ordinary numerical analysis and molecular simulations, appropriate treatments are necessary for the boundary conditions at the outer boundary surface of the simulation region and the material or particle surfaces. One mole of gas is composed of $6.02 \times 10^{23}$ molecules, and consequently it is not possible to conduct simulations for a real system composed of such a large number of molecules. This implies that simulations of a large system is computationally prohibitive even nowadays, and therefore we are required to treat a significantly restricted system, which is representative of the real experimental system, such as a simulation region (box). If we consider such a simulation box where molecules or dispersed particles move, then it is desirable to treat the boundary surface of the simulation box in such a way that the existence of this artificial boundary does not significantly influence the behavior of the particle motion. If there are no particles or molecules outside the region of the simulation box, the motion of the particles is fatally influenced by the boundary surface, and this kind of simulations can never imply a realistic treatment of a physical system. This problem associated with a small system is basically overcome by introducing replicas of the simulation box and arranging these replicated boxes around the simulation region as shown in Fig. 7.2(b). The particles are then allowed to interact with the virtual particles in replicated boxes, and consequently the influence of the boundary surface on the particle motion can largely be removed using this method. This treatment of the boundary surface is called the periodic boundary condition, as shown in Fig. 7.2(b), and is inevitably employed for almost all cases of thermodynamic equilibrium simulations.

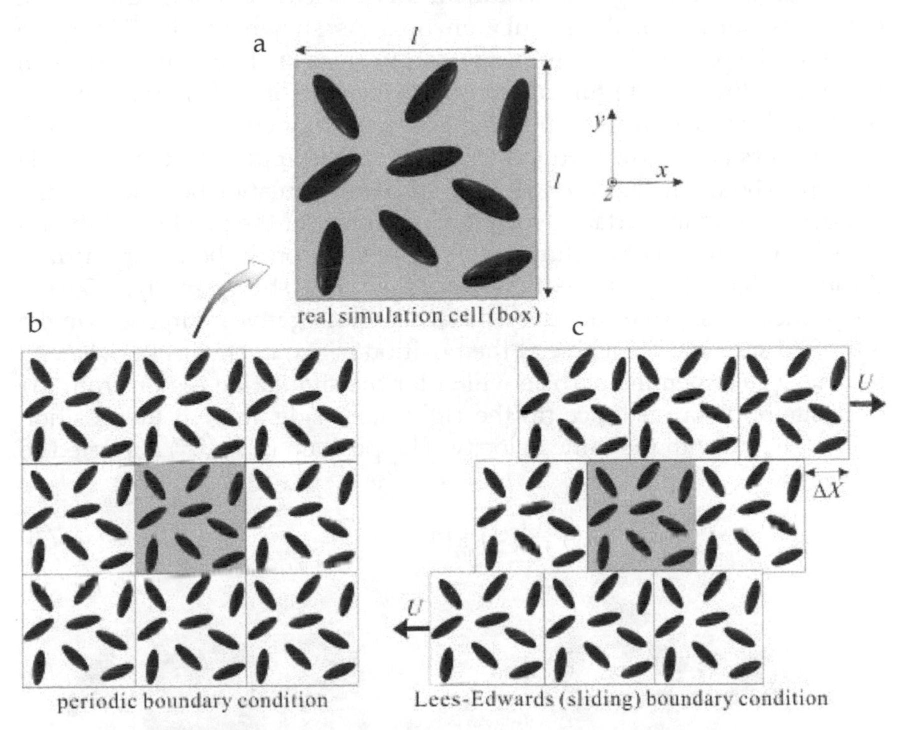


**Figure 7.2.** Simulation cell and the boundary conditions: (a) the simulation cell (box), (b) the usual periodic boundary condition and (c) the Lees-Edwards (sliding) boundary condition.

In the fields of colloid science and fluid engineering, a simple shear flow is frequently addressed in order to investigate the rheological characteristics of a particle suspension. Generation of a simple shear flow in a simulation is straightforwardly accomplished by modifying the above-mentioned periodic boundary condition. As shown in Fig. 7.2(c), the upper replicated simulation boxes are moved with a certain speed $U$ in the right direction and the lower replicated boxes are moved in the opposite direction with the same speed $U$. If the side length of the simulation box is denoted by $l$, then a shear flow will be induced with shear rate $\dot{\gamma} = U/l$. This sliding boundary condition is called the Lees-Edwards boundary condition and is widely used for generating a simple shear flow in non-equilibrium molecular dynamics simulations or in particle-based simulations for analyzing dynamic properties. In the following we discuss two typical treatments of particles, which are indispensable for successful application of the sliding boundary condition.

The first is regarding the treatment of particles crossing through a boundary surface of the simulation box. As shown in Fig. 7.3(a), we consider the particle in the upper region to cross the boundary surface in the upper direction. In this motion of the particle in the simulation box, its virtual particle in the replicated box naturally enters the simulation box by crossing the lower boundary surface. This implies that any particle crossing a boundary surface will enter into the simulation box through the opposite boundary surface. However, treatment of the position of the re-entered particle and its modified velocity depends on the boundary surface through which the particle of interest crosses. As shown in Fig. 7.3(b), if the particle crosses the boundary surface in the negative $x$-direction (in the left-hand side area) to arrive at the position $\boldsymbol{r} = (x, y, z)$ with the velocity $\boldsymbol{v} = (v_x, v_y, v_z)$, then the particle will enter the simulation region from the opposite boundary surface (in the right-hand side area) at the position $\boldsymbol{r}' = (x', y', z')$ with the same velocity. The position $(x', y', z')$ is related to $(x, y, z)$ as

$$(x', y', z') = (x + l, y, z) \tag{7.3}$$

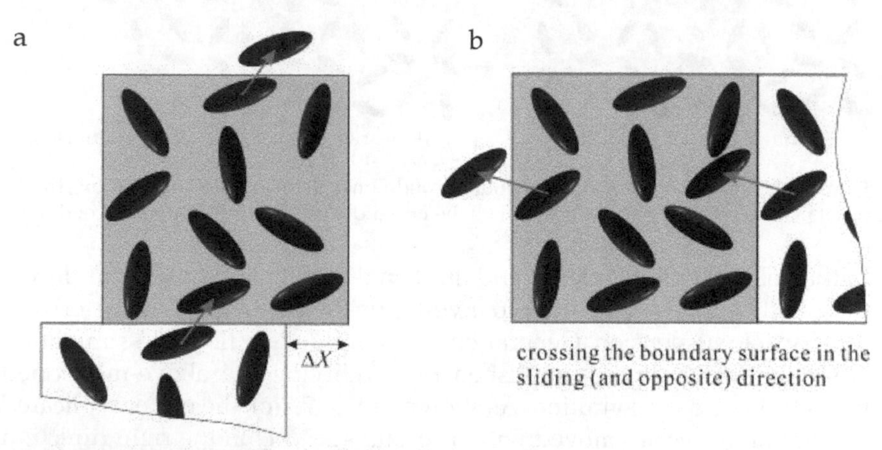


**Figure 7.3.** Treatment regarding particles crossing the boundary surfaces: (a) crossing the boundary surface in the direction normal to the sliding direction and (b) crossing the boundary surface in the sliding or the opposite direction.

This is a typical treatment used in the method of the periodic boundary condition. This procedure is also applied for the particles crossing the surfaces normal to the $z$-axis direction in Fig. 7.2(c). If the particles crosses the upper boundary surface and enter the simulation region from the lower surface as shown in Fig. 7.3(a), then the new position $\boldsymbol{r}' = (x', y', z')$ is related to the original position $\boldsymbol{r} = (x, y, z)$ just outside the upper surface as

$$(x', y', z') = (x - \Delta X, y - l, z) \tag{7.4}$$

in which the lower replicated box is assumed to be shifted in the negative $y$-direction by distance $\Delta X$. Furthermore, in this case, the original velocity $\boldsymbol{v} = (v_x, v_y, v_z)$ has to be changed into $\boldsymbol{v}' = (v_x', v_y', v_z')$, expressed as

$$(v_x', v_y', v_z') = (v_x - U, v_y, v_z) \tag{7.5}$$

in which $U$ is the sliding speed of the upper or lower replicated boxes. This treatment is conducted in a similar manner if the particles cross the boundary surface in the opposite direction.

The second treatment is regarding the calculation of the interaction energies or forces between particles. We consider the interaction of particle $i$ with particle $j$. As shown in Fig. 7.4, particle $i$ may interact with the replicated particles of particle $j$. In general, the size of the simulation box is taken as $l > 2r_{coff}$, where $r_{coff}$ is the cutoff radius for evaluating an interaction energy or force between two particles. This implies that it is sufficient to take into account only one particle among the real and virtual particles of particle $j$ as an object for evaluating an interaction energy or force. For specific discussion, we employ the arrangement of the particles shown in Fig. 7.4. If particle $j$ is apart from particle $i$ by distance $x_{ji}$ in the positive $x$-direction, and this distance is larger than the half side length $l/2$, then the particle as an object of interaction has to be switched from the real particle to the replicated virtual particle that is in the left-hand side area of particle $i$. Hence, the new relative position vector $\boldsymbol{r}_{ji}' = (x_{ji}', y_{ji}', z_{ji}')$ is expressed using the original relative position $\boldsymbol{r}_{ji} = (x_{ji}, y_{ji}, z_{ji})$ as

$$(x_{ji}', y_{ji}', z_{ji}') = (x_{ji} - l, y_{ji}, z_{ji}) \tag{7.6}$$

If the virtual particle in the upper sliding box is an object for interaction, as shown in Fig. 7.4, the new relative position vector $\boldsymbol{r}_{ji}' = (x_{ji}', y_{ji}', z_{ji}')$ is expressed as

$$(x_{ji}', y_{ji}', z_{ji}') = (x_{ji} + \Delta X, y_{ji} + l, z_{ji}) \tag{7.7}$$

Furthermore, if $x_{ji}'$ is larger than the half side length $l/2$, $x_{ji}'$ is modified as $(x_{ji}' - l)$, and similarly, if $x_{ji}'$ is smaller than $-l/2$, $x_{ji}'$ is modified as $(x_{ji}' + l)$.

Finally, it is noted that the method used in the periodic boundary condition is applicable to the present sliding boundary condition regarding the treatment of particles crossing the boundary surface and interaction with replicated particles in the $z$-axis direction.

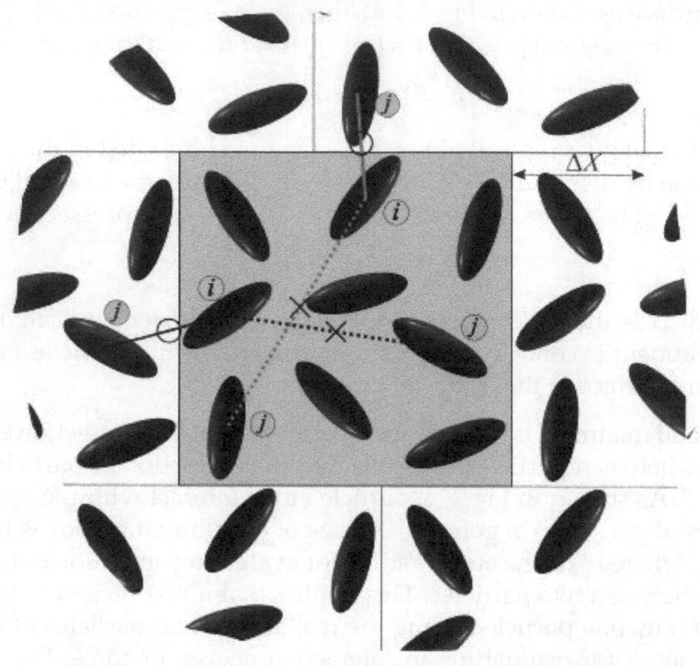


**Figure 7.4.** Treatment regarding the interactions with the neighboring particles and their replicated particles.

## 7.3 Analysis of the formation of clusters

In a strongly-interacting suspension system, magnetic particles generally aggregate to form clusters that are significantly dependent on a variety of factors such as the magnetic field strength and the magnetic interaction between particles. The cluster formation may be characterized by the internal structure of the aggregates and also by the cluster size distribution. The former characteristic is described using the radial distribution function, the pair correlation function, the orientational pair correlation function, and so forth. The latter cluster size distribution yields a direct understanding as to how large clusters are dominantly formed in the system. Moreover, the assessment of the cluster formation is necessary for performing the cluster-moving Monte Carlo method that is a desirable approach for treating a strongly-interacting system of spherical particles.

In the present section, we discuss the way of assessing the cluster formation of particles from the viewpoint of developing a simulation program.

A sophisticated method for assessing the cluster formation is the cluster analysis method [1, 2, 3]. In this method, the energy analysis of possible sets of clusters is performed in determining the clusters that can be regarded as being formed by a physically binding force. This method may be a desirable procedure for defining the formation of a cluster but unfortunately it is a time-consuming procedure. Hence, a simpler method based on the criterion of using the particle-particle distance alone is frequently used and may be preferred. This criterion for assessing cluster formation may be straightforward and physically reasonable because any particles without attractive forces which become accidentally located in the range of a cluster will quite easily drift away again with advancing time steps. Therefore, from a simulation point of view, the assessment of cluster formation using the particle-particle distance is understood to be a useful technique and in the following we will focus on this method for developing a simulation code.

In the assessment procedure for the cluster formation, we use the following variables:

*Unscan_p_name*(*) : saves the name of particles which are not yet scanned in the assessment procedure and therefore do not belong to any cluster.

*N_unscan_p_name* : saves the number of these un-scanned particles.

Figure 7.5 shows an example of a snapshot of $N = 15$ number of particles for explaining the procedure for assessing cluster formation. The main idea of the assessment procedure is first to take a particle from the table of un-scanned particles which are saved in the variable *Unscan_p_name*(*), to regard this particle as a pivot, and then to search for the particles that constitute a cluster from this pivot particle. After the name table of these particles constituting the cluster is constructed, the pivot particle is shifted to one of the cluster-constituting particles, and the same procedure is repeated. In this procedure, the name table of the cluster-constituting particles is updated if a new particle is found to be related to the pivot particle. The above-mentioned procedure for the assessment of the cluster formation is repeated until all the cluster-constituting particles have been used as a pivot for assessing the cluster formation.

In the following we apply the cluster assessment procedure to the example of a system snapshot shown in Fig. 7.5.

(I) Initial assignment of the name table of un-scanned particles:

*Unscan_p_name(i)=i* (*i*=1, 2, ..., *N*, where *N* is 15 in this example)

(II) Assessment of the cluster formation for cluster 1:

(II)-1 Set particle 1 (=*Unscan_p_name*(1)) as a pivot and find the related particles constituting cluster 1 from the pivot particle:

Table of related particles is 1, 2 and 9

(II)-2 Set particle 2 as a pivot and find the related particles:

Table of related particles is 1, 2, 9 and 7

(II)-3 Set particle 9 as a pivot and find the related particles:

Table of related particles is 1, 2, 9 and 7 (no related particle is found)

(II)-4 Set particle 7 as a pivot and find the related particles:

Table of related particles is 1, 2, 9 and 7 (no related particle is found)

(II)-5 Put the name table in order:

Table of related particles is 1, 2, 7 and 9

(II)-6 Final result regarding the formation of cluster 1 is obtained as

*Cls_const_p_name*(1,1)=1, *Cls_const_p_name*(2,1)=2,
*Cls_const_p_name*(3,1)=7, *Cls_const_p_name*(4,1)=9
*N_cls_const_*p(1)=4

in which the variable *Cls_const_p_name*(*, *j*) saves the name of constituting particles for cluster *j*, and *N_cls_const_p*(*j*) saves the number of constituting particles.

(III) Update the name table of un-scanned particles:

*Unscan_p_name*(1)=3, *Unscan_p_name*(2)=4,
*Unscan_p_name*(3)=5,
*Unscan_p_name*(4)=6, *Unscan_p_name*(5)=8,
*Unscan_p_name*(6)=10,
*Unscan_p_name*(7)=11, *Unscan_p_name*(8)=12,
*Unscan_p_name*(9)=13,
*Unscan_p_name*(10)=14, *Unscan_p_name*(11)=15
*N_unscan_p_name*=11 (=15–4)

(IV) Assessment of the cluster formation for cluster 2:

(IV)-1 Set particle 3 (=*Unscan_p_name*(1)) as a pivot and find the related particles constituting cluster 2 from the pivot particle:

Table of related particles is 3, 6, 8, 10 and 11

(IV)-2 Set particle 6 as a pivot and find the related particles:

Table of related particles is 3, 6, 8, 10, 11 and 5

(IV)-3 Set particle 8 as a pivot and find the related particles:

Table of related particles is 3, 6, 8, 10, 11 and 5 (no related particle is found)

(IV)-4 Set particle 10 as a pivot and find the related particles:

Table of related particles is 3, 6, 8, 10, 11 and 5 (no related particle is found)

(IV)-5 Set particle 11 as a pivot and find the related particles:

Table of related particles is 3, 6, 8, 10, 11 and 5 (no related particle is found)

(IV)-6 Set particle 5 as a pivot and find the related particles:

Table of related particles is 3, 6, 8, 10, 11, 5 and 15

(IV)-7 Set particle 15 as a pivot and find the related particles:

Table of related particles is 3, 6, 8, 10, 11, 5 and 15 (no related particle is found)

(IV)-8 Put the name table in order:

Table of related particles is 3, 5, 6, 8, 10, 11 and 15

(IV)-9 Final result regarding the formation of cluster 2 is obtained as

*Cls_const_p_name*(1,2)=3, *Cls_const_p_name*(2,2)=5,

*Cls_const_p_name*(3,2)=6, *Cls_const_p_name*(4,2)=8,

*Cls_const_p_name*(5,2)=10, *Cls_const_p_name*(6,2)=11,

*Cls_const_p_name*(7,2)=15

*N_cls_const_p*(2)=7

(V) Update the name table of un-scanned particles:

*Unscan_p_name*(1)=4, *Unscan_p_name*(2)=12,

*Unscan_p_name*(3)=13,

*Unscan_p_name*(4)=14

*N_unscan_p_name*=4 (=11–7)

(VI) Assessment of the cluster formation for cluster 3:

(VI)-1 Set particle 4 (=*Unscan_p_name*(1)) as a pivot and find the related particles constituting cluster 3 from the pivot particle:

Table of related particles is 4, 12 and 14

(VI)-2 Set particle 12 as a pivot and find the related particles:

Table of related particles is 4, 12 and 14 (no related particle is found)

(VI)-3 Set particle 14 as a pivot and find the related particles:

Table of related particles is 4, 12 and 14 (no related particle is found)

(VI)-4 Put the name table in order:

Table of related particles are 4, 12 and 14

(VI)-5 Final result regarding the formation of cluster 3 is obtained as

*Cls_const_p_name*(1,3)=4, *Cls_const_p_name*(2,3)=12,

*Cls_const_p_name*(3,3)=14

*N_cls_const_p*(3)=3

(VII) Update the name table of un-scanned particles:

*Unscan_p_name*(1)=13

*N_unscan_p_name*=1 (=4–3)

(VIII) Assessment of the cluster formation for cluster 4:

(VIII)-1 Set particle 13 (=*Unscan_p_name*(1)) as a pivot and find the related particles constituting cluster 4 from the pivot particle:

Table of related particles is 13 (no related particle is found)

(VIII)-2 Put the name table in order:

Table of related particles is 13

(VIII)-3 Final result regarding the formation of cluster 4 is obtained as

*Cls_const_p_name*(1,4)=13

*N_cls_const_p*(4)=1

(IX) Update the name table of un-scanned particles:

*Unscan_p_name*(1)=None

*N_unscan_p_name*=0 (=1–1)

(X) The cluster assessment is now completed and the final result is obtained as

cluster 1: 1, 2, 7, 9

cluster 2: 3, 5, 6, 8, 10, 11, 15

cluster 3: 4, 12, 14

cluster 4: 13

The number of the clusters is 4 and saved in the variable *N_cls* as *N_cls*=4.

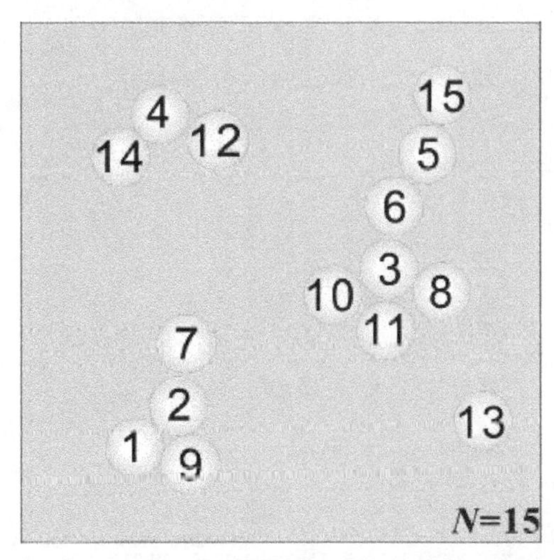


**Figure 7.5.** An example of a snapshot for assessment of the cluster formation.

For instance, if particle 1 is a pivot particle, the virtual (replicated) particles of 6, 5, 15 and so forth are objects for evaluating the distance between particle 1 and the partner particle. The specification whether a real or its virtual particle is a member of the cluster is significantly important in moving the cluster according to the Metropolis method in the cluster-moving Monte Carlo method, which will be explained in the following section.

Finally, we show an example of a subroutine for making the name table of cluster-constituting particles, which is written in FORTRAN language. This subroutine is directly applicable to a suspension of particles with any shape such as sphere, spherocylinder, spheroid or cube, because the procedure of evaluation of the distance between the two particles of interest is carried out in another subroutine and the present subroutine just uses this result as to whether or not these particles are within the criterion distance for the cluster formation.

```
001 C*********************************************************************
002 C                                                                    *
003 C    THE FOLLOWING SUBROUTINE IS FOR ASSESSING THE CLUSTER FORMATION *
004 C                                                                    *
005 C                                               WRITTEN BY A. SATOH *
006 C*********************************************************************
007 C       N       : NUMBER OF PARTICLES
008 C       RX(I),RY(I),RZ(I) : PARTICLE POSITION OF PARTICLE I
009 C
010 C       RCOFF  : CUTOFF RADIUS FOR CALCULATION OF INTERACTION ENERGIES
011 C       XL,YL,ZL : DIMENSIONS OF SIMULATION REGION
012 C       RCLSTR   : A CRITERION DISTANCE FOR ASSESSING THE CLUSTER
013 C                  FORMATION
014 C       COLORX(N), COLORY(N), COLORZ(N): ARE USED IN CLUSTER MOVEMENT
015 C
016 C       CLSN     : NAME OF CLUSTER (*) OR NUMBER OF CLUSTERS FORMED
017 C                  IN THE SYSTEM
018 C       CLSMX(*) : NUMBER OF CLUSTER-CONSTITUTING PARTICLES
019 C                  OF *-TH CLUSTER
020 C       CLS(+,*) : NAME OF CLUSTER-CONSTITUTING PARTICLES OF *-TH CLUSTER
021 C                                             + : 1,2,3,...,CLSMX(*)
022 C
023 C       PNUM1(#): NAME TABLE OF UN-SCANNED PARTICLES
024 C                                             # : 1,2,3,...,PMX1
025 C       PMX1     : NUMBER OF UN-SCANNED PARTICLES
026 C*********************************************************************
027
028       SUBROUTINE CLSFORM( N )
029 C
030       IMPLICIT REAL*8 (A-H,O-Z), INTEGER (I-N)
031 C
032       COMMON /BLOCK1/  RX   , RY   , RZ
033       COMMON /BLOCK2/  XL   , YL   , ZL
034       COMMON /BLOCK5/  CLSN , CLSMX , CLS , RCLSTR
035       COMMON /BLOCK6/  COLORX , COLORY , COLORZ
036 C
037       INTEGER      NN , TT
038       PARAMETER( NN=1000 , TT=1000 )
039 C
040       REAL*8    RX(NN) , RY(NN) , RZ(NN)
041 C
042       INTEGER  CLSN   , CLSMX(NN) , CLS(TT,NN)
043       INTEGER  COLORX(NN), COLORY(NN), COLORZ(NN)
044 C
045       INTEGER  PMX1   , PNUM1(NN)
046       INTEGER  N1 , PTCL , PTCLMN , IPTCLMN
047       INTEGER  III, KK , JJ , JJJ , JJE , II1 , JJ1
048       INTEGER  CLRX0 , CLRY0 , CLRZ0 , CLRXJ , CLRYJ , CLRZJ
049       REAL*8   RXI, RYI, RZI, RXJ, RYJ, RZJ  , RCLSTRSQ
050 C
051       LOGICAL  CLSTRUE
052 C                                 -------------------------------------
053 C                                   ASSESSMENT OF CLUSTER FORMATION
054 C                                 -------------------------------------
055       DO 10 I=1,N
056         COLORX(I) = 0
057         COLORY(I) = 0
058         COLORZ(I) = 0
059    10 CONTINUE
060 C
061       RCLSTRSQ = RCLSTR**2
062       N1    = N + 1
063       CLSN = 0
064       PMX1 = N
065       DO 499 J=1,N
066         PNUM1(J) = J
067   499 CONTINUE
068 C
069 C
060       DO 600 III=1,N
071 C
072 C
073         CLSN          = CLSN + 1
074         I             = PNUM1(1)
075         CLSMX(CLSN)   = 1
076         CLS(1,CLSN)   = I
077 C                   --- PARTICLE I IS SET AS AN INITIAL PIVOT ---
078         KK     = 0
079   500   KK     = KK + 1
080         PTCL   = CLS(KK,CLSN)
```

· COLORX(*), COLORY(*) and COLORZ(*) are used for the cluster-moving Monte Carlo method, and therefore are not referred to in the present subroutine.

· The name list of the un-scanned particles, PNUM1(*), is initially set as PNUM1(J)=J (J=1, 2, ..., N).

· The cluster, CLSN, is treated for the assessment of the cluster formation.
· The name of the first particle of the cluster is saved at CLS(1,CLSN).

· The present pivot particle is PTCL, which is taken from CLS(KK,CLSN).

```
081        CLRX0 = COLORX(PTCL)
082        CLRY0 = COLORY(PTCL)
083        CLRZ0 = COLORZ(PTCL)
084 C                                        --- PTCL IS THE PRESENT PIVOT ---
085 C
086        DO 550 J=1, N
087 C                                        --- ASSESSMENT PROCEDURE ---
088          IF( J .EQ. PTCL )  GOTO 550
089 C
090          DO 530 JJJJ=1,CLSMX(CLSN)
091            IF( J .EQ. CLS(JJJJ,CLSN) ) GOTO 550
092   530    CONTINUE
093 C
094          RXI = RX(PTCL)
095          RYI = RY(PTCL)
096          RZI = RZ(PTCL)
097          RXJ = RX(J)
098          RYJ = RY(J)
099          RZJ = RZ(J)
100          CALL CLSJUDGE( RXI, RYI, RZI , RXJ, RYJ, RZJ,
101      &                  RCLSTR, CLSTRUE, XL, YL, ZL,
102      &                  CLRX0, CLRY0, CLRZ0, CLRXJ, CLRYJ, CLRZJ )
103 C
104          IF( CLSTRUE ) THEN
105            CLSMX(CLSN)              = CLSMX(CLSN) + 1
106            CLS(CLSMX(CLSN),CLSN)    = J
107            COLORX(J) = CLRXJ
108            COLORY(J) = CLRYJ
109            COLORZ(J) = CLRZJ
110          END IF
111 C
112   550  CONTINUE
113 C
114 C
115        IF( KK .NE. CLSMX(CLSN) )  GOTO 500
116 C
117 C                                              -----
118 C
119 C                         --- SET THE DATA OF CLS(*,CLSN) IN ORDER ---
120 C
121        IF( CLSMX(CLSN) .LE. 2 )  GOTO 580
122        DO 570 JJ=2, CLSMX(CLSN)-1
123 C
124          PTCLMN  = CLS(JJ,CLSN)
125          IPTCLMN = JJ
126          DO 560 JJJ=JJ+1, CLSMX(CLSN)
127            IF( CLS(JJJ,CLSN) .LT. PTCLMN ) THEN
128              PTCLMN  = CLS(JJJ,CLSN)
129              IPTCLMN = JJJ
130            END IF
131   560    CONTINUE
132          CLS(IPTCLMN,CLSN) = CLS(JJ,CLSN)
133          CLS(JJ,CLSN)      = PTCLMN
134 C
135   570  CONTINUE
136 C
137 C                         ---- UPDATE TH
138   580  JJE  = PMX1
139        PMX1 = PMX1 - CLSMX(CLSN)
140        IF( PMX1 .EQ. 0 )  GOTO 700
141 C
142        II1 = 1
143        JJ1 = 0
144        DO 590 JJ=1,JJE
145          IF( PNUM1(JJ) .LT. CLS(II1,CLSN) ) THEN
146            JJ1          = JJ1 + 1
147            PNUM1(JJ1) = PNUM1(JJ)
148          ELSE
149            II1          = II1 + 1
150            IF( II1 .EQ. CLSMX(CLSN)+1 )   CLS(II1,CLSN) = N1
151          END IF
152   590  CONTINUE
153 C
154 C
155   600 CONTINUE
156   700                                                   RETURN
157                                                         END
```

- If particle J is already a constituent particle of the cluster, the assessment procedure is not necessary.

- The subroutine CLSJUDGE returns CLSTRUE=.TRUE. if particle J is assessed as a member of cluster CLSN.

- The number of the constituent particles of the cluster CLSN is saved at CLSMX(CLSN).
- The name of new constituent particle J is saved at CLS (CLSMX(CLSN), CLSN).

- If KK is not equal to CLSMX(CLSN), there remain particles as a pivot for the assessment.

- If there are only the two constituent particles of the cluster, the name table is in order and thus this procedure is not necessary.

- The minimum number (particle name) and the position of the particle in the list CLS(*,CLSN) are saved at PTCLMN and IPTCLMN, respectively.
- The particle CLS(JJ,CLSN) is saved at CLS(IPTCLMN, CLSN), and the particle PTCLMN is saved at CLS(JJ,CLSN).

- The name list of the un-scanned particles is updated by removing the constituent particles of the cluster CLSN from the table PNUM1(*).

- N1 is required to be an arbitrary number greater than N.

## 7.4 Attempt of cluster movement in Monte Carlo simulations

As stated in Section 7.3, dispersed particles generally aggregate to form clusters if particle-particle interactions are much stronger than the thermal energy. In a strongly-interacting system, it is basically difficult to obtain physically reasonable aggregate structures from ordinary Monte Carlo simulations, especially for a spherical particle suspension. This may be explained by focusing on the characteristic time of the motion of particles themselves and that of the clusters. If a cluster is composed of a large number of particles and can be regarded as moving as a unitary particle because of the strong connections between the particles, the characteristic time of motion of this cluster is certainly much longer than that of single-moving particles. The ordinary Monte Carlo method only treats the motion (i.e., movement) of single-moving particles, and therefore this implies that it treats a physical phenomenon described within the time range of the characteristic time of the motion of a single-moving particle. Hence, if the mechanism for inducing the movement of clusters does not function in the ordinary Monte Carlo method, physical reasonable aggregates cannot be obtained from the simulation of a strongly-interacting suspension. This problem does not significantly arise for a suspension of non-spherical particles such as rod-like particles because the above-mentioned weak point is significantly improved by the rotational motion of the non-spherical particles. That is, the rotational motion gives rise to sufficient movement for clusters to join other growing clusters, which leads to the formation of further larger clusters. Hence, ordinary Monte Carlo simulations generally capture physically reasonable aggregate formation in the case of a non-spherical particle system. From this background, the cluster-moving Monte Carlo method [4, 5] is a powerful technique for simulating a suspension composed of magnetic spherical particles in a strong applied magnetic field because large clusters are expected to be formed. In the following, we describe in more detail the background and the concrete treatments of the cluster-moving method from the viewpoint of developing a simulation program.

Figure 7.6 shows the feasibility of the translational movement of a linear cluster by rotation as a whole, which is induced by the movement of its constituent particles themselves, in the situation of either a weak or no applied magnetic field. If the external magnetic field is significantly weak, the magnetic moment of a particle is not restricted to the alignment with the field direction, so that the particles are not necessarily located along the magnetic field direction. This implies that a constituent particle can move in a direction normal to the magnetic field without dissociation from the cluster and the integrated motion of all the constituent particles may result in the rotation of the whole cluster as shown in Fig. 7.6. Hence, a series of such rotational motions may give rise to the translational motion of a cluster, which enables clusters in the system to further aggregate to form thick chain-like clusters in a weak or no external magnetic field.

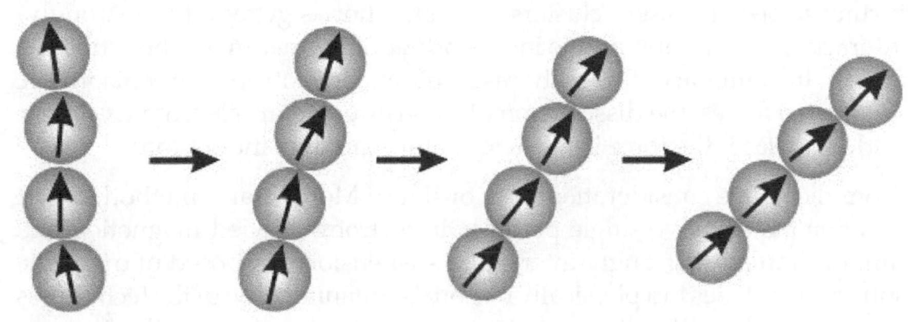

**Figure 7.6.** Rotational mode of a linear chain-like cluster in the situation of no external magnetic field.

Figure 7.7 illustrates how there is almost no feasibility of any translational movement of a linear cluster in the situation of a strong external magnetic field. In this situation, the magnetic moment of each constituent particle is strongly restricted to lie in the field direction, and therefore a linear chain-like cluster along the magnetic field direction is a preferred configuration for attaining a minimum magnetic interaction energy among the constituent particles. This implies also that an arbitrary constituent particle cannot move in directions normal to the field direction, as shown in Fig. 7.7, without the interaction energy significantly increasing. This strongly suggests that once a linear chain-like cluster is formed in a strong magnetic field, it cannot perform translational movements as a single unit in order to contribute to the growth of neighboring clusters.

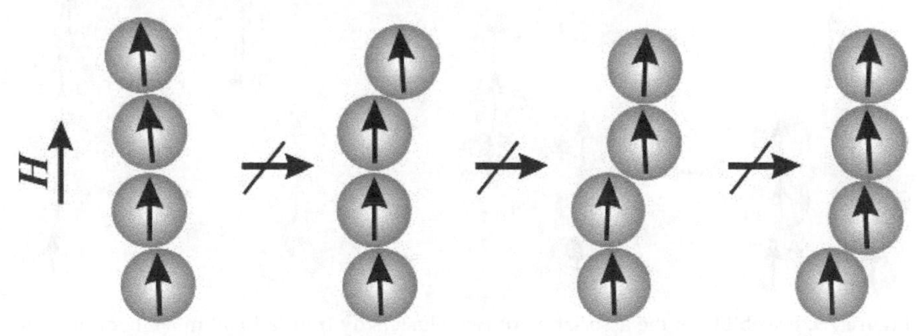


**Figure 7.7.** No feasibility of the translational movement in a strong magnetic field situation.

Figure 7.8 illustrates there is no feasibility of dissociation of a constituent particle from a cluster in a strong magnetic field situation. If a strong attractive interaction acts between particles, a constituent particle of a cluster is not able to dissociate from the cluster and this implies that once particles have aggregated to form a short cluster and there are no longer single-moving particles around the cluster, then these clusters cannot grow further to become larger clusters. This situation is general for a strongly-interacting suspension and is independent of the nature of the attractive forces. In summary, for both cases of either isotropic or anisotropic (magnetic) forces, the dissociation of a constituent particle from its cluster seldom arises if the force is much stronger than the random force.

From the above considerations, the ordinary Monte Carlo method, based on attempts to move single particles in a strong applied magnetic field, cannot simulate a strongly-interacting suspension composed of magnetic spherical particles in a physically reasonable manner. One of the techniques to overcome this difficulty may be to introduce the step of attempting to move the clusters themselves as a whole in addition to moving individual particles. Figure 7.9 schematically shows the process of growth of a longer chain-like cluster in a strong external magnetic field by the movement of clusters.

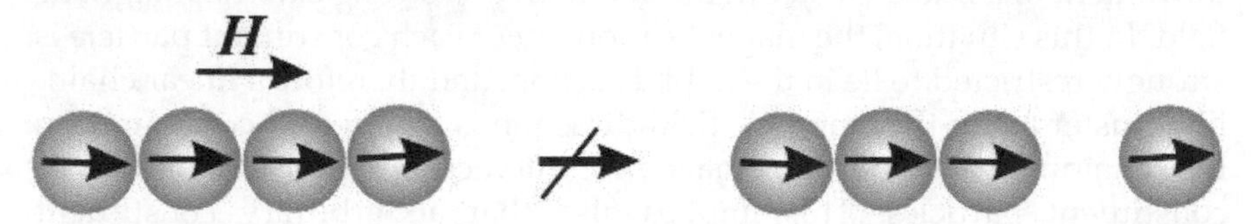


**Figure 7.8.** No feasibility of dissociation of a particle from the cluster in a strong magnetic field situation.

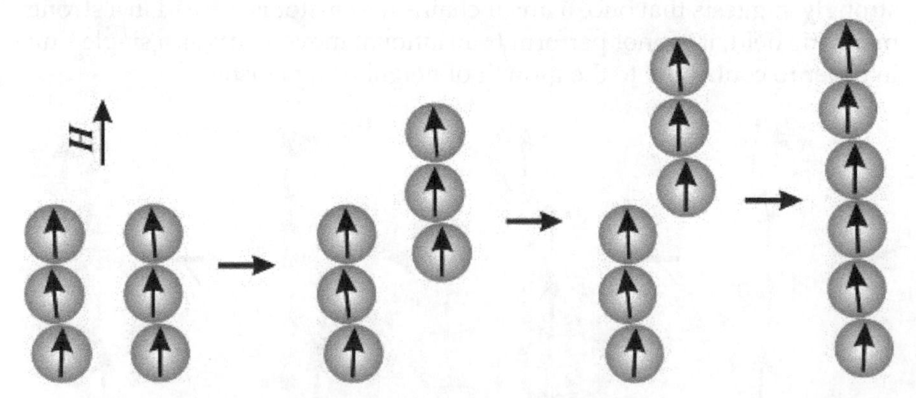


**Figure 7.9.** Feasibility of the association of two clusters by translational movement of each cluster in a strong magnetic field situation.

In this process, the clusters strongly tend to move to a position where their interaction energy attains to a lower value. When this process is repeatedly conducted, a long chain-like cluster will be formed along the magnetic field direction and this formation can basically be obtained with a characteristic time of motion within the range of the single-moving particles. In the cluster-moving Monte Carlo method [4, 5], the cluster-moving procedure is combined with the ordinary Monte Carlo steps for particle movement. Acceptance or rejection regarding the attempt of movement of a cluster is determined by the usual Metropolis method [6]. In the following, we discuss a practical treatment for the cluster movement that is useful for developing a simulation program.

The definition of a cluster has already been addressed above in Section 7.3. We therefore explain the cluster moving procedure under the assumption that several clusters have already been formed in a system. Figure 7.10(a) shows a snapshot that may be typical for magnetic or non-magnetic particles in a strongly-interacting system. Since the periodic boundary condition is generally used for a thermodynamic equilibrium situation, cluster 1 in Fig. 7.10(a) is understood to be composed of upper and lower aggregates. The formation of the cluster has been assessed using the particle shown darker in the figure as the pivot particle. When this cluster is moved to a new position, the particles in the upper part are transformed to virtual particles that interact with the real particles belonging to the part of the pivot particle, as shown in Fig. 7.10(b). Then, the cluster is moved to a new position according to the Metropolis procedure [6], as shown in Fig. 7.10(c), and the particles of the moved cluster that are outside the simulation region are transformed back into real particles in consideration with the periodic boundary condition, as shown in Fig. 7.10(d). Similarly, the rotational movement of the cluster may be carried out about the center of mass of the cluster. In this case the center of mass of the cluster in Fig. 7.10(b) has to be evaluated for the treatment of the rotation. However, the treatment of the rotation has to be carefully conducted in the situation where long clusters are formed from one end of the simulation region to the other. We exemplify this difficulty in the simulation of the rotation of clusters in the following.

Figure 7.11 shows a snapshot where a long and thick chain-like cluster is formed from the bottom of the region to the top. We consider the appearance of an overlap of particles in the rotational procedure of this cluster. In consideration of the periodic boundary condition, the ends of cluster 1 are assumed to be located at the shown broken line, which is quite understandable by referring to the relationship between Figs. 7.10(a) and (b). The center of mass of the cluster has been evaluated and shown

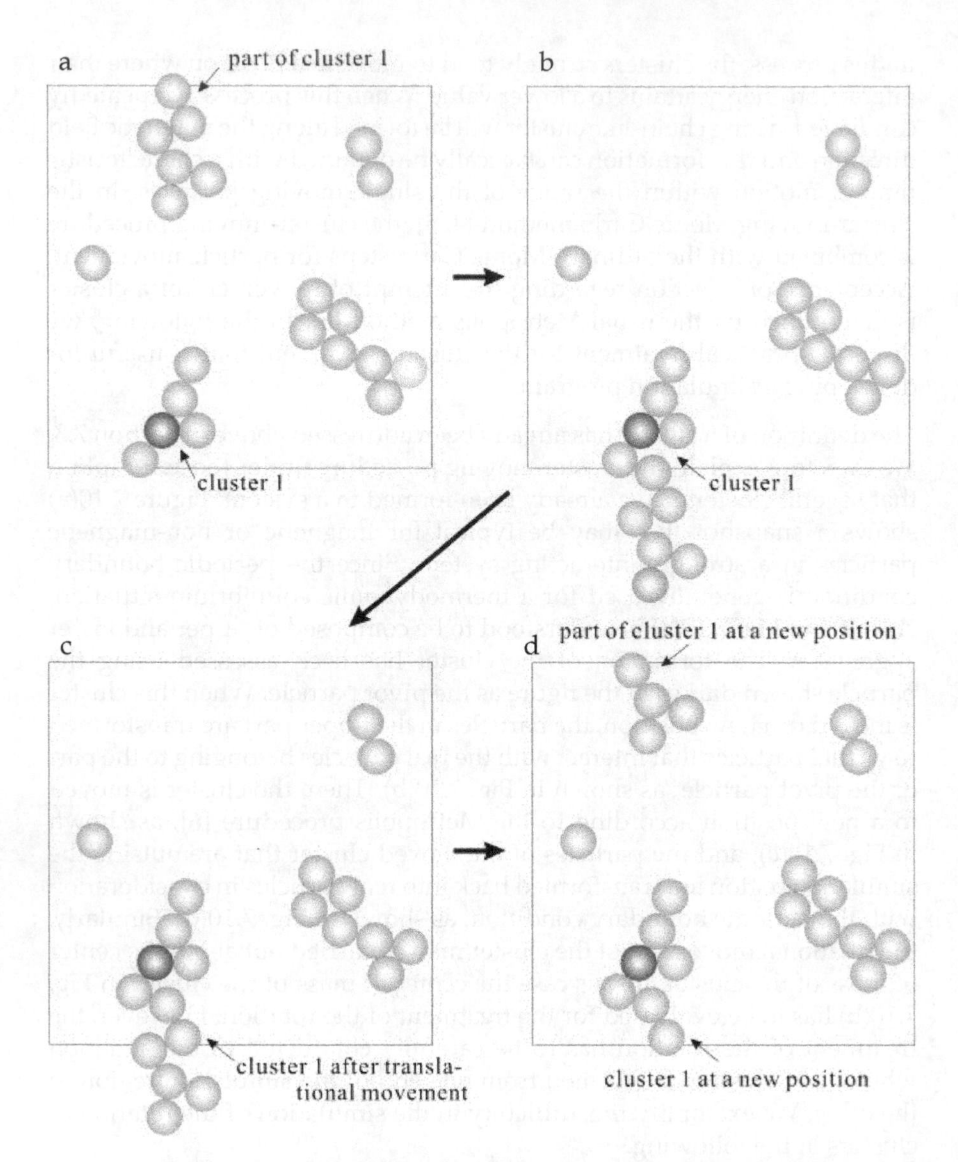


**Figure 7.10.** Treatment of the translational movement of a cluster: (a) the present snapshot, (b) real image of cluster 1 by reflecting the periodic BC, (c) new position of cluster 1 after the movement, and (d) reflection of the periodic BC (switch of virtual particles into real particles).

as a small solid circle in Fig. 7.11. An attempt is made to rotate the cluster about the center of mass, and this rotation is accepted if in the new position there is no overlap between the constituent particles of the cluster and also no overlap with any of the ambient neighboring particles. This procedure is important to ensure that the rotated cluster does not overlap with the neighboring particles. However, as shown in 7.11(b), both the end particles of the cluster may overlap after this rotation and this situation will appear more frequently for thicker chain-like clusters if they are formed from one end of the simulation box to the other. Hence, the rotational attempt in the cluster-moving Monte Carlo method has to be carefully conducted when large clusters are expected to be formed in the system. Fortunately, for the case of a spherical magnetic particle system in a strong applied magnetic field, physically reasonable chain-like clusters can be obtained by the introduction of only the translational movement of clusters, as explained in Fig. 7.9. Moreover, for a weak magnetic field, large clusters may be reasonably obtained by both the translational movement of the clusters and the rotational movement arising from the motion of only the constituent particles, as shown in Fig. 7.6.

The following program is an example, written in FORTRAN language, of a subroutine for performing the cluster-moving procedure in a cluster-moving Monte Carlo simulation. This subroutine calls another subroutine, not shown here, that calculates the interaction energy between two particles. Therefore, the subroutine below is directly applicable to a suspension of particles with any shape such as sphere, spherocylinder, spheroid or cube.

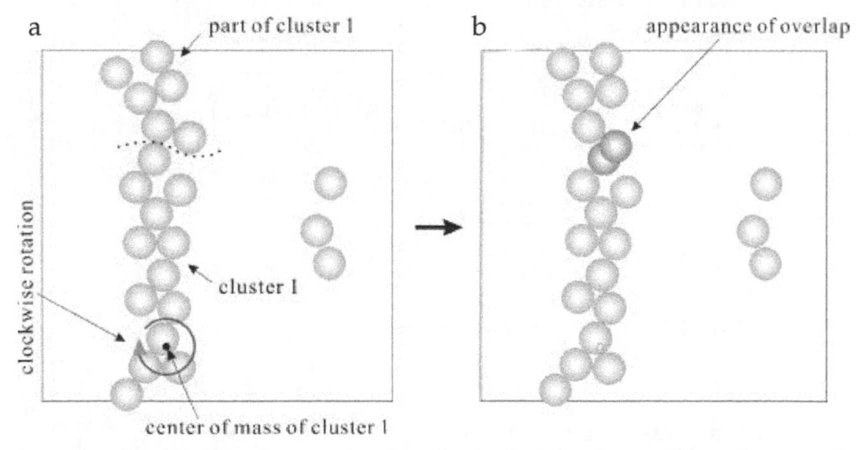


**Figure 7.11.** Appearance of an overlap between both the end particles in cluster 1 after the rotation about the center of mass of the cluster: (a) the present snapshot and (b) a new position of cluster 1 after the rotation.

```
001 C*********************************************************************
002 C   THE FOLLOWING SUBROUTINES ARE FOR THE PROCEDURE                   *
003 C                                          OF CLUSTER MOVEMENTS.      *
004 C                                              WRITTEN BY A. SATOH    *
005 C*********************************************************************
006 C
007 C      N       : NUMBER OF PARTICLES
008 C      RX(I),RY(I),RZ(I) : PARTICLE POSITION OF PARTICLE I
009 C      EX(I),EY(I),EZ(I) : DIRECTION OF MAGNETIC MOMENT OF PARTICLE I
010 C
011 C      RCOFF : CUTOFF RADIUS FOR CALCULATION OF INTERACTION ENERGIES
012 C      XL,YL,ZL : DIMENSIONS OF SIMULATION REGION
013 C      RCLSTR   : A CRITERION DISTANCE FOR ASSESSING THE CLUSTER
014 C                 FORMATION
015 C
016 C      DELR2    : MAXIMUM MOVEMENT DISTANCE FOR CLUSTER-MOVING
017 C
018 C      CLSN     : NAME OF CLUSTER (*) OR NUMBER OF CLUSTERS FORMED
019 C                 IN THE SYSTEM
020 C      CLSMX(*) : NUMBER OF CLUSTER-CONSTITUTING PARTICLES
021 C                 OF *-TH CLUSTER
022 C      CLS(+,*) : NAME OF CLUSTER-CONSTITUTING PARTICLES OF *-TH CLUSTER
023 C                                                + : 1,2,3,...,CLSMX(*)
024 C      COLORX(I), COLORY(I), COLORZ(I): ARE USED IN CLUSTER MOVEMENT
025 C
026 C      RAN(**)  : RANDOM NUMBERS RANGING FROM 0 TO 1
027 C                                                  ** : 1,2,...,NRANMX
028 C
029 C      ++++ -XL/2<RX(I)<XL/2 , -YL/2< RY(I)<YL/2, -ZL/2<RZ(I)<ZL/2 ++++
030 C---------------------------------------------------------------------
031 C
032 C**** SUB CLSMOV *****
033        SUBROUTINE CLSMOV( N RCOFF2 )
034 C
035        IMPLICIT REAL*8 (A-H,O-Z), INTEGER (I-N)
036 C
037        COMMON /BLOCK1/  RX   , RY   , RZ
038        COMMON /BLOCK2/  EX   , EY   , EZ
039        COMMON /BLOCK3/  XL   , YL   , ZL
040 C
041        COMMON /BLOCK11/ E    , ENEW , EOLD
042        COMMON /BLOCK12/ DELR2
043 C
044        COMMON /BLOCK21/ CLSN , CLSMX, CLS , RCLSTR
045        COMMON /BLOCK22/ COLORX, COLORY, COLORZ
046 C
047        COMMON /BLOCK31/ NRAN , RAN   , IXRAN
048 C
049 C      ---------------------------------------------------------------
050        INTEGER     NN , TT , NRANMX
051        PARAMETER( NN=1500 , TT=1500 , NRANMX=1000000 )
052        PARAMETER( PI=3.141592653589793D0 ) )
053 C      ----------------------------------
054 C
055        REAL*8   RX(NN) , RY(NN) , RZ(NN)
056        REAL*8   EX(NN) , EY(NN) , EZ(NN)
057        REAL*8   E(NN) , ENEW , EOLD
058 C
059        INTEGER  CLSN  , CLSMX(NN) , CLS(TT,NN)
060        INTEGER  COLORX(NN), COLORY(NN), COLORZ(NN)
061 C
062        REAL     RAN(NRANMX)
063        INTEGER  NRAN , IXRAN , NRANCHK
064 C
065        LOGICAL  OVRLAP
066 C      ----------------------------------
067 C
068        INTEGER  NN10
069        PARAMETER( NN10=500 )
060 C
071        INTEGER   CLSTR , ITH , ISTREET
072        INTEGER   NCLRMV , CLRMV(NN10)
073        INTEGER   CLRMVX(NN10) , CLRMVY(NN10) , CLRMVZ(NN10)
```

```
074        INTEGER   CLRMVX0      , CLRMVY0      , CLRMVZ0
075        REAL*8    RX0(NN), RY0(NN), RZ0(NN)
076        REAL*8    RXI    , RYI    , RZI    , EXI    , EYI    , EZI
077        REAL*8    RXCAN0    , RYCAN0    , RZCAN0
078        REAL*8    RXCAN(NN) , RYCAN(NN) , RZCAN(NN)
079        REAL*8    XL2 , YL2 , ZL2 , C1 , C2 , C3
080        LOGICAL   CLRMVCHK
081 C                                   -------------------------------------
082 C                                      MOVEMENT OF CLUSTERS
083 C                                                   BY MONTE CARLO METHOD
084 C                                   -------------------------------------
085        IF( ( CLSN .EQ. 1 ) .OR. ( CLSN .GE. N ) )  RETURN
086 C
087        XL2 = XL/2.D0
088        YL2 = YL/2.D0
089        ZL2 = ZL/2.D0
090 C
091 C
092        DO 1000 CLSTR=1,CLSN
093 C
094 C         -----------------------------------------------------------
095 C         ------------------------  ATTEMPT OF TRANSLATIONAL MOTION ---
096 C         -----------------------------------------------------------
097 C
098 C                                  ######  (1) CLUSTER'S ENERGY(OLD) #####
099 C
100          ETOTOLD =0.D0
101 C
102          DO 200 ITH=1,CLSMX(CLSTR)
103 C
104            I = CLS(ITH,CLSTR)
105            RXI    = RX(I)
106            RYI    = RY(I)
107            RZI    = RZ(I)
108            EXI    = EX(I)
109            EYI    = EY(I)
110            EZI    = EZ(I)
```

• CLSMX(CLSTR) is the number of the constituent particles of the cluster CLSTR.
• CLS(ITH,CLSTR) is the name of the constituent particle of the cluster CLSTR, which is saved at the ITH-position in the name table CLS(#,+).

```
111 C
112            DO 150 J=1,N
113 C
114              DO 100 III=1,CLSMX(CLSTR)
115                IF( J .EQ. CLS(III,CLSTR) )     GOTO 150
116  100         CONTINUE
```

• The particle J should not belong to the cluster of interest, CLSTR.

```
117 C
118 C                                   ++ ISTREET=0 FOR FULL CAL      ++
119 C                                   ++ ISTREET=1 FOR MAG.ENE.ONLY  ++
120 C                                   ++ ISTREET=2 FOR CLSTR MOV     ++
121 C
122              ISTREET = 2
123              CALL ENECAL( I , RXI, RYI, RZI, EXI, EYI, EZI,
124       &                   EOLD0, OVRLAP, ISTREET, J )
125 C
126              ETOTOLD = ETOTOLD + EOLD0
127 C
128  150       CONTINUE
129 C
130  200     CONTINUE
```

• In the subroutine ENECAL, the interaction energy between particles I and J is calculated.

```
131 C
132 C                                  ######  (2) CLUSTER'S ENERGY(CAN) #####
133 C
134          NRAN = NRAN + 1
135          C1   = DELR2*( 1.D0-2.D0*DBLE(RAN(NRAN)) )
136          NRAN = NRAN + 1
137          C2   = DELR2*( 1.D0-2.D0*DBLE(RAN(NRAN)) )
138          NRAN = NRAN + 1
139          C3   = DELR2*( 1.D0-2.D0*DBLE(RAN(NRAN)) )
```

• All the constituent particles of the cluster are attempted to be moved by the same small distance (C1, C2, C3) in the *x*-, *y*-, and *z*-direction, respectively.

```
140 C
141          NCLRMV  = 0
142          ETOTCAN =0.D0
143 C
144          DO 400 ITH=1,CLSMX(CLSTR)
145 C
146            I = CLS(ITH,CLSTR)
147            RXCAN0 = RX(I) + C1
148            RYCAN0 = RY(I) + C2
149            RZCAN0 = RZ(I) + C3
150            EXI    = EX(I)
151            EYI    = EY(I)
```

• (RXCAN0, RYCAN0, RZCAN0) is a new candidate position of particle I.

```
152           EZI      = EZ(I)
153           CLRMVCHK = .FALSE.
154           CLRMVX0  = COLORX(I)
155           CLRMVY0  = COLORY(I)
156           CLRMVZ0  = COLORZ(I)
157 C
158          IF( RXCAN0 .GT. XL2 ) THEN
159            RXCAN0   = RXCAN0  - XL
160            CLRMVX0  = CLRMVX0 - 1
161            CLRMVCHK = .TRUE.
162          ELSE IF( RXCAN0 .LE. -XL2 ) THEN
163            RXCAN0   = RXCAN0  + XL
164            CLRMVX0  = CLRMVX0 + 1
165            CLRMVCHK = .TRUE.
166          END IF
167C
168          IF( RYCAN0 .GT. YL2 ) THEN
169            RYCAN0   = RYCAN0  - YL
170            CLRMVY0  = CLRMVY0 - 1
171            CLRMVCHK = .TRUE.
172          ELSE IF( RYCAN0 .LE. -YL2 ) THEN
173            RYCAN0   = RYCAN0  + YL
174            CLRMVY0  = CLRMVY0 + 1
175            CLRMVCHK = .TRUE.
176          END IF
177C
178          IF( RZCAN0 .GT. ZL2 ) THEN
179            RZCAN0   = RZCAN0  - ZL
180            CLRMVZ0  = CLRMVZ0 - 1
181            CLRMVCHK = .TRUE.
182          ELSE IF( RZCAN0 .LE. -ZL2 ) THEN
183            RZCAN0   = RZCAN0  + ZL
184            CLRMVZ0  = CLRMVZ0 + 1
185            CLRMVCHK = .TRUE.
186          END IF
187C
188          IF ( CLRMVCHK ) THEN
189            NCLRMV = NCLRMV + 1
190            CLRMV( NCLRMV) = I
191            CLRMVX(NCLRMV) = CLRMVX0
192            CLRMVY(NCLRMV) = CLRMVY0
193            CLRMVZ(NCLRMV) = CLRMVZ0
194          END IF
195C
196          RXCAN(ITH) = RXCAN0
197          RYCAN(ITH) = RYCAN0
198          RZCAN(ITH) = RZCAN0
199C
200          DO 300 J=1,N
201C
202            DO 250 III=1,CLSMX(CLSTR)
203              IF( J .EQ. CLS(III,CLSTR) )     GOTO 300
204   250      CONTINUE
205C
206C                                   ++ ISTREET=0 FOR FULL CAL    . ++
207C                                   ++ ISTREET=1 FOR MAG.ENE.ONLY ++
208C                                   ++ ISTREET=2 FOR CLSTR MOV    ++
209C
210            ISTREET = 2
211            CALL ENECAL( I , RXCAN0, RYCAN0, RZCAN0, EXI, EYI, EZI,
212      &                  ECAN0, OVRLAP, ISTREET, J )
213C
214            IF( OVRLAP )  GOTO 1000
215C
216            ETOTCAN = ETOTCAN + ECAN0
217C
218   300    CONTINUE
219C
220   400  CONTINUE
221C
222C              ------ ASSESSMENT DUE TO METROPOLIS METHOD ------
223C
224        C3 = ETOTCAN - ETOTOLD
225        IF( C3 .GE. 0.D0 )THEN
226          NRAN = NRAN + 1
227          IF( DBLE(RAN(NRAN)) .GE. DEXP(-C3) ) THEN
228            GOTO 1000
229          END IF
230        END IF
231C                                        ++++++++++++++++++++++
232C                                        CANDIDATES ARE ACCEPTED
233C                                        ++++++++++++++++++++++
```

• COLORX(*), COLORY(*) and COLORZ(*) are used for the rotational procedure, and therefore are not referred to in the present subroutine.

• The treatment of the periodic boundary condition is conducted for the *x*-direction.

• The treatment of the periodic boundary condition is conducted for the *y*-direction.

• The treatment of the periodic boundary condition is conducted for the *z*-direction.

• CLSMVX(*), CLSMVY(*) and CLSMVZ(*) express whether the real or virtual position of particle I is an object for treatment in the *x*-, *y*- and *z*-direction, respectively.

• The particle J should not belong to the cluster CLSTR.

• IF an overlap between the present and neighboring particles arises, the treatment is terminated for the present cluster.

• Acceptance or rejection of the new position is determined by the ordinary Metropolis method.

```
234         DO 450 ITH=1,CLSMX(CLSTR)
235           I     = CLS(ITH,CLSTR)
236           RX(I) = RXCAN(ITH)
237           RY(I) = RYCAN(ITH)
238           RZ(I) = RZCAN(ITH)
239  450    CONTINUE
240         ETOTOLD = ETOTCAN
241C
242          IF( NCLRMV .GE. 1 ) THEN
243            DO 470 ITH=1, NCLRMV
244              I = CLRMV(ITH)
245              COLORX(I) = CLRMVX(ITH)
246              COLORY(I) = CLRMVY(ITH)
247              COLORZ(I) = CLRMVZ(ITH)
248  470       CONTINUE
249          END IF
250C
251C
252 1000 CONTINUE
253                                                  RETURN
254                                                  END
```

• The positions of all the constituent particles of the cluster CLSTR are updated if the candidate positions are accepted.

# Bibliography

[1] Coverdale, G. N., Chantrell, R. W., Hart, A. and Parker, D. 1993. A 3-D Simulation of a particulate dispersion. J. Magn. Magn. Mater. 120: 210–212.

[2] Coverdale, G. N., Chantrell, R. W., Hart, A. and Parker, D. 1994. A Computer simulation of the microstructure of a particulate dispersion. J. Appl. Phys. 75: 5574–5576.

[3] Coverdale, G. N., Chantrell, R. W., Martin, G. A. R., Bradbury, A., Hart, A. and Parker, D. 1998. Cluster analysis of the microstructure of colloidal dispersions using the maximum entropy technique. J. Magn. Magn. Mater. 188: 41–51.

[4] Satoh, A., Chantrell, R. W., Kamiyama, S. and Coverdale, G. N. 1996. Two-dimensional Monte Carlo simulations to capture thick chainlike clusters of ferromagnetic particles in colloidal dispersions. J. Colloid Interface Sci. 178: 620–627.

[5] Satoh, A., Chantrell, R. W., Kamiyama, S. and Coverdale, G. N. 1996. Three-dimensional Monte Carlo simulations of thick chainlike clusters composed of ferromagnetic fine particles. J. Colloid Interface Sci. 181: 422–428.

[6] Metropolis, N., Rosenbluth, A. W., Rosenbluth, M. N. and Teller, A. 1953. Equation of state calculations by fast computing machines. J. Chem. Phys. 21: 1087–1092.

CHAPTER 8

# Description of System Characteristics

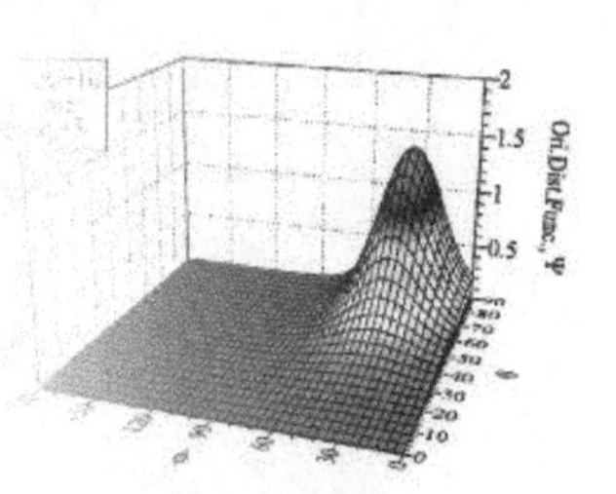

In a particle suspension where magnetic or non-magnetic particles strongly interact with each other to aggregate to form clusters, it is necessary to employ effective methods for describing characteristics regarding the system and the internal structure of aggregates. The most desirable method may be snapshots that can provide a direct visual image of aggregate structures formed in the system and are quite useful for discussing the characteristics of the system from a qualitative point of view. In order to discuss system characteristics at a deeper level, it is desirable to use the distribution functions such as the radial distribution function, the pair correlation function and the orientational distribution function as these quantities are quite appropriate for a quantitative discussion of the internal structure of aggregates. In the following we address these quantities, including their definitions, and demonstrate how they effectively characterize the system by using example snapshots obtained from Monte Carlo simulations for various cases of external magnetic field strength and magnetic particle-particle interaction strength.

## 8.1 Radial distribution function

The radial distribution function is the most common and useful quantity for characterizing the structure of particle aggregates in a suspension of magnetic or non-magnetic particles. It has originally been used for discussing the internal structure of a pure molecular system such a gas, liquid or solid state because they exhibit completely different features and the radial distribution function can characterize how densely the ambient particles are stochastically located around an arbitrary particle. We discuss the definition of the radial distribution function in the following.

In this discussion we consider a suspension composed of $N$ number of spherical particles in a thermodynamic equilibrium situation but it should be noted that the radial distribution function is not restricted to thermodynamic equilibrium. If the particles are uniformly distributed without aggregation, the average of a local number density $n$ at an arbitrary position in the system is equal to a given number density $n_0$, which is given by $n_0 = N/V$ where $V$ is the volume of the system. If there are aggregates of particles in the system, the mean local number density $n(r)$ at the distance $r$ from an arbitrary particle will deviate from the given number density $n_0$, i.e., $n(r) \neq n_0$. Hence, the ratio of $n(r)$ to $n_0$ becomes a useful index for characterizing the local internal structure of the system and the ratio $g(r) = n(r)/n_0$ is called the radial distribution function. If the notation $\Delta N_i(r)$ denotes the number of particles located in the infinitesimally shell volume $\Delta V = 4\pi r^2 \Delta r$ with radial width $\Delta r$ at the radial distance $r$ from an arbitrary particle $i$, then the radial distribution function $g(r)$ is evaluated as

$$g(r) = \frac{n(r)}{n_0} = \frac{1}{N/V} \cdot \left\langle \frac{1}{N} \sum_{i=1}^{N} \frac{\Delta N_i(r)}{\Delta V} \right\rangle = \frac{1}{n_0 N} \cdot \left\langle \sum_{i=1}^{N} \frac{\Delta N_i(r)}{4\pi r^2 \Delta r} \right\rangle \tag{8.1}$$

This expression is valid for a three-dimensional system and a slightly different form is used for a two-dimensional system as

$$g(r) = \frac{n(r)}{n_0} = \frac{1}{N/S} \cdot \left\langle \frac{1}{N} \sum_{i=1}^{N} \frac{\Delta N_i(r)}{\Delta S} \right\rangle = \frac{1}{n_0 N} \cdot \left\langle \sum_{i=1}^{N} \frac{\Delta N_i(r)}{2\pi r \Delta r} \right\rangle \tag{8.2}$$

in which $\Delta N_i(r)$ denotes the number of particles located in the infinitesimally shell (rim) with area $\Delta S = 2\pi r \Delta r$ of the radial width $\Delta r$ at the radial distance $r$ from an arbitrary particle $i$. The bracket enclosure $\langle - \rangle$ denotes the time average or ensemble average of the quantity of interest. The distance interval $\Delta r$ corresponding to the resolution of the radial distribution function is usually used with a sufficiently small value compared with the particle diameter $d$ and it may be reasonable that $\Delta r$ is taken as, for instance, $\Delta r/d = 0.1$ or $0.2$.

If a system is a dilute suspension, there are no aggregates in the system, so that the local number density $n(r)$ should tend to $n_0$ and $g(r)$ should be reasonably constant at unity. If there are particle aggregates in a system in a solid-like manner, the radial distribution function will tend to exhibit high peaks, where particle are located in a crystal-like structure, and the radial distribution function may resemble the Dirac's delta function at the positions with approximately zero value between these positions. In the following, we demonstrate how the internal structure of the aggregates is expressed by the characteristics of the radial distribution function.

Figure 8.1 shows snapshots of aggregate structures in a two-dimensional system composed of magnetic spherical particles in no external magnetic field, for the three cases of the magnetic interaction strength, $\lambda = 1, 5$ and 10. The results of the corresponding radial distribution functions are shown in Fig. 8.2. The snapshot in Fig. 8.1(a) is in gas-like structure without particle aggregation and so the curve of the radial distribution function shown in Fig. 8.2 exhibits the typical characteristic obtained for the gas state of a molecular system. The distribution, as expected for a dilute system, has an almost constant value of unity in a range over approximately double the particle diameter. In the vicinity of the situation of particle-particle contact, slightly larger values than unity are obtained, which clearly suggests that there is a tendency for two particles to weakly combine with each other to form small clusters and this expectation is supported by the snapshot shown in Fig. 8.1. As the magnetic interaction strength is increased from $\lambda = 1$ to 10, Fig. 8.1 clearly shows that there is a tendency for long chain-like clusters to be formed and the radial distribution function changes from exhibiting a gas-like to a liquid-like characteristic with high peaks at the positions of an integer times the particle diameter (= 1.3 including the steric layer) with an oscillatory tail around unity. These liquid-like characteristics will appear more significantly with increasing magnetic particle-particle interaction strength because more stable and longer chain-like clusters are expected to be formed.

From these examples, we understand that the radial distribution function can properly describe the essential characteristics of the internal structure of particle aggregates in the system of a strongly-interacting suspension.

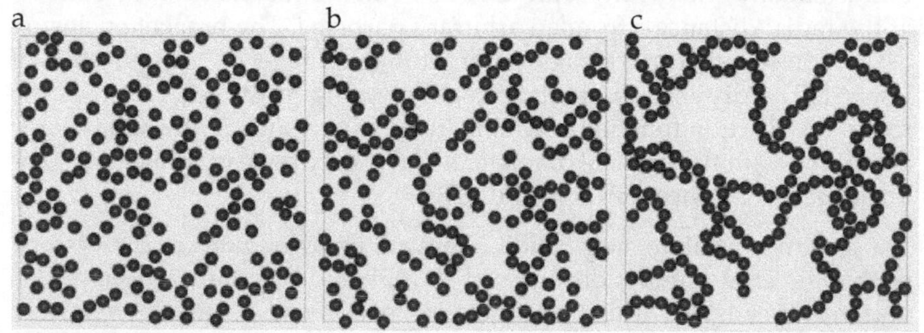


**Figure 8.1.** Snapshots for different magnetic interaction strengths in no external magnetic field: (a) $\lambda = 1$, (b) $\lambda = 5$ and $\lambda = 10$.

Finally, we show an example of a subroutine for evaluating the radial distribution function and the pair correlation function, which is written in FORTRAN language. In this evaluation, the position of each particle is necessary for calculating the distance between each pair of particles.

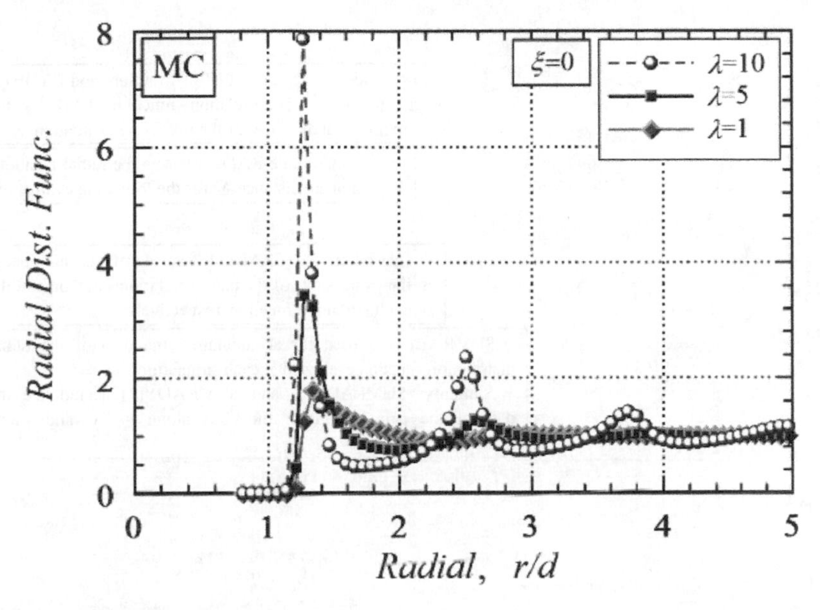


**Figure 8.2.** Radial distribution functions for different magnetic interaction strengths in no external magnetic field: $\lambda = 1$, $\lambda = 5$ and $\lambda = 10$.

```
001 C***********************************************************************
002 C                                                                      *
003 C   THE FOLLOWING SUBROUTINE IS FOR EVALUATING THE RADIAL               *
004 C         DISTRIBUTION FUNCTION AND THE PAIR CORRELATION FUNCTION      *
005 C                                                                      *
006 C                                                    WRITTEN BY A. SATOH *
007 C***********************************************************************
008 C      N      : NUMBER OF PARTICLES
009 C      D      : DIAMETER OF PARTICLE INCLUDING SURFACTANT LAYER
010 C                ( =1 FOR THIS CASE )
011 C      DS     : DIAMETER OF SOLID PARTICLE WITHOUT STERIC LAYER
012 C      DEL    : THICKNESS OF STERIC LAYER
013 C      VDENS  : VOLUMETRIC FRACTION OF PARTICLES
014 C      RCOFF  : CUTOFF RADIUS FOR CALCULATION OF MAG. FORCES
015 C      XL,YL  : DIMENSIONS OF SIMULATION REGION
016 C
017 C      RX(I),RY(I)  : PARTICLE POSITION OF PARTICLE I
018 C      NX(I),NY(I)  : DIRECTION OF MAGNETIC MOMENT
019 C
020 C      DR       : RADIAL INTERVAL FOR CALCULATING
021 C                 RADIAL DISTRIBUTION FUNC. AND PAIR CORRELATION FUNC.
022 C      DTHETA : THETA  INTERVAL FOR CALCULATING PAIR CORRELATION FUNC.
023 C      RMX      : MAXIMUM RADIAL
024 C      R0(*)    : IS USED FOR DETERMINING THE UNIT TO WHICH A PARTICLE IS
025 C                 BELONG ( 1=< * =< NR0MX )
026 C      R(*)     : REPRESENTATIVE POINT FOR EACH UNIT
027 C                   ( 1=< * =< NRMX )
028 C      RADDIST(*)  : RAD. DIST. FUNC.
029 C      SUMRAD(*)   : SUMMATION OF RAD. DIST. FUNC.
030 C
031 C      RADIALX(*)  : CORRELATION FUNCTION ALONG X-AXIS
```

```
032 C      RADIALY(*) : CORRELATION FUNCTION ALONG Y-AXIS
033 C      SUMRADX(*) : SUMMATION OF X-CORRELATION FUNCTION
034 C      SUMRADY(*) : SUMMATION OF Y-CORRELATION FUNCTION
035 C
036        NRADCNTR   : NUMBER OF SAMPLING
037 C
038 C     +++++++  -XL/2 <RX(I) <XL/2 , -YL/2 <RY(I)< YL/2  ++++++++
039 C--------------------------------------------------------------------
040 C

041                         ...
042                         ...
043 C                                                 --- PARAMETER (1) ---
044        DR      = 1.D0/DBLE( 20 )
045        DTHETA  = 10.D0/180.D0*PI
046        RMX     = 4.D0
047 C
048        CALL RADIALR( R0, R )
049 C
050        R0CHKSQ = R0(NR0MX)**2
051        SN      = DSIN( DTHETA/2.D0 )
052 C
053        CRADIAL = ( XL*YL/DBLE(N**2) ) / ( DR*DTHETA )
054        CRADIAL2= ( XL*YL/DBLE(N**2) ) / ( 2.D0*PI*DR )
055 C
056                         ...
057                         ...
058 C
059        NRADCNTR= 0
060        DO 50 I=1,NRMX
061          SUMRADX(I) = 0.D0
062          SUMRADY(I) = 0.D0
063          SUMRAD( I) = 0.D0
064     50 CONTINUE
065                         ...
066                         ...
067 C
068 C     -------------------------------------------------------------
069 C     ---------------    START OF MONTE CARLO PROCEDURE  ------------
060 C     -------------------------------------------------------------
071 C
072                         ...
073                         ...
074 C                                            ----- RADIAL DISTRIBUTION -----
075 C                                            ----- PAIR CORRELATION    -----
076          NRADCNTR = NRADCNTR + 1
077 C                                               --- ALONG X- AND Y-AXIS ---
078          CALL RADIALCA( N, XL, YL, P, R0CHKSQ, SN )
079 C
080          DO 500 J=1,NRMX

081            SUMRAD( J) = SUMRAD( J) + RADDIST(J)
082            SUMRADY(J) = SUMRADY(J) + RADIALY(J)
083            SUMRADX(J) = SUMRADX(J) + RADIALX(J)
084    500   CONTINUE
085                         ...
086                         ...
087 C
088 C        -------------------------------------------------------------
089 C        -----------------  END OF MONTE CARLO PROCEDURE -------------
090 C        -------------------------------------------------------------
091 C
092 C                                            ----- RADIAL DISTRIBUTION -----
093 C                                            ----- PAIR CORRELATION    -----
094        DO 1100 I=1,NRMX
095          RADIALX(I) = SUMRADX(I)/DBLE(NRADCNTR)
096          RADIALY(I) = SUMRADY(I)/DBLE(NRADCNTR)
097          RADDIST(I) = SUMRAD( I)/DBLE(NRADCNTR)
098   1100 CONTINUE
099                         ...
100                         ...
101                                                                    STOP
102                                                                    END
103 C*******************************************************************
104 C*************************    SUBROUTINE    ***********************
105 C*******************************************************************
106 C
107 C**** RADIALR ****
108        SUBROUTINE RADIALR( R0, R )
109 C
110        IMPLICIT REAL*8 (A-H,O-Z), INTEGER (I-N)
111 C
```

• DR implies $\Delta r$ for the radial distribution, and DTHETA is $\Delta\theta$ for the pair correlation function. RMX is the maximum radial distance for the following evaluation.

• The subroutine RADIALR sets the radial positions by increment distance $\Delta r$ for the following evaluation.

• CRADIAL and CRADIAL2 are constants that appear in the expressions of the pair correlation function and the radial distribution function, respectively.

• SUMRAD(*) is used for accumulating the data of the radial distribution function evaluated at each simulation step.
• Similarly, SUMRADX(*) and SUMRADY(*) accumulate the data of the pair correlation functions along the $x$- and $y$-axis directions, respectively.

• The values of the rad. dist. func. and the pair corr. func. are calculated in the subroutine RADIALCA.

• The final results are obtained by dividing the accumulated values by the number of sampling, NRADCNTR.

• The subroutine RADIALR sets the radial positions by increment distance $\Delta r$.

```
112       COMMON /BLOCK16/ DR , DTHETA , RMX , CRADIAL , NRMX , NR0MX
113 C
114       INTEGER    SS
115       PARAMETER( SS=200 )
116 C
117       REAL*8  R0(SS) , R(SS) , C0 , C1
118 C                                                     --- CAL. R(*) ---
119       C0     = DR/2.D0
120       R0(1)  = 0.6D0
121       R(1)   = R0(1) + C0
122 C
123       DO 100 I=2,200
124         R0(I) = R0(1) + DBLE(I-1)*DR
125         C1     = R0(I) + C0
126         IF( C1 .GT. RMX )THEN
127           NR0MX = I
128           NRMX  = I-1
129           RETURN
130         END IF
131         R(I) = C1
132   100 CONTINUE
133                                                              RETURN
134                                                              END
135 C**** RADIALCA ****
136       SUBROUTINE RADIALCA( N, XL, YL, P, R0CHKSQ, SN )
137 C
138       IMPLICIT REAL*8 (A-H,O-Z), INTEGER (I-N)
139 C
140       COMMON /BLOCK1/  RX   , RY
141       COMMON /BLOCK3/  GRPX, GRPY
142       COMMON /BLOCK5/  TMX , TABLE
143 C
144       COMMON /BLOCK16/ DR , DTHETA , RMX , CRADIAL , NRMX , NR0MX
145       COMMON /BLOCK17/ R   , R0
146       COMMON /BLOCK18/ RADIALX , RADIALY
147       COMMON /BLOCK19/ RADDIST , SUMRAD , CRADIAL2
148 C
149       INTEGER      NN , PP2 , TT , SS
150       PARAMETER( NN=1000 , PP2=225 , TT=500 , SS=200 )
151 C
152       INTEGER    P , GRPX(NN) , GRPY(NN)
153       INTEGER    TMX(PP2) , TABLE(TT,PP2)
154       REAL*8     RX(NN)    , RY(NN)
155       REAL*8     RADIALX(SS) , RADIALY(SS) , R(SS) , R0(SS)
156       REAL*8     RADDIST(SS) , SUMRAD(SS)
157 C
158       INTEGER    GX , GY , GP
159       REAL*8     RXIJ , RYIJ , RIJ , RIJSQ , MODX , MODY
160 C
161       INTEGER    NR0 , NRMX , NR0MX
162       REAL*8     C1  , C2
163       LOGICAL    LGX , LGY
164 C
165       DO 10 I=1,NRMX
166         RADDIST(I) = 0.D0
167         RADIALX(I) = 0.D0
168         RADIALY(I) = 0.D0
169    10 CONTINUE
170 C
171       DO 200 I=1,N
172 C                                              ---------------------------------
173 C                                              ---  INTERACTING CELL (GY,GX) ---
174 C                                              ---------------------------------
175         DO 150 JJ=-1,1
176           GY = GRPY(I) + JJ
177           IF( GY .EQ. 0 ) THEN
178             GY    = P
179             MODY =-YL
180             GOTO 110
181           END IF
182           IF( GY .EQ. P+1 ) THEN
183             GY    = 1
184             MODY = YL
185             GOTO 110
186           END IF
187           MODY = 0.D0
188 C
189   110     DO 150 II=-1,1
190             GX = GRPX(I) + II
191             IF( GX .EQ. 0 ) THEN
192               GX    = P
193               MODX =-XL
```

> • R0(*) is used for determining the radial shell site to which each particle belongs.
> • R(*) is the representative position of each radial shell site.

> • The subroutine RADIALCA evaluates the rad. dist. func. and the pair corr. func. by calculating the distance between each pair of particles.

> • These procedures are conducted by treating the periodic boundary condition.
> • The cell index method is used, where the simulation box is divided into the P number of cells in each axis direction.

```
194                 GOTO 120
195               END IF
196               IF( GX .EQ. P+1 ) THEN
197                 GX   = 1
198                 MODX = XL
199                 GOTO 120
200               END IF
201               MODX = 0.D0
202 C
203   120         GP = GX + (GY-1)*P
204               IF( TMX(GP) .EQ. 0 ) GOTO 150
205               DO 140 JJJ=1,TMX(GP)
206                 J  =TABLE(JJJ,GP)
207                 IF( J .EQ. I ) GOTO 140
208 C
209                 RXIJ = RX(J) + MODX - RX(I)
210                 RYIJ = RY(J) + MODY - RY(I)
211                 RIJSQ= RXIJ**2 + RYIJ**2
212 C                                    +++ OVER CHECK AREA(RADIAL) +++
213                 IF( RIJSQ .GE. ROCHKSQ ) GOTO 140
214                 RIJ = DSQRT(RIJSQ)
215 C
216 C               ------------------------ (1) RADIAL DISTRIBUTION FUNC. ---
217                 DO 125  NR0=2,NROMX
218                   IF( R0(NR0).GT.RIJ ) THEN
219                     RADDIST(NR0-1) = RADDIST(NR0-1) + 1.D0
220                     GOTO 127
221                   END IF
222   125           CONTINUE
223 C
224 C               ------------------------ (2) PAIR CORRELATION FUNCTION ---
225   127           C2 = DABS(RXIJ)/RIJ
226                 C1 = DABS(RYIJ)/RIJ
227                 IF( C2.GT.SN ) THEN
228                   LGY = .FALSE.
229                 ELSE
230                   LGY = .TRUE.
231                 END IF
232                 IF( C1.GT.SN ) THEN
233                   LGX = .FALSE.
234                 ELSE
235                   LGX = .TRUE.
236                 END IF
237 C                                    +++ OVER CHECK AREA(ANGLE)  +++
238                 IF( .NOT.(LGY .OR. LGX) )  GOTO 140
239 C
240                 DO 130  NR0=2,NROMX
241                   IF( R0(NR0).GT.RIJ ) THEN
242 C                                             +++ ACCUMULATION +++
243                     IF( LGX ) RADIALX(NR0-1) = RADIALX(NR0-1) + 1.D0
244                     IF( LGY ) RADIALY(NR0-1) = RADIALY(NR0-1) + 1.D0
245                     GOTO 140
246                   END IF
247   130           CONTINUE
248 C
249   140         CONTINUE
250
251   150       CONTINUE
252 C
253 C
254   200 CONTINUE
255 C                             ---- MULTIPLY THEM BY THE COEFFICIENT ---
256       DO 210 I=1,NRMX
257 C
258         RADDIST(I) = CRADIAL2 * RADDIST(I)/R(I)
259
260 C
261         C1 = CRADIAL * RADIALX(I)/R(I)
262         C2 = CRADIAL * RADIALY(I)/R(I)
263 C                             +++ MODIFICATION DUE TO THE COUNT +++
264 C                             +++ IN THE OPPOSITE DIRECTION     +++
265         RADIALX(I) = C1/2.D0
266         RADIALY(I) = C2/2.D0
267 C
268   210 CONTINUE
269                                                              RETURN
270                                                              END
```

• MODX and MODY are used for treating the periodic boundary condition.

• This IF condition determines the radial shell site to which particle J belongs.

• This IF condition determines whether or not particle J is within the angle range in the *y*-axis direction.

• This IF condition determines whether or not particle J is within the angle range in the *x*-axis direction.

• If the particle is within the angle range, the number of particle belonging to the radial shell site is increased by one.

• The final result of the rad. dist. func. is obtained both by multiplying it by the coefficient CRADIAL2 and by dividing it by the radial distance R(I).

• The pair corr. functions are obtained both by multiplying them by the coefficient CRADIAL and by dividing them by R(I).

## 8.2 Pair correlation function

As explained in the previous section, the radial distribution function provides the information as to how many ambient particles are stochastically situated around an arbitrary particle and is a function of the radial distance from the particle. This implies that the radial distribution function is a quantity that has been averaged in the tangential, $\theta$-direction, at an arbitrary radial distance. In a strong applied magnetic field, the magnetic spherical particles have a strong tendency to aggregate to form linear chain-like clusters along the magnetic field direction. Hence, the averaging procedure along the tangential direction weakens the inherent detail of the internal structure of the clusters. In order to overcome this drawback of the radial distribution function, the pair correlation function may be employed to describe the detailed characteristics along the magnetic field direction. This is because it is a function of both the radial distance $r$ from an arbitrary particle and an angle $\theta$ measured for instance from the external magnetic field direction. This allows the pair correlation function to capture the inherent detailed information regarding the stochastic location of the ambient particles in all the directions around an arbitrary particle. In addition to a magnetic spherical particle suspension, this correlation function is also applicable to a suspension composed of magnetic non-spherical particles such as the rod-like and disk-like particles. In the following, we show the definition of the pair correlation function and also how this function can describe the internal structure of the aggregates formed in a model suspension composed of magnetic spherical particles in a strong external magnetic field.

We first consider a two-dimensional system of $N$ number of magnetic spherical particles with the same diameter. Similar to the radial distribution function explained in Section 8.1, the pair correlation function is the ratio of the local number density $n(r, \theta)$ of particles at a point of the radial distance $r$ and the angle $\theta$ to the given number density $n_0 = N/S$ ($S$ is the area of the simulation region) for a uniform distribution of particles. An external magnetic field is applied in the $y$-axis direction and the angle $\theta$ is defined from the $y$-axis in the clock-wise direction. Employing these symbols, any point $\boldsymbol{r}$ from an arbitrary particle is expressed as $\boldsymbol{r} = (r, \theta)$, and the pair correlation function $g^{(2)}$ is a function of the point $\boldsymbol{r}$ as $g^{(2)} = g^{(2)}(\boldsymbol{r}) = g^{(2)}(r, \theta)$. With the notation $\Delta N_i(r,\theta)$ for the number of particles located in the infinitesimal area $\Delta S = r\Delta\theta\Delta r$ of the range $(r,\theta) \sim (r+\Delta r, \theta+\Delta\theta)$, then the pair correlation function $g^{(2)}(\boldsymbol{r})$ is evaluated by the following expression:

$$g^{(2)}(\boldsymbol{r}) = g^{(2)}(r,\theta) = \frac{n(r,\theta)}{n_0} = \frac{1}{N/S}\cdot\left\langle \frac{1}{N}\sum_{i=1}^{N}\frac{\Delta N_i(r,\theta)}{\Delta S}\right\rangle = \frac{1}{n_0 N}\cdot\left\langle \sum_{i=1}^{N}\frac{\Delta N_i(r,\theta)}{r\Delta\theta\Delta r}\right\rangle \tag{8.3}$$

in which ⟨–⟩ is the time average or ensemble average of the quantity of interest. If the dispersed particles are uniformly distributed in a system, then the local number density $n(r, \theta)$ is equal to the specified density $n_0$ and the pair correlation function will have the value of unity at any position except in the range of particle overlap.

In the case of a three-dimensional system, the pair correlation function $g^{(2)} = g^{(2)}(\boldsymbol{r}) = g^{(2)}(r, \theta, \varphi)$ at any point $\boldsymbol{r} = (r, \theta, \varphi)$ is evaluated using the following expression:

$$\begin{aligned} g^{(2)}(\boldsymbol{r}) = g^{(2)}(r,\theta,\varphi) &= \frac{n(r,\theta,\varphi)}{n_0} = \frac{1}{N/V}\cdot\left\langle \frac{1}{N}\sum_{i=1}^{N}\frac{\Delta N_i(r,\theta,\varphi)}{\Delta V}\right\rangle \\ &= \frac{1}{n_0 N}\cdot\left\langle \sum_{i=1}^{N}\frac{\Delta N_i(r,\theta,\varphi)}{r^2\sin\theta\Delta r\Delta\theta\Delta\varphi}\right\rangle \end{aligned} \tag{8.4}$$

in which $\theta$ and $\varphi$ are the polar angle measured from the $y$-axis or the magnetic field direction and the azimuthal angle measured from the $x$-axis direction, respectively. The term $\Delta N_i(r,\theta,\varphi)$ is the number of particles located in the infinitesimal volume $\Delta V = r^2\sin\theta\Delta r\Delta\theta\Delta\varphi$ of the range $(r,\theta,\varphi)$ ~ $(r+\Delta r,\theta+\Delta\theta,\varphi+\Delta\varphi)$. Similar to procedure for the radial distribution function, in simulations it may be reasonable to take the distance interval $\Delta r$ as $\Delta r/d = 0.1$ or 0.2, and to take the angle interval $(\Delta\theta, \Delta\varphi)$ as $(\Delta\theta, \Delta\varphi) = (\pi/90, \pi/90)$ or $(\pi/180, \pi/180)$. In the following, we demonstrate how the internal structure of chain-like clusters along the field direction are described by the characteristics of the pair correlation function.

Figure 8.3 shows snapshots of aggregate structures of spherical magnetic particles in a strong applied magnetic field, applied in the upward direction of the paper, in a two-dimensional system. Figures 8.3(a), (b) and (c) are three cases for the magnetic particle-particle interaction strength $\lambda = 1$, 5 and 10, respectively. Although the pair correlation function can be calculated for the whole radial and angle range from an arbitrary particle, we will focus on the pair correlation function in the $x$- and $y$-directions. It is noted that long chain-like clusters are expected to be formed in the magnetic field direction, i.e., in the $y$-axis direction, and therefore the $x$-axis is normal to the general orientation of the clusters. Figures 8.4(a) and (b), therefore, have been chosen to show the results of the pair correlation functions in the $y$- and $x$-directions, respectively. The snapshots and pair correlation functions were actually obtained from the results of Monte Carlo simulations for a thermodynamics equilibrium situation.

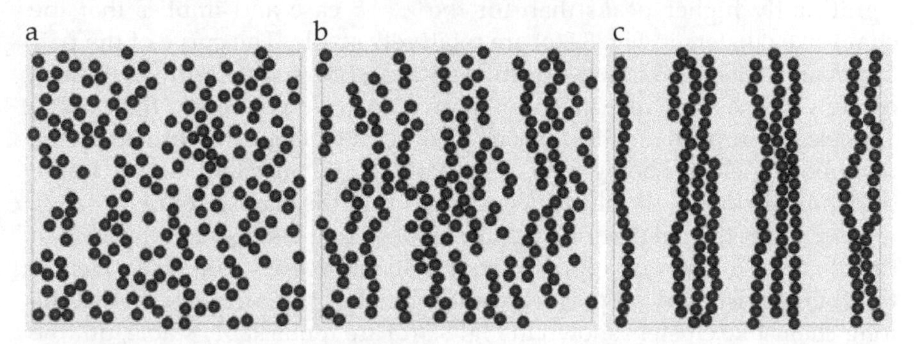


**Figure 8.3.** Snapshots for different magnetic interaction strengths in the external magnetic field strength $\xi = 20$: (a) $\lambda = 1$, (b) $\lambda = 5$ and (c) $\lambda = 10$: the magnetic field is applied in the upward direction of the paper.

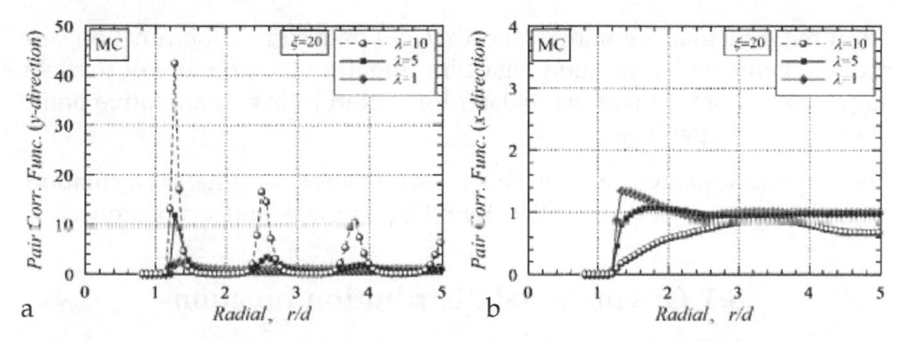


**Figure 8.4.** Pair correlation function for different magnetic interaction strengths in the magnetic field strength $\xi = 20$: $g^{(2)}(r)$ is (a) for the $y$-axis direction and (b) for the $x$-axis direction.

Figure 8.3(a) shows that there are almost no clusters formed in the system, therefore the pair correlation functions in both the $y$- and $x$-directions show the characteristics of a gas system, that is, the distributions exhibit the constant value of unity past the particle-particle contact range. In the case of relatively short clusters being formed in the field direction, shown in Fig. 8.3(b), high peaks appear in the pair correlation function in the $y$-direction at the positions of an integer times the particle diameter $d = 1.3$ including the uniform steric layer. In contrast, the pair correlation function in the $x$-direction still exhibits gas-like structure, so that this implies that these chain-like clusters are not thick as can be seen in the snapshot in Fig. 8.3(b). In the case of an increased magnetic interaction strength, shown in Fig. 8.3(c), long chain-like clusters are formed and these clusters aggregate further to form thick chain-like clusters in the field direction. The pair correlation function in the $y$-direction exhibits significantly higher peaks than for the $\lambda$ = 5 case and implies that the chain-like clusters in Fig. 8.3(c) are relatively stable. The curve of the pair correlation function in the $x$-direction shows a characteristic different from dilute gas-like structure, that is, it shows much lower values than unity. This clearly implies that the thick chain-like clusters are not formed by the neighboring cluster being merely shifted by the particle diameter in the direction normal to the cluster direction. From the snapshot in Fig. 8.3(c), it is seen that the neighboring cluster is located at a staggered distance in the cluster direction of approximately half the particle diameter. Hence, the characteristics of the pair correlation functions clearly suggest that the long chain-like clusters shown in Fig. 8.3(c) are remarkably stable, but the thick formation is not significantly stable. In other words, the neighboring long chain-like clusters retain a slight random motion that is normal to their own and the magnetic field direction.

From these results, we understand that the pair correlation function can provide valuable information regarding the internal structure of particle aggregates in order to discuss cluster formation from a quantitative point of view at a deeper level.

The subroutine shown above in Section 8.1 illustrates a suitable calculation procedure for the pair correlation functions in the $y$- and $x$-directions.

## 8.3 Orientational distribution function

As explained above in Sections 8.1 and 8.2, the radial distribution and pair correlation functions can provide the information as to how the ambient particles are stochastically located around an arbitrary particle, but do not yield the information regarding the orientational characteristics of the particle or the magnetic moment. In the case of non-magnetic spherical particles, the orientational characteristics of the particle are not expected to give rise to a significant influence on the physical properties of a particle suspension. On the other hand, the orientational characteristics can provide important and valuable information regarding the internal structure of aggregates in the case of non-spherical particles for obtaining their essential features at a deeper level. The orientational distribution function is also indispensable for investigating rheological or magneto-rheological properties of a particle suspension because these properties are significantly influenced by the orientational characteristics of the system. Hence, for a suspension composed of non-spherical particle or particles with non-symmetrical properties, it is particularly important to address the orientational distribution function for discussing the internal structure of aggregates. In the following, we discuss the definition of the orientational distribution function and demonstrate how this function can describe the orientational characteristics of a magnetic rod-like particle suspension.

We consider the orientation of a magnetic rod-like particle in an external magnetic field. The orientation of the particle is denoted by the unit vector $\boldsymbol{e}$, and this vector is expressed in terms of the polar angle $\theta$ and the azimuthal angle $\varphi$ as $\boldsymbol{e} = (\sin\theta\cos\varphi, \sin\theta\sin\varphi, \cos\theta)$ in the normal coordinate system. Employing these symbols, the orientational distribution function $\Psi(\theta, \varphi)$ is defined in such a way that the probability of the particle orientation being found in the range between $(\theta, \varphi)$ and $(\theta + d\theta, \varphi + d\varphi)$ is $\Psi(\theta,\varphi) \sin\theta d\theta d\varphi$. From this definition, the orientational distribution is recognized as a probability density function, and therefore it is required to satisfy the following probability condition:

$$\int_0^{2\pi}\int_0^{\pi} \Psi(\theta,\varphi)\sin\theta d\theta d\varphi = 1 \tag{8.5}$$

If the particle is able to incline uniformly to any point on a spherical surface and does not have a specific favored direction, then $\Psi(\theta, \varphi)$ becomes an equilibrium distribution that is expressed as

$$\Psi(\theta,\varphi) = \frac{1}{4\pi} \tag{8.6}$$

It is noted that a larger or smaller value of the orientational distribution function implies a larger or smaller probability of the particle inclining in that direction.

Using the function $\Psi$, the average of an arbitrary quantity $G(\theta, \varphi)$ is written as

$$\langle G \rangle = \int_0^{2\pi}\int_0^{\pi} G(\theta,\varphi)\Psi(\theta,\varphi)\sin\theta d\theta d\varphi \tag{8.7}$$

This expression may be used for evaluating a viscosity arising from the particle orientation in the simulation of particle suspensions.

When an unsteady flow such as a sinusoidal shear flow is treated, the orientational distribution function becomes dependent on time as $\Psi(\theta, \varphi, t)$ and the change in the orientational characteristics can be captured by this time-dependent orientational distribution function at an arbitrary simulation time step. For reference, we refer to the continuum condition that the orientational distribution function is required to satisfy:

$$\frac{\partial \Psi}{\partial t} = -\frac{\partial}{\partial \boldsymbol{e}} \cdot (\dot{\boldsymbol{e}} \Psi) \tag{8.8}$$

in which $\boldsymbol{e}$ is the particle orientation, $\dot{\boldsymbol{e}}$ is its time differential, and $\partial / \partial \boldsymbol{e}$ is the nabla operator with respect to the angles $\theta$ and $\varphi$. This condition is an important expression in the analytical investigation of the orientational characteristics of non-spherical particles based on the orientational distribution function. In the following, we discuss the procedure for evaluating the orientational distribution function in simulations.

We first discretize the space defined by the angles $\theta$ and $\varphi$ using a small angle interval $\Delta\theta$ and $\Delta\varphi$ in the polar and the azimuthal direction, respectively. Hence, $\theta$ is discretized from $\theta_0(=0)$ as $\theta_1(=\Delta\theta)$, $\theta_2(=2\Delta\theta)$, … , $\theta_i(=i\Delta\theta)$, … , $\theta_P(=P\Delta\theta=\pi)$, and similarly, $\varphi$ is discretized from $\varphi_0(=0)$ as $\varphi_1(=\Delta\varphi)$, $\varphi_2(=2\Delta\varphi)$, … , $\varphi_j(=j\Delta\varphi)$, … , $\varphi_Q(=Q\Delta\varphi=2\pi)$, where the number of discretization points ranges from 1 to $P$ for the polar angle and from 1 to $Q$ for the azimuthal angle. If the number of particles that incline in the direction range between $(\theta_i, \varphi_j)$ and $(\theta_{i+1}, \varphi_{j+1})$ is denoted by $f_{ij}^{(sum)}$ after $N_{smpl}$ number of sampling in a simulation, then the orientational distribution function $\Psi(\theta_i, \varphi_j) = \Psi_{ij}$ is obtained from the following expression:

$$\Psi_{ij} = \frac{f_{ij}^{(sum)} / (\sin\theta_i \Delta\theta\Delta\varphi)}{\sum_{i=1}^{P}\sum_{j=1}^{Q} f_{ij}^{(sum)}} \tag{8.9}$$

This expression is shown to satisfy the probability condition in Eq. (8.5) after substitution of the quantity in Eq. (8.9) into the discretized version of the integral in Eq. (8.5) to yield

$$\begin{aligned}\int_0^{2\pi}\int_0^{\pi} \Psi(\theta,\varphi)\sin\theta d\theta d\varphi &\approx \sum_{i=1}^{P}\sum_{j=1}^{Q} \Psi_{ij}\sin\theta_i \Delta\theta\Delta\varphi \\ &= \sum_{i=1}^{P}\sum_{j=1}^{Q} \frac{f_{ij}^{(sum)} / (\sin\theta_i \Delta\theta\Delta\varphi)}{\sum_{i=1}^{P}\sum_{j=1}^{Q} f_{ij}^{(sum)}} \sin\theta_i\Delta\theta\Delta\varphi = 1\end{aligned} \tag{8.10}$$

Hence, it is seen that the employment of sufficient small values for $\Delta\theta$ and $\Delta\varphi$ ensures satisfaction of the probability condition in Eq. (8.5).

Figure 8.5 shows results of the orientational distribution function of magnetic rod-like particles that perform rotational Brownian motion in a simple shear flow with the $x$-direction $((\theta, \varphi) = (90°, 0°))$ under an external magnetic field with the $y$-direction $((\theta, \varphi) = (90°, 90°))$. The three graphs in this figure were obtained for a strong shear flow with Peclet number (shear rate) $Pe = 10$ and exhibit the dependence of the orientational distribution function on the magnetic field strength $\xi$. For the weakest field strength $\xi = 1$, shown in Fig. 8.5(a), the particle tends to incline slightly in the shear flow direction that results in a low peak at $\varphi \simeq 22°$ with a wide foot. As the magnetic field strength is increased, shown for $\xi = 5$ in Fig. 8.5(b), the peak becomes higher with a narrower foot and its position is shifted toward the magnetic field direction $\varphi = 90°$. For the stronger field strength of $\xi = 10$, shown in Fig. 8.5(c), the above trend becomes more significant and a higher peak and a more significant alignment to the magnetic field are obtained. These orientational characteristics shown by the orientational distribution function are quite reasonable and straightforwardly understood from the dependence on various factors such as the external magnetic field strength as obtained from theoretical and simulation approaches.

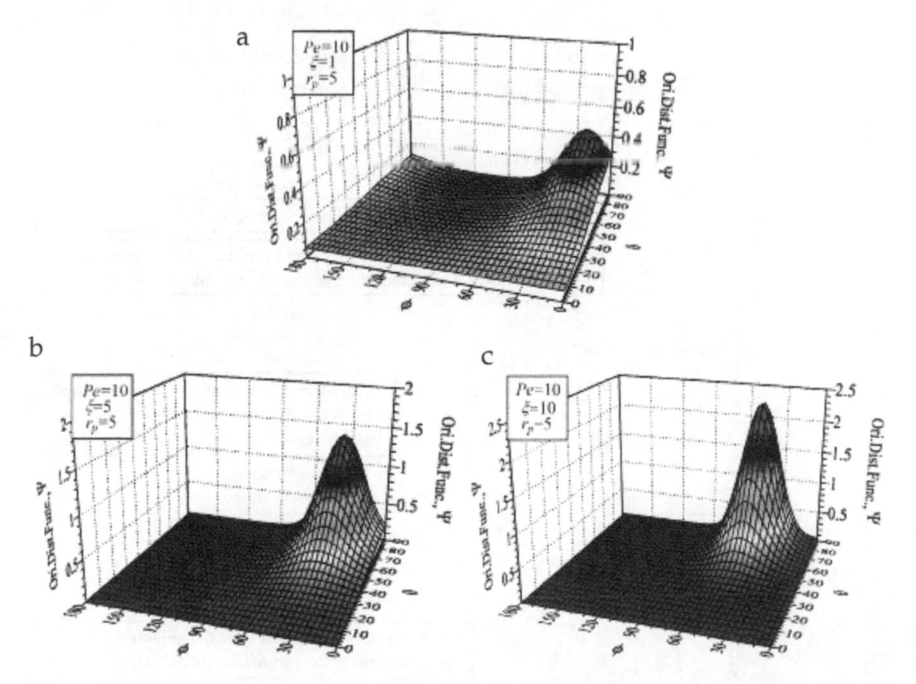


**Figure 8.5.** Orientational distribution function for the Peclet number (sear rate) $Pe = 10$: (a) $\xi = 1$, (b) $\xi = 5$ and (c) $\xi = 10$.

From this consideration, we understand that the orientational distribution function is a useful quantity for discussing the rotational motion of magnetic rod-like particles in a variety of circumstances of flow field and external magnetic field.

Finally, we show an example of a subroutine for evaluating the orientational distribution function, which is written in FORTRAN language. In this subroutine, the orientation of each particle is necessary for calculating the distribution function.

```
001 C*********************************************************************
002 C                                                                     *
003 C  THE FOLLOWING SUBROUTINE IS FOR EVALUATING                         *
004 C                      THE ORIENTATIONAL DISTRIBUTION FUNCTION        *
005 C                                                                     *
006 C                                                 WRITTEN BY A. SATOH *
007 C*********************************************************************
008 C
009 C     N    : NUMBER OF PARTICLES
010 C     EX(N),EY(N),EZ(N) : PARTICLE AXIS DIRECTION
011 C
012 C     THETA(*)  : REPRESENTATIVE ANGLE FOR EACH RANGE FOR ORI.DIST.FUN.
013 C     PHAI(*)   : REPRESENTATIVE ANGLE FOR EACH RANGE FOR ORI.DIST.FUN.
014 C
015 C     DTHETA(*): RANGE WIDTH FOR THETA ANGLE
016 C     DPHAI(*) : RANGE WIDTH FOR PHI ANGLE
017 C
018 C     ORIDIST(*,#): ORI.DIST.FUNC. AT (THETA,PHAI) POINT
019 C                               ( 0=< * <= NTHETA   , 0=< # <= NPHAI )
020 C     SUMRAD((*,#): SUMMATION OF ORI.DIST.FUNC.
021 C
022 C     COSTHCHK(*)  : ARE USED FOR DETERMINING THE POLAR ANGLE SITE
023 C     SINTHETA(*)
024 C     SINPHCHK(#)  : IS USED FOR DETERMINING THE AZIMUTHAL ANGLE SITE
025 C
026 C---------------------------------------------------------------------
027 C
028                      ...
029                      ...
030 C                                               --- PARAMETER (1) ---
031       NTHETA=  90
032       NPHAI  = 180
033 C
034       CALL ORIDINIT
035 C
036                      ...
037                      ...
038 C                                               --- INITIALIZATION ---
039       NORICTR = 0
040       DO 45 ITHETA=0, NTHETA
041       DO 45 IPHAI =0, NPHAI
042         SUMORI( ITHETA, IPHAI ) = 0.D0
043    45 CONTINUE
044 C
045                      ...
046                      ...
047 C
048 C     ----------------------------------------------------------------
049 C     ------------------    START OF MAIN LOOP     --------------------
050 C     ----------------------------------------------------------------
051 C
052                      ...
053                      ...
054 C                                 --- ORIENTATIONAL DISTRIBUTION ---
055          NORICTR = NORICTR + 1
056 C
057          CALL ORIDISTCA( N )
058 C
059          DO 200 ITHETA=0, NTHETA
060          DO 200 IPHAI =0, NPHAI-1
061            SUMORI( ITHETA,IPHAI) = SUMORI(  ITHETA,IPHAI)
062      &                             + ORIDIST(  ITHETA,IPHAI)
```

• $P$ is denoted by NTHETA and $Q$ by NPHAI.

• The space defined by the angles $\theta$ and $\varphi$ is discretized for evaluating the orientational distribution function in the subroutine ORIDISTCA.

• SUMORI(*,#) is used for accumulating the data of the orientational distribution function evaluated at each simulation step.

• The values of the orientational distribution function are calculated in the subroutine ORIDISTCA.

```
063   200      CONTINUE
064                     ...
065                     ...
066 C
067 C     ------------------------------------------------------------
068 C     ---------------------    END OF  MAIN LOOP   ---------------
069 C     ------------------------------------------------------------
060 C
071 C                                 -------- CAL. O.D.F.(THETA,PHAI) ---
072       SUMORI00     = 0.D0
073       DO 1230 ITHETA=1, NTHETA-1
074       DO 1230 IPHAI =0, NPHAI-1
075         SUMORI00 = SUMORI00 + SUMORI(ITHETA,IPHAI)
076         C00   = DTHETA(ITHETA)*DPHAI(IPHAI)*SINTHETA(ITHETA)
077         IF( C00 .LT. 1.D-20 )  C00 = 1.D30
078         SUMORI( ITHETA,IPHAI) = SUMORI( ITHETA,IPHAI)/C00
079  1230 CONTINUE
080       IF( SUMORI00 .LT. 1.D-20 )  SUMORI00 = 1.D30
```

- The calculated data are accumulated in the variable SUMORI(*,#).
- This DO loop calculates the numerator in Eq. (8.9).
- SUMORI00 implies the denominator in Eq. (8.9).

```
081       DO 1232 ITHETA=1, NTHETA-1
082       DO 1232 IPHAI =0, NPHAI-1
083         ORIDIST( ITHETA,IPHAI) = SUMORI( ITHETA,IPHAI)/SUMORI00
084  1232 CONTINUE
085 C
086 C                          --- SYMMETRY OF ORI. DIST. (THETA,PHAI) ---
087 C
088       CALL SYMMETCAL( NTHETA, NPHAI, THETA )
089                     ...
090                     ...
091                                                               STOP
092                                                               END
093 C*********************************************************************
094 C**************************    SUBROUTINE    ************************
095 C*********************************************************************
096 C
097 C**** ORIDINIT ****
098       SUBROUTINE ORIDINIT
099 C
100       IMPLICIT REAL*8 (A-H,O-Z) , INTEGER (I-N)
101 C
102       COMMON /BLOCK24/ NTHETA, NPHAI
103       COMMON /BLOCK25/ THETA, DTHETA, COSTHCHK, SINTHETA
104       COMMON /BLOCK26/ PHAI , DPHAI , SINPHCHK
105 C
106 C     ----------------------------------
107       INTEGER    NN
108       PARAMETER( NN=1000 , NNTHETA=91 , NNPHAI=181 )
109       PARAMETER( PI=3.141592653589793D0 )
110 C     ----------------------------------
111 C
112       REAL*8   THETA(   0:NNTHETA) , DTHETA(  0:NNTHETA)
113       REAL*8   COSTHCHK(0:NNTHETA) , SINTHETA(0:NNTHETA)
114       REAL*8   PHAI(     0:NNPHAI)  , DPHAI(    0:NNPHAI )
115       REAL*8   SINPHCHK(0:NNPHAI)
116 C
117       REAL*8   CTHETA, CDTHETA, CDTHETA2
118       REAL*8   CPHAI , CDPHAI , CDPHAI2
119 C
120 CCCC  NTHETA  =  90
```

- The subroutine SYMMETCAL may be necessary in a certain situation.
- The subroutine ORIDINIT discretizes the space defined by the angles $\theta$ and $\varphi$ for evaluating the orientational distribution function.

```
121 CCCC  NPHAI   = 180
122       CDTHETA =  PI          / DBLE( NTHETA )
123       CDPHAI  = (PI*2.D0) / DBLE( NPHAI  )
124 C     --------------------------------- CAL. THETA(*), SINTHETA(*) ---
125       DO 20 I=0, NTHETA
126         THETA(I)    = CDTHETA*DBLE(I)
127         SINTHETA(I) = DSIN( THETA(I) )
128    20 CONTINUE
129 C                                              --- CAL. DTHETA(*) ---
130       DTHETA(0)        = CDTHETA/2.D0
131       DTHETA(NTHETA)   = CDTHETA/2.D0
132       DO 40 I=1, NTHETA-1
133         DTHETA(I) = CDTHETA
134    40 CONTINUE
135 C                                            --- CAL. COSTHCHK(*) ---
136       CDTHETA2 = CDTHETA/2.D0
137       DO 60 I=1, NTHETA
138         CTHETA     = THETA(I) - CDTHETA2
139         COSTHCHK(I) = DCOS( CTHETA )
140    60 CONTINUE
141 C
```

- THETA(*) is the representative angle of each polar angle site.
- COSTHCHK(*) is used for determining the polar angle site to which each particle aligns.

```
142 C      ------------------------------------------ CAL. PHAI(*) ---
143        DO 120 I=0, NPHAI
144          PHAI(I) = CDPHAI*DBLE(I)
145    120 CONTINUE
146 C
147        DO 140 I=0, NPHAI
148          DPHAI(I) = CDPHAI
149    140 CONTINUE
150 C                                      --- CAL. SINPHCHK(*) ---
151        CDPHAI2 = CDPHAI/2.D0
152        DO 160 I=1, NPHAI
153          CPHAI = PHAI(I) - CDPHAI2
154          SINPHCHK(I) = DSIN( CPHAI )
155    160 CONTINUE
156
157                                                          END
158 C**** ORIDISTCA ****
159        SUBROUTINE ORIDISTCA( N )
160 C
161        IMPLICIT REAL*8 (A-H,O-Z) , INTEGER (I-N)
162 C
163        COMMON /BLOCK3/  EX    , EY    , EZ
164 C
165        COMMON /BLOCK24/ NTHETA, NPHAI
166        COMMON /BLOCK25/ THETA, DTHETA, COSTHCHK, SINTHETA
167        COMMON /BLOCK26/ PHAI , DPHAI , SINPHCHK
168        COMMON /BLOCK27/ ORIDIST , SUMORI
169 C
170 C      --------------------------------
171        INTEGER    NN
172        PARAMETER( NN=1000 , NNTHETA=91 , NNPHAI=181 )
173        PARAMETER( PI=3.141592653589793D0 )
174 C      --------------------------------
175 C
176        REAL*8   EX(NN) , EY(NN) , EZ(NN)
177 C
178        REAL*8   THETA(   0:NNTHETA) , DTHETA(  0:NNTHETA)
179        REAL*8   COSTHCHK(0:NNTHETA) , SINTHETA(0:NNTHETA)
180        REAL*8   PHAI(    0:NNPHAI)  , DPHAI(   0:NNPHAI )
181        REAL*8   SINPHCHK(0:NNPHAI)
182 C
183        REAL*8   ORIDIST(  0:NNTHETA , 0:NNPHAI )
184        REAL*8   SUMORI(   0:NNTHETA , 0:NNPHAI )
185 C
186        REAL*8   CTHETA, CDTHETA, CDTHETA2
187        REAL*8   CPHAI , CDPHAI , CDPHAI2
188 C
189        REAL*8   EXI, EYI, EZI
190        REAL*8   CS1, SN1, CS2, SN2
191        INTEGER  ITHETA, IPHAI
192 C
193        DO 500 I=1,N
194 C
195          EXI = EX(I)
196          EYI = EY(I)
197          EZI = EZ(I)
198 C                                      --- CS1,SN1,CS2,SN2 ---
199          CS1 = EZI
200          SN1 = DSQRT( 1.D0 - CS1**2 )

201          IF( SN1 .LT. 0.00001D0 ) THEN
202 CCCCCC      CS2 = 1.D0
203 CCCCCC      SN2 = 0.D0
204            CS2 = EXI/SN1
205            SN2 = EYI/SN1
206            WRITE(6,*) '------- 1330 in sub ORIDISTCA --------'
207          ELSE
208            CS2 = EXI/SN1
209            SN2 = EYI/SN1
210          END IF
211 C                                                --- ITHETA ---
212          DO 100 II=1, NTHETA
213            IF( COSTHCHK(II) .LT. CS1 ) THEN
214              ITHETA = II - 1
215              GOTO 120
216            END IF
217    100   CONTINUE
218          ITHETA = NTHETA
219 C                                                 --- IPHAI ---
220    120   DO 200 II=1, NPHAI/4
221            IF( SINPHCHK(II) .GT. DABS(SN2) ) THEN
222              IPHAI = II - 1
223              GOTO 220
224            END IF
```

• PHAI(*) is the representative angle of each azimuthal angle site.

• SINPHCHK(*) is used for determining the azimuthal angle site to which each particle aligns.

• The subroutine ORIDISTCA calculates the values of the orientational distribution function.

• (EXI,EYI,EZI) implies the orientation of particle $i$.

• The orientation $\boldsymbol{e}_i$ of particle $i$ is $\boldsymbol{e}_i=(e_{xi},e_{yi},e_{zi}) = (\sin\theta\cos\varphi, \sin\theta\sin\varphi, \cos\theta)$.

• CS1 implies $\cos\theta=e_{zi}$, and SN1 is $\sin\theta = (1-\cos\theta)^{1/2}$.

• When the singularity of $\sin\theta=0$ appears, a warning statement is printed out.

• CS2=$\cos\varphi$ is calculated from $e_{xi}/\sin\theta$ and SN2=$\sin\varphi$ is calculated from $e_{yi}/\sin\theta$.

• The polar angle site to which each particle aligns is determined using COSTHCHK(*).

• The azimuthal angle site to which each particle aligns is determined using SINPHCHK(*).

```
225    200     CONTINUE
226            IPHAI = NPHAI/4
227 C
228    220     IF( SN2 .GE. 0.D0 ) THEN
229              IF( CS2 .LT. 0.D0 )  IPHAI = NPHAI/2 - IPHAI
230            ELSE
231              IF( CS2 .GE. 0.D0 )  THEN
232                IPHAI = NPHAI - IPHAI
233              ELSE
234                IPHAI = NPHAI/2 + IPHAI
235              END IF
236            END IF
237 C
238            IF( IPHAI .EQ. NPHAI ) IPHAI = 0
239 C                                                  --- SAMPLING ---
240            ORIDIST(ITHETA, IPHAI) = ORIDIST(ITHETA, IPHAI) + 1.D0
241 C
242 C
243     500 CONTINUE
244                                                                RETURN
245                                                                END
```

• Any point in the first quadrant satisfies the relationships of $\sin\varphi > 0$ and $\cos\varphi > 0$.

• Any point in the second quadrant satisfies the relationships of $\sin\varphi > 0$ and $\cos\varphi < 0$.

• The third and the fourth quadrant satisfy the relationships of $\sin\varphi < 0$ and $\cos\varphi < 0$, and $\sin\varphi < 0$ and $\cos\varphi > 0$, respectively.

## 8.4 Orientational pair correlation function

As explained above in Section 8.2, the pair correlation function $g^{(2)}(\boldsymbol{r})$ describes how the ambient particles are stochastically located, that is, describes stochastic relationship of the ambient particles around an arbitrary particle. Hence, this function can provide information regarding the correlation of the position of the particles in a system but cannot provide the orientational correlation of the particle directions. If we consider a suspension composed of rod-like particles in a flow field with an external magnetic field, information regarding the correlation of particle orientations will surely be a significant aid in understanding the microstructure of particle aggregates. Similar to the pair correlation function, the orientational correlation function describes a measure of the degree regarding the correlation between particle orientations. In a suspension of magnetic non-spherical particles, it is significantly important to clarify the direction of alignment of the magnetic moment of each particle belonging to the same cluster, and the correlation of the particle or magnetic moment orientations of the clusters gives rise to information regarding the magnetic properties and microstructure of aggregates. This type of information is an important aid in understanding the dependence of the magneto-rheological characteristics of a suspension on the microstructure of the particle aggregates. In the following, we show the definition of the orientational pair correlation function and then demonstrate how this function can represent the characteristics of the orientational correlation properties of the particles in the case of a suspension of magnetic rod-like particles.

We consider a suspension composed of $N$ number of rod-like particles in a three-dimensional system with system volume $V$. If we employ the notation $\theta_{ij}$ for an angle between the orientations of particles $i$ and $j$,

$n_0 = N/V$ for a given number density, $\delta(-)$ for Dirac's delta function and $P_2(\cos\theta_{ij}) = (3\cos^2\theta_{ij} - 1)/2$ for the second Legendre polynomial, then the orientational pair correlation function $f_{o.c.}^{(2)}(\boldsymbol{r})$ is defined as [1]

$$f_{o.c.}^{(2)}(\boldsymbol{r}) = \left\langle \sum_{i=1}^{N} \sum_{j=1(\neq i)}^{N} P_2(\cos\theta_{ij})\delta(\boldsymbol{r}_{ji} - \boldsymbol{r}) \right\rangle \Big/ \left\langle \sum_{i=1}^{N} \sum_{j=1(\neq i)}^{N} \delta(\boldsymbol{r}_{ji} - \boldsymbol{r}) \right\rangle \tag{8.11}$$

in which $\langle - \rangle$ is the time or ensemble average. Similar to the radial distribution function, if we focus on the characteristics at a radial distance $r = |\boldsymbol{r}|$ (not angle-dependent), the orientational pair correlation function $f_{o.c.}(r)$, which is now a function of only the radial distance $r$, is evaluated from the following expression:

$$f_{o.c.}(r) = \frac{1}{n_0 g(r)} \cdot \frac{1}{N} \left\langle \sum_{i=1}^{N} \frac{\sum' P_2(\cos\psi_{ij}^{(p)})}{4\pi r^2 \Delta r} \right\rangle \tag{8.12}$$

in which $\sum'$ implies the summation of the contribution from particle $j$ that is located in the infinitesimal shell volume $\Delta V = 4\pi r^2 \Delta r$ at the radial $r$. From the expression of the radial distribution function in Eq. (8.1), it is seen that if all the particles in a system incline in the same or in the opposite direction, $P_2(\cos\theta_{ij})$ is equal to unity for all the pairs of particles, so that Eq. (8.12) gives rise to unity, i.e., $f_{o.c.}(r) = 1$ under the assumption of $g(r) = 1$. On the other hand, if all the pair of particles incline perpendicularly, then the orientational pair correlation function yields $f_{o.c.}(r) = -1/2$ and the case of no orientational correlation between two particles leads to $f_{o.c.}(r) = 0$.

Hematite particles are generally magnetized in a direction different from that of the particle axis [2–4], so that it may be valuable to address the orientational correlation function of the magnetic moment directions of two magnetic particles, $f_{o.c.}^{(m)}(r)$. An equation for evaluating this function is straightforwardly obtained by replacing the angle $\theta_{ij}$ with $\theta_{ij}^{(m)}$ that is an angle between the orientations of the magnetic moments of particles $i$ and $j$. In the following, we demonstrate how the orientational pair correlation function can describe the orientational correlation characteristics of the constituent particles in their clusters, using sample snapshots that were obtained from actual Monte Carlo simulations.

Figure 8.6 shows snapshots of the aggregate structures of magnetic rod-like particles that are magnetized in the particle axis direction with a point dipole moment at the particle center and are in thermodynamic equilibrium. Figures. 8.6(a) and (b) are snapshots for the magnetic particle-particle interaction strength $\lambda = 10$ and 50, respectively, in the situation of no external magnetic field. The corresponding results of the orientational correlation function $f_{o.c.}^{(p)}(r)$ $(= f_{o.c.}(r))$ of the particle orientations are shown in Fig. 8.7. Since the rotational Brownian motion of the rod-like particles has a significant influence on the stability of the cluster formation, there are no clusters formed in the system for the case of $\lambda = 10$, as shown in Fig. 8.6(a), whereas particles significantly aggregate to form raft-like clusters for $\lambda = 50$, as shown in Fig. 8.6(b). These results suggest that in the former snapshot there is almost no correlation of the particle orientations of the rod-like particles, and in contrast, for the case of $\lambda = 50$, the raft-like clusters have a strong orientational correlation between the directions of the constituent particles in each raft-like cluster. These observations are clearly exhibited in a quantitative manner by the characteristics of the orientational pair correlation function shown in Fig. 8.7. That is, in the case of $\lambda = 10$, the correlation function shows an almost zero value except in the close vicinity area, whereas in the case of $\lambda = 50$, it exhibits a curve that attains to much larger values and shows a series of high peaks with an oscillatory tail. The high peaks at the positions of an integer times the particle diameter (= 1.3, including the surfactant layer) result from the raft-like cluster formation while the intermediate peaks, for instance, at $r/d \simeq 1.9$ come from other structures such as the crystal-like structure that is observed in the upper-right and upper-left areas in Fig. 8.6(b).

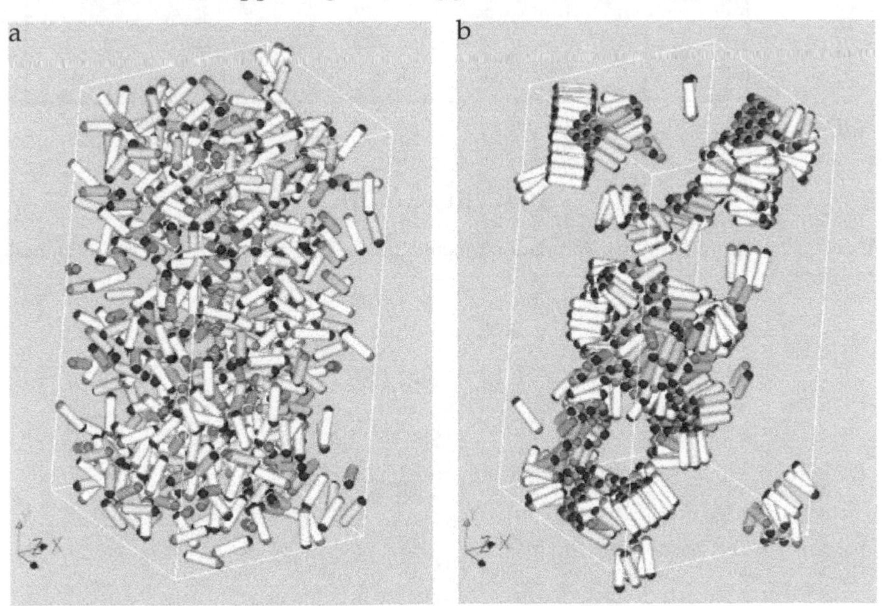


**Figure 8.6.** Snapshots for the two different cases of the magnetic particle-particle interaction strength in no external magnetic field: (a) $\lambda = 10$ and (b) $\lambda = 50$.

From these examples, we have illustrated that the orientational pair correlation function is a significantly appropriate quantity for clarifying the essence of the microstructure of the clusters from the viewpoint in regard to the characteristics of the orientational correlation of the constituent particles of the clusters.

Finally, we show an example of a subroutine for evaluating the orientational pair correlation function, which is written in FORTRAN language. In this subroutine, the position and orientation of each particle are necessary for calculating the correlation function.

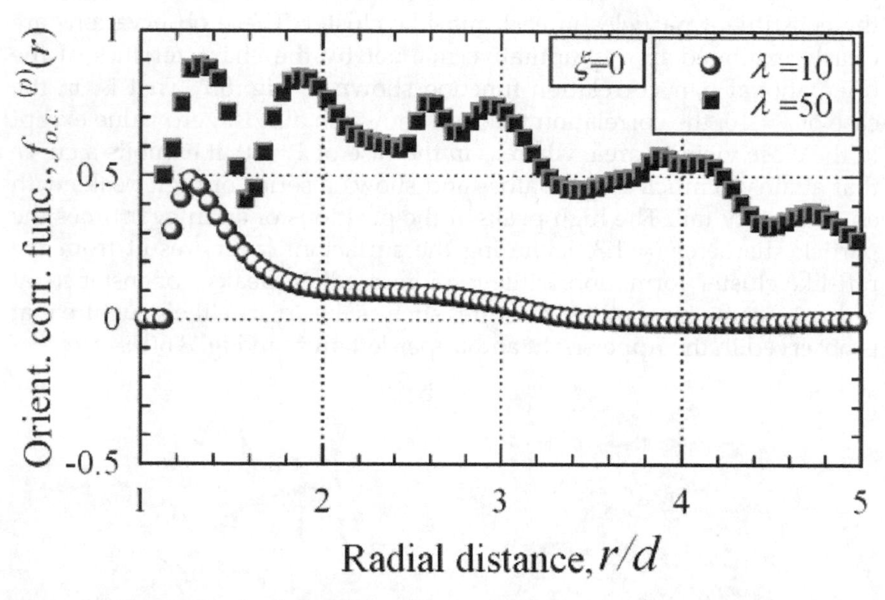


**Figure 8.7.** Orientational pair correlation function for the previous snapshots in no external magnetic field.

```
001 C**********************************************************************
002 C                                                                     *
003 C    THE FOLLOWING SUBROUTINE IS FOR EVALUATING                       *
004 C                       THE PAIR ORIENTATIONL CORRELATION FUNCTION    *
005 C                                                                     *
006 C                                                    WRITTEN BY A. SATOH *
007 C**********************************************************************
008 C      N           : NUMBER OF ROD-LIKE PARTICLES
009 C      RP          : ASPECT RATIO OF ROD-LIKE PARTICLES
010 C      XL,YL,ZL : DIMENSIONS OF SIMULATION REGION
011 C      RX(I),RY(I),RZ(I) : POSITION OF PARITCLE I
012 C      EX(I),EY(I),EZ(I) : DIRECTION OF RODLIKE PARTICLE I
013 C
014 C      DR          : RADIAL INTERVAL FOR CALCULATING RAD.DIST.FUNC.
015 C                    AND ORIENTATIONAL PAIR CORRELATION FUNCTION
016 C      RMX         : MAXIMUM RADIAL
017 C      RRATIO      : IS USED TO SAVE SAMPLING DATA EFFICIENTLY
018 C      RRAD(*)     : REPRESENTATIVE POINT FOR EACH UNIT FOR RAD.DIST.FUNC.
019 C                    AND ORIENTATIONAL PAIR CORRELATION FUNCTION
020 C                    ( NRADST=< * =< NRADED )
021 C
```

```
022 C      RADDIST(*)  : RAD. DIST. FUNC. AND TREATED AS KNOWN VALUES
023 C                    IN THIS SUBROUTINE
024 C
025 C      ORICORR(*)  : ORIENTATIONAL PAIR CORRELATION FUNCTION
026 C      SUMORICO(*) : SUMMATION OF ORI. PAIR CORR. FUNC.
027 C
028 C      ORDER       : ORDER PARAMETER OF PARTICLE DIRECTION
029 C
030 C      NRADCNTR    : NUMBER OF SAMPLING
031 C
032 C      +++++++  -XL/2 <RX(I) <XL/2 , -YL/2 <RY(I)< YL/2  ++++++++
033 C      +        -ZL/2 <RZ(I) <ZL/2                              +
034 C----------------------------------------------------------------------
035                         ...
036                         ...
037 C
```

• RRATIO is used for determining the index I where the ori. corr. func. is saved in the variable ORICORR(I).

```
038        RP          = 5.D0
039        RRATIO      = 20.D0
040        RMX         = 2.D0*RP + 1.D0
```

• The subroutine RADIALR sets the radial positions for the following evaluation.

```
041        CALL RADIALR
042        CRAD  = ( XL*YL*ZL/DBLE(N*N) ) / ( 4.D0*PI*DR )
043 C
044                         ...
045                         ...
046 C
```

• CRAD is a constant that appears in the expression of the orientational pair correlation function.

```
047        NRADCTR   = 0
048        ORDER     = 0.D0
049        DO 40 IRAD=1, NRADED
050          SUMORICO(IRAD) = 0.D0
051     40 CONTINUE
052 C
053                         ...
054                         ...
```

• SUMORICO(*) is used for accumulating the data of the orientational pair correlation function.

```
055 C
056 C      ------------------------------------------------------------------
057 C      --------------------    START OF MAIN LOOP   -------------------
058 C      ------------------------------------------------------------------
059                         ...
060                         ...
061 C
062 C                            ---ORIENTATIONAL PAIR CORRELATION FUNCTION ---
063        NRADCTR = NRADCTR + 1
064 C
065        CALL ORICORR( N )
066 C
```

• The values of the ori. pair corr. func. are calculated in the subroutine ORICORR.

```
067        DO 460 IRAD=NRADST, NRADED
068          SUMORICO(IRAD) = SUMORICO(IRAD) + ORICORR(IRAD)
069    460 CONTINUE
060                         ...
071                         ...
072 C
```

• The values of the ori. pair corr. func. are accumulated in the variable SUMORICO(*).

```
073 C      ------------------------------------------------------------------
074 C      ----------------------  END OF MAIN LOOP   ---------------------
075 C      ----------------------------------------
076 C
077 C                            ---ORIENTATIONAL PAIR CO
078 C
```

• The pre-final results are obtained by dividing the accumulated values by the number of sampling, NRADCNTR.

```
079        DO 1220 IRAD=NRADST, NRADED
080          ORICORR(IRAD) = SUMORICO(IRAD)/DBLE(NRADCTR)
081          IF( RADDIST(IRAD) .GE. 1.D-10 ) THEN
082            ORICORR(IRAD) = ORICORR(IRAD) / RADDIST(IRAD)
083          ELSE
084            ORICORR(IRAD) = 0.D0
085          END IF
086   1220 CONTINUE
```

• The final results are obtained by dividing the data by the radial distribution function.

```
087 C                            ---ORDER PARAMETER OF PARTICLE ORIENTATION ---
088 C
089        ORDER     = ORDER      / DBLE(NRADCTR)
090                         ...
091                         ...
092                                                                STOP
093                                                                END
```

• The order parameter is obtained by dividing the data by the number of sampling.

```
094 C**********************************************************************
095 C****************************    SUBROUTINE   *************************
096 C**********************************************************************
097 C
098 C**** RADIALR ****
099        SUBROUTINE RADIALR )
100 C
101        IMPLICIT REAL*8 (A-H,O-Z), INTEGER (I-N)
102 C
103        COMMON /BLOCK45/ DR, RMX, CRAD, NRADST, NRADED, RRATIO, RRAD
```

• The subroutine RADIALR sets the radial positions for the following evaluation.

```
104 C
105 C       -------------------------------------------------------------------
106         INTEGER    NN , SS
107         PARAMETER( NN=1500 , SS=10000 , PI=3.141592653589793D0 )
108 C       ---------------------------------
109 C
110         REAL*8   RRAD(SS) , C1 , C2
111 C
112         NRADST = IDNINT( RRATIO * 0.5D0 ) + 1
113         NRADED = IDNINT( RRATIO * RMX )
114 C                                                    --- CAL. RRAD(*) ---
115         DO 100 I=NRADST, NRADED
116           C1 = DBLE(I-1) / RRATIO
117           C2 = DBLE(I  ) / RRATIO
118           RRAD(I) = (C1+C2)/2.D0
119   100   CONTINUE
120         DR = RRAD(NRADST+1) - RRAD(NRADST)
121                                                                  RETURN
122                                                                  END
123 C**** ORICORR ****
124         SUBROUTINE ORICORR( N )
125 C
126         IMPLICIT REAL*8 (A-H,O-Z), INTEGER (I-N)
127 C
128         COMMON /BLOCK1/  RX   , RY   , RZ
129         COMMON /BLOCK2/  EX   , EY   , EZ
130 C
131         COMMON /BLOCK11/ XL   , YL   , ZL
132 C
133         COMMON /BLOCK45/ DR, RMX, CRAD, NRADST, NRADED, RRATIO, RRAD
134         COMMON /BLOCK46/ RADDIST
135         COMMON /BLOCK47/ ORICORR , SUMORICO
136         COMMON /BLOCK48/ ORDER
137 C
138 C       -------------------------------------------------------------------
139         INTEGER    NN , SS
140         PARAMETER( NN=1500 , SS=10000 , PI=3.141592653589793D0 )
141 C       ---------------------------------
142 C
143         REAL*8   RX(NN) , RY(NN) , RZ(NN)
144         REAL*8   EX(NN) , EY(NN) , EZ(NN)
145 C
146         REAL*8   RRAD(SS)     , RADDIST(SS)
147         REAL*8   ORICORR(SS) , SUMORICO(SS)
148         REAL*8   ORDER
149 C
150         REAL*8   RXIJ , RYIJ , RZIJ , RIJ , RIJSQ , RMX2
151         REAL*8   C1 , CP2 , CSUM1
152         INTEGER  IRAD , NPAIR
153 C
154         RMX2 = RMX**2
155         DO 10 IRAD =NRADST, NRADED
156           ORICORR(IRAD) = 0.D0
157    10   CONTINUE
158         NPAIR = 0
159         CSUM1 = 0.D0
160 C                                    --- CAL. ORI. PAIR CORR. FUNC. ---
161         DO 200 I=1,N
162 C
163           DO 150 J=1,N
164             IF( J .EQ. I ) GOTO 150
165 C
166             RXIJ  = RX(I)  - RX(J)
167             RXIJ  = RXIJ - DNINT(RXIJ/XL)*XL
168             IF( DABS(RXIJ) .GE. RMX )        GOTO 150
169             RYIJ  = RY(I)  - RY(J)
170             RYIJ  = RYIJ - DNINT(RYIJ/YL)*YL
171             IF( DABS(RYIJ) .GE. RMX )        GOTO 150
172             RZIJ  = RZ(I)  - RZ(J)
173             RZIJ  = RZIJ - DNINT(RZIJ/ZL)*ZL
174             IF( DABS(RZIJ) .GE. RMX )        GOTO 150
175 C
176             RIJSQ= RXIJ**2 + RYIJ**2 + RZIJ**2
177 C                                    +++ OVER CHECK AREA(RADIAL) +++
178             IF( RIJSQ .GE. RMX2 )            GOTO 150
179 C
180             NPAIR= NPAIR + 1
181             RIJ  = DSQRT(RIJSQ)
182             IRAD = IDINT( RRATIO*RIJ) + 1
183 CCCCC       RADDIST(IRAD) = RADDIST(IRAD) + 1.D0
184 C
```

• The start of the index in the variable ORICORR(*) is NRADST, and the final is NRADED.

• RRAD(*) is used for determining the radial shell site to which each particle belongs.

• DR implies the increment distance $\Delta r$.

• The subroutine ORICORR evaluates the ori. pair corr. func. by calculating the correlation between the directions of a pair of particles.

• These procedures are conducted by treating the periodic boundary condition.

• The index IRAD of the variable ORICORR(*) where the result should be saved is simply obtained from this algebraic expression.

• ORICORR(*) saves the result of $(3(\boldsymbol{e}_i\cdot\boldsymbol{e}_j)^2-1)/2$.

• CSUM1 is used for evaluating the order parameter of particle orientations.

```
185          C1    = EX(I)*EX(J) + EY(I)*EY(J) + EZ(I)*EZ(J)
186          CP2   = ( 3.D0*C1**2 - 1.D0 ) / 2.D0
187          ORICORR(IRAD) = ORICORR(IRAD) + CP2
188          CSUM1 = CSUM1 + CP2
189 C
190   150   CONTINUE
191 C
192   200 CONTINUE
193 C                                --- DIVIDE THE DATA BY SMALL
194       DO 210 IRAD=NRADST, NRADED
195         ORICORR(IRAD) = CRAD * ORICORR(IRAD)/RRAD(IRAD)**2
196   210 CONTINUE
197       IF( NPAIR .LE. 0 )  NPAIR = 1000000000
198       ORDER     = ORDER     + CSUM1/DBLE(NPAIR)
199                                                  RETURN
200                                                  END
```

• The ori. pair corr. func. at this step is obtained using the factors of the coefficient CRAD and the radial RRAD(*).

• The order parameter is obtained by dividing it by the number of a pair of particles.

## 8.5 Order parameters

In the previous sections of the present chapter, we have discussed the quantities by which the characteristics of the clusters can be quantitatively described to obtain the essential features of their microstructure. In this section, we focus on quantities that can describe characteristics of a system as a whole. For this objective, a scalar quantity may be desirable and should be significantly sensitive to the change in the characteristics of the whole system such as a phase change. The order parameter is one of the most well-known quantities or index for describing system characteristics, especially in the field of liquid crystals where the orientational order of the molecules plays a significantly important role for characterizing their performance in application. This order parameter is also an important index to quantify the orientational order of magnetic non-spherical particles because the order of alignment of the magnetic moments to a certain direction has a significant influence on the magneto-rheological characteristics of a particle suspension. Moreover, an appropriate expression of the orientational order parameter regarding magnetic moments of particles gives information relating to the magnetization of the whole system of magnetic particles. Hence, we may obtain information of both the microstructure and magnetic properties of a particle suspension as a whole rather than just the local characteristics of a suspension. In the following we discuss the definition of several order parameters that are widely used and then show how these order parameters can represent the system characteristics, using sample snapshots of magnetic rod-like particles.

We consider a suspension of $N$ number of rod-like particles in thermodynamic equilibrium. Using the notation $\theta_{ij}$ to denote an angle between the orientations of particles $i$ and $j$, then the orientational order parameter of particle orientations, $S^{(e)}$, is evaluated from the following definition:

$$S^{(e)} = \frac{1}{N_{pair}} \left\langle \sum_{i=1}^{N} \sum_{j=1(j>i)}^{N} P_2(\cos\theta_{ij}) \right\rangle \tag{8.13}$$

in which $P_2(\cos\theta_{ij}) = (3\cos^2 \theta_{ij} - 1)/2$ is the second Legendre polynomial, $N_{pair}$ ($= N(N-1)/2$) is the number of pairs of particles, and $\langle - \rangle$ is the time or ensemble average. It is noted that this order parameter is a scalar quantity and independent of the position of each particle. Similar to the orientational pair correlation function described in Section 8.4, if all the particles incline in the same direction, $P_2$ is equal to unity and therefore the order parameter gives rise to the value of $S^{(e)} = 1$. On the other hand, if the particles randomly incline without correlation between the particle orientations, the order parameter yields zero as $S^{(e)} = 0$. From these considerations, it is seen that we can quantitatively recognize the degree of order in the orientational characteristics of the system as a whole. However, the actual direction of the ordered orientation of the whole system cannot be obtained from this quantity, and thus another order parameter is necessary in order to discuss the characteristics regarding the alignment of the particles to a specific favored direction.

As we mainly address a suspension composed of magnetic particles with a variety of shapes, an applied magnetic field is one of the main factors that determine the characteristics of a magnetic particle suspension, in particular the magneto-rheological properties. Hence, it is valuable and important to consider the degree of the alignment of magnetic moments to the magnetic field direction. For simplicity, in this discussion the magnetic moment is assumed to be in the same direction of the particle axis direction, that is, the particle is magnetized in the particle axis direction. Hence, the direction of the magnetic moment of particle $i$ is equal to the particle direction $\boldsymbol{e}_i$. If an external magnetic field is applied in the $y$-axis direction, the order parameter of the alignment of magnetic moments to the field direction, $S^{(m)}$, is defined by the following expression:

$$S^{(m)} = \frac{1}{N} \left\langle \sum_{i=1}^{N} P_2(\boldsymbol{e}_i \cdot \boldsymbol{\delta}_y) \right\rangle \tag{8.14}$$

in which $\boldsymbol{\delta}_y$ is the unit vector denoting the $y$-axis direction. Similar to the characteristics of $S^{(e)}$, in the case of all the particles inclining in the field direction, or in the completely opposite direction, $S^{(m)}$ gives rise to a value of unity, and in the case of random orientations the order parameter yields a value of zero. Hence, this quantity is expected to be a significantly good index for quantifying the degree of the alignment of the magnetic moments to the magnetic field direction.

The normalized magnetization relative to the saturation magnetization, $n^{(m)} = M/M_0$, may be another quantity or index for capturing the order of the magnetic moment orientations. This quantity is defined by the following expression:

$$n^{(m)} = \frac{1}{N}\left\langle \sum_{i=1}^{N} (\boldsymbol{m}_i / m)\cdot \boldsymbol{\delta}_y \right\rangle = \frac{1}{N}\left\langle \sum_{i=1}^{N} \boldsymbol{n}_i \cdot \boldsymbol{\delta}_y \right\rangle \tag{8.15}$$

in which $\boldsymbol{m}_i$ $(= m\boldsymbol{n}_i)$ is the magnetic moment vector of particle $i$, and $m$ is the magnitude of the magnetic moment, expressed as $m = |\boldsymbol{m}_i|$. Similar to $S^{(m)}$, in the case of the complete alignment of the particles to the field direction the order parameter gives rise to $n^{(m)} = 1$ and in the case of their random orientation the order parameter yields a value of $n^{(m)} = 0$. It is expected that the order parameter $S^{(m)}$ varies more steeply than the order parameter $n^{(m)}$. This is because the order quantity $S^{(m)}$ has the factor of the square of $\cos\theta_i$, where $\theta_i$ is an angle of particle $i$ from the magnetic field direction.

We now illustrate how these order parameters represent the order characteristics regarding the degree of the particle orientations as a whole. To do so, we consider a suspension of $N$ number of magnetic rod-like particles that have a dipole moment in the particle axis direction at the particle center. This suspension is assumed to be in thermodynamic equilibrium under the circumstance of a uniform applied magnetic field.

Figure 8.8 shows aggregates of magnetic rod-like particles for a relatively weak magnetic interaction strength $\lambda = 10$ and the snapshots in Figs. 8.8(a), 8.8(b) and 8.8(c) are for the magnetic field strength $\xi = 0.1$, 5 and 20, respectively. Figures 8.9, 8.10(a) and 8.10(b) show results of the corresponding order parameters $S^{(e)}$, $S^{(m)}$ and $n^{(m)}$, respectively. In the case of almost zero magnetic field, as shown in Fig. 8.8(a), the particles randomly incline in any direction. In the case of a relatively strong magnetic field, as shown in Fig. 8.8(b), the particles tend to incline in the magnetic field direction, i.e., in the upward direction, and this orientational characteristic becomes more significant for a strong magnetic field, as shown in Fig. 8.8(c). Therefore a stronger magnetic field induces a strong tendency for the alignment of the magnetic particles to the magnetic field direction.

As expected, the result of $S^{(e)}$ at $\xi = 0.1$, shown in Fig. 8.9, yields zero, and increases with values of magnetic field strength $\xi$. Since there are no clusters formed in the system in this situation, the order parameter exhibits a monotonic increase, however, a more complex dependence on the field strength is to be expected if there are numerous clusters formed in the system. This is because these clusters will interact with each other to resist the alignment of their constituent particles to the magnetic field.

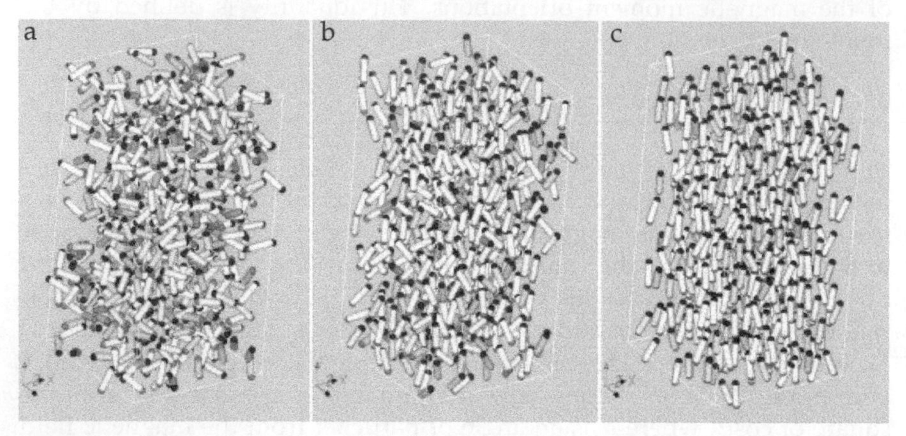


**Figure 8.8.** Snapshots for the case of the magnetic particle-particle interaction strength $\lambda = 10$: (a) $\xi = 0.1$, (b) $\xi = 5$ and (c) $\xi = 20$: an external magnetic field is applied in the upward direction.

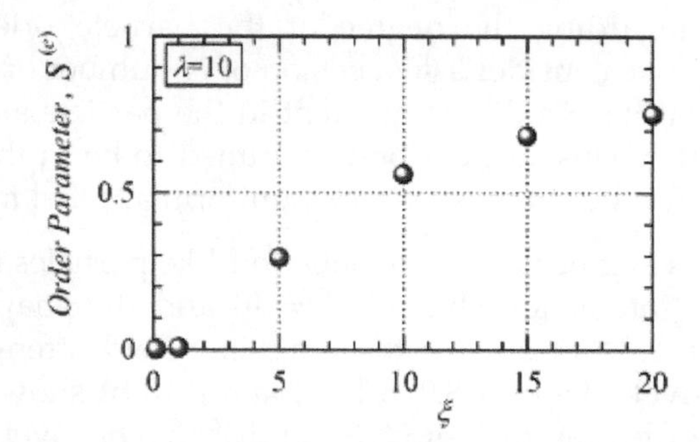


**Figure 8.9.** Order parameter of particle orientations.

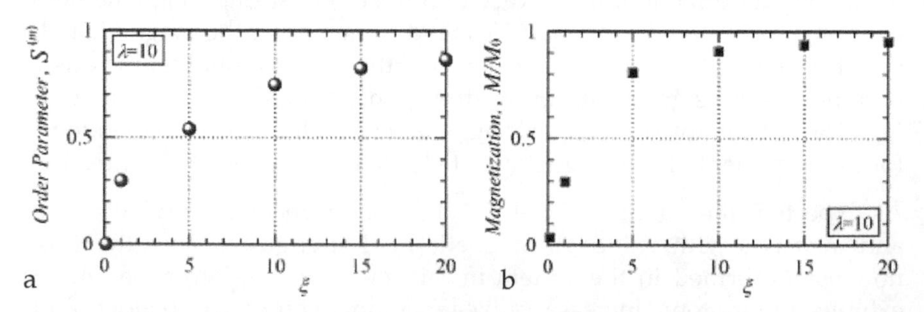


**Figure 8.10.** Order parameters regarding the alignment of magnetic moments to the magnetic field direction: (a) $S^{(m)}$ and (b) the normalized magnetization $n^{(m)}$ (=$M/M_0$).

The magnetic particles tend to incline more significantly in the magnetic field direction with increasing magnetic field strength and as a result the magnitude of the order parameters $S^{(m)}$ and $n^{(m)}$, as shown in Fig. 8.10, monotonically increase with increasing field strength and exhibit similar characteristics to the order parameter $S^{(e)}$. The curve of $S^{(m)}$ is closer to that of $S^{(e)}$ than that of $n^{(m)}$, but the results of $S^{(e)}$ yield smaller values than those of $S^{(m)}$. This seems to be reasonable because the order parameter $S^{(e)}$ is not restricted to the degree of the alignment of particles to a specific direction. Figure 8.10(b) shows results of the order parameter $n^{(m)}$ (= $M/M_0$), which implies the ratio of the magnetization $M$ of the system relative to the saturation magnetization $M_0$. The curve of $n^{(m)}$ increases more rapidly with field strength than that of $S^{(m)}$ and implies that the order parameter $n^{(m)}$ is a weaker index for representing the orientational order of the system in strong magnetic field cases, for instance, in the range of $\xi \gtrsim 5$.

From these results, we understand that the order parameters $S^{(e)}$, $S^{(m)}$ and $n^{(m)}$ are good indexes for quantifying the orientational characteristics of magnetic rod-like particles in a non-equilibrium system as well as a system in thermodynamic equilibrium. If the non-spherical particles are magnetized in a direction different from the particle axis direction, an expression similar to Eq. (8.13) is used for evaluating the corresponding order parameter of the magnetic moments as a whole system.

The subroutine shown previously in Section 8.4 also has the calculation procedure of the order parameters explained in the present section.

## 8.6 Cluster size distribution

In a strongly-interacting system where stable clusters of dispersed particles are formed, the radial distribution or pair correlation function can provide indirect information regarding the formation of clusters. However, more direct information of cluster formation may be desirable and the size distribution of the clusters is a more effective quantity to investigate the microstructure of a particle suspension for this objective. The cluster size distribution is expressed by the number of clusters, $N_S$, which are composed of $S$ particles and it is noted that the quantity $N_S$ is a time or ensemble averaged quantity. If there are no clusters formed in the system, then $N_1$ is equal to the number $N$ of particles in the system, and the other values are given by $N_S = 0$ for $S = 2, 3, \ldots, N$. If larger clusters are formed in the system, then $N_S$ shows large values for larger values of $S$, whereas $N_S$ becomes almost zero for the smaller values of $S$. Thus, the cluster size distribution seems to be a quite appropriate quantity for capturing a picture of how various size clusters are distributed in the system.

If the cluster size distribution has been evaluated and thus already known, we can then calculate the mean size of the clusters, $S^{(ave)}$, as

$$S^{(ave)} = \sum_{S=1} SN_S \Big/ \sum_{S=1} N_S = N \Big/ \sum_{S=1} N_S \tag{8.16}$$

Similar to the previous order parameters, this scalar quantity can provide the approximate image of the situation of the cluster formation.

In some situations, instead of the quantity $N_S$, the following ratio $\hat{N}_S$ of the product of $N_S$ and $S$ to the total number $N$ may be more effectively addressed as

$$\hat{N}_S = SN_S / N, \quad (S = 1, 2, 3, ...) \tag{8.17}$$

It is noted that this quantity takes a value ranging from zero to unity, therefore the quantity $\hat{N}_S$ can describe more clearly (sharply) the existence of large clusters in a system. For instance, if only one large cluster is formed in a system, $\hat{N}_N$ is equal to unity and the other values are zero, i.e., $\hat{N}_S = 0$ for $S = 1, 2, ..., N-1$.

The definition for a cluster has been discussed previously in Section 7.3. In a simulation, the cluster size distribution is evaluated at every certain number of simulation steps and accumulated using certain program variables. After the completion of the main loop, the accumulated data are divided by the total sampling number and this gives rise to the above-mentioned number distribution $N_S$ ($S = 1, 2, ..., N$) of each cluster size. The programing procedure is quite straightforward for this calculation and therefore we do not show a sample subroutine for the evaluation of the cluster size distribution.

In the following, we illustrate how the cluster size distribution can describe the microstructure of the clusters formed in a system by using sample snapshots and their corresponding cluster size distribution, which were obtained from Monte Carlo simulations for a magnetic rod-like particle suspension.

We consider the snapshots that were already shown on Fig. 8.6: the snapshots for $\lambda = 20$ and 40 are quite similar to those shown in Figs. 8.6(a) and (b) for $\lambda = 10$ and 50, respectively. Figure 8.11 shows results for the corresponding cluster size distribution of three different cases of the magnetic particle-particle interaction, $\lambda = 20$, 30 and 40. If the magnetic interaction is relatively weak as $\lambda = 20$, the magnetic rod-like particles do not aggregate to form raft-like clusters, as shown in Fig. 8.6. Hence, the curve for $\lambda = 20$ in Fig. 8.11 clearly shows that single-moving particles and small clusters significantly dominate the cluster size distribution where the number of single-moving particles is approximately 300 in the situation of the total number of system particles $N = 512$. In contrast, the snapshot shown previously in Fig. 8.6(b) clearly exhibits several large clusters formed in a system with relatively strong magnetic interaction as $\lambda = 50$ and the snapshot shows that some clusters have raft-like microstructure and some clusters are crystal-like. These characteristics of the clusters are clearly supported by the curve for $\lambda = 40$ shown in Fig. 8.11 for a similar strong interaction and in contrast, the number of single-moving particles has decreased to as low as $N_1 \simeq 10$ and to the contrary, large clusters such as a 20-constituing-particle cluster almost continuously appears in a system. From these results, it is seen that the cluster size distribution in the situation of a system with stable clusters being formed is significantly different from that of a system where no clusters are formed.

We therefore understand from the above that the cluster size distribution is a good quantity for capturing the number and range of clusters that are predominantly formed in a strongly-interacting system.

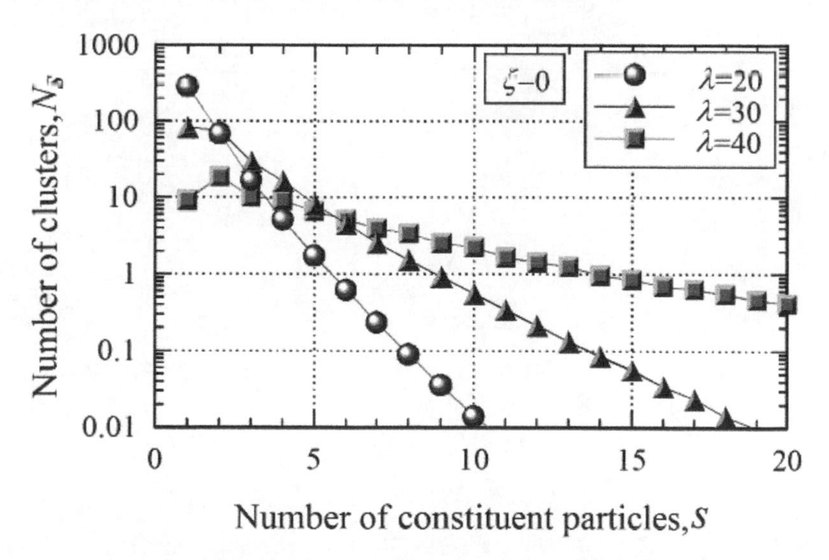


**Figure 8.11.** Cluster size distributions for the three cases of the magnetic particle-particle interaction strength $\lambda = 20$, 30 and 40 in no external magnetic field $\xi = 0$.

# Bibliography

[1] Löwen, H. 1994. Brownian dynamics of hard spherocylinders. Phys. Rev. E. 50: 1232–1242.

[2] Ozaki, M., Senna, M., Koishi, M. and Honda, H. 2000. Particles of specific functions. pp. 662–682. *In*: T. Sugimoto (ed.). Fine Particles. Marcel Dekker, New York.

[3] Van der Beek, D., Reich, H., Van der Schoot, P., Dijkstra, M., Schilling, T., Vink, R., Schmidt, M., Van Roij, R. and Lekkerkerker, H. 2006. Isotropic-nematic interface and wetting in suspensions of colloidal platelets. Phys. Rev. Lett. 97: 087801.

[4] Ozaki, M., Ookoshi, N. and Matijević, E. 1990. Preparation and magnetic properties of uniform hematite platelets. J. Colloid Interface Sci. 137: 546–549.

CHAPTER 9

# Several Examples of Simulations

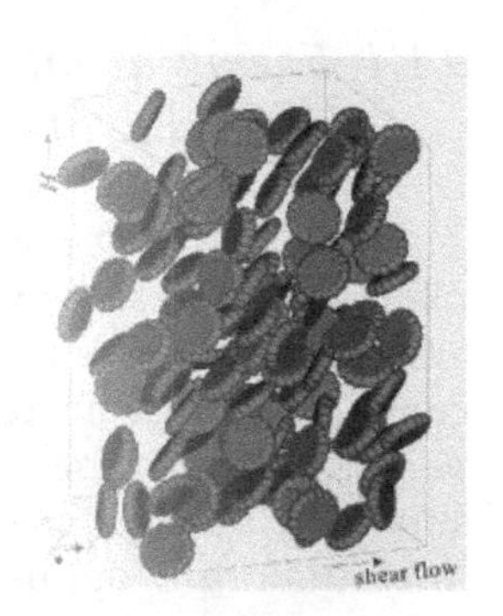


## 9.1 Monte Carlo simulations of cube-like particles

In the following sections we demonstrate molecular simulations of a suspension composed of several magnetic particle models that have been explained in the preceding chapters. In the present section, we address a suspension composed of magnetic cube-like particles in thermodynamic equilibrium in order to discuss the aggregation phenomena of these particles in an external magnetic field. The Monte Carlo method is employed as a simulation tool because it has a significant advantage as a simulation technique for thermodynamic equilibrium in that a series of Monte Carlo simulation procedures can be conducted even if the mathematical expression for the interaction of the repulsive layer covering each particle has not been obtained. This is because in Monte Carlo simulations, a solid particle model without a repulsive layer can be used as a first approximation, which yields a significant advantage to Monte Carlo method. In contrast, an explicit expression of the repulsive interaction is indispensable for dynamic simulation methods such as molecular dynamics and Brownian dynamics methods. From the viewpoint of developing a technology of changing surface quality, we focus here on a two-dimensional system where cube-like particles move with their bottom plane in contact with a plane representing the material surface. The orientation and configuration of the particles will be influenced by a variety of factors such as the magnetic particle-particle and particle-field interactions, which are treated in the discussion of the simulation results. Qualitative discussion is made in terms of snapshots that may provide readers with a direct image of the particle aggregate formation, and quantitative discussion is made using the order parameter and the cluster size distribution that describe the characteristics of the microstructure of the aggregates.

### 9.1.1 Formulation of the present 2D phenomenon and modeling of magnetic cube-like particles

As shown in Fig. 9.1, we consider the motion of cube-like particles on the plane surface, which may be regarded as the material surface. The particles are assumed to move on the surface plane with the bottom plane of each cube-like particle in contact with the surface plane. This assumption may be reasonable in the situation of a sufficiently strong gravitational field situation. In the present simulation, an external magnetic field is assumed to be uniformly applied along the bottom plane which is normal to the direction of the gravitational field.

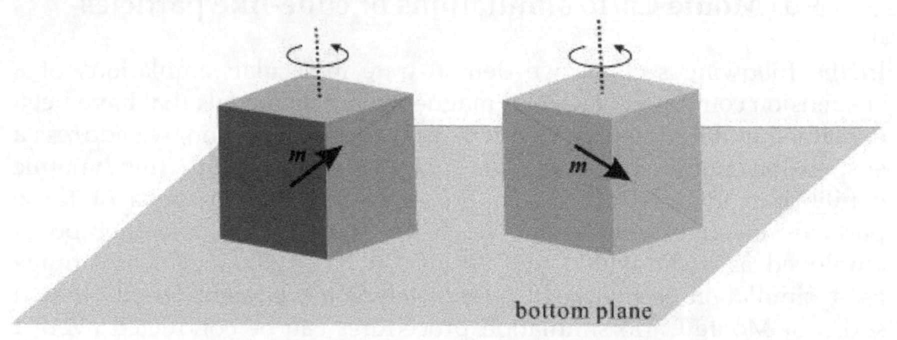


**Figure 9.1.** Particle model with two types of magnetic moment direction, (1) upward diagonal direction and (2) downward diagonal direction relative to the bottom plane; these magnetic cube-like particles move on a plane surface (or on the surface of a material) with the bottom surface of each cube in contact with the material surface.

The magnetic cube-like particle is modeled as a cube with a magnetic dipole moment at the center of the cube pointing in a diagonal direction, as shown in Fig. 3.21. This particle model may be regarded as a first approximation of the magnetic cube-like hematite particles which are experimentally synthesized [1]. In the present situation, the particles have a magnetic moment pointing either in the upward diagonal direction or in the downward diagonal direction relative to the bottom surface, which are schematically shown in Fig. 9.1. Hence, arbitrary motion of the particles on the bottom surface can be expressed as a combination of the translational motion parallel to the bottom surface and the rotational motion about a line, through the particle center, normal to the bottom surface. The orientation and configuration of the magnetic particles will be mainly determined by the two factors of the magnetic dipole-dipole interaction strength and the applied magnetic field strength.

We set the absolute orthogonal coordinate system *xyz* in such a way that the *z*-axis is taken in the positive direction normal to the bottom-surface plane. An external magnetic field $\boldsymbol{H}$ is assumed to be applied in the *y*-axis direction as $\boldsymbol{H} = H\boldsymbol{h} = (0, H, 0)$. The cube-like particles have side length $d$ and the system is assumed to be composed of equal-sized cube-like particles. The situation of particle $i$ is described by the position vector $\boldsymbol{r}_i = (x_i, y_i, d/2)$ and the magnetic moment direction $\boldsymbol{n}_i = (n_{ix}, n_{iy}, n_{iz})$. The particle direction $\boldsymbol{e}_i^b$ of an arbitrary particle $i$ in the body-fixed or particle coordinate system is necessarily expressed as $\boldsymbol{e}_i^b = (0,0,1)$. If the magnetic moment of particle $i$ is denoted by $\boldsymbol{m}_i = m\boldsymbol{n}_i$, where $\boldsymbol{n}_i$ is the unit vector describing the magnetic moment direction, then the magnetic dipole-dipole interaction $u_{ij}^{(m)}$ between particles $i$ and $j$ is expressed from Eq. (2.27) as

$$u_{ij}^{(m)} = \frac{\mu_0 m^2}{4\pi d^3 (r_{ij}/d)^3}\left\{ \boldsymbol{n}_i \cdot \boldsymbol{n}_j - 3(\boldsymbol{t}_{ij} \cdot \boldsymbol{n}_i)(\boldsymbol{t}_{ij} \cdot \boldsymbol{n}_j) \right\} \tag{9.1}$$

Also, the particle-field interaction energy $u_i^{(H)}$ is expressed from Eq. (2.9) as

$$u_i^{(H)} = -\mu_0 m H \boldsymbol{n}_i \cdot \boldsymbol{h} = -\mu_0 m H n_{iy} \tag{9.2}$$

In these equations, $\mu_0$ is the permeability of free space, $\boldsymbol{r}_{ij}(= \boldsymbol{r}_i - \boldsymbol{r}_j)$ is the relative position of particle $i$ to particle $j$, and $\boldsymbol{t}_{ij}$ is the unit vector denoting the direction of particle $i$ relative to particle $j$, expressed as $\boldsymbol{t}_{ij} = \boldsymbol{r}_{ij}/r_{ij}$, where $r_{ij} = |\boldsymbol{r}_{ij}|$. If energies are non-dimensionalized by the thermal energy $kT$, where $k$ is Boltzmann's constant and $T$ is the liquid temperature, then Eqs. (9.1) and (9.2) are expressed in non-dimensional form as in Eqs. (3.3) and (3.2), respectively. Hence, the present phenomenon is characterized by the two non-dimensional parameters $\xi = \mu_0 mH/(kT)$ and $\lambda = \mu_0 m^2/(4\pi d^3 kT)$, in which $\xi$ and $\lambda$ represent the strengths of magnetic particle-field and particle-particle interactions relative to the thermal energy, respectively.

Moreover, particles dispersed in a base liquid are generally covered by a repulsive layer such as a steric or an electric double layer, which prevents the particles from aggregating to obtain a stable particle suspension. However, at the current time, the interaction energy due to the overlap of the repulsive layers has not been expressed as a mathematical expression for the case of two cube-like particles. This may be a fatal problem for a molecular-dynamics-like simulation method, but this difficult problem may straightforwardly be overcome in a Monte Carlo simulation by adopting the solid particle model without a repulsive layer. This solid particle model will be a reasonable first approximation to advance studies regarding challenging subjects, and therefore we employ it here and do not take into account the repulsive layer.

As previously described in Section 3.5, the orientation of the magnetic moment of particle $i$, $\boldsymbol{n}_i$, is defined using the face vectors $\boldsymbol{a}_{i1}$, $\boldsymbol{a}_{i2}$ and $\boldsymbol{a}_{i3}$, shown in Fig. 3.21. In the present situation, $\boldsymbol{a}_{i3}$ is necessarily equal to $\boldsymbol{a}_{i3}$ = (0,0,$d/2$) or (0,0, – $d/2$) in the absolute coordinate system because the cubes move with their bottom face in contact with the bottom plane surface.

## 9.1.2 *Monte Carlo method*

We adopt here the Monte Carlo method for simulating a suspension composed of magnetic cube-like particles in thermodynamic equilibrium. In performing Monte Carlo simulations, we only need to treat the interactions between particles and the interaction between an applied magnetic field and the particles. The forces and torques acting on particles are not required in order to advance the Monte Carlo procedure, as explained in Section 6.1. In the present system, the number of magnetic particles, $N$, the volume of the system, $V$, and the system temperature $T$ are prescribed and constant in the simulation, and therefore we will use the *NVT* Monte Carlo method or the canonical ensemble Monte Carlo method. From the viewpoint of introducing the simulation methodology, we do not include the cluster-moving procedure into the ordinary Monte Carlo method, because this procedure makes the simulation program much more complex for readers' understanding. Fortunately, even if the cluster-moving Monte Carlo method is not used for a cube-like particle suspension, reasonable aggregate structures may be obtained as a first approximation.

There are two main Monte Carlo procedures in the present simulations. These are the attempt of translational movement of a particle on the bottom surface and the attempt of rotational motion of a particle about a line normal to the bottom surface which corresponds to the rotation of the magnetic moment. The procedure employed for the attempt of the translational and the rotational motion will be explained in detail in the following section.

The total energy of the system, $U$, is the sum of the particle-particle and the particle-field interaction energies, i.e.,

$$U = \sum_{i=1}^{N} u_i^{(H)} + \sum_{i=1}^{N} \sum_{j=1(j>i)}^{N} u_{ij}^{(m)} \tag{9.3}$$

This total energy of the system is used in the assessment procedure regarding the acceptance or rejection of a new position or a new orientation of a particle in the progression to a new microscopic state. In the following, we show the algorithm of this procedure in the present Monte Carlo simulation:

(1) An initial configuration of the particle position and orientation of all the cube-like particles is assigned

(2) Calculate the total interaction energy $U^{(old)}$ in Eq. (9.3)

(3) Choose a particle in order or at random and regard this particle as particle $a$

(4) Move particle $a$ to a random vicinity position using random numbers ranging from zero to unity

(5) Calculate the total interaction energy $U^{(new)}$ in Eq. (9.3)

(6) If $U^{(new)} \leq U^{(old)}$, this candidate position of particle $a$ is accepted, $U^{(new)}$ is regarded as $U^{(old)}$, and then go to step (8)

(7) If $U^{(new)} > U^{(old)}$, a random number $R$ is taken from a sequence of random numbers ranging from zero to unity and

(7)-1 If $\rho_0(=\exp\{-(U^{(new)} - U^{old)})/(kT)\}) \geq R$, then this candidate position of particle $a$ is accepted, $U^{(old)}$ is replaced with $U^{(new)}$, and then go to step (8)

(7)-2 If $\rho_0 < R$, then this candidate position of particle $a$ is rejected and then go to step (8)

(8) Rotate particle $a$ about a line, through the particle center, normal to the bottom surface by a small angle

(9) The total interaction energy is calculated and allotted to $U^{(new)}$ in Eq. (9.3)

(10) If $U^{(new)} \leq U^{(old)}$, this candidate orientation of particle $a$ is accepted, $U^{(old)}$ is replaced with $U^{(new)}$, and then go to step (3)

(11) If $U^{(new)} > U^{(old)}$, a random number $R$ is taken from a sequence of random numbers ranging from zero to unity

(11)-1 If $\rho_0(=\exp\{-(U^{(new)} - U^{old)})/(kT)\}) \geq R$, then the candidate orientation of particle $a$ is accepted, $U^{(old)}$ is replaced with $U^{(new)}$, and then go to step (3)

(11)-2 If $\rho_0 < R$, then the candidate orientation of particle $a$ is rejected and then return to step (3)

### 9.1.3 Attempt of translational movement of the particles and rotational motion of the magnetic moments

In Monte Carlo simulations, a series of microscopic states are required to be generated and this is accomplished using a sequence of pseudo-random numbers ranging from zero to unity. The generation of a new microscopic state is conducted regarding both the translational and the rotational configurations in the present non-spherical particle model. The translational generation is quite straightforward, and a new candidate position for an arbitrary particle $i$ is generated according to Eq. (6.5). In the present two-dimensional system, the new candidate position $\boldsymbol{r}_i^{(can)} = (x_i^{(can)}, y_i^{(can)}, z_i^{(can)})$ is expressed as

$$x_i^{(can)} = x_i + \Delta r(2R_1 - 1), \quad y_i^{(can)} = y_i + \Delta r(2R_2 - 1), \quad z_i^{(can)} = z_i = d/2 \tag{9.4}$$

in which $\boldsymbol{r}_i = (x_i, y_i, z_i)$ is the current position, $\Delta r$ is the maximum displacement of the translational movement and $R_1$ and $R_2$ are random numbers taken from the random number sequence. It is noted that a random number is only used once in the simulation and so different random numbers are used in each attempt of a translational position. The value of the maximum distance $\Delta r$ has to be carefully chosen for obtaining physically reasonable configurations and for the present cube-like particles with side length $d$, an appropriate value of $\Delta r$ must be sufficiently smaller than the side length $d$. If a particle is coated with a thin repulsive layer such as a surfactant layer, then $\Delta r$ must be much smaller than the thickness of the layer, but in the present simulation we do not take into account these repulsive layers.

We now briefly summarize the attempt of the rotation of the particles in the present two-dimensional system and it is noted that a detailed explanation has previously been given in Section 4.3 for the general three-dimensional case. The particles, shown schematically in Fig. 9.1, move on the bottom plane with their bottom face being in contact with the plane, and therefore only the attempt of the rotation about a line, the $z$-axis direction, through the particle center and normal to the bottom plane is required in the present case. The rotation matrix $\hat{\boldsymbol{R}}$ about the $z$-axis direction by a small angle $\Delta\theta$ is expressed from Eq. (4.15) as

$$\hat{\boldsymbol{R}} = \begin{bmatrix} \cos\Delta\theta & -\sin\Delta\theta & 0 \\ \sin\Delta\theta & \cos\Delta\theta & 0 \\ 0 & 0 & 1 \end{bmatrix} \tag{9.5}$$

Using this rotation matrix, the face vector $\boldsymbol{a}_k^b$ ($k$ = 1,2, ..., 6) in the body-fixed coordinate system, defined in Fig. 3.21 of Section 3.5, is rotated to give rise to the new vector $\boldsymbol{a}_k^{b\prime}$ as $\boldsymbol{a}_k^{b\prime} = \hat{\boldsymbol{R}} \cdot \boldsymbol{a}_k^b$ ($k$ = 1,2, ..., 6). As explained in Section 4.1, an arbitrary vector $\boldsymbol{c}^b$ in the body-fixed coordinate system is related to the vector $\boldsymbol{c}$ in the absolute coordinate system with the rotation matrix $\boldsymbol{R}$ by the expression $\boldsymbol{c}^b = \boldsymbol{R} \cdot \boldsymbol{c}$. Hence the face vector $\boldsymbol{a}_k'$ ($k$ = 1,2, ..., 6) in the absolute system is expressed as $\boldsymbol{a}_k' = \boldsymbol{R}^t \cdot \boldsymbol{a}_k^{b\prime}$. Then, the new rotation matrix $\boldsymbol{R}'$, which relates the vectors $\boldsymbol{a}_k'$ in the absolute coordinate system to the vectors $\boldsymbol{a}_k^b$ ($k$ = 1,2, ..., 6) (e.x., $\boldsymbol{a}_1^b = (d/2, 0, 0)$) in the body-fixed coordinate system, is obtained from the following relationship:

$$\boldsymbol{a}_k' = \boldsymbol{R}'^t \cdot \boldsymbol{a}_k^b \quad (k = 1,2,3) \tag{9.6}$$

The final result of the components $R_{ij}'$ ($i,j$ = 1,2,3) of the rotation matrix is obtained as

$$\begin{bmatrix} R_{11}' \\ R_{12}' \\ R_{13}' \end{bmatrix} = (2/d) \begin{bmatrix} a_{1x}' \\ a_{1y}' \\ a_{1z}' \end{bmatrix}, \quad \begin{bmatrix} R_{21}' \\ R_{22}' \\ R_{23}' \end{bmatrix} = (2/d) \begin{bmatrix} a_{2x}' \\ a_{2y}' \\ a_{2z}' \end{bmatrix}, \quad \begin{bmatrix} R_{31}' \\ R_{32}' \\ R_{33}' \end{bmatrix} = (2/d) \begin{bmatrix} a_{3x}' \\ a_{3y}' \\ a_{3z}' \end{bmatrix} \tag{9.7}$$

in which $\boldsymbol{a}_k' = (a_{kx}', a_{ky}', a_{kz}')$. Finally, the new direction $\boldsymbol{n}'$ of the magnetic moment is obtained from these new face vectors as

$$\boldsymbol{n}' = \frac{\boldsymbol{a}_1' + \boldsymbol{a}_2' + \boldsymbol{a}_3'}{|\, \boldsymbol{a}_1' + \boldsymbol{a}_2' + \boldsymbol{a}_3' \,|} = \frac{\boldsymbol{a}_1' + \boldsymbol{a}_2' + \boldsymbol{a}_3'}{\sqrt{3}d/2} \tag{9.8}$$

In the above equations, the subscript denoting the particle name is omitted for simplicity. The direction $\boldsymbol{n}'$ is a candidate magnetic moment direction $\boldsymbol{n}^{(can)}$ for generating a new configuration regarding the rotation of the particles in the Monte Carlo procedures.

### 9.1.4 *Assessment of the particle overlap and the cluster formation*

The assessment of the overlap between the two cube-like particles with the same size has been discussed in detail for the three-dimensional case in Section 5.4. In the present case, we treat a two-dimensional motion on the bottom surface, and therefore the assessment procedure described in Section 5.4.1 is directly relevant to the present situation. In a two-dimensional case with the motion on a plane surface, the overlap of the solid cubes is relatively straightforwardly assessed and based on the configuration of the two squares shown in Fig. 5.10. In this procedure, one of the main tasks is to find the face and the corner required for the assessment of the particle overlap and they are found according to the procedure explained in Section 5.4 using the quantities $\beta_n{}^i$ and $\gamma_m{}^i$ that are defined in Eqs. (5.36) and (5.37). Once the corner and the face of interest have been determined, the condition shown in Eq. (5.41) is used for assessing whether or not the two squares, and thus the two cubes, have overlapped.

During the above-mentioned procedure, the minimum distance $r_{min}{}^{(surf)}$ between the two particles is obtained and may be used for evaluating the formation of clusters, which gives rise to the cluster size distribution. Hence, the assessment procedure for the particle overlap is indispensable for evaluation of the cluster size distribution in addition to the essential procedure in the Monte Carlo and the cluster-moving Monte Carlo methods. Although there are more sophisticated methods available to determine cluster formation, such as the cluster analysis method [2–4], we employ here the simplest criterion that uses the minimum surface-to-surface distance $r_{min}{}^{(surf)}$. That is, if the distance $r_{min}{}^{(surf)}$ is smaller than a given criterion value $r_{clstr}$, these particles are regarded as forming a cluster. This criterion distance $r_{clstr}$ is usually taken much smaller than the particle size $d$, i.e., $r_{clstr} << d$.

## 9.1.5 Description of the characteristics of the system

In Chapter 8, we have discussed a variety of system quantities that can be used to describe the physical characteristics of a particle suspension. Snapshots of particle configurations are indispensable for visualizing the formation of clusters from a qualitative point of view. For a quantitative assessment of the cluster formation, the order parameters explained in Section 8.5 are a typical approach, but here we introduce the following modified order parameters in order to obtain more accurate features of the configurations of the particle aggregates.

As understandable from the aggregate structures that will be shown later, in the case of a weak magnetic particle-particle interaction the aggregates are typically formed in a small area and the cluster size will increase with a stronger interaction. This implies that the order parameter defined in Eq. (8.13) is less informative in a situation where small aggregates are formed and scattered throughout the whole system than in a situation where larger clusters are formed. This arises from averaging the quantity of interest for all the pairs of particles in the whole simulation area, and therefore the quantity defined in Eq. (8.13) will be able to describe the characteristics of the system more accurately if large clusters are formed in the system. This consideration suggests a modification of the order parameters so that they are evaluated as a function of the radial distance. That is, the radial distance is discretized and the averaging procedure is evaluated for the number of pairs of particles that are limited to being in a certain area or volume defined by the radial distance. This modification is also expected to give rise to information of the typical size of the cluster formed in the system.

Similar to Eq. (8.13), we address the following order parameter of the magnetic moment, $S_1(r)$:

$$S_1(r) = \frac{1}{N_{pair}(r)} \left\langle \sum_{i=1}^{N} \sum_{\substack{j=1 \\ (j>i)}}^{N} P_2(\cos\psi_{ij}^{(n)}) \right\rangle \tag{9.9}$$

in which $\psi_{ij}^{(n)}$ is the angle between the magnetic moment directions $\boldsymbol{n}_i$ and $\boldsymbol{n}_j$ of particles $i$ and $j$, expressed as $\cos\psi_{ij}^{(n)} = \boldsymbol{n}_i \cdot \boldsymbol{n}_j$, $\langle - \rangle$ is the ensemble average, the summation with respect to particles $i$ and $j$ treats all the pairs of particles being within the range of the radial distance $r$, and $N_{pair}(r)$ is the number of these pairs of particles.

As previously discussed in Section 3.5, aggregate configurations shown in Figs. 3.24 and 3.27 are preferable in the case of sufficiently strong magnetic particle–particle interactions and we address the question as to what order parameter has a quantitative value that can describe this kind of aggregate. The answer is accomplished by considering the projection of the magnetic moments on the bottom plane surface. The notation $\hat{\boldsymbol{n}}_i$ and $\hat{\boldsymbol{n}}_j$ are employed for the unit projection vectors of particles $i$ and $j$, respectively, and $\theta_{ij}^{(n)}$ is used for the angle between these unit vectors. If the aggregate structures in Figs. 3.24 and 3.27 are predominately formed in a system, then $\cos 4\theta_{ij}^{(n)}$ should give rise to a value tending to unity. Hence, we address the following another order parameter $S_2(r)$ for the present case:

$$S_2(r) = \frac{1}{N_{pair}(r)} \left\langle \sum_{i=1}^{N} \sum_{\substack{j=1 \\ (j>i)}}^{N} \cos 4\theta_{ij}^{(n)} \right\rangle \tag{9.10}$$

It is noted that as in the previous order parameter, $S_2$ is evaluated as a function of the radial distance $r$.

In order to assess whether or not stable clusters are formed in the current system, we will also treat the cluster size distribution previously explained in Section 8.6. This distribution $N_S$ ($S = 1,2,\ldots,N$) implies the number of the clusters that are composed of $S$ number of constituent particles. Hence, a larger number of clusters with $S$ particles formed in the system will give rise to a larger value of $N_S$.

We also address two other order parameters regarding the alignment of the magnetic moments to the applied magnetic field direction. The first order parameter $\langle \hat{n}_y \rangle$ corresponds to the normalized magnetization and is defined from Eq. (8.15) as

$$\langle \hat{n}_y \rangle = \frac{1}{N}\left\langle \sum_{i=1}^{N} (\hat{\boldsymbol{n}}_i \cdot \boldsymbol{h}) \right\rangle = \frac{1}{N}\left\langle \sum_{i=1}^{N} (\hat{n}_{iy}) \right\rangle \tag{9.11}$$

in which $\boldsymbol{h} = (0,1,0)$ is the unit vector denoting the magnetic field direction and $\hat{n}_{iy}$ is the $y$-component of the unit projected vector of the magnetic moment vector $\hat{\boldsymbol{n}}_i = (\hat{n}_{ix}, \hat{n}_{iy}, \hat{n}_{iz})$.

The second order parameter $S_{ny}$ is defined in a similar way as

$$S_{ny} = \frac{1}{N}\left\langle \sum_{i=1}^{N} P_2(\hat{\boldsymbol{n}}_i \cdot \boldsymbol{h}) \right\rangle = \frac{1}{N}\left\langle \sum_{i=1}^{N} \frac{3\cos^2(\hat{\boldsymbol{n}}_i \cdot \boldsymbol{h}) - 1}{2} \right\rangle \tag{9.12}$$

This order parameter is expected to describe the characteristics regarding the tendency of the alignment of the magnetic moments to the applied magnetic field direction with a more distinguishable variation than the former order parameter $\langle \hat{n}_y \rangle$.

In the following, we discuss the results obtained by Monte Carlo simulations from a quantitative point of view in terms of these four order parameters in addition to the cluster size distribution and from a qualitative point of view in terms of snapshots.

## *9.1.6 Specification of parameters for simulations*

Unless specifically noted, the simulation results shown in the following sections were obtained by adopting the following values for performing the simulations. The volumetric fraction of magnetic cube-like particles, $\phi_V$, is set as $\phi_V = 0.2$, and the number of particles $N$ is $N = 100$. We focus on a special case where half of the total number of particles have a magnetic moment in the upward diagonal direction, i.e., in the diagonal direction away from the plane bottom surface and the reminder has a magnetic moment in the downward diagonal direction, i.e., in the diagonal direction toward the bottom surface. The cube-like particles with these two types of the magnetic moment direction were initially located randomly in the system and the periodic boundary condition is applied for treatment at the boundary surfaces of the simulation box both in the $x$- and $y$-directions. As already mentioned, an external magnetic field is applied along the bottom plane surface in the $y$-axis direction. In the snapshots that will be shown below, the red or dark and the blue or light colored cubes have a magnetic moment in the upward and the downward diagonal directions, respectively. The cutoff distance $r_{coff}$ for the evaluation of interaction energies is set as $r_{coff} = 10d$, and the surface-to-surface criterion distance $r_{clstr}$ for the assessment of cluster formation is set as $r_{clstr} = 0.1d$. The maximum random translational distance $\Delta r$ and rotation angle $\Delta\theta$ in Monte Carlo procedures arc adopted as $\Delta r = 0.1d$ and $\Delta\theta = (2°/180°)\pi$, respectively. The non-dimensional parameters for the applied magnetic field strength $\xi$ and the magnetic particle-particle interaction strength $\lambda$ are set in a wide range with $\xi = 0 \sim 20$ and $\lambda = 0 \sim 20$. It is noted that for the case of $\lambda >> 1$ and $\xi >> 1$ the effect of either one is more dominant than the thermal motion and in the converse case with $\lambda \lesssim 1$ and $\xi \lesssim 1$ the effect of thermal motion is more dominant than either one. The total number of MC steps per simulation run, $N_{smplmx}$, is taken as $N_{smplmx} = 2{,}000{,}000$ and the last 90% of thc data were used for the averaging procedure.

Finally, it should be noted that although the present small system may be regarded as giving rise to acceptable results as a first approximation, in an academic context a larger system such as $N = 500$ or 1,000 may be desirable and thc cluster-moving method should be employed for conducting the simulation in a more accurate manner.

### 9.1.7 Results and discussion

#### (a) Dependence of aggregate structures on the magnetic particle-particle interaction strength

Figure 9.2 shows snapshots for a simulation with no applied magnetic field, $\xi = 0$, for the three different cases of the magnetic particle-particle interaction strength $\lambda$: Figs. 9.2(a), (b) and (c) are for $\lambda = 0$, 5 and 10, respectively. Similar snapshots are shown in Fig. 9.3 for a strong applied field strength $\xi = 10$. We first discuss results shown in Fig. 9.2 for the case of no applied magnetic field. In the case of no magnetic interaction

$\lambda = 0$, as shown in Fig. 9.2(a), thermal motion is absolutely dominant, therefore, the particles do not aggregate to form clusters, the magnetic moment of each particle does not incline in a specific favored direction, and the particles show an almost uniform directional characteristic. As the magnetic interaction strength is increased from $\lambda = 0$ in Fig. 9.2(a), the particles come to aggregate at $\lambda = 5$ in Fig. 9.2(b) and finally form significant aggregate structures at the magnetic interaction strength $\lambda = 10$, shown in Fig. 9.2(c). As expected from the discussion in Section 3.5 regarding favored aggregate structures from an energy point of view, it is evident that the large aggregate structures in Fig. 9.2(c) are a combination of the basic structures shown in Fig. 3.27(b). These aggregates are not chain-like clusters, but are designated as close-packed clusters. In the intermediary interaction case of $\lambda = 5$, the basic structures *e*1 and *e*2 in Fig. 3.25(a) are certainly observed in small clusters in Fig. 9.2(b).

Next, we consider results for the case of the strong magnetic field strength $\xi = 10$, shown in Fig. 9.3. In the case of no magnetic particle-particle interaction $\lambda = 0$, shown in Fig. 9.3(a), the particles do not aggregate, but the magnetic moment of each particle is strongly restricted to the magnetic field direction. In the case of a intermediary magnetic interaction $\lambda = 5$, shown in Fig. 9.3(b), chain-like clusters are observed to be formed in the magnetic field direction, and for the stronger case $\lambda = 10$, shown in Fig. 9.3(c), the particles exhibit long and thick chain-like clusters formed in the field direction. These clusters are formed based on the basic structure shown in Fig. 3.27(a), and appear to exhibit significant stability, which will be verified in discussing results of the order parameters and the cluster size distribution. In the intermediary interaction case $\lambda = 5$, the combination of the basic structure in Fig. 3.26(a) gives rise to short clusters in Fig. 9.3(b). From these results, we understand that application of an external magnetic field induces a significant tendency for the formation of thick chain-like clusters in the field direction if the magnetic particle-particle interaction strength is sufficiently stronger than the thermal energy.

Figure 9.4 shows results of the order parameters $S_1$ and $S_2$ for the case of no applied magnetic field case $\xi = 0$, where the curves are shown for the four different cases of the magnetic interaction strength, $\lambda = 5$, 7, 8 and 10. Similar results are shown in Fig. 9.5 for the strong applied magnetic field case $\xi = 10$. In the case of no applied magnetic field, shown in Fig. 9.4, the order parameter $S_1$ in Fig 9.4(a) shows features that are common to all four cases of the magnetic interaction strength in that significantly small negative values tend to zero with increasing radial distance $r^*(= r/d)$. In the case of no external magnetic field, the cube-like particles do not incline in a specific favored direction, so that it is quite reasonable

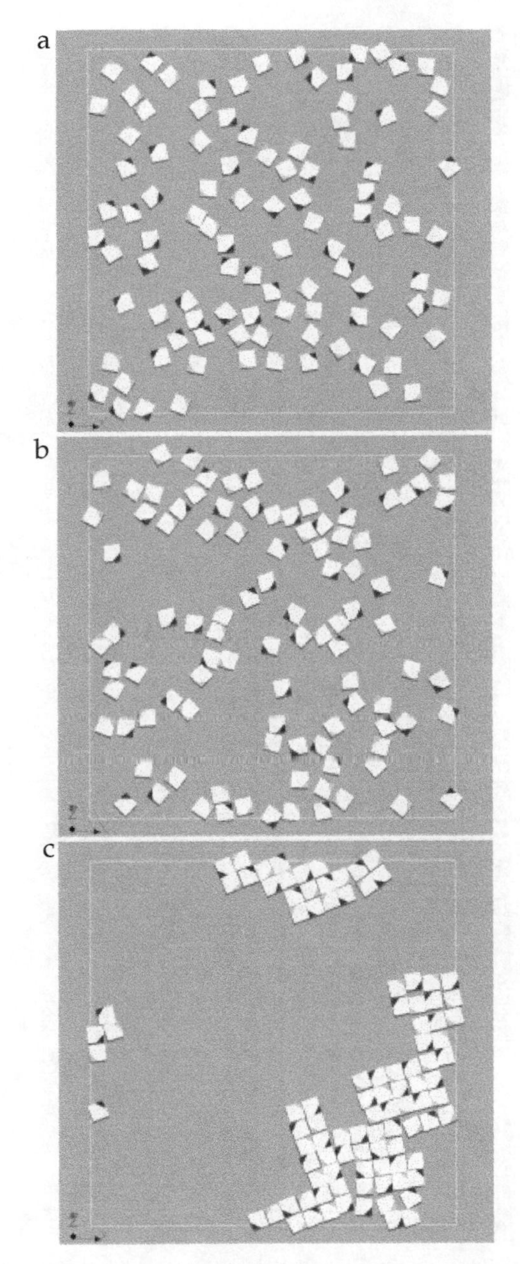


**Figure 9.2.** Aggregate structures of cubic particles on a plane under no applied magnetic field for the three cases of magnetic particle-particle interaction strengths: (a) $\lambda = 0$, (b) $\lambda = 5$ and (c) $\lambda = 10$.

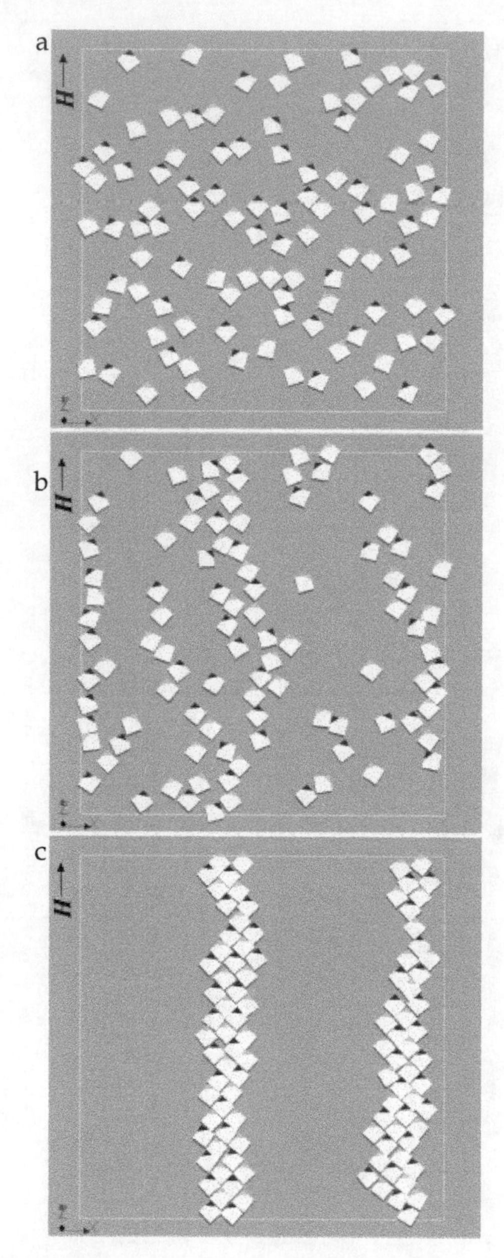


**Figure 9.3.** Aggregate structures of cubic particles on a plane for the magnetic field strength $\xi = 10$: (a) $\lambda = 0$, (b) $\lambda = 5$ and (c) $\lambda = 10$: an external magnetic field is applied in the upward direction.

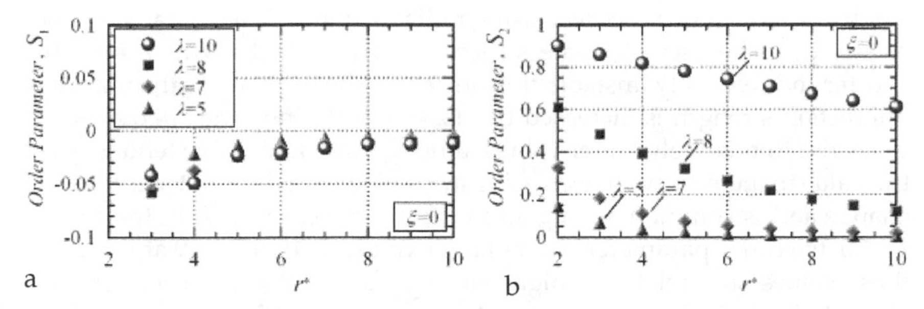


**Figure 9.4.** Order parameters as a function of the radial distance $r$ under no applied magnetic field situation: (a) order parameter $S_1$ and (b) order parameter $S_2$ are shown for the four different cases of magnetic particle-particle interaction strengths, $\lambda = 5, 7, 8$ and 10.

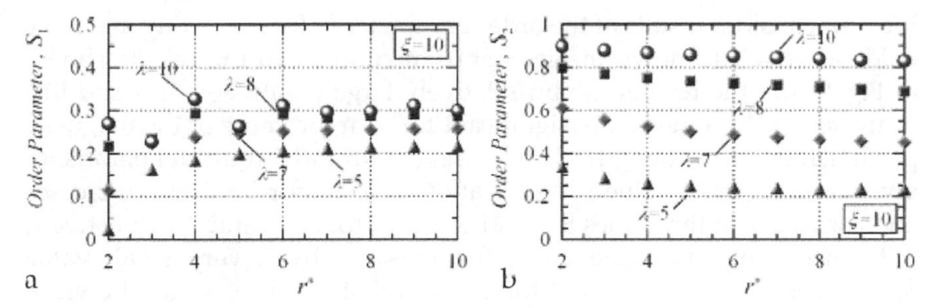


**Figure 9.5.** Order parameters (a) $S_1$ and (b) $S_2$ under the situation of the applied magnetic field strength $\xi = 10$ for the four different cases of $\lambda = 5, 7, 8$ and 10.

that in the averaging procedure $S_1$ converges to zero with the expanding area of an increasing radial distance. Prohibitive configurations of the cube-like particles are expected to arise from the particle geometrical shape (i.e., cubic shape) in a particle-particle distance area of the order of particle size $d$. This effect may be seen in the curves in Fig. 9.4(a), that is, the value for each case monotonically decreases with decreasing radial distance. From a quantitative point of view, this characteristic is strikingly similar among the four curves. Hence the significantly different aggregate structures, clearly shown in Fig. 9.2, which are strongly dependent on the magnetic particle interaction strength, cannot be distinguished in a reasonable manner in terms of the order parameter $S_1$. This drawback is remarkably improved by the order parameter $S_2$, shown in Fig. 9.4(b).

In contrast to the results of $S_1$ shown in Fig. 9.4(a), the curves of $S_2$ in Fig. 9.4(b) exhibit significantly different characteristics for each value of the magnetic interaction strength. In the case of $\lambda = 5$, the curve tends to be zero in value in large averaging areas and increases slightly to a small positive value with decreasing radial distance. These characteristics clearly describe the aggregate structures in Fig. 9.2(b), where only the shorter and relatively unstable clusters tend to be formed. As the magnetic interaction strength is increased from $\lambda = 5$ to 10, the order parameter $S_2$ gives rise to relatively larger values although the qualitative tendency for the value to increase with decreasing averaging area does not significantly change and is common among all four different cases of $\lambda$. In the case of $\lambda = 10$, the order parameter shows larger values with $S_2 \simeq 0.9$ at $r^* = 2$ and these values stay relatively high with $S_2 \simeq 0.6$ at $r^* = 10$. Therefore, the characteristics of the order parameter $S_2$ can reflect the formation of the large scaled clusters in the system and indirectly verify that the internal structure is expanded across the whole system, which certainly describes the situation of the large packed clusters shown in Fig. 9.2(c).

We now consider results of the order parameter $S_1$ for the strong magnetic field case $\xi = 10$, shown in Fig. 9.5(a). In contrast to the results shown in Fig. 9.4(a), the results exhibit relatively large positive values and this characteristic becomes more significant for a larger value of the magnetic particle interaction strength $\lambda$. In the cases of $\lambda = 5$ and 7, the order parameter gives rise to a small value $S_1 \simeq 0.02$ at $r^*(= r/d) = 2$ for $\lambda = 5$, and increases and converges to the values $S_1 \simeq 0.21$ and 0.25 for $\lambda = 5$ and 7, respectively, with increasing averaging area. The reason why a very small value $S_1 \simeq 0.02$ is observed at $r^* = 2$ for $\lambda = 5$ is that in this case as shown in Fig. 9.3(b), only unstable short chain-like clusters are formed in the field direction, and these short clusters have mainly the basic structure *e*1 shown in Fig. 3.25(a), but do not include the basic structure *a*1 or *b*1 in Fig. 3.22(a) that give rise to a much larger value of the order parameter, $S_1 \simeq 1$, than the basic structure *e*1 in Fig. 3.25(a). Therefore, it is seen that the increase in the results for $\lambda = 5$ and 7 with increasing averaging area is due to inclusion of the basic structure *a*1 in Fig. 3.22(a), which is supported by the weak cluster formation in Fig. 9.3(b). The magnetic particle-particle interaction increases the tendency of the cluster formation based on the basic structure *a*1 in Fig. 3.22(a) in a strong applied magnetic field situation. This tendency gives rise to larger values of the order parameter $S_1$ for stronger magnetic particle-particle interaction in Fig. 9.5(a). Since stable long and thick chain-like clusters are formed in the field direction across the simulation box for the cases of $\lambda = 8$ and 10, the basic structure *a*1 in Fig. 3.22(a) is significantly present in all the averaging areas. This leads to the relatively large value of $S_1 \simeq 0.3$, which is almost independent of the value of $r^*$, and it is noted that even at $r^* = 2$ the order parameter yields $S_1 \simeq 0.27$ and 0.21 for $\lambda = 10$ and 8, respectively. It is observed in the curves for $\lambda = 10$ and $\lambda = 8$ that the value of $S_1$ exhibits a zigzag type feature of $S_1$, that is, $S_1 \simeq 0.27$, 0.22,

0.32, 0.27 and 0.31 at $r^* = 2, 3, 4, 5$ and 6, respectively, for the case of $\lambda = 10$. This is because another basic structure, similar to e1 in Fig. 3.25(a), with the second particle being shifted in the magnetic field direction by about $\sqrt{2}d$, makes the order parameter decrease slightly because it is not included in the averaging area $r^* = 2$, leading to a larger value, but is included in $r^* = 3$, leading to a smaller value. This suggestion may be verified if the stable long and thick chain-like clusters composed of combination of the basic structure in Fig 3.27 (a) are formed in the field direction and the snapshot shown in Fig. 9.3(c) clearly exhibits this formation. From this discussion, the order parameter $S_1$ can describe the difference in the characteristics of the cluster formation, dependent on the magnetic interaction strength, in the situation of a strong applied magnetic field, but does not seem to be sufficient for understanding the essential microstructure at a physically reasonable level.

We now discuss results of the order parameter $S_2$ for the strong magnetic field case $\xi = 10$, shown in Fig. 9.5(b). In the case of $\lambda = 10$ and 8 stable long and thick chain-like clusters are formed, and this feature is significantly well describe by the order parameters $S_2$. That is, it shows the large values of $S_2 \simeq 0.9$ (nearly equal to unity) at $r^* = 2$ and remains almost constant with enlarging averaging areas with value $S_2 \simeq 0.83$ at $r^* = 10$. The result for $\lambda = 5$ shows much smaller values than those for $\lambda = 10$ and 8, and this characteristic seems to be quite reasonable because the unstable short chain-like clusters are formed as shown in Fig. 9.3(b). The reason why the curve shows relatively large values around $S_2 \simeq 0.2$ (not zero) even at $r^* = 10$ is that the correlation of the magnetic moments between arbitrary pairs of particles does not significantly decrease for a strong magnetic field situation such as $\xi = 10$. Whether stable clusters are formed in the system or not, this alignment of the magnetic moments does not essentially change, but may be exaggerated by the magnetic particle-particle interaction and this leads to larger values of $S_2$ for stronger magnetic interactions, as observed in Fig. 9.5(b).

The above characteristics of $S_2$ certainly enable one to conclude that stable long and thick chain-like clusters are formed between magnetic particle-particle interaction strengths $\lambda = 7$ and 8, which may be supported by the results of the snapshots in Fig. 9.3. From these results, it is evident that the order parameter $S_2$ is much more appropriate than the more common order parameter $S_1$ for describing how large and stable clusters are formed in a system.

The cluster size distribution may be a more direct quantity for assessing the formation of stable clusters in a system. Figure 9.6 shows results of the cluster size distribution for two cases of the magnetic field strength, $\xi = 0$ shown in Fig. 9.6(a) and $\xi = 10$ shown in Fig. 9.6(b). Each figure is for three different cases of the magnetic interaction strength, $\lambda = 7$, 8 and 10. As already discussed, we anticipate that the stable long and thick chain-like clusters shown in Fig. 9.3(c) are formed between $\lambda = 7$ and 8. Hence, we address the cluster size distribution for the present three cases of the magnetic interaction strength. In the case of no applied magnetic field $\xi = 0$, shown in Fig. 9.6(a), it is seen that the cluster size distribution dramatically changes between $\xi = 7$ and $\xi = 8$. That is, the distribution for $\lambda = 7$ rapidly decreases and converges to almost zero at $S \simeq 20$ with no larger clusters than that size. This implies that only relatively short clusters are formed, but there are no long clusters that are expanded to range from one side to the other across the simulation box. In contrast, the distribution for $\lambda = 8$ exhibits the appearance of larger clusters such as $S \simeq 50$, and shows that the increase in the magnetic interaction strength induces the formation of further larger clusters. It is seen in Fig. 9.6(a) for the case of $\lambda = 10$ that a larger cluster with $S = 100$ is observed with almost no single-moving particles.

We now discuss results of the cluster size distribution in a strong magnetic field case $\xi = 10$, shown in Fig. 9.6(b). In the case of $\lambda = 7$, there is no peak in the curve, similar to the case of $\xi = 0$, but the range of the distribution is expanded to slightly larger areas of $S$. This implies that an external magnetic field tends to induce the formation of clusters slightly although the main factor for the cluster formation is the magnetic particle-particle interaction. In the cases of $\lambda = 8$ and 10, a high peak appears at $S \simeq 66$ and 47 for $\lambda = 8$ and 10, respectively. In the case of $\lambda = 10$, the result shows that $N_s$ has almost zero value except at this peak range, so that only several large clusters are formed in the system with few other small clusters and few single-moving particles. The reason why a higher peak appears at a smaller value of $S$ for $\lambda = 10$ than for $\lambda = 8$ is that a small number of longer clusters are formed in the system for the case of $\lambda = 10$; this consideration is certainly supported by the snapshots in Fig. 9.3(c), where the two long and thick chain-like clusters are observed to be formed across the simulation box. In the case of the intermediary strength case $\lambda = 8$, the number of single moving particles and shorter clusters decreases in a strong magnetic field situation, and the distribution in the small range of $S$ is expanded to larger area such as $S \simeq 30$. As already pointed for the order parameter $S_2$, it is seen that the increase in the magnetic interaction strength induces the formation of further larger clusters and diminishes the number of single-moving particles and smaller clusters.

From this consideration, we understand that the cluster size distribution can more clearly and directly provide the information as to how large clusters are formed in the system, although it is necessary to address the order parameters in discussing the internal structure of clusters.

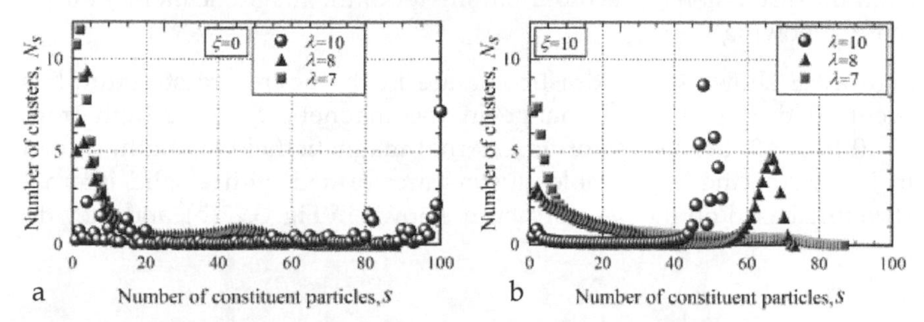


**Figure 9.6.** Number of clusters for each size cluster for the three different cases of magnetic particle-particle interaction strengths, $\lambda = 7, 8$ and 10: (a) $\xi = 0$ and (b) $\xi = 10$.

## *(b) Dependence of aggregate structures on the external magnetic field strength*

We discuss here the dependence of the aggregate structure of particles on the external magnetic field strength. From the discussion above, it is seen that an increase in the magnetic interaction strength induces the formation of further larger clusters. In the following, we discuss this characteristic from a different point of view using the order parameter of the magnetic moment itself rather than the correlation of the magnetic moments evaluated as $S_1$ and $S_2$.

Figure 9.7 shows results of the snapshots of particle aggregates for the magnetic interaction strength $\lambda = 5$ in the two cases of the magnetic field strengths, $\xi = 5$ shown in Fig. 9.7(a) and $\xi = 20$ shown in Fig. 9.7(b). Referring to these and the snapshots in Figs. 9.2(b) for $\xi = 0$ and 9.3(b) for $\xi = 10$ together with the present snapshots in Figs. 9.7, long and thick chain-like clusters tend to be formed in the magnetic field direction with increasing magnetic field strength. However, even at $\xi = 20$ shown in Fig. 9.7(b), the internal structure of these clusters is relatively loose and not as rigid as those in the snapshot in Fig. 9.3(c) because the magnetic interaction $\lambda = 5$ is not sufficient for forming such stable and rigid clusters. If the magnetic moment of each particle is strongly restricted to the magnetic field direction by a strong magnetic field, the particle configuration expressed by the basic structure *e*1 in Fig. 3.25(a) appears more frequently with greater possibility. This induces the formation of chain-like clusters with loose internal structure, as shown in Fig. 9.7(b), but if the magnetic interaction is much smaller than the thermal energy, then this tendency does not arise, even for increasing magnetic field strength. The criterion field strength for this tendency appearing significantly will be made clear from the discussion of the order parameter of the magnetic moment itself in the following.

Figure 9.8 shows that a drastic change in the aggregate structure has occurred during a small change in the magnetic field strength from $\xi = 0$ to $\xi = 2$. In the case of no external magnetic field strength, shown in Fig. 9.2(c), the large stable clusters are formed with a solid internal structure based on the basic structure shown in Fig. 3.27(b), and they do

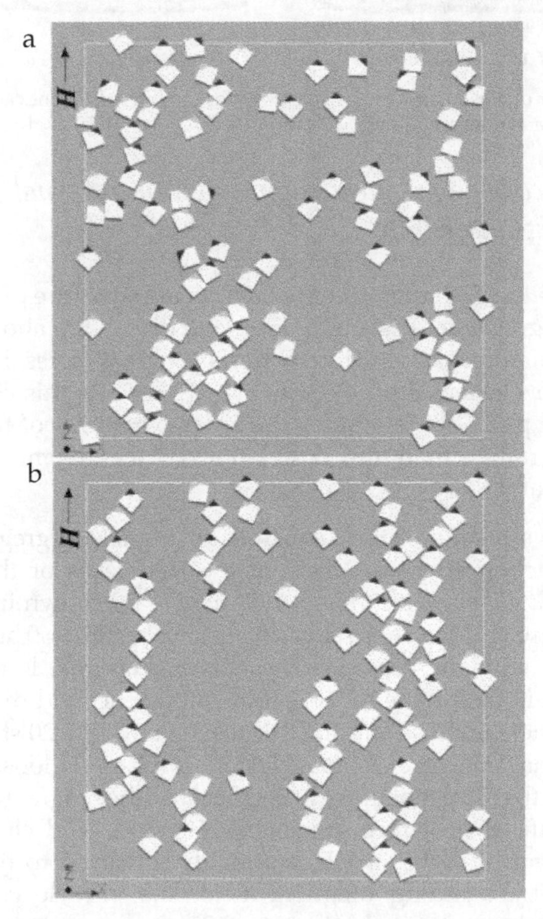


**Figure 9.7.** Dependence on the aggregate structures of cubic particles on the magnetic field strength for $\lambda = 5$: (a) $\xi = 5$ and (b) $\xi = 20$.

not incline in a specific favored direction. In this situation, application of even a weak external magnetic field such as $\xi = 1$ or $\xi = 2$ drastically induces the significant formation of long and thick chain-like clusters in the field direction. This effect induced by the field strength is seemingly beyond our previous expectation because the value of $\xi = 1$ or $\xi = 2$ implies that the random motion should be much more dominant than the magnetic interaction. However, the snapshots shown in Fig. 9.8 clearly show that this effect occurs for a small field strength. This may be explained in the following. For the case of the present cube-like particles, the internal structure of the aggregates is significantly restricted to the configuration of a face-to-face contact situation. For the constituent particles in the same cluster, this restriction exaggerates the alignment of the magnetic

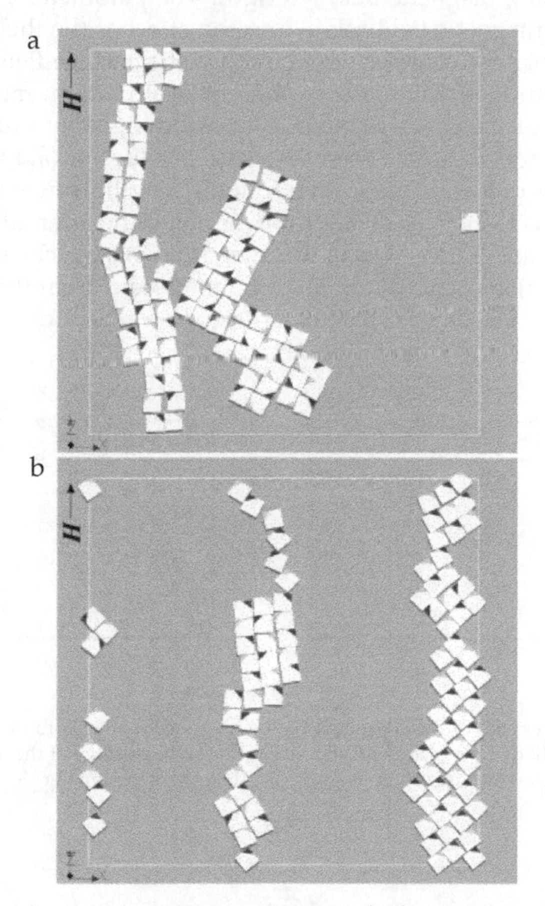


**Figure 9.8.** Dependence on the aggregate structures of cubic particles on the magnetic field strength for $\lambda = 10$: (a) $\xi = 1$ and (b) $\xi = 2$.

moment to the field direction. This alignment of the magnetic moments further exaggerates the magnetic particle-particle interaction between the constituent particles in the clusters, which leads to the appearance of the tendency of the clusters to incline in the magnetic field direction , even at a weak magnetic field with a strength such as $\xi = 1$ or 2.

Figure 9.9 shows results of the two order parameters as a function of the magnetic field strength $\xi$ where Fig. 9.9(a) is for the order parameter $\langle \hat{n}_y \rangle$ shown in Eq. (9.11) and Fig. 9.9(b) is for $S_{ny}$ shown in Eq. (9.12). The three different cases of the magnetic interaction strength $\lambda = 0$, 5 and 10 are in each figure. In the case of the order parameter $\langle \hat{n}_y \rangle$, all the curves steeply increase and tend to a sufficiently large value approaching unity at $\xi \simeq 10$ with increasing magnetic field strength. The parameter $\xi$ denotes the strength of the magnetic field relative to the effect of the thermal energy, so that a value of $\xi = 10$ implies that the magnetic particle-field interaction is significantly more dominant than the effect of random motion, hence, the attainment of the order parameter to approximately unity seems to be quite reasonable. It is also seen that larger values of $\langle \hat{n}_y \rangle$ are obtained for larger values of the magnetic interaction strength. However, this order parameter seems to be insufficient for describing the internal structure of aggregates because in considering the formation of such clusters as found in Figs. 9.3(a), (b) and (c), we are not able to discern from the differences in the curves in Fig. 9.9(a) that the internal structure of these particle configurations is completely different.

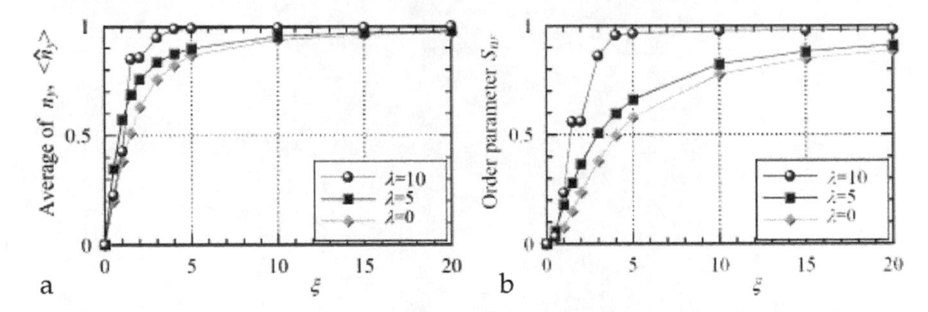


**Figure 9.9.** Characteristics of the alignment of the magnetic moments to the magnetic field direction as a function of the magnetic field strength: (a) the average of the $y$-component $\hat{n}_y$ of the magnetic moment direction $\hat{\boldsymbol{n}}$ and (b) the order parameter $S_{ny}$ regarding the $y$-component $\hat{n}_y$.

The order parameter $S_{ny}$ shown in Fig. 9.9(b) can describe more clearly the difference in the internal structure of the clusters than the previous order parameter $\langle \hat{n}_y \rangle$. Although the three curves show similar characteristics in that the results monotonically increase with increasing magnetic field strength, the curve for $\lambda = 10$ shows completely different features in a quantitative manner from those for $\lambda = 0$ and 5. That is, the former curve for $\lambda = 10$ increases more rapidly in value and converges to almost unity at $\xi =$ 5, but the latter curves clearly exhibit much smaller values over the same range, even in the large field areas such as $\xi = 20$. These characteristics of the order parameter $S_{ny}$ certainly reflect the difference in features of the internal structure of the clusters. As shown in Fig. 9.3, long and stable thick chain-like clusters are formed for $\lambda = 10$ in contrast to single-particle and short clusters with loose internal structure that are formed for $\lambda = 0$ and 5. From the curves of $S_{ny}$ in Fig. 9.9(b), it is seen that the magnetic particle-particle interaction significantly encourages the alignment of the magnetic moments to the magnetic field direction.

We understand from these considerations that the order parameter $S_{ny}$ can describe the difference in the characteristics of the alignment of the magnetic moments and the difference in the internal structure of aggregates more sensitively and clearly than the more common order parameter $\langle \hat{n}_y \rangle$.

### 9.1.8 Sample simulation program

Finally, we show the sample simulation program, written in FORTRAN language, that has been adapted from a full three-dimensional version of the simulation code in order to obtain the present results for the two-dimensional system composed of the current cube-like particles that have been discussed above. In consideration of the reader who desires to develop a three-dimensional system program in future, we did not completely remove the three-dimensional descriptions in the program to obtain the present two-dimensional version. For instance, the position vector $\boldsymbol{r}_i$ is still expressed as (RX(I), RY(I), RZ(I)) including the $z$-component, and also in subroutine PTCLMDL the variables are specified for describing the shape of a three-dimensional cube and therefore this subroutine is directly applicable to a three-dimensional simulation without any modifications. From a tutorial point of view, a number of non-essential subroutines, which may make the logical flow more complex, have been omitted from the original three-dimensional code. Readers may, therefore, be more able to understand the present two-dimensional version and more straightforwardly develop their own version of a three-dimensional simulation code. In this development, the main task would be to modify the ENECAL subroutine for analyzing the particle overlap in a three-dimensional configuration of a pair of cube-like particles and a variety of discussions made in Section 5.4 may be an important guide for readers to do so. It would also be straightforward for readers to add subroutines regarding the radial distribution, the pair correlation function and the orientation distribution functions to the present two-dimensional simulation program.

All the important variables are defined in the first comment area of the program before the start of the specification of parameters, so it is strongly recommended for readers to understand the meaning of these variables. It is noted that a non-dimensionalized system is used for variables in the program where for instance distances and energies are non-dimensionalized by the side length $d$ and the thermal energy $kT$, respectively. The important variables used in the program are explained in the following:

RX(I),RY(I),RZ(I) : $(x,y,z)$ coordinates of the position vector of magnetic particle $i$

NX(I),NY(I),NZ(I) : $(x,y,z)$ coordinates of the magnetic moment direction $\boldsymbol{n}_i$

EX(I),EY(I),EZ(I) : is the particle direction but not used in the present Monte Carlo procedures

N,NDENS,VDENS : number of particles $N$, number density $n^*$ and volumetric fraction $\phi_V$

(HX,HY,HZ) : unit vector denoting the magnetic field direction (= (0,1,0) in the present case)

D : is the side length of the cube and is equal to unity in the present non-dimensional system.

(XL,YL) : side lengths of the simulation box in the $(x, y)$-directions

RA,KU : non-dimensional parameters $\lambda$ and $\xi$

As an aid to understanding the program, explanatory descriptions have been added to the important procedures. The line numbers are only for the convenience of the reader and are not necessary for executing the FORTRAN program.

```
00010 C*********************************************************************
00020 C*                          mc_cube_1.f                               *
00030 C*                                                                    *
00040 C*     OPEN(9, FILE='@bbb1.dat', STATUS='UNKNOWN'); para,results      *
00050 C*     OPEN(11,FILE='bbb11.dat', STATUS='UNKNOWN'); parameters        *
00060 C*     OPEN(12,FILE='bbb21.dat', STATUS='UNKNOWN'); ORDER PARAMETERS  *
00070 C*     OPEN(13,FILE='bbb31.dat', STATUS='UNKNOWN'); CLSTR.DIST.       *
00080 C*     OPEN(14,FILE='bbb41.mgf', STATUS='UNKNOWN'); ANIME,MicroAVS    *
00090 C*     OPEN(21,FILE='bbb001.dat',STATUS='UNKNOWN'); PTCL POS.ORI      *
00100 C*     OPEN(22,FILE='bbb011.dat',STATUS='UNKNOWN'); PTCL POS.ORI      *
00110 C*     OPEN(23,FILE='bbb021.dat',STATUS='UNKNOWN'); PTCL POS.ORI      *
00120 C*     OPEN(24,FILE='bbb031.dat',STATUS='UNKNOWN'); PTCL POS.ORI      *
00130 C*     OPEN(25,FILE='bbb041.dat',STATUS='UNKNOWN'); PTCL POS.ORI      *
00140 C*                                                                    *
00150 C*     ---------------  MONTE CARLO SIMULATIONS   ---------------     *
00160 C*       TWO-DIMENSIONAL MONTE CARLO SIMULATIONS OF                   *
00170 C*       A DISPERSION COMPOSED OF HEMATITE CUBIC PARTICLES            *
00180 C*       IN THREMODYNAMIC EQUILIBRIUM.                                *
00190 C*                                                                    *
00200 C*      1. CUBIC PARTICLES.                                           *
00210 C*      2. HEMATITE PARTICLES ARE CONSIDERED.                         *
00220 C*      3. STERIC LAYER IS NOT CONSIDERED.                            *
00230 C*      4. NOTE THAT MAGNITUDE OF CUBEA(xyz,6) IS NOT 1, BUT 0.5.     *
00240 C*      5. NORMALIZATION SHOULD BE CONDUCTED EVERY MC STEP.           *
00250 C*        (OTHERWISE, OVERLAPS HAPPEN BETWEEN PARTICLES.)             *
00260 C*                                                                    *
00270 C*                                      VER.1  BY A.SATOH , '15 10/10 *
00280 C*********************************************************************
00290 C    N      : NUMBER OF PARTICLES (N=INISQUX*INISQUY)
00300 C    D      : LENGTH OF EACH LINE OF THE CUBE (=1)
00310 C    VP     : VOLUME OF THE PARTICLE (=1)
00320 C    NDENS  : NUMBER DENSITY
00330 C    VDENS  : VOLUMETRIC FRACTION
00340 C    RA     : NONDIMENSIONAL PARAMETER OF PARTICLE-PARTICLE INTERACT
00350 C    KU     : NONDIMENSIONAL PARAMETER OF PARTICLE-FIELD INTERACTION
00360 C    HX,HY,HZ : MAGNETIC FIELD DIRECTION (UNIT VECTOR)
00370 C    RCOFF  : CUTOFF RADIUS FOR CALCULATION OF INTERACTION ENERGIES
00380 C    XL,YL,ZL : DIMENSIONS OF SIMULATION REGION (ZL=D)
00390 C    INISQUX,INISQUY : NUMBER OF PARTICLES IN EACH DIRECTION
00400 C    BETA   : RATE OF NUMBER OF PTCLS WITH UPWARD OBLIQUE MAG. MOMENT
00410 C             RELATIVE TO THE TOTAL NUMBERS (=0.5)
00420 C    NUP    : NUMBER OF PTCLS WITH UPWARD OBLIQUE MAG. MOMENT
00430 C    NDWN   : NUMBER OF PTCLS WITH DOWNWARD OBLIQUE MAG. MOMENT
00440 C           (N = NUP + NDWN)
00450 C
00460 C                     INISQUX  INISQUY       N
00470 C                        2        2          4
00480 C                        3        3          9
00490 C                       ...      ...        ...
00500 C
00510 C                       10       10        100
00520 C                       ...      ...        ...
00530 C
00540 C                       30       30        900
00550 C
00560 C    RX(I),RY(I),RZ(I)     : PARTICLE POSITION (RZ(I)=D/2)
00570 C    EX(I),EY(I),EZ(I)     : DIRECTION OF PARTICLE AXIS DIRECTION
00580 C                            ( (0,0,1) OR (0,0,-1) )
00590 C    NX(I),NY(I),NZ(I)     : DIRECTION OF MAGNETIC MOMENT
00600 C    E(I)         : INTERACTION ENERGY OF PARTICLE I WITH THE OTHERS
00610 C    MOMX(**),MOMY(**)     : MAG. MOMENT OF SYSTEM AT EACH TIME STEP
00620 C    MOMZ(**)
00630 C
00640 C    RMAT(3,3,N)  : ROTATIONAL MATRIX
00650 C
00660 C    DELR      : MAXIMUM MOVEMENT DISTANCE
00670 C    DELT      : MAXIMUM MOVEMENT IN MAGNETIC MOMENT ORIENTATION
00680 C    DELDEG    : = DELT
00690 C    MCSMPLMX: MAXIMUM NUMBER OF MC STEP
00700 C
00710 C    MORDER, MORDER4 : ORDER PARAMETERS
00720 C    MORDERB(10), MORDER4B(10) : ORDER PARAMETERS DEPENDENT
00730 C                                                ON RADIAL DISTANCE
00740 C
00750 C    WIDTHCLS   : = 0.1 * D FOR ASSESSMENT OF CLSTR FORMATION
00760 C    RCLSLMT    : PTCLS BEYOND THIS RANGE ARE NOT REGARDED AS A MEMBER
00770 C                 OF A CLUSTER
00780 C    CLSNDIST(N,10)  : NUMBER DISTRIBUTION OF CLUSTERS
00790 C    ANSNDIST(N)     : AVERAGE OF NUM. DIST. OF CLUSTERS
00800 C    NDISTSMP(10)    : SAMPLING NUMBERS FOR EACH RANGE
```

• The position of particle $i$ is denoted by (RX(I), RY(I), RZ(I)).
• The orientation of the magnetic moment of particle $i$ is denoted by (NX(I), NY(I), NZ(I)).

• DELR, DELT and DELDEG are used in generating random displacements.

```
00810 C       ANSNDSMP          : TOTAL NUMBER OF SAMPLING NUMBERS
00820 C       NDISTCNT          : COUNTER FOR DATA SAVING (=1,2,...10)
00830 C       NDISTMAX          : NUMBER OF CONSTITUTING PARTICLES OF LARGEST
00840 C                           CLUSTER
00850 C
00860 C       ++++ -XL/2<RX(I)<XL/2 , -YL/2< RY(I)<YL/2, RZ(I)=D/2 ++++
00870 C-----------------------------------------------------------------------
00880         IMPLICIT REAL*8 (A-H,O-Z), INTEGER (I-N)
00890 C
00900         COMMON /BLOCK101/ RX    , RY    , RZ
00910         COMMON /BLOCK102/ EX    , EY    , EZ
00920         COMMON /BLOCK103/ NX    , NY    , NZ
00930         COMMON /BLOCK111/ XL    , YL    , ZL
00940         COMMON /BLOCK112/ HX    , HY    , HZ
00950         COMMON /BLOCK113/ RA    , KU
00960         COMMON /BLOCK115/ N     , NDENS, VDENS , NUP , NDWN
00970         COMMON /BLOCK116/ D     , VP    , D2    , DCHK
00980         COMMON /BLOCK118/ INISQUX , INISQUY , INISQUZ , BETA
00990 C
01000         COMMON /BLOCK130/ E     , ENEW , EOLD
01010         COMMON /BLOCK131/ RCOFF, DELR , DELT , DELDEG
01020         COMMON /BLOCK142/ NRAN , RAN  , IXRAN
01030         COMMON /BLOCK143/ MOMX , MOMY , MOMZ , MEANENE
01040 C
01050         COMMON /BLOCK160/ WIDTHCLS , RCLSLMT
01060         COMMON /BLOCK161/ CLSN  , CLSMX , CLS
01070         COMMON /BLOCK163/ CLSNDIST, ANSNDIST, NDISTSMP, ANSNDSMP
01080         COMMON /BLOCK164/ NDISTMAX, NDISTCNT
```

• These variables are used for evaluation of cluster size distribution.

```
01090 C
01100         COMMON /BLOCK171/ EDIRX , EDIRY , EDIRZ
01110         COMMON /BLOCK172/ MORDX , MORDY , MORDZ
01120         COMMON /BLOCK173/ EORDX , EORDY , EORDZ
01130         COMMON /BLOCK175/ MORDER , MORDER4 , MORDERB ,  MORDER4B
```

• These variables are used for evaluation of order parameters.

```
01140 C
01150         COMMON /BLOCK200/ RMAT
01160         COMMON /BLOCK201/ CUBEA    , ANEIGHBA , ANEIALPH , ANEIA
01170         COMMON /BLOCK203/ CUBEALPH, ALPHNEIG , ALNEIA
01180         COMMON /BLOCK205/ RMATI , CUBEAI , CUBEAJ , CUBEALI, CUBEALJ
01190 C
01200         COMMON /WORK116/  ENEMAGI
01210         COMMON /WORK201/  CEXI , CEYI , CEZI
01220         COMMON /WORK202/  CNXI , CNYI , CNZI
01230 C       ---------------------------------------------------------------
01240         INTEGER     NN , NNS , TT
01250         PARAMETER( NN=1000 , NNS=5000000 , TT=1000 )
01260         PARAMETER( NRANMX=1000000 , PI=3.141592653589793D0 )
01270 C       ---------------------------
01280         REAL*8   KU      , NDENS
01290         REAL*8   BETA
01300         REAL*8   RX(NN) , RY(NN) , RZ(NN)
01310         REAL*8   EX(NN) , EY(NN) , EZ(NN) , NX(NN) , NY(NN) , NZ(NN)
01320         REAL*8   E(NN)  , ENEW   , EOLD
01330 C       ---------------------------
01340         REAL     RAN(NRANMX)
01350         INTEGER  NRAN , IXRAN , NRANCHK
01360 C       ---------------------------
01370         REAL     MOMX(NNS), MOMY(NNS) , MOMZ(NNS)  , MEANENE(NNS)
01380         REAL     EDIRX(NNS) , EDIRY(NNS) , EDIRZ(NNS)
01390         REAL     MORDX(NNS) , MORDY(NNS) , MORDZ(NNS)
01400         REAL     EORDX(NNS) , EORDY(NNS) , EORDZ(NNS)
01410         REAL*8   MORDER , MORDER4, MORDERB(10) , MORDER4B(10)
01420 C       ---------------------------
01430         REAL*8   RMAT(3,3,NN)
01440         REAL*8   CUBEA(3,6,NN) , CUBEALPH(3,8,NN)
01450         REAL*8   RMATI(3,3)
01460         REAL*8   CUBEAI(3,6)   , CUBEALI(3,8)
01470         REAL*8   CUBEAJ(3,6)   , CUBEALJ(3,8)
01480         INTEGER  ANEIGHBA(4,6), ANEIALPH(4,6) , ANEIA(2,6)
01490         INTEGER  ALPHNEIG(3,8), ALNEIA(3,8)
01500 C       ---------------------------
01510         INTEGER  NDNSMX
01520         INTEGER  CLSN , CLSMX(NN) , CLS(TT,NN)
01530 C
01540         REAL     CLSNDIST(NN,10), ANSNDIST(NN)
01550         INTEGER  NDISTSMP(10)    , ANSNDSMP    , NDISTMAX , NDISTCNT
01560 C       ---------------------------------------------------------------
01570         REAL*8   RXI  , RYI  , RZI  , RXJ  , RYJ  , RZJ
01580         REAL*8   EXI  , EYI  , EZI  , EXJ  , EYJ  , EZJ
01590         REAL*8   NXI  , NYI  , NZI  , NXJ  , NYJ  , NZJ
01600         REAL*8   RXIJ , RYIJ , RZIJ , RXJI , RYJI , RZJI
```

```
01610       REAL*8    RIJ  , RIJSQ, RCOFF2
01620 C
01630       REAL*8    RXCAN, RYCAN, RZCAN
01640       REAL*8    EXCAN, EYCAN, EZCAN
01650       REAL*8    CEXI , CEYI , CEZI
01660       REAL*8    CNXI , CNYI , CNZI
01670       REAL*8    ECAN
01680 C
01690       REAL*8    XL2  , YL2  , ZL2
01700       REAL*8    C0   , C00  , C1  , C2  , C3
01710       REAL*8    C11  , C12  , C13 , C14 , C21 , C22 , C23
01720       REAL*8    C31  , C32  , C33 ,       C41 , C42 , C43
01730       REAL*8    EMAGOLD , EMAGCAN
01740       REAL*8    RMATIHAT(3,3) , CUBEAIB(3,6) , CUBEAIS(3,6)
01750       REAL*8    BALPH(3,4)
01760 C
01770       INTEGER   MCSMPL , MCSMPLMX , NGRAPH , DNSMPL , NP , NOPT
01780       INTEGER   NANIME , NANMCTR
01790       INTEGER   NORDER , DNORDR, NORDRCTR
01800       INTEGER   NCLSCHK, DNCLSCHK
01810       INTEGER   ISTREET, NSMPL, NSMPL1, NSMPL2
01820       INTEGER   K0 , IXYZ
01830       LOGICAL   OVRLAP , CLSTRUE
01840 C ---------------------------------------------------------------------
01850           OPEN(9, FILE='@alku20ra20a1.dat', STATUS='UNKNOWN')
01860           OPEN(11,FILE='alku20ra20a11.dat', STATUS='UNKNOWN')
01870           OPEN(12,FILE='alku20ra20a21.dat', STATUS='UNKNOWN')
01880           OPEN(13,FILE='alku20ra20a31.dat', STATUS='UNKNOWN')
01890           OPEN(14,FILE='alku20ra20a41.mgf', STATUS='UNKNOWN')
01900           OPEN(21,FILE='alku20ra20a001.dat',STATUS='UNKNOWN')
01910           OPEN(22,FILE='alku20ra20a011.dat',STATUS='UNKNOWN')
01920           OPEN(23,FILE='alku20ra20a021.dat',STATUS='UNKNOWN')
01930           OPEN(24,FILE='alku20ra20a031.dat',STATUS='UNKNOWN')
01940           OPEN(25,FILE='alku20ra20a041.dat',STATUS='UNKNOWN')
01950                                                         NP   = 9
01960                                                         NOPT = 20
01970 C
01980 C
01990       KU       = 20.0D0
02000       RA       = 20.0D0
02010       VDENS    =  0.2D0
02020 C
02030       D        =  1.D0
02040       VP       =  1.D0
02050       DSQ      =  D**2
02060       DCHK     =  DSQRT( 3.D0 )
02070       D2       =  D/2.D0
02080       NDENS    =  VDENS/VP
02090 C
02100       INISQUX  =  6
02110       INISQUY  = INISQUX
02120       N        = INISQUX * INISQUY
02130       BETA     = 0.5D0
02140       NUP      = IDNINT(BETA*DBLE(N))
02150       NDWN     = N - NUP
02160 C
02170       DELR   =  0.10D0
02180       DELT   =  (2.0D0/180.D0 )*PI
02190       DELDEG =  DELT
02200 C
02210       HX      = 0.D0
02220       HY      = 1.D0
02230       HZ      = 0.D0
02240 C
02250       RCOFF  = 10.D0
02260       RCOFF2 = RCOFF**2
02270 C
02280       MCSMPLMX = 2000000
02290       NGRAPH   = MCSMPLMX/5
02300       NANIME   = MCSMPLMX/200
02310       NORDER   = MCSMPLMX/10
02320       NCLSCHK  = MCSMPLMX/10
02330 C
02340       DNSMPL   = 10
02350       DNORDR   = 10
02360       DNCLSCHK = 10
02370 C
02380       WIDTHCLS = 0.1D0 * D
02390       RCLSLMT  = DCHK + WIDTHCLS
02400 C
```

• KU and RA are set as $\zeta$=20 and $\lambda$=20, respectively.
• DELR, DELT and DELDEG are used in generating random displacements.
• The volumetric fraction is $\phi_V$=0.2.

• It is noted that the side length of cubes is set as unity.

• INISQUX should be set as a sufficiently large value such as 20 for a research objective.
• Half of the particles have a magnetic moment in the upward diagonal direction, and the reminders do it in the lower diagonal direction.

• DELR and DELT specify the maximum translational and rotational displacements of particles, respectively, in generation of a new configuration for Monte Carlo assessment procedure.

• The cutoff distance is set as $r_{coff}$=10 for restricting ambient particles for energy calculation.
• NORDER and NCLSCHK imply that the first 10% data are not used for averaging procedure.
• DNSMPL, DNORDR and DNCLSCHK imply that data sampling is carried out every 10 MC steps.

• WIDTHCLS=0.1 implies that the two particles of interest are regarded as forming a cluster if the minimum surface-to-surface distance is smaller than 0.1.

```
02410          IXRAN   = 0
02420          CALL RANCAL( NRANMX, IXRAN, RAN )
02430          NRAN    = 1
02440          NRANCHK = NRANMX - 5*N
02450 C
02460 C        -----------------------------------------------------------------
02470 C        -----------------     INITIAL CONFIGURATION     -----------------
02480 C        -----------------------------------------------------------------
02490 C
02500 C                                                --- SET INITIAL CONFIG. ---
02510 CCCC     OPEN(19,FILE='a1ku20ra20a041.dat',STATUS='OLD')
02520 CCCC     READ(19,472)  N , NUP , NDWN , XL , YL , ZL , D
02530 CCCC     READ(19,474)  (RX(I),I=1,N),(RY(I),I=1,N),(RZ(I),I=1,N),
02540 CCCC &                 (EX(I),I=1,N),(EY(I),I=1,N),(EZ(I),I=1,N),
02550 CCCC &                 (NX(I),I=1,N),(NY(I),I=1,N),(NZ(I),I=1,N)
02560 CCCC     READ(19,474)  (  ( (RMAT(II,JJ,I),II=1,3), JJ=1,3 ), I=1,N  )
02570 CCCC     READ(19,474)  (  ( (CUBEA(    II,JJ,I),II=1,3), JJ=1,6 ), I=1,N  )
02580 CCCC     READ(19,474)  (  ( (CUBEALPH(II,JJ,I),II=1,3), JJ=1,8 ), I=1,N  )
02590 CCCC     CLOSE(19,STATUS='KEEP')
02600 CCCC     GOTO 7
02610 C
02620          CALL INITIAL
02630 C
```

· Subroutine INITIAL sets the position and orientation of each particle on the plane surface.

```
02640        7 IF( RCOFF .GE. XL/2.D0 ) THEN
02650              RCOFF = XL/2.D0 - 0.00001D0
02660          END IF
02670          IF( RCOFF .GE. YL/2.D0 ) THEN
02680             RCOFF = YL/2.D0 - 0.00001D0
02690          END IF
```

· The cutoff distance $r_{coff}$ should be shorter than the half side length of the square simulation region.

```
02700 C                                                   --- PARAMETER (9) ---
02710          RCOFF2 = RCOFF**2
02720          XL2 = XL/2.0D0
02730          YL2 = YL/2.0D0
02740 C
02750 C        ------------------
02760 C
02770          CALL PTCLMDL
02780 C
```

· In subroutine PTCLMDL, the relationships regarding vertex-to-vertex, plane-to-plane, vertex-to-plane, etc. of the cube are specified using various variables. These variables are used in assessment of the particle overlap of cube-like particles.

```
02790 C        ------------------------------------------------------ PRINT OUT ---
02800          WRITE(NP,12) N, NUP, NDWN, VDENS, NDENS, RA, KU
02810          WRITE(NP,13) D, PV
02820          WRITE(NP,15) HX, HY, HZ, RCOFF, BETA, XL,YL, ZL
02830          WRITE(NP,17) DELR, DELT, DELTN
02840          WRITE(NP,19) WIDTHCLS , RCLSLMT , NCLSCHK , DNCLSCHK
02850          WRITE(NP,21) MCSMPLMX, NGRAPH, NANIME, NORDER
02860          WRITE(NP,28) INISQUX , INISQUY
02870 C
02880 C        ------------------------------------------------ INITIALIZATION ---
02890          NORDRCTR = 0
02900          MORDER   = 0.D0
02910          MORDER4  = 0.D0
02920          DO 46 I=1,10
02930            MORDERB( I) = 0.D0
02940            MORDER4B(I) = 0.D0
02950     46 CONTINUE
```

· These variables are used for evaluating the order parameters.

```
02960 C
02970          NSMPL    = 0
02980          NANMCTR  = 0
02990 C
03000          NDISTCNT = 1
03010          ANSNDSMP = 0
03020          NDISTMAX = 0
03030          DO 50 J=1,10
03040            NDISTSMP(J) = 0
03050            DO 48 I=1,N
03060              IF( J .EQ. 1 )  ANSNDIST(I) = 0.0
03070              CLSNDIST(I,J) = 0.0
03080     48     CONTINUE
03090     50 CONTINUE
```

· These variables are used for evaluating the cluster size distribution.

```
03100 C
03110 C        -----------------------------------------------------------------
03120 C        -----------------------------------------------------------------
03130 C        -----------------     START OF MAIN LOOP     --------------------
03140 C        -----------------------------------------------------------------
03150 C        -----------------------------------------------------------------
03160 C
03170          DO 1000 MCSMPL = 1, MCSMPLMX
03180 C
03190            DO 500 I = 1,N
03200 C
```

```
03210 C                                             ----------------------------------
03220 C               ------------------------------ (1) TRANSLATIONAL MOTION ---
03230 C                                             ----------------------------------
03240 C
03250 C                                                            --- OLD ENERGY ---
03260               RXI = RX(I)
03270               RYI = RY(I)
03280               RZI = RZ(I)
03290               EXI = EX(I)
03300               EYI = EY(I)
03310               EZI = EZ(I)
03320               NXI = NX(I)
03330               NYI = NY(I)
03340               NZI = NZ(I)
03350 C
03360               DO 110 IX=1,3
03370               DO 110 IY=1,3
03380                 RMATI(IX,IY) = RMAT(IX,IY,I)
03390   110         CONTINUE
03400 C
03410               DO 115 II= 1,8
03420               DO 115 IX= 1,3
03430                 IF( II.LE.6) THEN
03440                   CUBEAI(IX,II) = CUBEA(IX,II,I)
03450                 END IF
03460                 CUBEALI(IX,II)   = CUBEALPH(IX,II,I)
03470   115         CONTINUE
```

• The face vector $\boldsymbol{a}_i$ from the center of cube $i$ is denoted by (CUBEAI(1,I), CUBEAI(2,I), CUBEAI(3,I)).

• The vertex vector $\boldsymbol{a}_i$ from the center of cube $i$ is denoted by (CUBEALI(1,I), CUBEALI(2,I), CUBEALI(3,I)).

```
03480 C                                  ++ ISTREET=0 FOR FULL CAL     . ++
03490 C                                  ++ ISTREET=1 FOR MAG.ENE.ONLY ++
03500 C                                  ++ ISTREET=2 FOR CLSTR MOV     ++
03510               ISTREET = 1
03520               CALL ENECAL( I, RXI, RYI, RZI, EXI, EYI, EZI,
03530      &                 NXI, NYI, NZI, EOLD, OVRLAP, ISTREET, J, CLSTRUE )
03540 C
03550               EMAGOLD = ENEMAGI
03560 C                                             ---------- (1A) CANDIDATE
```

• Interaction energy of particle $i$ is calculated in this subroutine.

```
03570               NRAN  = NRAN + 1
03580               RXCAN = RX(I) + DELR*( 1.D0 - 2.D0*DBLE(RAN(NRAN)) )
03590               IF( RXCAN .GE. XL2 ) THEN
03600                 RXCAN = RXCAN - XL
03610               ELSE IF( RXCAN .LT. -XL2 ) THEN
03620                 RXCAN = RXCAN + XL
03630               END IF
03640 C
03650               NRAN  = NRAN + 1
03660               RYCAN = RY(I) + DELR*( 1.D0 - 2.D0*DBLE(RAN(NRAN)) )
03670               IF( RYCAN .GE. YL2 ) THEN
03680                 RYCAN = RYCAN - YL
03690               ELSE IF( RYCAN .LT. -YL2 ) THEN
03700                 RYCAN = RYCAN + YL
03710               END IF
03720 C
03730               RZCAN = RZ(I)
03740 C
03750               DO 120 IX=1,3
03760               DO 120 IY=1,3
03770                 RMATI(IX,IY) = RMAT(IX,IY,I)
03780   120         CONTINUE
03790 C
03800               DO 125 II= 1,8
03810               DO 125 IX= 1,3
03820                 IF( II.LE.6) THEN
03830                   CUBEAI(IX,II) = CUBEA(IX,II,I)
03840                 END IF
03850                 CUBEALI(IX,II)   = CUBEALPH(IX,II,I)
03860   125         CONTINUE
03870 C
03880 C                                                            --- NEW ENERGY ---
03890 C                                  ++ ISTREET=0 FOR FULL CAL      ++
03900 C                                  ++ ISTREET=1 FOR MAG.ENE.ONLY ++
03910 C                                  ++ ISTREET=2 FOR CLSTR DIST.   ++
03920               ISTREET = 0
03930               CALL ENECAL( I , RXCAN, RYCAN, RZCAN, EXI, EYI, EZI,
03940      &                 NXI, NYI, NZI, ECAN, OVRLAP, ISTREET, J, CLSTRUE )
03950 C
03960               EMAGCAN = ENEMAGI
03970 C
03980               IF( OVRLAP ) THEN
03990                 GOTO 140
04000               END IF
```

• A candidate position is set using a random number RAN(NRAN) ranging from 0 to 1.
• The periodic boundary condition has to be applied.

• Interaction energy of particle $i$ is calculated in this subroutine for the candidate position (RXCAN, RYCAN, RZCAN).

• The logical variable OVRLAP has ".TRUE." if particle $i$ overlaps with an ambient particle. In this case, the following assessment procedure based on Metropolis method is not necessary.

```
04010 C                                          --------  (1B) ENERGY ASSESMENT --
04020 C
04030          C3 = ECAN - EOLD
04040          IF( C3 .GE. 0.D0 )THEN
04050            NRAN = NRAN + 1
04060            IF( DBLE(RAN(NRAN)) .GE. DEXP(-C3) )THEN
04070              GOTO 140
04080            END IF
04090          END IF
04100 C                                                  +++++++++++++++++++++
04110 C                                                  CANDIDATES ARE ACCEPTED
04120 C                                                  +++++++++++++++++++++
04130          RX(I) = RXCAN
04140          RY(I) = RYCAN
04150          RZ(I) = D2
04160          EOLD  = ECAN
04170          EMAGOLD = EMAGCAN
04180          E(I)  = ECAN
04190 C
04200 C
04210 C                                          ---------------------------------
04220 C        ---------------------------------  (2) ROTATIONAL MOTION ---
04230 C                                          ---------------------------------
04240 C
04250 C                                          +++++++++++++++++++++++++++++++++
04260 C                  ++++++++++++++++++++++++++ (2)-1 (EX,EY,EZ) +++
04270 C                                          +++++++++++++++++++++++++++++++++
04280 C
04290 C      --- (EX,EY,EZ) IS NOT CHANGED IN THE PRESENT 2D SYSTEM ---
04300 C
04310 C                                          +++++++++++++++++++++++++++++++++
04320 C                  ++++++++++++++++++++++++ (2)-2 ABOUT Z-AXIS +++
04330 C                                          +++++++++++++++++++++++++++++++++
04340   140    RXI  = RX(I)
04350          RYI  = RY(I)
04360          RZI  = RZ(I)
04370          EXI  = EX(I)
04380          EYI  = EY(I)
04390          EZI  = EZ(I)
04400          NXI  = NX(I)
04410          NYI  = NY(I)
04420          NZI  = NZ(I)
04430 C
04440          DO 205 IX=1,3
04450          DO 205 IY=1,3
04460            RMATI(IX,IY) = RMAT(IX,IY,I)
04470   205    CONTINUE
04480 C
04490          NRAN  = NRAN + 1
04500          C00   = 2.D0 * DBLE(RAN(NRAN)) - 1.D0
04510          CDEG  = C00 * DELDEG
04520          CCS   = DCOS( CDEG )
04530          CSN   = DSIN( CDEG )
04540 C
04550          C11   =  CCS
04560          C21   =  CSN
04570          C31   =  0.D0
04580          C12   = -CSN
04590          C22   =  CCS
04600          C32   =  0.D0
04610          C13   =  0.D0
04620          C23   =  0.D0
04630          C33   =  1.D0
04640 C
04650          RMATIHAT(1,1) =  C11
04660          RMATIHAT(2,1) =  C21
04670          RMATIHAT(3,1) =  C31
04680          RMATIHAT(1,2) =  C12
04690          RMATIHAT(2,2) =  C22
04700          RMATIHAT(3,2) =  C32
04710          RMATIHAT(1,3) =  C13
04720          RMATIHAT(2,3) =  C23
04730          RMATIHAT(3,3) =  C33
04740 C
04750          DO 215 K0 = 1,3
04760 C
04770            IF(      K0 .EQ. 1 ) THEN
04780              C1 = 0.5D0
04790              C2 = 0.D0
04800              C3 = 0.D0
```

• Assessment procedure based on Metropolis method.

• The candidate position is now accepted and therefore the information of the position is updated.

• The present interaction energy is saved in the variable EOLD.

• The cube-like particles are assumed to perform the translational and rotational motion on the plane surface with their bottom face in contact with this plane surface. Hence, the orientation of the magnetic moment of each particle is attempted to be rotated about the $z$-axis direction (or about the direction normal to the plane surface).

• CDEG is a rotation angle, specified by a random number, about the $z$-axis direction.

• RMATIHAT(*,+) is a rotation matrix about the $z$-axis direction with an angle CDEG.

```
04810              ELSE IF( K0 .EQ. 2 ) THEN
04820                C1 = 0.D0
04830                C2 = 0.5D0
04840                C3 = 0.D0
04850              ELSE IF( K0 .EQ. 3 ) THEN
04860                C1 = 0.D0
04870                C2 = 0.D0
04880                C3 = 0.5D0
04890              END IF
04900              DO 210 IX = 1,3
04910                CUBEAIB(IX,K0) =  RMATIHAT(IX,1)*C1 + RMATIHAT(IX,2)*C2
04920         &                                         + RMATIHAT(IX,3)*C3
04930    210       CONTINUE
04940              CUBEAIB(1,K0+3) = -CUBEAIB(1,K0)
04950              CUBEAIB(2,K0+3) = -CUBEAIB(2,K0)
04960              CUBEAIB(3,K0+3) = -CUBEAIB(3,K0)
04970 C
04980    215     CONTINUE
04990 C
05000            DO 220 K0 =1,3
05010              DO 217 IX =1,3
05020                CUBEAIS(IX,K0) =     RMATI(IX,1)  * CUBEAIB(1,K0)
05030         &                         + RMATI(IX,2)  * CUBEAIB(2,K0)
05040         &                         + RMATI(IX,3)  * CUBEAIB(3,K0)
05050    217       CONTINUE
05060              CUBEAIS(1,K0+3) = -CUBEAIS(1,K0)
05070              CUBEAIS(2,K0+3) = -CUBEAIS(2,K0)
05080              CUBEAIS(3,K0+3) = -CUBEAIS(3,K0)
05090    220     CONTINUE
05100 C
05110            DO 223 K0 =1,6
05120            DO 223 IX =1,3
05130              CUBEAI(IX,K0) = CUBEAIS(IX,K0)
05140    223     CONTINUE
05150 C
05160            DO 230 K0=1,3
05170            DO 230 IX=1,3
05180              RMATI(IX,K0) = 2.D0 * CUBEAI(IX,K0)
05190    230     CONTINUE
05200 C
05210            DO 235 IX=1,3
05220              CUBEALI(IX,1)=  CUBEAI(IX,1) + CUBEAI(IX,2) + CUBEAI(IX,3)
05230              CUBEALI(IX,2)= -CUBEAI(IX,1) + CUBEAI(IX,2) + CUBEAI(IX,3)
05240              CUBEALI(IX,3)=  CUBEAI(IX,1) - CUBEAI(IX,2) + CUBEAI(IX,3)
05250              CUBEALI(IX,4)=  CUBEAI(IX,1) + CUBEAI(IX,2) - CUBEAI(IX,3)
05260    235     CONTINUE
05270 C
05280            DO 240 K0 = 1,4
05290            DO 240 IXYZ = 1,3
05300              CUBEALI(IXYZ,K0+4) = -CUBEALI(IXYZ,K0)
05310    240     CONTINUE
05320 C
05330            CNXI = CUBEALI(1,1)
05340            CNYI = CUBEALI(2,1)
05350            CNZI = CUBEALI(3,1)
05360            C00  = DSQRT( CNXI**2 + CNYI**2 + CNZI**2 )
05370            CNXI = CNXI / C00
05380            CNYI = CNYI / C00
05390            CNZI = CNZI / C00
05400 C
05410            CEXI = RMATI(1,3)
05420            CEYI = RMATI(2,3)
05430            CEZI = RMATI(3,3)
05440            C00  = DSQRT( CEXI**2 + CEYI**2 + CEZI**2 )
05450            CEXI = CEXI / C00
05460            CEYI = CEYI / C00
05470            CEZI = CEZI / C00
05480 C
05490 C                                                --- NEW ENERGY ---
05500 C                                     ++ ISTREET=0 FOR FULL CAL     . ++
05510 C                                     ++ ISTREET=1 FOR MAG.ENE.ONLY ++
05520 C                                     ++ ISTREET=2 FOR CLSTR DIST.  ++
05530 C
05540            ISTREET = 0
05550            CALL ENECAL( I , RXI, RYI, RZI, CEXI, CEYI, CEZI,
05560         &          CNXI, CNYI, CNZI, ECAN, OVRLAP, ISTREET, J, CLSTRUE )
05570 C
05580            EMAGCAN = ENEMAGI
05590 C
05600            IF( OVRLAP ) THEN
```

• The face vector $a_k^b$ ($k$=1,2,…,6), in the body-fixed coordinate system, for the candidate orientation situation is evaluated from rotating the original vector using the above rotation matrix.

• The face vector $a_k^s$ ($k$=1,2,…,6), in the absolute coordinate system, is obtained from matrix manipulation of the face vector $a_k^b$ and the orientation matrix RMATI(*,+).

• The vertex vector $a_k^s$ ($k$=1,2,…, 8), in the absolute coordinate system, is obtained from the face vector $a_k^s$ ($k$=1,2,…,6) in the absolute coordinate system.

• Normalization procedure for ensuring the unit vector $n_i$.

• Interaction energy of particle $i$ is calculated in this subroutine for the candidate orientation of the magnetic moment (CNXI, CNYI, CNZI).

```
05610              GOTO 500
05620            END IF
05630 C                                           --------  (3) ENERGY ASSESMENT --
05640 C
05650            C3 = ECAN - EOLD
05660            IF( C3 .GE. 0.D0 )THEN
05670              NRAN = NRAN + 1
05680              IF( DBLE(RAN(NRAN)) .GE. DEXP(-C3) )THEN
05690                GOTO 500
05700              END IF
05710            END IF
05720 C                                                ++++++++++++++++++++++
05730 C                                                CANDIDATES ARE ACCEPTED
05740 C                                                ++++++++++++++++++++++
05750            EX(I) = CEXI
05760            EY(I) = CEYI
05770            EZ(I) = CEZI
05780            NX(I) = CNXI
05790            NY(I) = CNYI
05800            NZ(I) = CNZI
05810 C
05820            DO 250 IX=1,3
05830            DO 250 IY=1,3
05840              RMAT(IX,IY,I) = RMATI(IX,IY)
05850    250     CONTINUE
05860 C
05870            DO 255 K0= 1,8
05880            DO 255 IXYZ= 1,3
05890              IF( K0.LE.6) THEN
05900                CUBEA(IXYZ,K0,I) = CUBEAI(IXYZ,K0)
05910              END IF
05920              CUBEALPH(IXYZ,K0,I) = CUBEALI(IXYZ,K0)
05930    255     CONTINUE
05940 C
05950            EOLD    = ECAN
05960            EMAGOLD = EMAGCAN
05970            E(I)    = ECAN
05980 C
05990    500   CONTINUE
06000 C
06010 C        ////////////////////////////////////////////////////////////////
06020 C        ////////////////////////////////////////////////////////////////
06030 C
06040 C        ----------------------------------------------------------------
06050 C
06060 C                                                --- MOMENT OF SYSTEM ---
06070          IF( MOD(MCSMPL,DNSMPL) .EQ. 0 ) THEN
06080            NSMPL   = NSMPL + 1
06090            C11 = 0.D0
06100            C12 = 0.D0
06110            C13 = 0.D0
06120            C14 = 0.D0
06130            C21 = 0.D0
06140            C22 = 0.D0
06150            C23 = 0.D0
06160            C31 = 0.D0
06170            C32 = 0.D0
06180            C33 = 0.D0
06190            C41 = 0.D0
06200            C42 = 0.D0
06210            C43 = 0.D0
06220            DO 525 JJ=1,N
06230              C11 = C11 + NX(JJ)
06240              C12 = C12 + NY(JJ)
06250              C13 = C13 + NZ(JJ)
06260              C14 = C14 + E(JJ)
06270              C31 = C31 + EX(JJ)
06280              C32 = C32 + EY(JJ)
06290              C33 = C33 + EZ(JJ)
06300              C21 = C21 + (3.D0* NX(JJ)**2 - 1.D0) / 2.D0
06310              C22 = C22 + (3.D0* NY(JJ)**2 - 1.D0) / 2.D0
06320              C23 = C23 + (3.D0* NZ(JJ)**2 - 1.D0) / 2.D0
06330              C41 = C41 + (3.D0* EX(JJ)**2 - 1.D0) / 2.D0
06340              C42 = C42 + (3.D0* EY(JJ)**2 - 1.D0) / 2.D0
06350              C43 = C43 + (3.D0* EZ(JJ)**2 - 1.D0) / 2.D0
06360    525     CONTINUE
06370            MOMX(NSMPL) = REAL(C11)/REAL(N)
06380            MOMY(NSMPL) = REAL(C12)/REAL(N)
06390            MOMZ(NSMPL) = REAL(C13)/REAL(N)
06400            MEANENE(NSMPL)= REAL( C14 - KU*(C11*HX + C12*HY
```

• Assessment procedure based on Metropolis method.

• The candidate orientation is now accepted and therefore the information of the magnetic moment direction is updated.

• The information of the face vector $\boldsymbol{a}_k^s$ ($k$=1,2,…,6) and the vertex vector $\boldsymbol{\alpha}_k^s$ ($k$=1,2,…, 8) are updated.

• Sampling is conducted every DNSMPL steps for evaluating characteristics of the system.

```
06410      &                                            + C13*HZ) )/REAL(2*N)
06420            EDIRX(NSMPL) = REAL(C31)/REAL(N)
06430            EDIRY(NSMPL) = REAL(C32)/REAL(N)
06440            EDIRZ(NSMPL) = REAL(C33)/REAL(N)
06450            MORDX(NSMPL) = REAL(C21)/REAL(N)
06460            MORDY(NSMPL) = REAL(C22)/REAL(N)
06470            MORDZ(NSMPL) = REAL(C23)/REAL(N)
06480            EORDX(NSMPL) = REAL(C41)/REAL(N)
06490            EORDY(NSMPL) = REAL(C42)/REAL(N)
06500            EORDZ(NSMPL) = REAL(C43)/REAL(N)
06510          END IF
```

• MOMY(*) and MORDY(*) are used for describing the order parameters $\langle \hat{n}_{iy} \rangle$ and $S_{ny}$, respectively. In subroutine ORDERCA, the other order parameters are evaluated as a function of radial distance.

```
06520 C                                             --- ORDER PARAMETERS ---
06530          IF( (MCSMPL.GT.NORDER) .AND. (MOD(MCSMPL,DNORDR).EQ. 0) )THEN
06540            NORDRCTR = NORDRCTR + 1
06550            CALL ORDERCA( N , NUP , NDWN )
06560          END IF
06570 C                          --------------------------------------------
06580 C                           CHECKING PROCEDURE OF THE CLUSTER FORMATION
06590 C                          --------------------------------------------
06600          IF(    (MCSMPL.GT.NCLSCHK) .AND.
06610      &          (MOD(MCSMPL,DNCLSCHK).EQ.0)    ) THEN
06620 C                                              +++ CHECK CLSTR FORM. +++
06630            CALL CLSFORM( N )
06640 C
06650            CALL CLSDIST( N )
06660 C
```

• In CLSFORM, the formation of clusters is evaluated.
• In CLSDIST, the cluster size distribution is calculated.

```
06670            IF( MOD(MCSMPL,NCLSCHK) .EQ. 0 )  THEN
06680              DO 555 K=1,NDISTMAX
06690                IC1 = NDISTSMP(NDISTCNT)
06700                IF( IC1 .LE. 0 ) IC1 = 1000000000
06710                CLSNDIST(K,NDISTCNT) = CLSNDIST(K,NDISTCNT) / REAL(IC1)
06720   555        CONTINUE
06730              NDISTCNT = NDISTCNT + 1
06740            END IF
06750 C
06760          END IF
06770 C
06780 C
```

• Sub-average of the cluster size distribution is calculated every NCLSCHK MC steps. Ten sub-averages are obtained in one simulation run.

```
06790 C        ---------------------------- DATA OUTPUT (1) FOR GRAPHICS ---
06800 C
06810          IF( MOD(MCSMPL,NGRAPH) .EQ. 0 ) THEN
06820            NOPT = NOPT + 1
06830            WRITE(NOPT,472)  N , NUP , NDWN , XL , YL , ZL , D
06840            WRITE(NOPT,474)  (RX(I),I=1,N), (RY(I),I=1,N), (RZ(I),I=1,N),
06850      &                      (EX(I),I=1,N), (EY(I),I=1,N), (EZ(I),I=1,N),
06860      &                      (NX(I),I=1,N), (NY(I),I=1,N), (NZ(I),I=1,N)
06870            WRITE(NOPT,474) (  ( (RMAT(II,JJ,I),II=1,3), JJ=1,3 ),
06880      &                                                       I=1,N  )
06890            WRITE(NOPT,474) (  ( (CUBEA(    II,JJ,I),II=1,3), JJ=1,6 ),
06900      &                                                       I=1,N  )
06910            WRITE(NOPT,474) (  ( (CUBEALPH(II,JJ,I),II=1,3), JJ=1,8 ),
06920      &                                                       I=1,N  )
06930                                           CLOSE(NOPT,STATUS='KEEP')
06940          END IF
06950 C                                 --- DATA OUTPUT
06960          IF( MOD(MCSMPL,NANIME) .EQ. 0 ) THEN
06970            NANMCTR = NANMCTR + 1
06980            NOPT1   = 14
06990            CALL ANIMEOPT( NOPT1, NANMCTR, MCSMPLMX, NANIME )
07000          END IF
```

• Data output for making snapshots and animations by means of a commercial software MicroAVS.

```
07010 C
07020 C        ---------------------- CHECK OF THE SUM OF RANDOM NUMBERS ---
07030 C
07040          IF( NRAN .GE. NRANCHK )THEN
07050            CALL RANCAL( NRANMX, IXRAN, RAN )
07060            NRAN = 1
07070          END IF
07080 C
07090 C        -----------------------------------
07100          IF( MOD(MCSMPL,1) .EQ. 0 ) THEN
07110 C
07120            DO 650 I=1,N
```

• The following procedure is necessary for keeping the exact shape of the cube, since the cubic shape may slightly be distorted with advancing MC steps.

```
07130 C
07140 C               +++ TAKE alph_1 AS A CRITERION FOR NORMALIZATION +++
07150 C
07160              DO 605 K0 = 2,4
07170                BALPH(1,K0) = CUBEALPH(1,K0,I) - CUBEALPH(1,1,I)
07180                BALPH(2,K0) = CUBEALPH(2,K0,I) - CUBEALPH(2,1,I)
07190                BALPH(3,K0) = CUBEALPH(3,K0,I) - CUBEALPH(3,1,I)
07200                C0 = DSQRT( BALPH(1,K0)**2 + BALPH(2,K0)**2
```

• BALPH(*,2),BALPH(*,3) and BALPH(*,4) denote the three side lines from the vertex point 1 of the cube.

```
07210      &                               + BALPH(3,K0)**2 )
07220              BALPH(1,K0) = BALPH(1,K0) / C0
07230              BALPH(2,K0) = BALPH(2,K0) / C0
07240              BALPH(3,K0) = BALPH(3,K0) / C0
07250   605      CONTINUE
07260 C
07270 C             +++ TAKE balph_2 AS A CRITERION FOR NORMALIZATION +++
07280 C                                                   --- balph_2 ---
07290            C12 =  BALPH(1,2)
07300            C22 =  BALPH(2,2)
07310            C32 =  BALPH(3,2)
07320 C                                                   --- balph_3 ---
07330            C13 =  BALPH(1,3)
07340            C23 =  BALPH(2,3)
07350            C33 =  BALPH(3,3)
07360 C
07370            C1  =  C13*C12 + C23*C22 + C33*C32
07380            C13 =  C13 - C1 * C12
07390            C23 =  C23 - C1 * C22
07400            C33 =  C33 - C1 * C32
07410            C0  =  DSQRT(  C13**2 + C23**2 + C33**2  )
07420            K0  =  3
07430            BALPH(1,K0) = C13 / C0
07440            BALPH(2,K0) = C23 / C0
07450            BALPH(3,K0) = C33 / C0
07460 C                                                   --- balph_4 ---
07470 CCCCC      C14 =  BALPH(1,4)
07480 CCCCC      C24 =  BALPH(2,4)
07490 CCCCC      C34 =  BALPH(3,4)
07500            C14 = -(  C22*C33 - C32*C23  )
07510            C24 = -(  C32*C13 - C12*C33  )
07520            C34 = -(  C12*C23 - C22*C13  )
07530            C0  =  DSQRT(  C14**2 + C24**2 + C34**2  )
07540            K0  =  4
07550            BALPH(1,K0) = C14 / C0
07560            BALPH(2,K0) = C24 / C0
07570            BALPH(3,K0) = C34 / C0
07580 C
07590 C          ------------------------------------------------- alph_1 ---
07600            C0 = DSQRT( CUBEALPH(1,1,I)**2 + CUBEALPH(2,1,I)**2
07610      &                                 + CUBEALPH(3,1,I)**2 )
07620            C1 = ( DSQRT(3.D0)/2.D0 ) / C0
07630            CUBEALPH(1,1,I) = CUBEALPH(1,1,I)  * C1
07640            CUBEALPH(2,1,I) = CUBEALPH(2,1,I)  * C1
07650            CUBEALPH(3,1,I) = CUBEALPH(3,1,I)  * C1
07660 C                                  --- alph_2,alph_3,alph_4 ---
07670            DO 610 K0 = 2,4
07680              CUBEALPH(1,K0,I) =  BALPH(1,K0) + CUBEALPH(1,1,I)
07690              CUBEALPH(2,K0,I) =  BALPH(2,K0) + CUBEALPH(2,1,I)
07700              CUBEALPH(3,K0,I) =  BALPH(3,K0) + CUBEALPH(3,1,I)
07710              C0 = DSQRT( CUBEALPH(1,K0,I)**2 + CUBEALPH(2,K0,I)**2
07720      &                                      + CUBEALPH(3,K0,I)**2 )
07730              C1 = ( DSQRT(3.D0)/2.D0 ) / C0
07740              CUBEALPH(1,K0,I) = CUBEALPH(1,K0,I)  * C1
07750              CUBEALPH(2,K0,I) = CUBEALPH(2,K0,I)  * C1
07760              CUBEALPH(3,K0,I) = CUBEALPH(3,K0,I)  * C1
07770   610      CONTINUE
07780 C                              --- alph_5,alph_6,alph_7,alph_8 ---
07790            DO 615 K0 = 1,4
07800            DO 615 IXYZ = 1,3
07810              CUBEALPH(IXYZ,K0+4,I) = - CUBEALPH(IXYZ,K0,I)
07820   615      CONTINUE
07830 C          ------------------
07840            DO 620 IXYZ = 1,3
07850              CUBEA(IXYZ,1,I) = (   CUBEALPH(IXYZ,3,I)
07860      &                           + CUBEALPH(IXYZ,4,I)   ) / 2.D0
07870              CUBEA(IXYZ,2,I) = (   CUBEALPH(IXYZ,2,I)
07880      &                           + CUBEALPH(IXYZ,4,I)   ) / 2.D0
07890              CUBEA(IXYZ,3,I) = (   CUBEALPH(IXYZ,2,I)
07900      &                           + CUBEALPH(IXYZ,3,I)   ) / 2.D0
07910   620      CONTINUE
07920 C                                            --- a_4,a_5,a_6 ---
07930            DO 625 K0 = 1,3
07940            DO 625 IXYZ = 1,3
07950              CUBEA(IXYZ,K0+3,I) = - CUBEA(IXYZ,K0,I)
07960   625      CONTINUE
07970 C
07980            DO 627 K0 = 1,6
07990              C0 = DSQRT( CUBEA(1,K0,I)**2 + CUBEA(2,K0,I)**2
08000      &                                 + CUBEA(3,K0,I)**2 )
```

· These vectors have to be unit vectors.

· BALPH(*,3) has to be normal to BALPH(*,2).

· BALPH(*,4) has to be normal to BALPH(*,2) and BALPH(*,3).

· The vertex vector $\boldsymbol{\alpha}_1^s$ is adopted as a criterion for modification.

· The vertex vector $\boldsymbol{\alpha}_k^s$ ($k$=2,3,4) is made using $\boldsymbol{\alpha}_1^s$ and BALPH(*,+).

· The face vector $\boldsymbol{a}_k^s$ ($k$=1,2,3) is made using $\boldsymbol{\alpha}_k^s$ ($k$=2,3,4) .

```
08010                 C1 = 0.5D0 / C0
08020                 CUBEA(1,K0,I) = CUBEA(1,K0,I) * C1
08030                 CUBEA(2,K0,I) = CUBEA(2,K0,I) * C1
08040                 CUBEA(3,K0,I) = CUBEA(3,K0,I) * C1
08050   627         CONTINUE
08060 C             ---------------------------------------------------
08070               DO 630 K0 = 1,3
08080               DO 630 IXYZ = 1,3
08090                 RMAT(IXYZ,K0,I) = 2.D0 * CUBEA(IXYZ,K0,I)
08100   630         CONTINUE
08110 C             --------------------------------(RX,EY,EZ),(NX,NY,NZ) ---
08120               C1 = CUBEA(1,3,I)
08130               C2 = CUBEA(2,3,I)
08140               C3 = CUBEA(3,3,I)
08150               C0 = DSQRT( C1**2 + C2**2 + C3**2 )
08160               EX(I) = C1 / C0
08170               EY(I) = C2 / C0
08180               EZ(I) = C3 / C0
08190 C
08200               C1 = CUBEALPH(1,1,I)
08210               C2 = CUBEALPH(2,1,I)
08220               C3 = CUBEALPH(3,1,I)
08230               C0 = DSQRT( C1**2 + C2**2 + C3**2 )
08240               NX(I) =  C1 / C0
08250               NY(I) =  C2 / C0
08260               NZ(I) =  C3 / C0
08270 C
08280   650       CONTINUE
08290 C
08300           END IF
08310 C
08320 C
08330  1000 CONTINUE
08340 C
08350 C     -------------------------------------------------------------
08360 C     ---------------  END OF MONTE CARLO SIMULATIONS  -------------
08370 C     -------------------------------------------------------------
08380 C
08390 C                                                --- PRINT OUT (2) ---
08400       WRITE(NP,1011)
08410       NSMPL1 = 1
08420       NSMPL2 = NSMPL
08430       CALL PRNTDAT( NSMPL1 , NSMPL2 , NP )
08440       WRITE(NP,1013) NSMPL1 , NSMPL2
08450 C
08460       MORDER  = MORDER   / DBLE(NORDRCTR)
08470       MORDER4 = MORDER4  / DBLE(NORDRCTR)
08480       DO 1021 I=1,10
08490         MORDERB( I) = MORDERB( I) / DBLE(NORDRCTR)
08500         MORDER4B(I) = MORDER4B(I) / DBLE(NORDRCTR)
08510  1021 CONTINUE
08520 C                                   --- PRINT OUT (3) ORDER PARA ---
08530       WRITE(NP,1031) NORDRCTR , MORDER , MORDER4
08540       DO 1035 I=1,10
08550         WRITE(NP,1033) I, MORDERB(I), MORDER4B(I)
08560  1035 CONTINUE
08570 C
08580 C                                    --- DATA OUT (3) ORDER PARA ---
08590       WRITE(12,1043) NORDRCTR , MORDER , MORDER4
08600       DO 1047 I=1,10
08610         WRITE(12,1045) I, MORDERB(I), MORDER4B(I)
08620  1047 CONTINUE
08630 C
08640 C                                             --- DATA OUTPUT (4) ---
08650       WRITE(11,1111) N, NUP, NDWN, VDENS, NDENS, RA, KU
08660       WRITE(11,1114) D, PV
08670       WRITE(11,1116) HX, HY, HZ, RCOFF, BETA, XL,YL, ZL
08680       WRITE(11,1117) DELR, DELT, DELTN
08690       WRITE(11,1119) WIDTHCLS , RCLSLMT , NCLSCHK , DNCLSCHK
08700       WRITE(11,1121) MCSMPLMX, NGRAPH, NANIME, NORDER
08710       WRITE(11,1128) INISQUX , INISQUY
08720       WRITE(11,1130) NSMPL
08730       WRITE(11,1132) ( MOMX(I),I=NSMPL1, NSMPL2)
08740      &               ,( MOMY(I),I=NSMPL1, NSMPL2)
08750      &               ,( MOMZ(I),I=NSMPL1, NSMPL2)
08760       WRITE(11,1134) ( MEANENE(I),I=NSMPL1, NSMPL2)
08770 C
08780 C                               ------------ CAL. NO. CLS DIST. ---
08790       NDISTCNT = NDISTCNT - 1
08800       J1       = 2
```

• The face vector $\boldsymbol{a}_k^s$ ($k$=1,2,..,6) has to be a vector with length 0.5.

• The rotation matrix is made from the face vectors $\boldsymbol{a}_k^s$ ($k$=1,2,3).

• The unit vector of the orientation of the magnetic moment is obtained using the vertex vector $\boldsymbol{a}_1^s$.

• MORDERB(*) and MORDER4B(*) denote the order parameter $S_1$ and $S_2$ as a function of the radial distance.

```
08810          JMAX      = NDISTCNT
08820          DO 1310 K=1,NDISTMAX
08830 C                                                  +++ JMAX MUST BE TEN +++
08840            DO 1305 J=J1,JMAX
08850              IF( K .EQ. 1 ) THEN
08860                ANSNDSMP = ANSNDSMP + NDISTSMP(J)
08870              END IF
08880              ANSNDIST(K)= ANSNDIST(K) + CLSNDIST(K,J)*REAL( NDISTSMP(J) )
08890  1305      CONTINUE
08900  1310  CONTINUE
08910 C
08920          DO 1315 K=1,NDISTMAX
08930            IF( ANSNDSMP .LE. 0 )  GOTO 1315
08940            IC1 = ANSNDSMP
08950            ANSNDIST(K) = ANSNDIST(K) / REAL( IC1 )
08960  1315  CONTINUE
08970 C                                            --- PRINT OUT CLS.DIST. (4) ---
08980          WRITE(NP,1331) (ANSNDIST(I), I=1,NDISTMAX)
08990 C
09000 C                                            --- DATA OUTPUT CLS.DIST. (5) ---
09010          WRITE(13,1333) NDISTMAX
09020          DO 1337 I= 1, NDISTMAX
09030            WRITE(13,1335) I, ANSNDIST(I)
09040  1337  CONTINUE
09050 C
09060          WRITE(13,1341) NDISTCNT
09070          DO 1350 J=1, NDISTCNT
09080            WRITE(13,1343) J , NDISTSMP(J)
09090            WRITE(13,1345) ( CLSNDIST(I,J), I=1,NDISTMAX )
09100  1350  CONTINUE
09110                                              CLOSE(9, STATUS='KEEP')
09120                                              CLOSE(11,STATUS='KEEP')
09130                                              CLOSE(12,STATUS='KEEP')
09140                                              CLOSE(13,STATUS='KEEP')
09150                                              CLOSE(14,STATUS='KEEP')
09160 C      -------------------------- FORMAT ------------------------------
09170     12 FORMAT(/1H ,'-------------------------------------------------'
09180        &       /1H ,'-                   MONTE CARLO METHOD          -'
09190        &       /1H ,'-------------------------------------------------'
09200        &      //1H ,'N=',I4, 2X,'NUP=',I4, 2X, 'NDWN=',I4, 2X,
09210        &            'VDENS=',F8.5, 2X ,'NDENS=',F9.6
09220        &       /1H ,'RAM=',F6.2, 1X, 'KU=',F9.2)
09230     13 FORMAT(/1H ,'D=',F5.2, 2X, 'PV=',F7.3)
09240     15 FORMAT(/1H ,'HX=',F5.2, 2X, 'HY=',F5.2, 2X, 'HZ=',F5.2, 2X,
09250        &            'RCOFF=',F6.2, 2X, 'BETA=',F5.2
09260        &       /1H ,'XL=',F8.3, 2X, 'YL=',F8.3, 2X, 'ZL=',F8.3)
09270     17 FORMAT(/1H ,'DELR=',F7.4, 2X, 'DELT=',F7.4, 2X, 'DELTN=',F7.4)
09280     19 FORMAT(/1H ,'WIDTHCLS=',F8.3, 2X, 'RCLSLMT=',F8.3, 2X,
09290        &            'NCLSCHK=',I8, 2X, 'DNCLSCHK=',I8)
09300     21 FORMAT(/1H ,'MCSMPLMX=',I8, 2X, 'NGRAPH=',I8,2X,'NANIME=',I8,2X,
09310        &            'NORDER=',I8 )
09320     28 FORMAT(/1H ,'INISQUX=',I3,2X, 'INISQUY=',I3/)
09330 C
09340    472 FORMAT( 3I5 , 3F9.4 , F8.4 )
09350    474 FORMAT( (5F16.10) )
09360    476 FORMAT( (6F13.10) )
09370   1011 FORMAT(/1H ,'+++++++++++++++++++++++++++++'
09380        &       /1H ,'    MONTE CARLO SIMULATIONS    '
09390        &       /1H ,'+++++++++++++++++++++++++++++'/)
09400   1013 FORMAT(///1H ,18X, 'START OF MC SAMPLING STEP=',I7
09410        &        /1H ,18X, 'END   OF MC SAMPLING STEP=',I7/)
09420   1031 FORMAT(1H ,'********************** NORDRCTR=', I9, 2X,
09430        &          'MORDER=', F7.4, 2X, 'MORDER4=', F7.4/)
09440   1033 FORMAT(1H ,'I=', I3, 2X, 'MORDERB(*)=', F7.4, 2X,
09450        &                        'MORDER4B(*)=', F7.4)
09460   1043 FORMAT( I9, 2E13.5 )
09470   1045 FORMAT( I4, 2E13.5 )
09480 C
09490   1111 FORMAT( 3I5 , 2F7.4 , 2F9.3 )
09500   1114 FORMAT( 2F10.4 )
09510   1116 FORMAT( 3F6.2 , 2F7.2, 3F7.2 )
09520   1117 FORMAT( 3F8.5 )
09530   1119 FORMAT( 2F10.4, 2I8 )
09540   1121 FORMAT( 4I9 )
09550   1128 FORMAT( 2I8 )
09560   1130 FORMAT( I9 )
09570   1132 FORMAT( (10F8.4) )
09580   1134 FORMAT( (6E13.5) )
09590 C
09600   1331 FORMAT(/'*************** NUMBER OF CLUSTERS **************'
```

• The average of the cluster size distribution is calculated using ten sub-average data; NDISTMAX=10.

```
09610       &         //'N_CLS(1), N_CLS(2), N_CLS(3), .....................'
09620       &          /( (10F8.3) ) )
09630  1333 FORMAT( I8 )
09640  1335 FORMAT( I5 , F10.6 )
09650  1341 FORMAT( I8 )
09660  1343 FORMAT( 2I8 )
09670  1345 FORMAT( (10F8.3) )
09680 C
09690                                                                          STOP
09700                                                                          END
09710 C*********************************************************************
09720 C***************************   SUBROUTINE   *************************
09730 C*********************************************************************
09740 C
```

· An initial configuration regarding the particle positions and the magnetic moment directions is assigned.

```
09750 C**** SUB INITIAL ****
09760       SUBROUTINE INITIAL
09770 C
09780       IMPLICIT REAL*8 (A-H,O-Z), INTEGER (I-N)
09790 C
09800       COMMON /BLOCK101/ RX    , RY    , RZ
09810       COMMON /BLOCK102/ EX    , EY    , EZ
09820       COMMON /BLOCK103/ NX    , NY    , NZ
09830       COMMON /BLOCK111/ XL    , YL    , ZL
09840       COMMON /BLOCK115/ N     , NDENS, VDENS , NUP , NDWN
09850       COMMON /BLOCK116/ D     , VP    , D2    , DCHK
09860       COMMON /BLOCK118/ INISQUX , INISQUY , INISQUZ , BETA
09870 C
09880       COMMON /BLOCK200/ RMAT
09890       COMMON /BLOCK201/ CUBEA     , ANEIGHBA , ANEIALPH , ANEIA
09900       COMMON /BLOCK203/ CUBEALPH, ALPHNEIG , ALNEIA
09910 C     -----------------------------------------------------------------
09920       INTEGER     NN , NNS , TT
09930       PARAMETER( NN=1000 , NNS=5000000 , TT=1000 )
09940       PARAMETER( NRANMX=1000000 , PI=3.141592653589793D0 )
09950 C     ---------------------------
09960       REAL*8    KU       , NDENS
09970       REAL*8    BETA
09980       REAL*8    RX(NN) , RY(NN) , RZ(NN)
09990       REAL*8    EX(NN) , EY(NN) , EZ(NN) , NX(NN) , NY(NN) , NZ(NN)
10000 C     ---------------------------
10010       REAL*8    RMAT(3,3,NN)
10020       REAL*8    CUBEA(3,6,NN) , CUBEALPH(3,8,NN)
10030       INTEGER   ANEIGHBA(4,6), ANEIALPH(4,6) , ANEIA(2,6)
10040       INTEGER   ALPHNEIG(3,8), ALNEIA(3,8)
10050 C     ---------------------------
10060       INTEGER   PTCL , ICNTR , INIPX , INIPY , N0
10070       REAL*8    NXI , NYI , NZI
10080       REAL*8    XLUNT , YLUNT ,  RAN1 , RAN2
10090       REAL*8    CUBEAX(3) , CUBEAY(3) , CUBEAZ(3)
10100       REAL*8    ROT11, ROT21, ROT31, ROT12, ROT22, ROT32
10110       REAL*8    ROT13, ROT23, ROT33
10120       REAL*8    C0 , C1 , C2 , C3 , CS1 , SN1
10130 C     -----------------------------------------------------------------
10140 C
10150 CCC   N    = INIPX*INIPY
10160 CCC   NUP  = IDNINT(BETA*DBLE(N))
10170 CCC   NDWN = N - NUP
10180 C
10190       INIPX = INISQUX
10200       INIPY = INISQUY
10210 C
10220       XL = (  DBLE(INIPX*INIPY) / ( VDENS )  )**(1./2.)
10230       YL = XL
10240 C
10250       XLUNT = XL/DBLE(INIPX)
10260       YLUNT = XLUNT
10270 C                                         -------------------- POSITION -----
```

· An initial position of each particle is assigned using quasi-random numbers ranging from 0 to 1.

```
10280       RAN1 = DSQRT( 2.D0 )
10290       RAN2 = DSQRT( 7.D0 )
10300       C0   = 1.D-3
10310       PTCL = 0
10320       DO 10 I=0, INIPX-1
10330       DO 10 J=0, INIPY-1
10340         PTCL = PTCL + 1
10350         C1 = RAN1*DBLE(PTCL)
10360         C1 = C1 - DINT(C1)
10370         C1 = C1 - 0.5D0
10380         C2 = RAN2*DBLE(PTCL)
10390         C2 = C2 - DINT(C2)
10400         C2 = C2 - 0.5D0
```

```
10410         RX(PTCL) = DBLE(I)*XLUNT + C1*C0 + 0.5D0 - XL/2.D0
10420         RY(PTCL) = DBLE(J)*YLUNT + C2*C0 + 0.5D0 - YL/2.D0
10430         RZ(PTCL) = D2
10440    10 CONTINUE
10450 C                                    -------------------- ORIENTATION ----
10460       RAN1 = DSQRT( 2.D0 )
10470       RAN2 = DSQRT( 3.D0 )
10480       PTCL  =0
10490       DO 500 I=1,N
10500 C
10510         PTCL = PTCL + 1
10520         C1 = RAN2*DBLE( PTCL )
10530         C1 = C1 - DINT(C1)
10540         C1 = (C1 - 0.5D0)*2.D0
10550         C2 = ( PI*(90.D0/180.D0) ) * C1
10560 C
10570         CS1   = DCOS(C2)
10580         SN1   = DSIN(C2)
10590         ROT11 =  CS1
10600         ROT21 =  SN1
10610         ROT31 =  0.D0
10620         ROT12 = -SN1
10630         ROT22 =  CS1
10640         ROT32 =  0.D0
10650         ROT13 =  0.D0
10660         ROT23 =  0.D0
10670         ROT33 =  1.D0
10680 C
10690         IF( I .LE. NUP ) THEN
10700 C                 --- (A) FOR PTCLS WITH UPWARD OBLIQUE DIRECTION ---
10710 C
10720           NXI = ROT11 * 1.D0 + ROT12 * 1.D0 + ROT13 * 1.D0
10730           NYI = ROT21 * 1.D0 + ROT22 * 1.D0 + ROT23 * 1.D0
10740           NZI = ROT31 * 1.D0 + ROT32 * 1.D0 + ROT33 * 1.D0
10750 C
10760           C0 = DSQRT( NXI**2 + NYI**2 + NZI**2 )
10770           NX(I) = NXI / C0
10780           NY(I) = NYI / C0
10790           NZ(I) = NZI / C0
10800 C
10810           CUBEAX(1) = ROT11 * 0.5D0 + ROT12 * 0.0D0 + ROT13 * 0.0D0
10820           CUBEAY(1) = ROT21 * 0.5D0 + ROT22 * 0.0D0 + ROT23 * 0.0D0
10830           CUBEAZ(1) = ROT31 * 0.5D0 + ROT32 * 0.0D0 + ROT33 * 0.0D0
10840 C
10850           CUBEAX(2) = ROT11 * 0.0D0 + ROT12 * 0.5D0 + ROT13 * 0.0D0
10860           CUBEAY(2) = ROT21 * 0.0D0 + ROT22 * 0.5D0 + ROT23 * 0.0D0
10870           CUBEAZ(2) = ROT31 * 0.0D0 + ROT32 * 0.5D0 + ROT33 * 0.0D0
10880 C
10890           CUBEAX(3) = ROT11 * 0.0D0 + ROT12 * 0.0D0 + ROT13 * 0.5D0
10900           CUBEAY(3) = ROT21 * 0.0D0 + ROT22 * 0.0D0 + ROT23 * 0.5D0
10910           CUBEAZ(3) = ROT31 * 0.0D0 + ROT32 * 0.0D0 + ROT33 * 0.5D0
10920 C
10930           DO 110 K0 = 1, 3
10940             CUBEA(1,K0,I) =  CUBEAX(K0)
10950             CUBEA(2,K0,I) =  CUBEAY(K0)
10960             CUBEA(3,K0,I) =  CUBEAZ(K0)
10970   110     CONTINUE
10980 C
10990           DO 120 K0 = 1, 3
11000             RMAT(1,K0,I) = 2.D0 * CUBEAX(K0)
11010             RMAT(2,K0,I) = 2.D0 * CUBEAY(K0)
11020             RMAT(3,K0,I) = 2.D0 * CUBEAZ(K0)
11030   120     CONTINUE
11040 C
11050         ELSE
11060 C               --- (B) FOR PTCLS WITH DOWNWARD OBLIQUE DIRECTION ---
11070 C
11080           NXI = ROT11 * 1.D0 + ROT12 * (-1.D0) + ROT13 * (-1.D0)
11090           NYI = ROT21 * 1.D0 + ROT22 * (-1.D0) + ROT23 * (-1.D0)
11100           NZI = ROT31 * 1.D0 + ROT32 * (-1.D0) + ROT33 * (-1.D0)
11110 C
11120           C0 = DSQRT( NXI**2 + NYI**2 + NZI**2 )
11130           NX(I) = NXI / C0
11140           NY(I) = NYI / C0
11150           NZ(I) = NZI / C0
11160 C
11170           CUBEAX(1) = ROT11* 0.5D0 + ROT12* 0.0D0   + ROT13* 0.0D0
11180           CUBEAY(1) = ROT21* 0.5D0 + ROT22* 0.0D0   + ROT23* 0.0D0
11190           CUBEAZ(1) = ROT31* 0.5D0 + ROT32* 0.0D0   + ROT33* 0.0D0
11200 C
```

• The original face vectors $\boldsymbol{a}_1^s$ =(0.5,0,0), $\boldsymbol{a}_2^s$ =(0,0.5,0) and $\boldsymbol{a}_3^s$ =(0,0,0.5) for the particles with the upward diagonal moment. Similarly, $\boldsymbol{a}_1^s$ =(0.5,0,0), $\boldsymbol{a}_2^s$ =(0,−0.5,0) and $\boldsymbol{a}_3^s$ =(0,0, −0.5) for the particles with the downward diagonal moment.

• An initial direction of each magnetic moment is randomly assigned in the range of −90° to 90° about the $z$-axis direction.

• The original magnetic moment direction $\boldsymbol{n}_i^s$ is $\boldsymbol{n}_i^s$ =(1/ $\sqrt{3}$ ,1/ $\sqrt{3}$ ,1/ $\sqrt{3}$ ) in the absolute coordinate system for particles with a magnetic moment in the upward diagonal direction.

• The new magnetic moment direction $\boldsymbol{n}_i^s$ is obtained by multiplying the rotation matrix and the original direction.

• The new face vector $\boldsymbol{a}_k^s$ ($k$=1,2,3) is obtained by multiplying the rotation matrix and the original one.

• The original magnetic moment direction $\boldsymbol{n}_i^s$ is $\boldsymbol{n}_i^s$ =(1/ $\sqrt{3}$ , −1/ $\sqrt{3}$ ,−1/ $\sqrt{3}$ ) for particles with a magnetic moment in the downward diagonal direction.

• The new magnetic moment direction $\boldsymbol{n}_i^s$ is obtained by multiplying the rotation matrix and the original direction.

```
11210             CUBEAX(2) = ROT11* 0.0D0 + ROT12*(-0.5D0) + ROT13* 0.0D0
11220             CUBEAY(2) = ROT21* 0.0D0 + ROT22*(-0.5D0) + ROT23* 0.0D0
11230             CUBEAZ(2) = ROT31* 0.0D0 + ROT32*(-0.5D0) + ROT33* 0.0D0
11240 C
11250             CUBEAX(3) = ROT11* 0.0D0 + ROT12* 0.0D0   + ROT13*(-0.5D0)
11260             CUBEAY(3) = ROT21* 0.0D0 + ROT22* 0.0D0   + ROT23*(-0.5D0)
11270             CUBEAZ(3) = ROT31* 0.0D0 + ROT32* 0.0D0   + ROT33*(-0.5D0)
11280 C
11290             DO 130 K0 = 1, 3
11300               CUBEA(1,K0,I) =  CUBEAX(K0)
11310               CUBEA(2,K0,I) =  CUBEAY(K0)
11320               CUBEA(3,K0,I) =  CUBEAZ(K0)
11330   130       CONTINUE
11340 C
11350             DO 140 K0 = 1, 3
11360               RMAT(1,K0,I) = 2.D0 * CUBEAX(K0)
11370               RMAT(2,K0,I) = 2.D0 * CUBEAY(K0)
11380               RMAT(3,K0,I) = 2.D0 * CUBEAZ(K0)
11390   140       CONTINUE
11400 C
11410           END IF
11420 C
11430           C1 = CUBEA(1,3,I)
11440           C2 = CUBEA(2,3,I)
11450           C3 = CUBEA(3,3,I)
11460           C0 = DSQRT( C1**2 + C2**2 + C3**2 )
11470           EX(I) = C1 / C0
11480           EY(I) = C2 / C0
11490           EZ(I) = C3 / C0
11500 C
11510           DO 200 IXYZ = 1, 3
11520             CUBEA(IXYZ,4,I) = -CUBEA(IXYZ,1,I)
11530             CUBEA(IXYZ,5,I) = -CUBEA(IXYZ,2,I)
11540             CUBEA(IXYZ,6,I) = -CUBEA(IXYZ,3,I)
11550   200     CONTINUE
11560 C
11570           DO 220 IXYZ = 1, 3
11580             CUBEALPH(IXYZ,1,I) =  CUBEA(IXYZ,1,I) + CUBEA(IXYZ,2,I)
11590      &                                           + CUBEA(IXYZ,3,I)
11600             CUBEALPH(IXYZ,2,I) = -CUBEA(IXYZ,1,I) + CUBEA(IXYZ,2,I)
11610      &                                           + CUBEA(IXYZ,3,I)
11620             CUBEALPH(IXYZ,3,I) =  CUBEA(IXYZ,1,I) - CUBEA(IXYZ,2,I)
11630      &                                           + CUBEA(IXYZ,3,I)
11640             CUBEALPH(IXYZ,4,I) =  CUBEA(IXYZ,1,I) + CUBEA(IXYZ,2,I)
11650      &                                           - CUBEA(IXYZ,3,I)
11660   220     CONTINUE
11670 C
11680           DO 240 IXYZ = 1, 3
11690             CUBEALPH(IXYZ,5,I) = -CUBEALPH(IXYZ,1,I)
11700             CUBEALPH(IXYZ,6,I) = -CUBEALPH(IXYZ,2,I)
11710             CUBEALPH(IXYZ,7,I) = -CUBEALPH(IXYZ,3,I)
11720             CUBEALPH(IXYZ,8,I) = -CUBEALPH(IXYZ,4,I)
11730   240     CONTINUE
11740 C
11750 C
11760 CCCCC     C1 = CUBEALPH(1,1,I)
11770 CCCCC     C2 = CUBEALPH(2,1,I)
11780 CCCCC     C3 = CUBEALPH(3,1,I)
11790 CCCCC     C0 = DSQRT( C1**2 + C2**2 + C3**2 )
11800 CCCCC     NX(I) =  C1 / C0
11810 CCCCC     NY(I) =  C2 / C0
11820 CCCCC     NZ(I) =  C3 / C0
11830 C
11840 C
11850   500 CONTINUE
11860                                                              RETURN
11870                                                              END
11880 C**** SUB PTCLMDL *****
11890       SUBROUTINE PTCLMDL
11900 C
11910       IMPLICIT REAL*8 (A-H,O-Z), INTEGER (I-N)
11920 C
11930       COMMON /BLOCK201/ CUBEA    , ANEIGHBA , ANEIALPH , ANEIA
11940       COMMON /BLOCK203/ CUBEALPH, ALPHNEIG , ALNEIA
11950 C     ------------------------------------------------------------------
11960       INTEGER    NN , NNS , TT
11970       PARAMETER( NN=1000 , NNS=5000000 , TT=1000 )
11980       PARAMETER( NRANMX=1000000 , PI=3.141592653589793D0 )
11990 C     --------------------------
12000       REAL*8   CUBEA(3,6,NN), CUBEALPH(3,8,NN)
```

• The new face vector $\boldsymbol{a}_k^s$ ($k$=1,2,3) is obtained by multiplying the rotation matrix and the original one.

• The vertex vector $\boldsymbol{\alpha}_k^s$ ($k$=1,2,…,8), in the absolute coordinate system, is obtained from the face vector $\boldsymbol{a}_k^s$ ($k$=1,2,…,6) in the absolute coordinate system.

• In subroutine PTCLMDL, the variables characterizing the relationships regarding vertex-to-vertex, plane-to-plane, vertex-to-plane, etc. of the cube are specified.

```
12010         INTEGER  ANEIGHBA(4,6), ANEIALPH(4,6), ANEIA(2,6)
12020         INTEGER  ALPHNEIG(3,8), ALNEIA(3,8)
12030 C
12040 C       ------------------------------------------- FACE VECTOR A ---
12050         ANEIGHBA(1,1) = 2
12060         ANEIGHBA(2,1) = 3
12070         ANEIGHBA(3,1) = 5
12080         ANEIGHBA(4,1) = 6
12090         ANEIGHBA(1,2) = 1
12100         ANEIGHBA(2,2) = 3
12110         ANEIGHBA(3,2) = 4
12120         ANEIGHBA(4,2) = 6
12130         ANEIGHBA(1,3) = 1
12140         ANEIGHBA(2,3) = 2
12150         ANEIGHBA(3,3) = 4
12160         ANEIGHBA(4,3) = 5
12170         ANEIGHBA(1,4) = ANEIGHBA(1,1)
12180         ANEIGHBA(2,4) = ANEIGHBA(2,1)
12190         ANEIGHBA(3,4) = ANEIGHBA(3,1)
12200         ANEIGHBA(4,4) = ANEIGHBA(4,1)
12210         ANEIGHBA(1,5) = ANEIGHBA(1,2)
12220         ANEIGHBA(2,5) = ANEIGHBA(2,2)
12230         ANEIGHBA(3,5) = ANEIGHBA(3,2)
12240         ANEIGHBA(4,5) = ANEIGHBA(4,2)
12250         ANEIGHBA(1,6) = ANEIGHBA(1,3)
12260         ANEIGHBA(2,6) = ANEIGHBA(2,3)
12270         ANEIGHBA(3,6) = ANEIGHBA(3,3)
12280         ANEIGHBA(4,6) = ANEIGHBA(4,3)
12290 C
```

• The name of the neighboring faces of face $k$ ($k$=1,2,…,6) is saved in the variable ANEIGHBA($l$,$k$); the number of neighboring faces, $l$, is $l$=4.

```
12300         ANEIALPH(1,1) = 1
12310         ANEIALPH(2,1) = 4
12320         ANEIALPH(3,1) = 3
12330         ANEIALPH(4,1) = 6
12340         ANEIALPH(1,2) = 2
12350         ANEIALPH(2,2) = 1
12360         ANEIALPH(3,2) = 7
12370         ANEIALPH(4,2) = 4
12380         ANEIALPH(1,3) = 8
12390         ANEIALPH(2,3) = 2
12400         ANEIALPH(3,3) = 3
12410         ANEIALPH(4,3) = 1
12420         ANEIALPH(1,4) = 5
12430         ANEIALPH(2,4) = 8
12440         ANEIALPH(3,4) = 7
12450         ANEIALPH(4,4) = 2
12460         ANEIALPH(1,5) = 6
12470         ANEIALPH(2,5) = 5
12480         ANEIALPH(3,5) = 3
12490         ANEIALPH(4,5) = 8
12500         ANEIALPH(1,6) = 4
12510         ANEIALPH(2,6) = 6
12520         ANEIALPH(3,6) = 7
12530         ANEIALPH(4,6) = 5
12540 C
12550         ANEIA(1,1) = 2
12560         ANEIA(2,1) = 3
12570         ANEIA(1,2) = 3
12580         ANEIA(2,2) = 4
12590         ANEIA(1,3) = 4
12600         ANEIA(2,3) = 5
12610         ANEIA(1,4) = 5
12620         ANEIA(2,4) = 6
12630         ANEIA(1,5) = 6
12640         ANEIA(2,5) = 1
12650         ANEIA(1,6) = 1
12660         ANEIA(2,6) = 2
12670 C
```

• The name of the related vertices of face $k$ ($k$=1,2,…,6) is saved in the variable ANEIALPH($l$,$k$); the number of neighboring vertices, $l$, is $l$=4, as shown in Table 5.2.

```
12680 C       -------------------------------------- CORNER VECTOR ALPHA ---
12690         ALPHNEIG(1,1) = 2
12700         ALPHNEIG(2,1) = 3
12710         ALPHNEIG(3,1) = 4
12720         ALPHNEIG(1,2) = 1
12730         ALPHNEIG(2,2) = 7
12740         ALPHNEIG(3,2) = 8
12750         ALPHNEIG(1,3) = 1
12760         ALPHNEIG(2,3) = 6
12770         ALPHNEIG(3,3) = 8
12780         ALPHNEIG(1,4) = 1
12790         ALPHNEIG(2,4) = 6
12800         ALPHNEIG(3,4) = 7
```

• The name of the neighboring vertices of vertex $k$ ($k$=1,2,…,8) is saved in the variable ALPHNEIG($l$,$k$); the number of neighboring vertices, $l$, is $l$=3, as shown in Table 5.1.

```
12810         ALPHNEIG(1,5) = 6
12820         ALPHNEIG(2,5) = 7
12830         ALPHNEIG(3,5) = 8
12840         ALPHNEIG(1,6) = 3
12850         ALPHNEIG(2,6) = 4
12860         ALPHNEIG(3,6) = 5
12870         ALPHNEIG(1,7) = 2
12880         ALPHNEIG(2,7) = 4
12890         ALPHNEIG(3,7) = 5
12900         ALPHNEIG(1,8) = 2
12910         ALPHNEIG(2,8) = 3
12920         ALPHNEIG(3,8) = 5
12930 C
12940         ALNEIA(1,1) = 1
12950         ALNEIA(2,1) = 2
12960         ALNEIA(3,1) = 3
12970         ALNEIA(1,2) = 2
12980         ALNEIA(2,2) = 3
12990         ALNEIA(3,2) = 4
13000         ALNEIA(1,3) = 1
13010         ALNEIA(2,3) = 3
13020         ALNEIA(3,3) = 5
13030         ALNEIA(1,4) = 1
13040         ALNEIA(2,4) = 2
13050         ALNEIA(3,4) = 6
13060         ALNEIA(1,5) = 4
13070         ALNEIA(2,5) = 5
13080         ALNEIA(3,5) = 6
13090         ALNEIA(1,6) = 1
13100         ALNEIA(2,6) = 5
13110         ALNEIA(3,6) = 6
13120         ALNEIA(1,7) = 2
13130         ALNEIA(2,7) = 4
13140         ALNEIA(3,7) = 6
13150         ALNEIA(1,8) = 3
13160         ALNEIA(2,8) = 4
13170         ALNEIA(3,8) = 5
13180                                                               RETURN
13190                                                               END
```

• The name of the related faces of vertex $k$ ($k$=1,2,…,8) is saved in the variable ALNEIA($l$,$k$); the number of neighboring faces, $l$, is $l$=3, as shown in Table 5.1.

• In subroutine PRNTDAT, the ten sub-average of MOMY(*) and MORDY(*) are calculated and printed out; these describe the order parameters $\langle \hat{n}_y \rangle$ and $S_{ny}$, respectively.

```
13200 C**** SUB PRNTDAT ****
13210       SUBROUTINE PRNTDAT( MCSST, MCSMX, NP )
13220 C
13230       IMPLICIT REAL*8 (A-H,O-Z), INTEGER (I-N)
13240 C
13250       COMMON /BLOCK143/ MOMX , MOMY , MOMZ , MEANENE
13260       COMMON /BLOCK171/ EDIRX , EDIRY , EDIRZ
13270       COMMON /BLOCK172/ MORDX , MORDY , MORDZ
13280       COMMON /BLOCK173/ EORDX , EORDY , EORDZ
13290 C     -----------------------------------------------------------------
13300       INTEGER    NN , NNS , TT
13310       PARAMETER( NN=1000 , NNS=5000000 , TT=1000 )
13320       PARAMETER( NRANMX=1000000 , PI=3.141592653589793D0 )
13330 C     --------------------------
13340       REAL      MOMX(NNS), MOMY(NNS) , MOMZ(NNS)  , MEANENE(NNS)
13350       REAL      EDIRX(NNS) , EDIRY(NNS) , EDIRZ(NNS)
13360       REAL      MORDX(NNS) , MORDY(NNS) , MORDZ(NNS)
13370       REAL      EORDX(NNS) , EORDY(NNS) , EORDZ(NNS)
13380 C     --------------------------
13390       INTEGER   MCSST     , MCSMX     , NP
13400 C     --------------------------
13410       REAL      AMOMX(10) , AMOMY(10)  , AMOMZ(10)   , C0
13420       REAL      AEDIRX(10), AEDIRY(10) , AEDIRZ(10)
13430       REAL      AMORDX(10), AMORDY(10) , AMORDZ(10)
13440       REAL      AEORDX(10), AEORDY(10) , AEORDZ(10)
13450       INTEGER   IC , IMC(0:10) , JS , JE
13460 C                                  ----- TIME STEP AVERAGE -----
13470       IC = ( MCSMX-MCSST+1 )/10
13480       DO 30 I=0,10
13490         IMC(I) = MCSST - 1 + IC*I
13500         IF( I .EQ. 10 ) IMC(I) =MCSMX
13510    30 CONTINUE
13520 C
13530 C
13540       DO 35 I=1,10
13550         AMOMX( I) = 0.
13560         AMOMY( I) = 0.
13570         AMOMZ( I) = 0.
13580         AEDIRX(I) = 0.
13590         AEDIRY(I) = 0.
13600         AEDIRZ(I) = 0.
```

```
13610          AMORDX(I) = 0.
13620          AMORDY(I) = 0.
13630          AMORDZ(I) = 0.
13640          AEORDX(I) = 0.
13650          AEORDY(I) = 0.
13660          AEORDZ(I) = 0.
13670    35 CONTINUE
13680 C
13690       DO 50 I=1,10
13700          JS = IMC(I-1) + 1
13710          JE = IMC(I)
13720          DO 40 J=JS,JE
13730             AMOMX(I)  = AMOMX(I)  + MOMX( J)
13740             AMOMY(I)  = AMOMY(I)  + MOMY( J)
13750             AMOMZ(I)  = AMOMZ(I)  + MOMZ( J)
13760             AEDIRX(I) = AEDIRX(I) + EDIRX(J)
13770             AEDIRY(I) = AEDIRY(I) + EDIRY(J)
13780             AEDIRZ(I) = AEDIRZ(I) + EDIRZ(J)
13790             AMORDX(I) = AMORDX(I) + MORDX(J)
13800             AMORDY(I) = AMORDY(I) + MORDY(J)
13810             AMORDZ(I) = AMORDZ(I) + MORDZ(J)
13820             AEORDX(I) = AEORDX(I) + EORDX(J)
13830             AEORDY(I) = AEORDY(I) + EORDY(J)
13840             AEORDZ(I) = AEORDZ(I) + EORDZ(J)
13850    40    CONTINUE
13860    50 CONTINUE
13870 C
13880       DO 70 I=1,10
13890          C0        = REAL( IMC(I)-IMC(I-1) )
13900          AMOMX(I)  = AMOMX(I)  /C0
13910          AMOMY(I)  = AMOMY(I)  /C0
13920          AMOMZ(I)  = AMOMZ(I)  /C0
13930          AEDIRX(I) = AEDIRX(I) /C0
13940          AEDIRY(I) = AEDIRY(I) /C0
13950          AEDIRZ(I) = AEDIRZ(I) /C0
13960          AMORDX(I) = AMORDX(I) /C0
13970          AMORDY(I) = AMORDY(I) /C0
13980          AMORDZ(I) = AMORDZ(I) /C0
13990          AEORDX(I) = AEORDX(I) /C0
14000          AEORDY(I) = AEORDY(I) /C0
14010          AEORDZ(I) = AEORDZ(I) /C0
14020    70 CONTINUE
14030 C                                                ----- STEP HEIKIN INSATU -----
14040       WRITE(NP,75)
14050       DO 90 I=1,10
14060          WRITE(NP,80) I, IMC(I-1)+1, IMC(I),
14070     &                  AMOMX(I) , AMOMY(I) , AMOMZ(I),
14080     &                  AEDIRX(I), AEDIRY(I), AEDIRZ(I),
14090     &                  AMORDX(I), AMORDY(I), AMORDZ(I),
14100     &                  AEORDX(I), AEORDY(I), AEORDZ(I)
14110    90 CONTINUE
14120 C
14130    75 FORMAT(//1H ,'-----------------------------------------------'
14140     &        /1H ,'                    TIME AVERAGE                '
14150     &        /)
14160    80 FORMAT(1H ,'I=',I2, 2X ,'SMPLMN=',I8, 2X ,'SMPLMX=',I8
14170     &  /1H ,9X ,'MOMX =',F7.4, 2X,'MOMY =',F7.4, 2X, 'MOMZ  =',F7.4
14180     &  /1H ,9X ,'EDIRX=',F7.4, 2X,'EDIRY=',F7.4, 2X, 'EDIRZ =',F7.4
14190     &  /1H ,9X ,'MORDX=',F7.4, 2X,'MORDY=',F7.4, 2X, 'MORDZ =',F7.4
14200     &  /1H ,9X ,'EORDX=',F7.4, 2X,'EORDY=',F7.4, 2X, 'EORDZ =',F7.4/)
14210                                                                  RETURN
14220                                                                  END
14230 C**** SUB ANIMEOPT ****
14240       SUBROUTINE ANIMEOPT( NOPT1, NANMCTR, MCSMPLMX, NANIME )
14250 C
14260       IMPLICIT REAL*8 (A-H,O-Z), INTEGER (I-N)
14270 C
14280       COMMON /BLOCK101/ RX    , RY    , RZ
14290       COMMON /BLOCK102/ EX    , EY    , EZ
14300       COMMON /BLOCK103/ NX    , NY    , NZ
14310       COMMON /BLOCK111/ XL    , YL    , ZL
14320       COMMON /BLOCK112/ HX    , HY    , HZ
14330       COMMON /BLOCK113/ RA    , KU
14340       COMMON /BLOCK115/ N     , NDENS, VDENS, NUP , NDWN
14350       COMMON /BLOCK116/ D     , VP    , D2    , DCHK
14360 C
14370       COMMON /BLOCK200/ RMAT
14380       COMMON /BLOCK201/ CUBEA    , ANEIGHBA , ANEIALPH , ANEIA
14390       COMMON /BLOCK203/ CUBEALPH, ALPHNEIG , ALNEIA
14400       COMMON /BLOCK205/ RMATI , CUBEAI , CUBEAJ , CUBEALI, CUBEALJ
```

· For a commercial software MicroAVS which makes snapshots and animations.

```
14410 C     ----------------------------------------------------------------
14420         INTEGER    NN , NNS , TT
14430         PARAMETER( NN=1000 , NNS=5000000 , TT=1000 )
14440         PARAMETER( NRANMX=1000000 , PI=3.141592653589793D0 )
14450 C     --------------------------
14460         REAL*8   KU     , NDENS
14470         REAL*8   RX(NN) , RY(NN) , RZ(NN)
14480         REAL*8   EX(NN) , EY(NN) , EZ(NN) , NX(NN) , NY(NN) , NZ(NN)
14490 C     --------------------------
14500         REAL*8   RMAT(3,3,NN)
14510         REAL*8   CUBEA(3,6,NN) , CUBEALPH(3,8,NN)
14520         REAL*8   RMATI(3,3)
14530         REAL*8   CUBEAI(3,6)   , CUBEALI(3,8)
14540         REAL*8   CUBEAJ(3,6)   , CUBEALJ(3,8)
14550         INTEGER  ANEIGHBA(4,6), ANEIALPH(4,6) , ANEIA(2,6)
14560         INTEGER  ALPHNEIG(3,8), ALNEIA(3,8)
14570 C     ----------------------------------------------------------------
14580         REAL*8   CX1 , CY1 , CZ1 , CX2 , CY2 , CZ2 , CX3 , CY3 , CZ3
14590         REAL*8   RXI , RYI , RZI , CZ00
14600 C     --------------------------
14610 C
14620         IF( NANMCTR .EQ. 1 ) THEN
14630           WRITE(NOPT1,181) (MCSMPLMX/NANIME)
14640         END IF
14650 C
14660         IF( (NANMCTR.GE.1) .AND. (NANMCTR.LE.9) ) THEN
14670           WRITE(NOPT1,183) NANMCTR
14680         ELSE IF( (NANMCTR.GE.10) .AND. (NANMCTR.LE.99) ) THEN
14690           WRITE(NOPT1,184) NANMCTR
14700         ELSE IF( (NANMCTR.GE.100) .AND. (NANMCTR.LE.999) ) THEN
14710           WRITE(NOPT1,185) NANMCTR
14720         ELSE IF( (NANMCTR.GE.1000) .AND. (NANMCTR.LE.9999) ) THEN
14730           WRITE(NOPT1,186) NANMCTR
14740         END IF
14750 C
14760 C     ---------------------------------------------------- CUBE (1) ---
14770 C         ---------------------------------------------------------
14780 C           AVS CRNR  1 - SIM CRNR  2      AVS CRNR  5 - SIM CRNR  7
14790 C           AVS CRNR  2 - SIM CRNR  8      AVS CRNR  6 - SIM CRNR  5
14800 C           AVS CRNR  3 - SIM CRNR  3      AVS CRNR  7 - SIM CRNR  6
14810 C           AVS CRNR  4 - SIM CRNR  1      AVS CRNR  8 - SIM CRNR  4
14820 C         ---------------------------------------------------------
14830 C
14840         N0 = N
14850         DO 250 I= 1, N0
14860 C
14870           WRITE(NOPT1,210)
14880 C
14890           DO 230 KCRNR = 1, 8
14900             IF( KCRNR.EQ.1 )  KAVS = 2
14910             IF( KCRNR.EQ.2 )  KAVS = 8
14920             IF( KCRNR.EQ.3 )  KAVS = 3
14930             IF( KCRNR.EQ.4 )  KAVS = 1
14940             IF( KCRNR.EQ.5 )  KAVS = 7
14950             IF( KCRNR.EQ.6 )  KAVS = 5
14960             IF( KCRNR.EQ.7 )  KAVS = 6
14970             IF( KCRNR.EQ.8 )  KAVS = 4
14980 C
14990             RXI= RX(I) + CUBEALPH( 1, KAVS,I )
15000             RYI= RY(I) + CUBEALPH( 2, KAVS,I )
15010             RZI= RZ(I) + CUBEALPH( 3, KAVS,I )
15020 C
15030             WRITE(NOPT1,220)  RXI, RYI, RZI
15040   230     CONTINUE
15050 C
15060           WRITE(NOPT1,240)
15070 C
15080   250   CONTINUE
15090 C
15100 C     ---------------------------------------- RIGHT TRIANGLE (2) ---
15110 C                                               --- FOR MAG MOMENT ---
15120         DO 450 I=1,N
15130 C
15140           WRITE(NOPT1,410)
15150 C
15160           IF( I .LE. NUP ) THEN
15170             CZ00 = 0.01D0
15180           ELSE
15190             CZ00 = 0.01D0 + 1.D0
15200           END IF
```

```
15210 C
15220         CX1 = RX(I) + CUBEALPH(1,1,I)
15230         CY1 = RY(I) + CUBEALPH(2,1,I)
15240         CZ1 = RZ(I) + CUBEALPH(3,1,I) + CZ00
15250         CX2 = CX1   - CUBEA(1,1,I)
15260         CY2 = CY1   - CUBEA(2,1,I)
15270         CZ2 = CZ1   - CUBEA(3,1,I)
15280         CX3 = CX1   - CUBEA(1,2,I)
15290         CY3 = CY1   - CUBEA(2,2,I)
15300         CZ3 = CZ1   - CUBEA(3,2,I)
15310 C
15320         IF( I .LE. NUP ) THEN
15330           WRITE(NOPT1,420) CX1, CY1, CZ1, 1.0, 0.0, 0.0
15340           WRITE(NOPT1,420) CX2, CY2, CZ2, 1.0, 0.0, 0.0
15350           WRITE(NOPT1,420) CX3, CY3, CZ3, 1.0, 0.0, 0.0
15360         ELSE
15370           WRITE(NOPT1,420) CX1, CY1, CZ1, 0.6, 0.8, 1.0
15380           WRITE(NOPT1,420) CX2, CY2, CZ2, 0.6, 0.8, 1.0
15390           WRITE(NOPT1,420) CX3, CY3, CZ3, 0.6, 0.8, 1.0
15400         END IF
15410 C
15420         WRITE(NOPT1,440)
15430 C
15440   450 CONTINUE
15450 C
15460 C     -------------------------------------- SIM.REGEON LINES  (3) ---
15470       XL2 = XL/2.0
15480       YL2 = YL/2.0
15490       ZL2 = ZL/2.0
15500       WRITE(NOPT1,510)  17
15510       WRITE(NOPT1,520)  0.-XL2 ,  0.-YL2 ,  0.-ZL2
15520       WRITE(NOPT1,520)  XL-XL2 ,  0.-YL2 ,  0.-ZL2
15530       WRITE(NOPT1,520)  XL-XL2 ,  YL-YL2 ,  0.-ZL2
15540       WRITE(NOPT1,520)  0.-XL2 ,  YL-YL2 ,  0.-ZL2
15550       WRITE(NOPT1,520)  0.-XL2 ,  0.-YL2 ,  0.-ZL2
15560       WRITE(NOPT1,520)  0.-XL2 ,  0.-YL2 ,  ZL-ZL2
15570       WRITE(NOPT1,520)  XL-XL2 ,  0.-YL2 ,  ZL-ZL2
15580       WRITE(NOPT1,520)  XL-XL2 ,  YL-YL2 ,  ZL-ZL2
15590       WRITE(NOPT1,520)  0.-XL2 ,  YL-YL2 ,  ZL-ZL2
15600       WRITE(NOPT1,520)  0.-XL2 ,  0.-YL2 ,  ZL-ZL2
15610       WRITE(NOPT1,520)  0.-XL2 ,  0.-YL2 ,  0.-ZL2
15620       WRITE(NOPT1,520)  0.-XL2 ,  YL-YL2 ,  0.-ZL2
15630       WRITE(NOPT1,520)  0.-XL2 ,  YL-YL2 ,  ZL-ZL2
15640       WRITE(NOPT1,520)  XL-XL2 ,  YL-YL2 ,  ZL-ZL2
15650       WRITE(NOPT1,520)  XL-XL2 ,  YL-YL2 ,  0.-ZL2
15660       WRITE(NOPT1,520)  XL-XL2 ,  0.-YL2 ,  0.-ZL2
15670       WRITE(NOPT1,520)  XL-XL2 ,  0.-YL2 ,  ZL-ZL2
15680 C
15690 C     -------------------------- FORMAT ---------------------------
15700   181 FORMAT('# Micro AVS Geom:2.00'
15710      &       /'# Animation of MC simulation results'
15720      &       /I4)
15730   183 FORMAT('step',I1)
15740   184 FORMAT('step',I2)
15750   185 FORMAT('step',I3)
15760   186 FORMAT('step',I4)
15770   210 FORMAT( 'polyhedron'/'cube'/'facet'/'vertex'/'8')
15780   220 FORMAT( 3F11.4 )
15790   240 FORMAT( '6'/'4'/' 1 2 3 4'/'4'/' 1 5 6 2'/'4'/' 2 6 7 3'
15800      &            /'4'/' 3 7 8 4'/'4'/' 1 4 8 5'/'4'/' 5 8 7 6')
15810   410 FORMAT( 'polyhedron'/'triangle'/'facet'/'color'/'3')
15820   420 FORMAT( 3F11.4 , 3F7.1 )
15830   440 FORMAT( '1'/'3'/' 1 2 3' )
15840   510 FORMAT( 'polyline'/'pline_sample'/'vertex'/I3 )
15850   520 FORMAT( 3F10.3 )
15860                                                                RETURN
15870                                                                END
15880 C**** ORDERCA ****
15890       SUBROUTINE ORDERCA( N , NUP , NDWN )
15900 C
15910       IMPLICIT REAL*8 (A-H,O-Z), INTEGER (I-N)
15920 C
15930       COMMON /BLOCK101/ RX    , RY    , RZ
15940       COMMON /BLOCK103/ NX    , NY    , NZ
15950       COMMON /BLOCK111/ XL    , YL    , ZL
15960 C
15970       COMMON /BLOCK175/ MORDER , MORDER4 , MORDERB ,  MORDER4B
15980 C     -----------------------------------------------------------------
15990       INTEGER     NN , NNS , TT
16000       PARAMETER( NN=1000 , NNS=5000000 , TT=1000 )
```

• The order parameters $S_1$ and $S_2$ are evaluated as a function of radial distance.

```
16010         PARAMETER( NRANMX=1000000 , PI=3.141592653589793D0 )
16020 C       ----------------------------
16030         REAL*8    RX(NN) , RY(NN) , RZ(NN)
16040         REAL*8    NX(NN) , NY(NN) , NZ(NN)
16050         REAL*8    MORDER , MORDER4, MORDERB(10) , MORDER4B(10)
16060 C       ----------------------------
16070         REAL*8    NX1 , NY1 , NZ1 , NX2 , NY2 , NZ2
16080         REAL*8    XL2 , YL2 , ZL2 , RXIJ, RYIJ, RZIJ, RIJ
16090         REAL*8    C1  , C2  , C4  , CSUM1 , CSUM2 , COP1 , COP2
16100         REAL*8    CSUMB(10) , CSUM4B(10)
16110         INTEGER   NPAIR , NPAIRB0(10)
16120 C
16130         XL2 = XL/2.D0
16140         YL2 = YL/2.D0
16150         ZL2 = ZL/2.D0
16160 C
16170         NPAIR = 0
16180         CSUM1 = 0.D0
16190         CSUM2 = 0.D0
16200         DO 10 I=1,10
16210           NPAIRB0(I) = 0
16220           CSUMB( I) = 0.D0
16230           CSUM4B( I) = 0.D0
16240    10 CONTINUE
16250 C                                      --- ORDER PARAMETERS OF MAG. MOM. ---
16260         DO 200 I=1, N-1
16270         DO 150 J=I+1, N
16280 C
16290           NPAIR= NPAIR + 1
16300 C                                          --------------- (1) MORDER ---
16310           C1   = NX(I)*NX(J) + NY(I)*NY(J) + NZ(I)*NZ(J)
16320           COP1 = ( 3.D0*C1**2 - 1.D0 ) / 2.D0
16330           CSUM1 = CSUM1 + COP1
16340 C
16350           NX1 = NX(I)
16360           NY1 = NY(I)
16370           NZ1 = 0.D0
16380           C0 = DSQRT( NX1**2 + NY1**2 + NZ1**2)
16390           NX1 = NX1/C0
16400           NY1 = NY1/C0
16410           NZ1 = NZ1/C0
16420 C
16430           NX2 = NX(J)
16440           NY2 = NY(J)
16450           NZ2 = 0.D0
16460           C0 = DSQRT( NX2**2 + NY2**2 + NZ2**2)
16470           NX2 = NX2/C0
16480           NY2 = NY2/C0
16490           NZ2 = NZ2/C0
16500 C                                          --------------- (2) MORDER4 ---
16510           C1   = NX1*NX2 + NY1*NY2 + NZ1*NZ2
16520           C2   = C1*C1
16530           C4   = C2*C2
16540           COP2  = 8.D0*C4 - 8.D0*C2 + 1.D0
16550           CSUM2 = CSUM2 + COP2
16560 C
16570 C                                    --- (3) MORDERB, MORDER4B(*) ---
16580           RXIJ = RX(I)   - RX(J)
16590           IF( RXIJ .GT. XL2 ) THEN
16600             RXIJ = RXIJ - XL
16610           ELSE IF( RXIJ .LT. -XL2 ) THEN
16620             RXIJ = RXIJ + XL
16630           END IF
16640           IF( DABS(RXIJ) .GE. 10.D0 )       GOTO 150
16650 C
16660           RYIJ = RY(I)   - RY(J)
16670           IF( RYIJ .GT. YL2 ) THEN
16680             RYIJ = RYIJ - YL
16690           ELSE IF( RYIJ .LT. -YL2 ) THEN
16700             RYIJ = RYIJ + YL
16710           END IF
16720           IF( DABS(RYIJ) .GE. 10.D0 )       GOTO 150
16730 C
16740           RIJ = DSQRT( RXIJ**2 + RYIJ**2 )
16750           IF( DABS(RIJ) .GE. 10.D0 )        GOTO 150
16760 C
16770           II = IDINT( RIJ ) + 1
16780 C
16790           NPAIRB0(II) = NPAIRB0(II) + 1
16800           CSUMB( II) = CSUMB( II) + COP1
```

- The variables CSUMB(*) and CSUM4B(*) are used for the order parameters $S_1$ and $S_2$, respectively.

- The order parameter $S_1$ is calculated.

- The order parameter $S_2$ is calculated.

- DBLE(II) corresponds to the radial distance.

- The results of the order parameters $S_1$ and $S_2$ are saved in the corresponding variables which are for the radial circle shell between DINT(RIJ)+1 and DINT(RIJ).

```
16810           CSUM4B( II) = CSUM4B( II) + COP2
16820 C
16830    150 CONTINUE
16840    200 CONTINUE
16850 C
16860 C                                    --- AVERAGING AND ACCUMULATING DATA ---
16870 C                                             +++ MORDER, MORDER4 +++
16880        MORDER     = MORDER  + CSUM1/DBLE(NPAIR)
16890        MORDER4    = MORDER4 + CSUM2/DBLE(NPAIR)
16900 C                                       +++ MORDERB(*), MORDER4B(*) +++
16910        NPAIRB0(1) = 0
16920        CSUMB(  1) = 0.D0
16930        CSUM4B( 1) = 0.D0
16940 C
16950        DO 320 I=2, 10
16960          NPAIRB0(I) = NPAIRB0( I-1) + NPAIRB0(I)
16970          CSUMB(  I) = CSUMB(   I-1) + CSUMB  ( I)
16980          CSUM4B( I) = CSUM4B(  I-1) + CSUM4B( I)
16990    320 CONTINUE
17000 C
17010        DO 340 I=2, 10
17020          IC0 = NPAIRB0(I)
17030          IF( IC0 .LE. 0 ) IC0 = 99999999
17040          CSUMB( I) = CSUMB( I) / DBLE( IC0 )
17050          CSUM4B(I) = CSUM4B(I) / DBLE( IC0 )
17060    340 CONTINUE
17070 C
17080        MORDERB( 1) = 0.D0
17090        MORDER4B(1) = 0.D0
17100        DO 360 I=2, 10
17110          MORDERB( I) = MORDERB( I) + CSUMB( I)
17120          MORDER4B(I) = MORDER4B(I) + CSUM4B(I)
17130    360 CONTINUE
17140                                                          RETURN
17150                                                          END
17160 C**** SUB ENECAL *****
17170        SUBROUTINE ENECAL( I, RXI, RYI, RZI, EXI, EYI, EZI, NXI, NYI,
17180     &                   NZI, ECAN, OVRLAP, ISTREET, JPTCL0, CLSTRUE )
17190 C
17200        IMPLICIT REAL*8 (A-H,O-Z), INTEGER (I-N)
17210 C
17220        COMMON /BLOCK101/ RX    , RY    , RZ
17230        COMMON /BLOCK102/ EX    , EY    , EZ
17240        COMMON /BLOCK103/ NX    , NY    , NZ
17250        COMMON /BLOCK111/ XL    , YL    , ZL
17260        COMMON /BLOCK112/ HX    , HY    , HZ
17270        COMMON /BLOCK113/ RA    , KU
17280        COMMON /BLOCK115/ N     , NDENS, VDENS , NUP , NDWN
17290        COMMON /BLOCK116/ D     , VP    , D2    , DCHK
17300        COMMON /BLOCK131/ RCOFF, DELR , DELT , DELDEG
17310 C
17320        COMMON /BLOCK160/ WIDTHCLS , RCLSLMT
17330 C
17340        COMMON /BLOCK201/ CUBEA    , ANEIGHBA , ANEIALPH , ANEIA
17350        COMMON /BLOCK203/ CUBEALPH, ALPHNEIG , ALNEIA
17360        COMMON /BLOCK205/ RMATI , CUBEAI , CUBEAJ , CUBEALI, CUBEALJ
17370 C
17380        COMMON /WORK116/  ENEMAGI
17390 C      ---------------------------------------------------------------
17400        INTEGER    NN , NNS , TT
17410        PARAMETER( NN=1000 , NNS=5000000 , TT=1000 )
17420        PARAMETER( NRANMX=1000000 , PI=3.141592653589793D0 )
17430 C      --------------------------
17440        REAL*8    KU     , NDENS
17450        REAL*8    RX(NN) , RY(NN) , RZ(NN)
17460        REAL*8    EX(NN) , EY(NN) , EZ(NN) , NX(NN) , NY(NN) , NZ(NN)
17470 C      --------------------------
17480        REAL*8    CUBEA(3,6,NN) , CUBEALPH(3,8,NN)
17490        REAL*8    RMATI(3,3)
17500        REAL*8    CUBEAI(3,6)   , CUBEALI(3,8)
17510        REAL*8    CUBEAJ(3,6)   , CUBEALJ(3,8)
17520        INTEGER   ANEIGHBA(4,6), ANEIALPH(4,6) , ANEIA(2,6)
17530        INTEGER   ALPHNEIG(3,8), ALNEIA(3,8)
17540 C
17550        INTEGER   ISTREET
17560        LOGICAL   OVRLAP ,  CLSTRUE
17570 C      ---------------------------------------------------------------
17580        REAL*8    RXI  , RYI  , RZI  , RXJ  , RYJ  , RZJ
17590        REAL*8    RXIJ , RYIJ , RZIJ , RXJI , RYJI , RZJI
17600        REAL*8    RIJ  , RIJSQ, RIJ2 , RIJ3
```

• The order parameters $S_1$ and $S_2$ in the range of each radial range are evaluated using the above calculated values for the radial circle shell. For instance, CSUMB(I) is for the order parameter $S_1$ within the range of the radial DBLE(I).

• The data of the order parameters $S_1$ and $S_2$ are accumulated in the variables MORDERB(*) and MORDER4B(*) for the final averaging procedure after the main loop in the main program.

• Energy of the interaction between particle $i$ and the other particles is calculated in this subroutine.

• If this subroutine is used for assessment of the cluster formation, only particles I and JPTCL0 are addressed for evaluation.

```
17610          REAL*8   NXI  , NYI  , NZI  , NXJ  , NYJ  , NZJ
17620          REAL*8   EXI  , EYI  , EZI  , EXJ  , EYJ  , EZJ
17630 C
17640          REAL*8   ECAN , RCOFF2
17650          REAL*8   XL2 , YL2 , ZL2
17660          REAL*8   C1  , C00 , C01 , C02
17670 C        ---------------------------
17680          INTEGER  IXYZ , KXYZ , K0
17690          INTEGER  JPTCL
17700          INTEGER  IFACE, JFACE, ICRNR , JCRNR
17710          INTEGER  IICRNR, JJFACE, JJCRNR1, JJCRNR2, IC00, JC00, II, II0
17720          INTEGER  IUPDWN, JUPDWN, IIUPDWN, JJUPDWN
17730 C
17740          REAL*8   SIJ , SIJRHD
17750          REAL*8   BETAI(8) , BETAJ(8) , GAMMAI(6) , GAMMAJ(6)
17760          REAL*8   CUBEAII(3,6)  , CUBEALII(3,8)
17770          REAL*8   CUBEAJJ(3,6)  , CUBEALJJ(3,8)
17780          REAL*8   BJJ0(3,2) , BII0(3)
17790          REAL*8   CQJ(3) , CQJ1 , CQJ2 , CQJ3
17800 C        ---------------------------------------------------------------
17810 C                                      +++ NEVER USE I AS A COUNTER +++
17820 C
17830          OVRLAP  = .FALSE.
17840          CLSTRUE = .FALSE.
17850 C
17860          ECAN    = - KU*( NXI*HX + NYI*HY + NZI*HZ )
17870          ENEMAGI = ECAN
17880 C
17890          XL2     = XL/2.D0
17900          YL2     = YL/2.D0
17910          ZL2     = ZL/2.D0
17920          RCOFF2  = RCOFF**2
17930 C
17940 C        ------------------------------------------------- MAIN LOOP START
17950 C
17960 C                            +++++++++++++++++++++++++++++++++++++++++++++
17970 C                             I
17980 C                             RXI, RYI, RZI, EXI, EYI, EZI, NXI, NYI, NZI
17990 C                             RCOFF2, ECAN, OVRLAP, ISTREET
18000 C                            +++++++++++++++++++++++++++++++++++++++++++++
18010          DO 1000 JPTCL=1, N
18020 C                                       +++ ISTREET=0 FOR FULL CAL     +++
18030 C                                       +++ ISTREET=1 FOR MAG.ENE.ONLY +++
18040 C                                       +++ ISTREET=2 FOR CLUSTER-SIZE +++
18050            IF( ISTREET .EQ. 2 ) THEN
18060              J = JPTCL0
18070              IF( JPTCL .GE. 2 ) RETURN
18080            ELSE
18090              J = JPTCL
18100            END IF
18110 C
18120            IF( J .EQ. I )  GOTO 1000
18130 C
18140            RXJ  = RX(J)
18150            RYJ  = RY(J)
18160            RZJ  = RZ(J)
18170            EXJ  = EX(J)
18180            EYJ  = EY(J)
18190            EZJ  = EZ(J)
18200            NXJ  = NX(J)
18210            NYJ  = NY(J)
18220            NZJ  = NZ(J)
18230 C                               +++ IUPDWN IS 1 AND -1 FOR UPWARD AND  +++
18240 C                               +++ DOWNWARD MAG. MOMENT, RESPECTIVELY +++
18250            IF( I .LE. NUP ) THEN
18260              IUPDWN =  1
18270            ELSE
18280              IUPDWN = -1
18290            END IF
18300            IF( J .LE. NUP ) THEN
18310              JUPDWN =  1
18320            ELSE
18330              JUPDWN = -1
18340            END IF
18350 C
18360            RXIJ = RXI  - RXJ
18370            IF( RXIJ .GT. XL2 ) THEN
18380              RXIJ = RXIJ - XL
18390              RXJ  = RXJ  + XL
18400            ELSE IF( RXIJ .LT. -XL2 ) THEN
```

• The variable IUPDWN is used for describing the magnetic moment direction; IUPDWN is 1 or -1 for particle $i$ with a magnetic moment in the upward or downward diagonal direction, respectively.

• Treatment of the periodic boundary condition.

```
18410             RXIJ = RXIJ + XL
18420             RXJ  = RXJ  - XL
18430           END IF
18440           IF( ISTREET .EQ. 2 ) THEN
18450             IF( DABS(RXIJ) .GE. RCLSLMT ) THEN
18460               CLSTRUE = .FALSE.
18470               RETURN
18480             END IF
18490           ELSE
18500             IF( DABS(RXIJ) .GE. RCOFF )       GOTO 1000
18510           END IF
18520 C
18530           RYIJ = RYI  - RYJ
18540           IF( RYIJ .GT. YL2 ) THEN
18550             RYIJ = RYIJ - YL
18560             RYJ  = RYJ  + YL
18570           ELSE IF( RYIJ .LT. -YL2 ) THEN
18580             RYIJ = RYIJ + YL
18590             RYJ  = RYJ  - YL
18600           END IF
18610           IF( ISTREET .EQ. 2 ) THEN
18620             IF( DABS(RYIJ) .GE. RCLSLMT ) THEN
18630               CLSTRUE = .FALSE.
18640               RETURN
18650             END IF
18660           ELSE
18670             IF( DABS(RYIJ) .GE. RCOFF )       GOTO 1000
18680           END IF
18690 C
18700           RZIJ = 0.D0
18710           RZJ  = D2
18720 C
18730           RIJSQ= RXIJ**2 + RYIJ**2
18740           IF( ISTREET .EQ. 2 ) THEN
18750             IF( RIJSQ .GE. RCLSLMT**2 ) THEN
18760               CLSTRUE = .FALSE.
18770               RETURN
18780             END IF
18790             IF( RIJSQ .LE. (D+WIDTHCLS)**2 ) THEN
18800               CLSTRUE = .TRUE.
18810               RETURN
18820             END IF
18830           ELSE
18840             IF( RIJSQ .GE. RCOFF2 )           GOTO 1000
18850             IF( RIJSQ .LT. 1.D0 ) THEN
18860               OVRLAP = .TRUE.
18870               RETURN
18880             END IF
18890           END IF
18900 C
18910 C
18920           RIJ  = DSQRT(RIJSQ)
18930           RIJ3 = RIJ*RIJSQ
18940 C
18950 C         ------------------------------- START OF MAGNETIC ENERGY ---
18960 C
18970           RXJI = -RXIJ
18980           RYJI = -RYIJ
18990           RZJI =  0.D0
19000 C
19010           IF( ISTREET .NE. 2 ) THEN
19020             C00  = NXI*NXJ  + NYI*NYJ  + NZI*NZJ
19030             C01  = NXI*RXIJ + NYI*RYIJ
19040             C02  = NXJ*RXIJ + NYJ*RYIJ
19050             C1   = (RA/RIJ3)*( C00 - 3.D0*C01*C02/RIJSQ )
19060 C
19070             ECAN    = ECAN    + C1
19080             ENEMAGI = ENEMAGI + C1
19090           END IF
19100 C
19110 C         ----------------------------------- END OF MAGNETIC ENERGY ---
19120 C
19130 C                                    +++ ISTREET=0 FOR FULL CAL       ++
19140 C                                    +++ ISTREET=1 FOR MAG.ENE.ONLY   ++
19150 C                                    +++ ISTREET=2 FOR CLUSTER-SIZE +++
19160           IF( ISTREET .NE. 2 ) THEN
19170             IF( ISTREET .EQ. 1 )  GOTO 1000
19180             IF( RIJ .GE. DCHK )   GOTO 1000
19190           END IF
19200 C
```

• Assessment of cluster formation is conducted.

• The magnetic interaction energy between particles $i$ and $j$ is calculated.

```
19210 C        *****************************************************************
19220 C        *********    CHECK THE OVERLAP OF PARTICLES I AND J    *******
19230 C        *****************************************************************
19240 C
19250          DO 30 KO= 1, 8
19260          DO 30 IXYZ=1, 3
19270            IF( KO.LE.6 ) THEN
19280 CCCC         CUBEAI( IXYZ,KO ) = CUBEA(     IXYZ,KO,I )
19290              CUBEAJ( IXYZ,KO ) = CUBEA(     IXYZ,KO,J )
19300            END IF
19310 CCCC       CUBEALI(  IXYZ,KO ) = CUBEALPH( IXYZ,KO,I )
19320            CUBEALJ(  IXYZ,KO ) = CUBEALPH( IXYZ,KO,J )
19330    30    CONTINUE
19340 C
19350          DO 35 KO= 1, 3
19360            GAMMAI(KO)   = CUBEAI(1,KO) * RXJI + CUBEAI(2,KO) * RYJI
19370            GAMMAI(KO+3)= -GAMMAI(KO)
19380 C
19390            GAMMAJ(KO)   = CUBEAJ(1,KO) * RXIJ + CUBEAJ(2,KO) * RYIJ
19400            GAMMAJ(KO+3)= -GAMMAJ(KO)
19410    35    CONTINUE
19420 C
19430          DO 40 KO= 1, 4
19440            BETAI(KO)   = CUBEALI(1,KO) * RXJI + CUBEALI(2,KO) * RYJI
19450            BETAI(KO+4)= -BETAI(KO)
19460 C
19470            BETAJ(KO)   = CUBEALJ(1,KO) * RXIJ + CUBEALJ(2,KO) * RYIJ
19480            BETAJ(KO+4)= -BETAJ(KO)
19490    40    CONTINUE
19500 C                 -------------------------------------------------
19510 C                 RXIJ, RYIJ, RZIJ, RXJI, RYJI, RZJI, RIJ, RIJSQ
19520 C                 CUBEAI(3,6), CUBEAJ(3,6), CUBEALI(3,8), CUBEALJ(3,8)
19530 C                 GAMMAI(6), GAMMAJ(6), BETAI(8), BETAJ(8)
19540 C                 -------------------------------------------------
19550 C        -----------------------------------------------------------
19560 C
19570          RIJ2 = RIJ / 2.D0
19580 C
19590 C                          ---------- (A) LOOK FOR MAXIMUM CORNER ---
19600 C                                                     +++ PTCL I +++
19610          C00 = 0.D0
19620          DO 105 KO= 1, 8
19630            IF((KO.EQ.4).OR.(KO.EQ.5).OR.(KO.EQ.6).OR.(KO.EQ.7)) GOTO 105
19640            IF( BETAI(KO)  .LT.  0.D0 ) GOTO 105
19650            IF( BETAI(KO)  .GE.  C00 ) THEN
19660              C00   = BETAI(KO)
19670              IC00 = KO
19680            END IF
19690   105    CONTINUE
19700          ICRNR = IC00
19710 C                                                     +++ PTCL J +++
19720          C00 = 0.D0
19730          DO 115 KO= 1, 8
19740            IF((KO.EQ.4).OR.(KO.EQ.5).OR.(KO.EQ.6).OR.(KO.EQ.7)) GOTO 115
19750            IF( BETAJ(KO)  .LT.  0.D0 ) GOTO 115
19760            IF( BETAJ(KO)  .GE.  C00 ) THEN
19770              C00   = BETAJ(KO)
19780              JC00 = KO
19790            END IF
19800   115    CONTINUE
19810          JCRNR = JC00
19820 C                             +++ RESULT  I: ICRNR, J: JCRNR +++
19830 C
19840 C
19850 C                           ----------- (B) LOOK FOR MAXIMUM FACE ---
19860 C                                                     +++ PTCL I +++
19870          C00 = 0.D0
19880          DO 125 KO= 1, 6
19890            IF( (KO.EQ.3)  .OR.  (KO.EQ.6) ) GOTO 125
19900            IF( GAMMAI(KO)  .LT.  0.D0 ) GOTO 125
19910            IF( GAMMAI(KO)  .GE.  C00 ) THEN
19920              C00   = GAMMAI(KO)
19930              IC00 = KO
19940            END IF
19950   125    CONTINUE
19960          IFACE = IC00
19970 C                                                     +++ PTCL J +++
19980          C00 = 0.D0
19990          DO 135 KO= 1, 6
20000            IF( (KO.EQ.3)  .OR.  (KO.EQ.6) ) GOTO 135
```

• From now on, the calculation of steric interactions or assessment of cluster formation is conducted.

• CUBEAI(*,+) and CUBEALI(*,+) were calculated before transferring to this subroutine.

• $\gamma_k^i$ ($k$=1,2,…,6) is calculated for face $k$ of particle $i$; similarly, for particle $j$.

• $\beta_k^i$ ($k$=1,2,…,8) is calculated for vertex $k$ of particle $i$.

• $\beta_k^j$ ($k$=1,2,…,8) is calculated for vertex $k$ of particle $j$.

• The name of the vertex giving rise to the maximum $\beta_k^i$ ($k$=1,2,…,8) is searched for particle $i$.

• The vertices 4 to 7 are not necessary to be checked in the present 2D system.

• The name of the vertex giving rise to the maximum $\beta_k^j$ ($k$=1,2,…,8) is searched for particle $j$.

• The name of the face giving rise to the maximum $\gamma_k^i$ ($k$=1,2,…,6) is searched for particle $i$.

• The faces 3 and 6 are not necessary to be checked in the present 2D system.

```
20010             IF( GAMMAJ(K0) .LT. 0.D0 ) GOTO 135
20020             IF( GAMMAJ(K0) .GE. C00 ) THEN
20030               C00  = GAMMAJ(K0)
20040               JC00 = K0
20050             END IF
20060     135   CONTINUE
20070           JFACE = JC00
20080 C                                   +++ RESULT  I: IFACE, J: JFACE +++
20090 C
```

• The face and the vertex of interest are determined in the following procedure.

```
20100           IF( GAMMAJ(JFACE) .GT. GAMMAI(IFACE) ) THEN
20110             JJFACE  = JFACE
20120             IICRNR  = ICRNR
20130             IIUPDWN = IUPDWN
20140             JJUPDWN = JUPDWN
20150             DO 137 K0= 1, 8
20160             DO 137 IXYZ=1, 3
20170               IF( K0.LE.6 ) THEN
20180                 CUBEAII( IXYZ,K0 ) = CUBEAI(   IXYZ,K0 )
20190                 CUBEAJJ( IXYZ,K0 ) = CUBEAJ(   IXYZ,K0 )
20200               END IF
20210               CUBEALII(  IXYZ,K0 ) = CUBEALI(  IXYZ,K0 )
20220               CUBEALJJ(  IXYZ,K0 ) = CUBEALJ(  IXYZ,K0 )
20230     137     CONTINUE
20240           ELSE
20250             JJFACE  = IFACE
20260             IICRNR  = JCRNR
20270             IIUPDWN = JUPDWN
20280             JJUPDWN = IUPDWN
20290             DO 138 K0= 1, 8
20300             DO 138 IXYZ=1, 3
20310               IF( K0.LE.6 ) THEN
20320                 CUBEAII( IXYZ,K0 ) = CUBEAJ(   IXYZ,K0 )
20330                 CUBEAJJ( IXYZ,K0 ) = CUBEAI(   IXYZ,K0 )
20340               END IF
20350               CUBEALII(  IXYZ,K0 ) = CUBEALJ(  IXYZ,K0 )
20360               CUBEALJJ(  IXYZ,K0 ) = CUBEALI(  IXYZ,K0 )
20370     138     CONTINUE
20380           END IF
20390 C                       +++ RESULT  FACE: JJFACE, CRNR: IICRNR +++
```

• The face and the vertex of interest are determined as JJFACE and IICRNR, respectively. Also, the related information is saved in the variables IIUPDWN, CUBEAII(*,+), CUBEALII(*.+), etc.

```
20400           II0= 0
20410           DO 140 II = 1,4
20420             K0 = ANEIALPH(II,JJFACE)
20430             IF((K0.EQ.4).OR.(K0.EQ.5).OR.(K0.EQ.6).OR.(K0.EQ.7)) GOTO 140
20440             II0 = II0 + 1
20450             IF( II0 .EQ. 1 ) JJCRNR1 = K0
20460             IF( II0 .EQ. 2 ) JJCRNR2 = K0
20470     140   CONTINUE
20480 C                                   ++++++++ name of face : JJFACE +++
20490 C                                   +++ name of corner  :IICRNR    +++
20500 C                              +++ name of corner  :JJCRNR1, JJCRNR2 +++
20510 C
```

• The names of the related vertices are JJCRNR1 and JJCRNR2.

• The number of these vertices is necessarily 2 in the present 2D system.

```
20520           DO 150 KXYZ= 1, 3
20530             BJJ0(KXYZ,1)= CUBEALJJ(KXYZ,JJCRNR1) - CUBEAJJ(KXYZ,3)
20540             BJJ0(KXYZ,2)= CUBEALJJ(KXYZ,JJCRNR2) - CUBEAJJ(KXYZ,3)
20550             BII0(KXYZ  )= CUBEALII(KXYZ,IICRNR ) - CUBEAII(KXYZ,3)
20560     150   CONTINUE
20570 C
```

• $\boldsymbol{b}_{p1}$ for IICRNR is evaluated and saved as BII0(*).

• $\boldsymbol{b}_{q2}$ and $\boldsymbol{b}_{q3}$ for JJFACE are saved as BJJ0(*,1) and BJJ0(*,2), respectively.

```
20580           DO 155 KXYZ= 1,3
20590             CQJ(KXYZ)=( BJJ0(KXYZ,1) + BJJ0(KXYZ,2) ) / 2.D0
20600     155   CONTINUE
20610 C                               --- RIGHT-HAND SIDE FOR ASSESSMENT ---
20620           C1 = CQJ(1)*BII0(1)+ CQJ(2)*BII0(2) + CQJ(3)*BII0(3)
20630           SIJRHD = DABS(C1)/D2 + D2
20640 C
```

• $\boldsymbol{c}_q$ is evaluated and saved as CQJ(*).

• The quantity on the right-hand side in Eq. (5.41) is evaluated and saved in SIJRHD.

```
20650           C1 = DSQRT( CQJ(1)**2 + CQJ(2)**2 + CQJ(3)**2 )
20660           CQJ1 = CQJ(1) / C1
20670           CQJ2 = CQJ(2) / C1
20680           CQJ3 = CQJ(3) / C1
20690 C
20700           SIJ =  RXIJ * CQJ1 + RYIJ * CQJ2
20710           SIJ = DABS( SIJ )
20720 C
20730 C
```

• $s_{ij}$ is $|\boldsymbol{r}_{ij}\cdot\boldsymbol{c}_q/|\boldsymbol{c}_q||$.

```
20740           IF( ISTREET .EQ. 2 ) THEN
20750             IF(  SIJ .LE. (SIJRHD + WIDTHCLS)  ) THEN
20760               CLSTRUE = .TRUE.
20770               RETURN
20780             ELSE
20790               CLSTRUE = .FALSE.
20800               RETURN
```

• The assessment procedure for the cluster formation is conducted according to the criterion similar to Eq. (5.41).

```
20810            END IF
20820          ELSE
20830            IF( SIJ .GE. SIJRHD ) THEN
20840 C
20850 CCCC          OVRLAP = .FALSE.
20860               GOTO 1000
20870            ELSE
20880 C
20890               OVRLAP = .TRUE.
20900               RETURN
20910            END IF
20920          END IF
20930 C
20940 C
20950  1000 CONTINUE
20960                                                                  RETURN
20970                                                                  END
20980 C*********************************************************************
20990 C                                                                   *
21000 C THE FOLLOWING SUBROUTINES ARE FOR CHECKING THE CLUSTER FORMATION    *
21010 C                                                                   *
21020 C*********************************************************************
21030 C**** CLSFORM ****
21040       SUBROUTINE CLSFORM( N )
21050 C
21060       IMPLICIT REAL*8 (A-H,O-Z), INTEGER (I-N)
21070 C
21080       COMMON /BLOCK101/ RX    , RY    , RZ
21090       COMMON /BLOCK102/ EX    , EY    , EZ
21100       COMMON /BLOCK103/ NX    , NY    , NZ
21110       COMMON /BLOCK111/ XL    , YL    , ZL
21120 C
21130       COMMON /BLOCK160/ WIDTHCLS , RCLSLMT
21140       COMMON /BLOCK161/ CLSN   , CLSMX , CLS
21150 C     -------------------------------------------------------------
21160       INTEGER     NN , NNS , TT
21170       PARAMETER( NN=1000 , NNS=5000000 , TT=1000 )
21180       PARAMETER( NRANMX=1000000 , PI=3.14159265358979300 )
21190 C     --------------------------
21200       REAL*8    RX(NN) , RY(NN) , RZ(NN)
21210       REAL*8    EX(NN) , EY(NN) , EZ(NN) , NX(NN) , NY(NN) , NZ(NN)
21220 C     ---------------------------
21230       INTEGER   NDNSMX
21240       INTEGER   CLSN , CLSMX(NN) , CLS(TT,NN)
21250 C     -------------------------------------------------------------
21260       INTEGER   PMX1 , PNUM1(NN)
21270       INTEGER   N1 , PTCL , PTCLMN , IPTCLMN
21280       INTEGER   III, KK , JJ , JJJ , JJE , II1 , JJ1 , ISTREET
21290       REAL*8    RXI, RYI, RZI, NXI, NYI, NZI, EXI, EYI, EZI
21300       REAL*8    ECAN
21310       LOGICAL   OVRLAP, CLSTRUE
21320 C                                    ------------------------------------
21330 C                                     EXAMINATION OF CLUSTER FORMATION
21340 C                                    ------------------------------------
21350       N1   = N + 1
21360       CLSN = 0
21370       PMX1 = N
21380       DO 499 J=1,N
21390         PNUM1(J) = J
21400   499 CONTINUE
21410 C
21420 C
21430       DO 600 III=1,N
21440 C                                 ++++++++++++++++++++++++++++++++++++
21450 C                                  CLSN      : NAME OF CLUSTER (*)
21460 C                                  CLSMX(*)  : NUMBER OF CONSISTING
21470 C                                              PARTICLES OF *-TH CLUSTER
21480 C                                  CLS(**,*): NAME OF CONSISTING
21490 C                                              PARTICLE  OF *-TH CLUSTER
21500 C                                  **        : 1,2,3,...,CLSMX(*)
21510 C                                 ++++++++++++++++++++++++++++++++++++
21520         CLSN          = CLSN + 1
21530         I             = PNUM1(1)
21540         CLSMX(CLSN)   = 1
21550         CLS(1,CLSN)   = I
21560 C                           --- PARTICLE I IS SET AS AN INITIAL PIVOT ---
21570         KK    = 0
21580   500   KK    = KK + 1
21590         PTCL  = CLS(KK,CLSN)
21600 C                               --- PTCL IS THE PRESENT PIVOT ---
```

- The assessment procedure according to Eq. (5.41).

- The main task for assessment of the cluster formation is conducted in the subroutine.

- The particles to be scanned are initially set in the variable PNUM1(*) as PNUM1(J)=J (J=1,2,...,N).

- The first particle to be checked is taken from PNUM1(1) and this particle is I. The present cluster of interest is CLSN.

- The particle to be checked is now PTCL. This particle may form cluster CLSN with particle I.

- In the following procedure, the particles forming cluster CLSN are searched.

```
21610 C
21620       DO 550 J=1, N
21630 C                                         --- ASSESSMENT PROCEDURE ---
21640         IF( J .EQ. PTCL )  GOTO 550
21650 C
21660         DO 530 JJJ=1,CLSMX(CLSN)
21670           IF( J .EQ. CLS(JJJ,CLSN) ) GOTO 550
21680   530   CONTINUE
21690 C
21700         RXI = RX(PTCL)
21710         RYI = RY(PTCL)
21720         RZI = RZ(PTCL)
21730         EXI = EX(PTCL)
21740         EYI = EY(PTCL)
21750         EZI = EZ(PTCL)
21760         NXI = NX(PTCL)
21770         NYI = NY(PTCL)
21780         NZI = NZ(PTCL)
21790 C                                                --- NEW ENERGY ---
21800 C                                   ++ ISTREET=0 FOR FULL CAL    . ++
21810 C                                   ++ ISTREET=1 FOR MAG.ENE.ONLY ++
21820 C                                   ++ ISTREET=2 FOR CLSTR DIST.  ++
21830         ISTREET = 2
21840         CALL ENECAL( PTCL, RXI, RYI, RZI, EXI, EYI, EZI,
21850      &         NXI, NYI, NZI, ECAN, OVRLAP, ISTREET, J, CLSTRUE )
21860
21870 C
21880         IF( CLSTRUE ) THEN
21890           CLSMX(CLSN)            = CLSMX(CLSN) + 1
21900           CLS(CLSMX(CLSN),CLSN) = J
21910         END IF
21920 C
21930   550 CONTINUE
21940 C
21950 C
21960       IF( KK .NE. CLSMX(CLSN) )  GOTO 500
21970 C
21980 C                                      ----- END OF CLSN-CLUSTER -----
21990 C
22000 C                         --- SET THE DATA OF CLS(*,CLSN) IN ORDER ---
22010 C
22020       IF( CLSMX(CLSN) .LE. 2 )   GOTO 580
22030       DO 570 JJ=2,CLSMX(CLSN)-1
22040 C
22050         PTCLMN  = CLS(JJ,CLSN)
22060         IPTCLMN = JJ
22070         DO 560 JJJ=JJ+1,CLSMX(CLSN)
22080           IF( CLS(JJJ,CLSN) .LT. PTCLMN ) THEN
22090             PTCLMN  = CLS(JJJ,CLSN)
22100             IPTCLMN = JJJ
22110           END IF
22120   560   CONTINUE
22130         CLS(IPTCLMN,CLSN) = CLS(JJ,CLSN)
22140         CLS(JJ,CLSN)      = PTCLMN
22150 C
22160   570 CONTINUE
22170 C
22180 C                                     --- UPDATE PNUM1(*) DATA ---
22190   580 JJE  = PMX1
22200       PMX1 = PMX1 - CLSMX(CLSN)
22210       IF( PMX1 .EQ. 0 )  GOTO 700
22220 C
22230       II1 = 1
22240       JJ1 = 0
22250       DO 590 JJ=1,JJE
22260         IF( PNUM1(JJ) .LT. CLS(II1,CLSN) ) THEN
22270           JJ1        = JJ1 + 1
22280           PNUM1(JJ1) = PNUM1(JJ)
22290         ELSE
22300           II1       = II1 + 1
22310           IF( II1 .EQ. CLSMX(CLSN)+1 )   CLS(II1,CLSN) = N1
22320         END IF
22330   590 CONTINUE
22340 C
22350   600 CONTINUE
22360 C
22370   700                                                          RETURN
22380                                                                END
22390 C**** SUB CLSDIST ****
22400       SUBROUTINE CLSDIST( N )
```

• Particle J=PTCL has already been a member of cluster CLSN.

• After the procedure of ENECAL, the logical variable CLSTRUE returns ".TRUE.", if particle J is regarded as being in the range of cluster formation from particle PTCL.

• Particle J is now regarded as a member of cluster CLSN.

• This procedure is a typical routine for setting the data in order; smaller values are saved in earlier variables.

• The particles that have already been scanned are removed from the list of the variable PNUM1(*).

• N1 has to be an arbitrary value greater than N.

• Accumulation of the data of the cluster size distribution for the final averaging procedure in the main program.

```
22410 C
22420       IMPLICIT REAL*8 (A-H,O-Z), INTEGER (I-N)
22430 C
22440       COMMON /BLOCK160/ WIDTHCLS , RCLSLMT
22450       COMMON /BLOCK161/ CLSN    , CLSMX , CLS
22460       COMMON /BLOCK163/ CLSNDIST, ANSNDIST, NDISTSMP, ANSNDSMP
22470       COMMON /BLOCK164/ NDISTMAX, NDISTCNT
22480 C     ----------------------------------------------------------------
22490       INTEGER     NN , NNS , TT
22500       PARAMETER( NN=1000 , NNS=5000000 , TT=1000 )
22510       PARAMETER( NRANMX=1000000 , PI=3.141592653589793D0 )
22520 C     -------------------------
22530       INTEGER  NDNSMX
22540       INTEGER  CLSN , CLSMX(NN) , CLS(TT,NN)
22550 C
22560       REAL     CLSNDIST(NN,10), ANSNDIST(NN)
22570       INTEGER  NDISTSMP(10)    , ANSNDSMP    , NDISTMAX , NDISTCNT
22580 C     ----------------------------------------------------------------
22590       INTEGER  ICLS , NCLSMX
22600 C
22610       NDISTSMP(NDISTCNT) = NDISTSMP(NDISTCNT) + 1
22620       IF( CLSN .LE. 0 )  RETURN
22630 C
22640       DO 1000 ICLS=1, CLSN
22650         NCLSMX = CLSMX(ICLS)
22660         CLSNDIST(NCLSMX,NDISTCNT) = CLSNDIST(NCLSMX,NDISTCNT) + 1.0
22670 C
22680         IF( NCLSMX .GT. NDISTMAX ) NDISTMAX = NCLSMX
22690  1000 CONTINUE
22700                                                                 RETURN
22710                                                                 END
22720 C*********************************************************************
22730 C   THIS SUBROUTINE IS FOR GENERATING UNIFORM RANDOM NUMBERS          *
22740 C    (SINGLE PRECISION) FOR 64-BIT COMPUTER.                          *
22750 C      N     : NUMBER OF RANDOM NUMBERS TO GENERATE                   *
22760 C      IXRAN : INITIAL VALUE OF RANDOM NUMBERS (POSITIVE INTEGER)     *
22770 C            : LAST GENERATED VALUE IS KEPT                           *
22780 C      X(N)  : GENERATED RANDOM NUMBERS (0<X(N)<1)                    *
22790 C*********************************************************************
22800 C**** SUB RANCAL ****
22810       SUBROUTINE RANCAL( N, IXRAN, X )
22820 C
22830       IMPLICIT REAL*8 (A-H,O-Z), INTEGER (I-N)
22840 C
22850       REAL      X(N)
22860       INTEGER INTEGMX, INTEG64, INTEGST, INTEG
22870 C
22880       DATA INTEGMX/2147483647/
22890 CCC   DATA INTEG64/2147483648/
22900       DATA INTEGST,INTEG/584287,48828125/
22910 C
22920       AINTEGMX = REAL( INTEGMX )
22930 CCC   AINTEGMX = REAL( INTEG64 )
22940 C
22950       IF ( IXRAN.LT.0 ) STOP
22960       IF ( IXRAN.EQ.0 ) IXRAN = INTEGST
22970       DO 30 I=1,N
22980    10   IXRAN = IXRAN*INTEG
22990 CCC     IXRAN = KMOD(IXRAN,INTEG64)
23000         IF (IXRAN .EQ. 0 ) THEN
23010           IXRAN = INTEGST
23020           GOTO 10
23030         ELSE IF (IXRAN .LT. 0 ) THEN
23040           IXRAN  = (IXRAN+INTEGMX)+1
23050           X(I)   = REAL(IXRAN)/AINTEGMX
23060         ELSE
23070           X(I)   = REAL(IXRAN)/AINTEGMX
23080         END IF
23090    30 CONTINUE
23100       RETURN
23110       END
```

• The largest cluster is composed of NDISTMAX number of particles.

• Generation of a sequence of pseudo-random numbers ranging from 0 to 1.

• This subroutine is for a 32-bit computer, but simple replacement of "CCC" makes the subroutine for a 64-bit computer.

• By printing out X(*) before starting the present simulation program, it is necessary to confirm that random values ranging from 0 to 1 are surely generated.

• In our experience, the state of IXRAN= 0 happened only once, and therefore, this risk should be removed for any chance. Otherwise, a plenty of time is to be consumed to search for the cause of a sudden divergence of the system at a certain Monte Carlo step.

## 9.2 Brownian dynamics simulations of disk-like particles

In this second example of particle-based simulations, we focus on the Brownian dynamics simulations of a suspension composed of magnetic disk-like particles in a simple shear flow. Different from the former Monte Carlo method, the Brownian dynamics method can investigate the dynamic characteristics of a particle suspension such as the rheological properties and the shear-dependent microstructure of the aggregates. For the case of a suspension composed of spherical particles, it is relatively straightforward to take into account multi-body hydrodynamic interactions among particles because the resistance functions or the mobility functions have already been obtained as mathematical expressions or numerical expressions for the two equal-sized spherical particles. In contrast, these functions are generally quite difficult to derive for a non-spherical particle system, so that only the friction terms between the particles and the ambient liquid may be combined with the basic equations for the Brownian dynamics. This simplified approach of omitting the multi-body hydrodynamic interactions among particles may be basically acceptable and frequently employed as a first approximation in the case of a non-dense suspension of either spherical or non-spherical particles. Moreover, it should be noted that the ordinary Brownian dynamics simulations can simulate the motion of suspended particles in a given linear flow field such as a simple shear flow and an elongational flow. If we have to simulate both the particle motion and the ambient flow field simultaneously, then other particle-based methods such as dissipative particle dynamics and lattice Boltzmann methods may be employed but in these methods it may be difficult to activate the particle Brownian motion at a physically reasonable level. The present ordinary Brownian dynamics method for a non-dense particle suspension requires the explicit analytical expressions for the translational and rotational diffusion or friction coefficients of the disk-like particles. Although the diffusion coefficients have not been derived for disk-like particles, this difficulty may be overcome by representing the disk-like particle as an oblate spheroid, for which both the translational and rotational diffusion coefficients may be expressed in mathematical form. As already pointed out in Section 9.1, another difficulty to be overcome in the employment of Brownian dynamics is treating the repulsive layer, i.e., the steric or electric double layer, covering each suspension particle in an appropriate manner. This treatment is significantly important because a substantially small time interval for performing a simulation must be employed due to a thin repulsive layer in order to obtain physically reasonable results without a system divergence.

Although there is no analytical expression for the interaction energy for the overlap of repulsive layers of our two current disk-like particles, an appropriate modelling in regard to the steric layers may be possible by applying the expression derived for two spherical particles to the present case of disk-like ones, which will be explained soon below.

From this background, in the present section we employ the ordinary Brownian dynamics method in order to discuss the behavior of the disk-like particles of a suspension in a simple shear flow under a uniform applied magnetic field. In this simulation method, the spin rotational Brownian motion about the particle axis of each disk-like particle is also taken into account in addition to the usual rotational and translational Brownian motion. Using the results obtained from the present Brownian dynamics simulation, we are able to discuss the dependence of the microstructure of the particle aggregates on the various factors such as magnetic field strength, the magnetic particle-particle interaction strength and the shear rate of a simple shear flow. As in the previous Monte Carlo simulations, snapshots are used for qualitative discussion and the radial distribution function and the order parameter of the magnetic moments are addressed for quantitative discussion regarding the aggregate structures. We do not address here the magneto-rheological characteristics of a suspension because we are focusing on the microstructure of the aggregates from an educational point of view. Since the magneto-rheological effect is dependent on the formation of aggregates in a complex manner, the first important step for clarifying the magneto-rheological characteristics is to investigate the microstructure of the suspended particles in a suspension under a variety of circumstances.

## 9.2.1 Formalization of the present phenomenon and modeling of magnetic disk-like particles

As shown in Fig. 9.10, we consider the motion of disk-like particles in a simple shear flow in the $x$-axis direction under a uniform applied magnetic field in the $y$-axis direction. The flow field of the present simple shear flow, $\boldsymbol{U}$, is expressed as $\boldsymbol{U} = (\dot{\gamma}y, 0, 0)$ with the shear rate $\dot{\gamma}$. The disk-like particles in the upper and the lower area will flow in the right and the left direction, respectively, whilst performing the translational and rotational Brownian motion in a simple shear flow situation. An external magnetic field $\boldsymbol{H}$ is uniformly applied in the $y$-axis direction, which is normal to the shearing plane of the simple shear flow, as $\boldsymbol{H} = (0, H, 0) = H\boldsymbol{h}$, where $\boldsymbol{h}$ is the unit vector denoting the applied magnetic field direction. A dilute suspension of these particles is considered here, so that only the friction forces and torques due to the interaction between the particles and their ambient fluid are taken into account and these are combined into the basic equation of motion of the particles in addition to the random Brownian forces and torques. These particles may move almost independently or may aggregate to form clusters in the simple shear flow. The formation of clusters will be governed by a variety of factors such as the magnetic field strength, the magnetic particle-particle interaction strength and the shear rate. The dependence of the cluster formation on these factors will be discussed in detail.

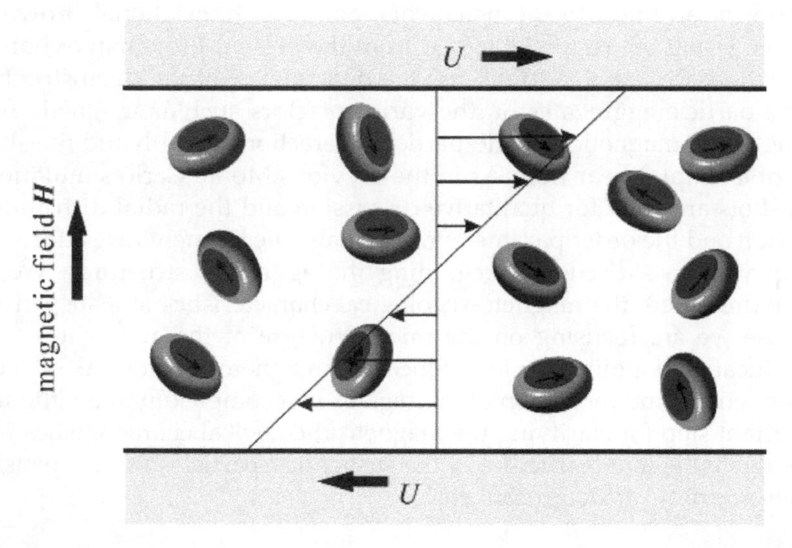


**Figure 9.10.** Brownian motion of disk-like particles in a simple shear flow under a uniform applied magnetic field.

As shown in Fig. 9.11, the magnetic disk-like particles are modeled as a short cylinder with diameter $d$ that is covered by a torus perimeter with the cross section of a semicircle: the diameter and the thickness of the disk-like particle are denoted by symbol $d_1$ and $b_1$, respectively. Here we consider hematite disk-like particles [5] and therefore assume that the particles have a magnetic point dipole moment at the particle center inclining along the plane of the disk and normal to the particle direction. The situation of an arbitrary disk-like particle $i$ is specified by the position vector $\boldsymbol{r}_i$, the unit vector $\boldsymbol{e}_i$ normal to the disk plane, denoting the particle orientation, and the magnetic moment vector $\boldsymbol{m}_i$. The magnetic moment $\boldsymbol{m}_i$ may be expressed as $\boldsymbol{m}_i = m\boldsymbol{n}_i$ using the unit vector $\boldsymbol{n}_i$ denoting the magnetic moment direction. These particles are assumed to be covered by a thin steric layer to prevent the particles from aggregating to form large aggregates that would sediment in the gravitational field. The treatment of the particle interaction due to the steric layers will be explained later in this section. The particle volume $V_p$ for the present disk-like particles shown in Fig. 9.11 is expressed as

$$V_p = \frac{\pi}{4} b_1^3 (r_p - 1)^2 + \frac{\pi^2}{8} b_1^3 (r_p - 1) + \frac{\pi}{6} b_1^3 \tag{9.13}$$

in which $r_p$ is the particle aspect ratio as $r_p = d_1/b_1$. The particle volume is used for specifying the volumetric fraction $\phi_V$ for initiating simulations.

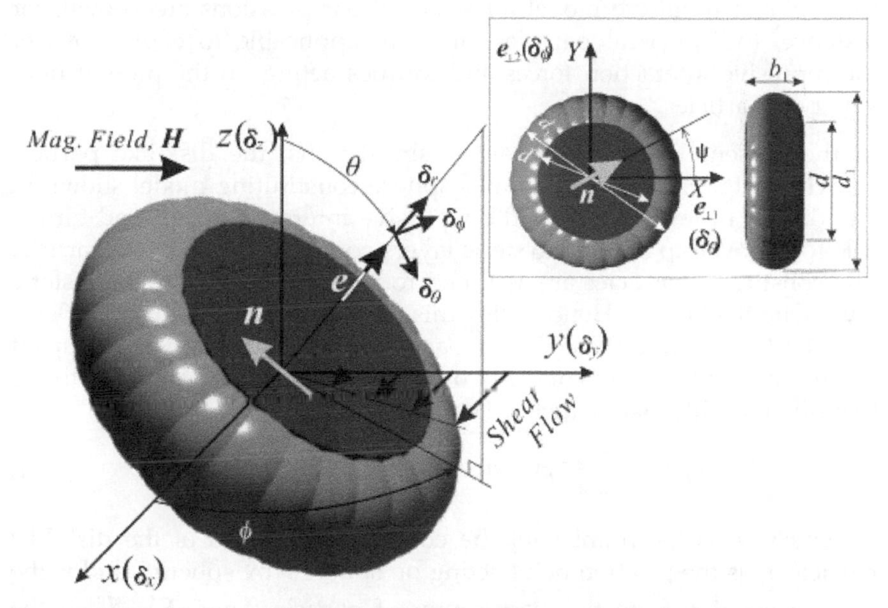


**Figure 9.11.** Model of magnetic disk-like particle with cross section of spherocylinder.

Since we employ the point dipole moment for magnetic properties of the particles, Eqs. (3.5), (3.7) and (3.8) are valid without any modifications for the force $\boldsymbol{F}_{ij}^{(m)*}$ and the torque $\boldsymbol{T}_{ij}^{(m)*}$ acting on particle $i$ by particle $j$, and the torque $\boldsymbol{T}_i^{(H)*}$ due to the applied magnetic field. However, it is noted that these equations are non-dimensionalized by the corresponding representative values based on thermal energy $kT$, and also noted that $\boldsymbol{t}_{ij}$ is the unit vector denoting the direction of the relative position $\boldsymbol{r}_{ij}$ ($= \boldsymbol{r}_i - \boldsymbol{r}_j$), expressed as $\boldsymbol{t}_{ij} = (\boldsymbol{r}_i - \boldsymbol{r}_j)/|\boldsymbol{r}_i - \boldsymbol{r}_j| = \boldsymbol{r}_{ij}/r_{ij}$. The non-dimensional parameters $\xi$ and $\lambda$, which are expressed in Eqs. (3.9), imply the strengths of the magnetic particle-field and the magnetic particle-particle interactions, respectively.

As already pointed out, the modeling of the repulsive layer covering the disk-like particles is indispensable in molecular-dynamics-like simulations. An appropriate choice of the time interval for the overlap between the repulsive layers of the particles will yield a physically reasonable force and torque, which may enable one to simulate a particle suspension without divergence of the system. If the analytical expressions of a force and a torque for the overlap between these repulsive layers are known, no difficult problems will be encountered to develop a simulation code for a molecular-dynamics-like method from a mathematical point of view. However, if these repulsive forces and torques are not able to be expressed as mathematical expressions, then we are necessitated to employ a repulsive model for which the expressions are known, for instance, for spherical particles, and are applicable to expressions of the repulsive interaction forces and torques acting on the present non-spherical particles.

In the present Brownian dynamics simulations, the disk-like particle shown in Fig. 9.11 is modeled as a sphere-constituting model shown in Fig. 3.17 in order to evaluate the repulsive interaction force and torque due to the overlap of the two steric layers coating each disk-like particle. The constituent particles are assumed to be covered by a uniform steric layer with thickness $\delta$. Hence, using this particle model, the total repulsive force $\boldsymbol{F}_{ij}^{(V)}$ and torque $\boldsymbol{T}_{ij}^{(V)}$ acting on particle $i$ by particle $j$ can be evaluated by summing all the pairs of the two constituent spheres belonging to the two different disk-like particles. That is,

$$\boldsymbol{F}_{ij}^{(V)} = \sum_{i_a=1}^{N_p}\sum_{j_b=1}^{N_p} \boldsymbol{F}_{i_a j_b}^{(sphere)}, \quad \boldsymbol{T}_{ij}^{(V)} = \sum_{i_a=1}^{N_p}\sum_{j_b=1}^{N_p} \boldsymbol{r}_{i_a} \times \boldsymbol{F}_{i_a j_b}^{(sphere)} \tag{9.14}$$

in which $N_p$ is the number of the constituent spheres of the disk-like particle, $\boldsymbol{r}_{ia}$ is the position point acting on sphere $i_a$ by sphere $j_b$ and is the relative position from the mass center of particle $i$, and $\boldsymbol{F}_{iajb}^{(sphere)}$ is the repulsive force of particle $i_a$ exerted by particle $j_b$ due to the overlap of the steric layers of the two spheres with the same diameter $b_1$. This expression is written in a similar manner to Eq. (3.6) as

$$\boldsymbol{F}_{i_a j_b}^{(sphere)} = kT\lambda_V \frac{1}{\delta} \boldsymbol{t}_{i_a j_b} \ln\left(\frac{b_1 + 2\delta}{r_{i_a j_b}}\right) \quad (b_1 \le r_{i_a j_b} \le b_1 + 2\delta) \tag{9.15}$$

in which $\lambda_V = \pi b_1^2 n_s/2$ is the non-dimensional parameter representing the strength of the steric interaction, $n_s$ is the number of surfactant molecules per unit area on the surface of the spherical particles, and $\boldsymbol{t}_{iajb}$ is the unit vector given by $t_{iajb} = r_{iajb} / r_{iajb}$, as previously defined, and $\delta$ is the thickness of the steric layer.

Generation of the random translational and rotational displacements in Brownian dynamics method requires explicit expressions for the translational and rotational diffusion coefficients or frictions coefficients of the present disk-like particle. There are certainly mathematical expressions for diffusion coefficients of the sphere and the spheroid, that is, both the prolate and oblate spheroid. However, both the translational and the rotational diffusion coefficients are not known for the present disk-like particles. Hence, as in the previous evaluation of the interaction of the repulsive layers, we are required to employ an appropriate modelling technique for applying the known diffusion coefficients to the present disk-like particle. It seems to be reasonable to use the expressions for the translational and rotational diffusion coefficients, expressed in Eqs. (3.25) and (3.26), for the oblate spheroid by using the correct aspect ratio for the disk-like particles, therefore these equations will be used in the present simulations. The detailed explanation regarding the diffusion coefficients has already been provided in Section 3.4, and therefore we do not add anything further here. It is noted that in the present particle model, we treat the spin Brownian motion about the particle axis line, which is not usually addressed in Brownian dynamics simulations.

### 9.2.2 *Basic equations of translational and rotational Brownian motion*

The basic equations of the Brownian motion of non-spherical (axisymmetric) particles are obtained by considering the translation motion and the spin rotational motion about the particle axis line in addition to the ordinary rotational motion about the line through the particle center and normal to the particle axis direction. The first translational motion gives rise to the equation for the particle position, the second motion yields the equation providing the magnetic moment direction, and the third rotational motion yields the equation providing the particle orientation. We consider here the axisymmetric disk-like particle, and therefore the translational equations can be decomposed into the equations for motion parallel and normal to the particle axis direction. Similarly, the equations providing the particle orientation and the magnetic moment direction correspond to the particle angular velocity, normal and parallel to the particle axis direction, respectively. As already explained sufficiently in Section 6.3, if the inertial terms are negligible, these basic equations are written in equations similar to Eqs. (6.38), (6.39), (6.48) and (6.49) as

$$\boldsymbol{r}_{\|}(t+\Delta t)=\boldsymbol{r}_{\|}(t)+\boldsymbol{U}_{\|}\Delta t+\frac{D_{\|}^{t}}{kT}\boldsymbol{F}_{\|}(t)\Delta t+\Delta r^{B}\boldsymbol{e}(t) \tag{9.16}$$

$$\boldsymbol{r}_{\perp}(t+\Delta t)=\boldsymbol{r}_{\perp}(t)+\boldsymbol{U}_{\perp}\Delta t+\frac{D_{\perp}^{t}}{kT}\boldsymbol{F}_{\perp}(t)\Delta t+\Delta r_{\perp 1}^{B}\boldsymbol{e}_{\perp 1}(t)+\Delta r_{\perp 2}^{B}\boldsymbol{e}_{\perp 2}(t) \tag{9.17}$$

$$\boldsymbol{e}(t+\Delta t)=\boldsymbol{e}(t)+\boldsymbol{\Omega}_{\perp}\times\boldsymbol{e}\Delta t+\frac{D_{\perp}^{r}}{kT}\boldsymbol{T}_{\perp}(t)\times\boldsymbol{e}\Delta t-\frac{Y^{H}}{Y^{C}}\big((\boldsymbol{\varepsilon}\cdot\boldsymbol{ee}):\boldsymbol{E}\big)\times\boldsymbol{e}\Delta t+\Delta\phi_{\perp 1}^{B}\boldsymbol{e}_{\perp 1}+\Delta\phi_{\perp 2}^{B}\boldsymbol{e}_{\perp 2} \tag{9.18}$$

$$\boldsymbol{n}(t+\Delta t)=\boldsymbol{n}(t)+\boldsymbol{\Omega}_{\|}\times\boldsymbol{n}(t)\Delta t+\frac{D_{\|}^{r}}{kT}\boldsymbol{T}_{\|}(t)\times\boldsymbol{n}(t)\Delta t+\Delta\phi_{\|}^{B}\,\boldsymbol{e}(t)\times\boldsymbol{n}(t) \tag{9.19}$$

These equations are valid for the motion of the present axisymmetric particle with a dipole moment normal to the particle axis direction in a simple shear flow situation.

As shown in Fig. 9.11, the vectors $\boldsymbol{e}_{\perp 1}$ and $\boldsymbol{e}_{\perp 2}$ are the unit vectors normal to each other and in the plane normal to the particle axis direction $\boldsymbol{e}$. The random displacements $\Delta r_{\|}^{B}$, $\Delta r_{\perp 1}^{B}$ and $\Delta r_{\perp 2}^{B}$ are related to the translational diffusion coefficients $D_{\|}^{t}$ and $D_{\perp}^{t}$ through the following stochastic characteristics:

$$\langle\Delta r_{\|}^{B}\rangle=\langle\Delta r_{\perp 1}^{B}\rangle=\langle\Delta r_{\perp 2}^{B}\rangle=0,\quad \langle(\Delta r_{\|}^{B})^{2}\rangle=2\Delta t D_{\|}^{t},\quad \langle(\Delta r_{\perp 1}^{B})^{2}\rangle=\langle(\Delta r_{\perp 2}^{B})^{2}\rangle=2\Delta t D_{\perp}^{t} \tag{9.20}$$

Similarly, the rotational random displacements $\Delta\phi_{\|}^{B}$, $\Delta\phi_{\perp 1}^{B}$ and $\Delta\phi_{\perp 2}^{B}$ are related to the rotational diffusion coefficients $D_{\|}^{r}$ and $D_{\perp}^{r}$ as

$$\langle\Delta\phi_{\|}^{B}\rangle=\langle\Delta\phi_{\perp 1}^{B}\rangle=\langle\Delta\phi_{\perp 2}^{B}\rangle=0,\quad \langle(\Delta\phi_{\|}^{B})^{2}\rangle=2\Delta t D_{\|}^{r},\quad \langle(\Delta\phi_{\perp 1}^{B})^{2}\rangle=\langle(\Delta\phi_{\perp 2}^{B})^{2}\rangle=2\Delta t D_{\perp}^{r} \tag{9.21}$$

As already pointed out, since the translational and rotational diffusion coefficients are not known for the present disk-like particle, we employ the diffusion coefficients for the oblate spheroid with the same particle aspect ratio $r_p$, which are already available as analytical expressions. The translational diffusion coefficients $D_{\|}^{t}$ and $D_{\perp}^{t}$ and the rotational diffusion coefficients $D_{\|}^{r}$ and $D_{\perp}^{r}$ have already been written in Eqs. (3.25) and (3.26) for the oblate spheroid. The ratio of the resistance functions, $Y^H/Y^C$, in Eq. (9.18) is expressed as

$$\frac{Y^H}{Y^C} = -\frac{\hat{s}^2}{2-\hat{s}^2} \tag{9.22}$$

Finally, it is noted that the quantities parallel and normal to the particle axis direction appearing in Eqs. (9.16) to (9.19) are shown in Eq. (6.51).

### 9.2.3 Normalization and non-dimensional parameters characterizing the phenomenon

Non-dimensionalization of the equations gives rise to the non-dimensional system parameters that characterize the physical phenomenon of interest. Dimensionless quantities shown below are denoted by superscript "*". The following representative values are used for the non-dimensionalization procedure: the particle thickness $b_1$ for distances, $1/\dot{\gamma}$ for time, $\dot{\gamma}b_1$ for velocities, $\dot{\gamma}$ for angular velocities, $3\pi\eta_s\dot{\gamma}b_1^2$ for forces, $\pi\eta_s\dot{\gamma}b_1^3$ for torques, $3\pi\eta_s\dot{\gamma}$ for stresses, the base liquid viscosity $\eta_s$ for viscosity, $kT/(3\pi\eta_s b_1)$ for translational diffusion coefficients, and $kT/(\pi\eta_s b_1{}^3)$ for rotational diffusion coefficients. Employing these representative values, non-dimensionalization of the basic equations gives rise to the following four non-dimensional parameters:

$$Pe_0 = \frac{\pi\eta_s\dot{\gamma}b_1^3}{kT}, \quad R_m = \frac{\mu_0 m^2}{4\pi^2\eta_s\dot{\gamma}b_1^6}, \quad R_V = \frac{kT\lambda_V}{3\pi\eta_s\dot{\gamma}b_1^3(\delta/b_1)}, \quad R_H = \frac{\mu_0 mH}{\pi\eta_s\dot{\gamma}b_1^3} \tag{9.23}$$

in which $\delta$ is the thickness of the steric layer as defined before, $Pe_0$ implies the strength of the influence of the viscous shear force relative to that of the rotational Brownian motion, $R_m$ and $R_V$ imply the strengths of the magnetic particle-particle force and the steric repulsive force relative to the viscous shear force, respectively, and $R_H$ implies the strength of the magnetic torque due to the particle-field interaction relative to the viscous shear torque. However, the following non-dimensional parameters may be useful for discussing the simulation results of a magnetic particle suspension:

$$\lambda = \frac{\mu_0 m^2}{4\pi b_1^3 kT}, \quad \xi = \frac{\mu_0 mH}{kT}, \quad \lambda_V = \frac{\pi b_1^2 n_s}{2} \tag{9.24}$$

in which $\lambda$ and $\xi$ imply the strengths of the magnetic particle-particle and the magnetic particle-field interaction relative to the thermal motion, respectively, and $\lambda_V$ implies the steric interaction strength relative to the thermal motion. The former parameters $R_m$, $R_V$ and $R_H$ are related to $\lambda$, $\lambda_V$, $\xi$ and $Pe_0$ as

$$R_m = \lambda / Pe_0, \quad R_V = \frac{\lambda_V}{3Pe_0(\delta / b_1)}, \quad R_H = \xi / Pe_0 \tag{9.25}$$

### 9.2.4 *Calculation of the repulsive interactions due to the steric layers and treatment of the particle overlap*

As described in Section 9.2.1, the interaction energy due to the overlap of the steric layers is evaluated using the sphere-constituting particle model shown in Fig. 3.17. In Section 3.4, the detailed explanation was made regarding the concept of this modelling technique for the evaluation of the steric interaction between two disk-like particles. In the present section, we mainly address the concrete treatment of the sphere-constituting particle model, which may be a useful aid for the reader in understanding the implementation of the calculation of the steric interaction forces and torques in the sample simulation program shown later. We consider the overlap between the two sphere-constituting particles $i$ and $j$, assumed to be the lower and the upper particles shown in Fig. 3.17. Several symbols used in the following are defined in Fig. 9.11. Each particle of the two sphere-connecting particles of interest has its own body-fixed two-dimensional coordinate system $XY$: the origins of these coordinate systems are set as the positions, shown as the darker particles in Fig. 3.17, which give rise to the minimum distance between the two disk-like particles. The positions giving rise to the minimum distance are denoted by $\boldsymbol{r}_i^{p(mn)}$ and $\boldsymbol{r}_j^{p(mn)}$ for particle $i$ and particle $j$, respectively. The positions $\boldsymbol{r}_j^{p(mn)}$ is used as the position of the criterion or the first constituent sphere, shown as sphere 1 in Fig. 9.12. The other constituent spheres are then added and located as shown in Fig 9.12 in close contact along the $Y$-axis direction, which is denoted by the unit vector $\boldsymbol{\delta}_j^{pY}$ in the absolute coordinate system, so as to construct a sphere-constituting particle. The lower disk-like particle $i$ in Fig. 3.17 is constructed in a similar manner, but in this case it is noted that the first criterion sphere is set at the position $\boldsymbol{r}_i^{p(mn)}$ and the other particles are added in the corresponding direction $\boldsymbol{\delta}_i^{pY}$ to construct the sphere-constituting particle for particle $i$. In the case of the configuration shown in Fig. 3.17, the unit vectors $\boldsymbol{\delta}_i^{pY}$ and $\boldsymbol{\delta}_j^{pY}$ are equal to the unit vectors $\boldsymbol{e}_i^s$ and $-\boldsymbol{e}_j^s$, respectively, defined by Eq. (5.27) and Fig. 5.5, and it is noted that these unit vectors are normal to the line of the intersection of the two planes to which the two disk-like particles belong. Each constituent sphere is assumed to be coated by a uniform steric layer with the thickness $\delta$. Hence, using the sphere-constituting particle, the interaction force and torque in Eq. (9.14) is evaluated by summing all the pairs of the corresponding constituent spheres in the two disk-like particles. The above discussion has been made under the assumption that the positions $\boldsymbol{r}_i^{p(mn)}$ and $\boldsymbol{r}_j^{p(mn)}$, giving rise to the minimum distance, for particles $i$ and $j$, respectively, have already been evaluated. In the following, therefore, we describe the concrete method to find the positions giving rise to the minimum distance, which has already been mentioned in the latter part of Section 5.3

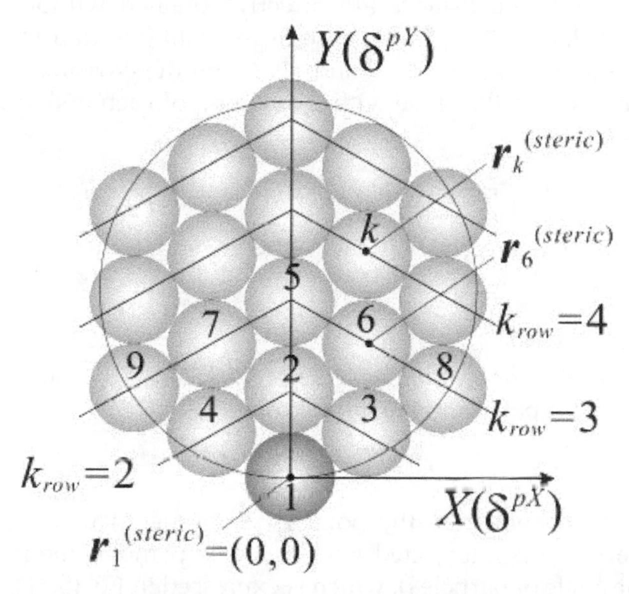


**Figure 9.12.** Modelling of the disk-like particle as a sphere-constituting particle for evaluation of the steric interaction due to the overlap of the steric layers of the two disk-like particles: each sphere is assumed to be coated with a uniform steric layer.

The present method employed to find the positions giving rise to the minimum distance between the two disk-like particles is made up of two procedures. In the first procedure, a guess is made regarding the approximate positions $\boldsymbol{r}_i^{p(orgn)}$ and $\boldsymbol{r}_j^{p(orgn)}$, for the given configuration of the two disk-like particles, which are near to the exact positions, and then these positions are used as the starting positions for the second procedure. In the second procedure, the exact positions $\boldsymbol{r}_i^{p(mn)}$ and $\boldsymbol{r}_j^{p(mn)}$, giving rise to the minimum distance, are evaluated using a continuous scanning method around the above starting positions $\boldsymbol{r}_i^{p(orgn)}$ and $\boldsymbol{r}_j^{p(orgn)}$, by narrowing (shortening) the scanning area over a period of several searching steps, as described in Section 5.3. We explain these two main procedures in detail in the following.

In the case of the configuration in Fig, 5.6(c), the positions $\boldsymbol{r}_i^{p(mn)}$ and $\boldsymbol{r}_j^{p(mn)}$ are given as

$$\boldsymbol{r}_i^{p(mn)} = \boldsymbol{r}_{i(j)}^{Q} = \boldsymbol{r}_j^{p(mn)} - k_{i(j)}^{Q}\,\boldsymbol{e}_i,\quad \boldsymbol{r}_j^{p(mn)} = \boldsymbol{r}_j + (d/2)\,\boldsymbol{e}_j^{s} \tag{9.26}$$

in which $\boldsymbol{r}_{i(j)}{}^{Q}$, $k_{i(j)}{}^{Q}$ and $\boldsymbol{e}_j^{s}$ are expressed in Eqs. (5.31), (5.32) and (5.27), respectively; if undefined quantities are used in the following, descriptions in Section 5.3 should be referred to. For this case, the positions giving rise to the minimum distance are directly obtained without using the positions $\boldsymbol{r}_i^{p(orgn)}$ and $\boldsymbol{r}_j^{p(orgn)}$. In the case of the configuration in Fig. 5.6(a), the approximate positions $\boldsymbol{r}_i^{p(orgn)}$ and $\boldsymbol{r}_j^{p(orgn)}$ are the crossing point of the intersection line and the circle with diameter $d$ of each disk-like particle, expressed as

$$\boldsymbol{r}_i^{p(orgn)} = \boldsymbol{r}_i + k_i^{s}\boldsymbol{e}_i^{s} - h_i^{Q}\,\boldsymbol{t}_{ij}^{s},\quad \boldsymbol{r}_j^{p(orgn)} = \boldsymbol{r}_j + k_j^{s}\boldsymbol{e}_j^{s} + h_j^{Q}\,\boldsymbol{t}_{ij}^{s} \tag{9.27}$$

in which $h_i^{Q}$ and $h_j^{Q}$ are given by the following equations:

$$h_i^{Q} = \sqrt{(d/2)^2 - (k_i^{s})^2},\quad h_j^{Q} = \sqrt{(d/2)^2 - (k_j^{s})^2} \tag{9.28}$$

In the cases of the configurations in Figs. 5.6(b) and (d), the approximate position $\boldsymbol{r}_j^{p(orgn)}$ for particle $j$ is first determined as

$$\boldsymbol{r}_j^{p(orgn)} = \boldsymbol{r}_j + (d/2)\,\boldsymbol{e}_j^{s} \tag{9.29}$$

Then, the line drawn from the point $\boldsymbol{r}_j^{p(orgn)}$ perpendicularly to the disk plane of particle $i$ is intersected with this disk plane as the position $\boldsymbol{r}_{i(j)}{}^{Q}$ (outside the circle of particle $i$), which is expressed in Eq. (5.31). Using this position $\boldsymbol{r}_{i(j)}{}^{Q}$, the approximate position $\boldsymbol{r}_i^{p(orgn)}$ for particle $i$ is regarded as

$$\boldsymbol{r}_i^{p(orgn)} = \boldsymbol{r}_i + h_i^{Q}(\boldsymbol{r}_{i(j)}^{Q} - \boldsymbol{r}_i) \tag{9.30}$$

in which $h_i^{Q}$ is given by the following equation:

$$h_i^{Q} = \frac{d}{2}\cdot\frac{1}{|\boldsymbol{r}_{i(j)}^{Q} - \boldsymbol{r}_i|} \tag{9.31}$$

If the two disk-like planes are in the same plane, the exact positions $\boldsymbol{r}_i^{p(mn)}$ and $\boldsymbol{r}_j^{p(mn)}$ are straightforwardly obtained as

$$\boldsymbol{r}_i^{p(mn)} = \boldsymbol{r}_i - (d/2)\boldsymbol{t}_{ij},\quad \boldsymbol{r}_j^{p(mn)} = \boldsymbol{r}_j + (d/2)\boldsymbol{t}_{ij} \tag{9.32}$$

in which in this case $\boldsymbol{t}_{ij} = (\boldsymbol{r}_i - \boldsymbol{r}_j)/|\boldsymbol{r}_i - \boldsymbol{r}_j|$. Similarly, in the case of the two disk-like particles being in the two parallel planes, the important position is the center of these two parallel particles. The projected point of this center position on the disk plane of particles $i$ and $j$ are denoted by $\boldsymbol{r}_i^p$ and $\boldsymbol{r}_j^p$, respectively. The position $\boldsymbol{r}_j^{p(mn)}$ is straightforwardly searched for from this point as the nearest position to the outer circle of particle $j$ in the direction $(\boldsymbol{r}_j^p - \boldsymbol{r}_j)$; then, the position $\boldsymbol{r}_i^{p(mn)}$ is obtained from this point as $\boldsymbol{r}_i^{p(mn)} = \boldsymbol{r}_i^p + (\boldsymbol{r}_j^{p(mn)} - \boldsymbol{r}_j)$.

Using the approximate or starting positions $\boldsymbol{r}_i^{p(orgn)}$ and $\boldsymbol{r}_j^{p(orgn)}$ which have been obtained above in the first main procedure, we are now ready to explain the procedure to search for the positions $\boldsymbol{r}_i^{p(mn)}$ and $\boldsymbol{r}_j^{p(mn)}$ that give rise to the minimum distance. Since the concept of this procedure based on scanning technique using quasi-random numbers is sufficiently explained in Section 5.3, in the present section we show how to implement the scanning procedure for these positions. The solution should be reasonably close to the approximate positions $\boldsymbol{r}_i^{p(orgn)}$ and $\boldsymbol{r}_j^{p(orgn)}$, and therefore, it is reasonable to scan in the vicinity of the neighboring region to find the exact positions $\boldsymbol{r}_i^{p(mn)}$ and $\boldsymbol{r}_j^{p(mn)}$. Since the scanning technique is the same for particles $i$ and $j$, we concentrate on the treatment for particle $i$. An arbitrary position around the point $\boldsymbol{r}_i^{p(orgn)}$ is to be taken on the line of the circle with diameter $d$, so that the unit vector $\boldsymbol{n}_i^Q$ that is tangential to the circle is convenient for specifying the neighboring points. This unit vector $\boldsymbol{n}_i^Q$ is obtained with Eq. (5.33), where vector $\boldsymbol{r}_i^Q$ is equal to $\boldsymbol{r}_i^{p(orgn)}$. If the scanning range is taken within the range $\pm\beta$ in angle on the circle line, then an arbitrary neighboring point $\boldsymbol{r}_{i(try)}^{(min)}$ is specified using a random number $R$, ranging from zero to unity, as Eq. (5.35), where $r_0$ denotes $d/2$. This scanning procedure is conducted to obtain ten neighboring points. The point giving rise to the minimum distance between $\boldsymbol{r}_{i(try)}^{(min)}$ and $\boldsymbol{r}_{j(try)}^{(min)}$ is regarded as the position $\boldsymbol{r}_i^{p(orgn)}$, which is then used in the next scanning stage with a new narrow angle range of $\pm\beta$. The new angle $\beta$ specifying the scanning range is specified using the first and third positions that yield the minimum and the third minimum distances. The second scanning stage is conducted to determine the position $\boldsymbol{r}_i^{p(mn)}$ and $\boldsymbol{r}_j^{p(mn)}$, which give rise to the minimum distance in the second scanning stage. A new narrow angle range is also specified using the first and the third positions that yield the minimum and the third minimum distances. The third scanning stage is conducted in a similar manner. It is, therefore, seen that the positions yielding the minimum distance of the disk-like particles is determined after the three stages of scanning procedure where each stage is made up of ten scanning attempts in the present simulations.

### 9.2.5 Modification of the magnetic moment direction in relation to the change in the particle direction

If, as is usual, the disk-like particle is rotated about a line normal to the particle axis direction, the magnetic moment direction, which is normal to the particle axis, will be automatically rotated due to the particle rotation. In the following, we discuss the modification of the magnetic moment direction, which must be taken into account due to the particle rotation. If the particle direction $\boldsymbol{e}(t)$ is changed to $\boldsymbol{e}(t + \Delta t)$ during the time interval $\Delta t$, the line about which the particle is rotated is expressed by the unit vector $\boldsymbol{\delta}_{\perp}$ written as

$$\boldsymbol{\delta}_{\perp} = \frac{\boldsymbol{e}(t)\times(\boldsymbol{e}(t+\Delta t)-\boldsymbol{e}(t))}{|\boldsymbol{e}(t)\times(\boldsymbol{e}(t+\Delta t)-\boldsymbol{e}(t))|} \tag{9.33}$$

For simplification of analysis, the magnetic moment direction $\boldsymbol{n}$ is decomposed into the vector components $\boldsymbol{n}_h$ and $\boldsymbol{n}_v$ in the direction normal and parallel to the unit vector $\boldsymbol{\delta}_{\perp}$, respectively. The vectors are expressed as

$$\boldsymbol{n}_v = (\boldsymbol{n}\cdot\boldsymbol{\delta}_{\perp})\boldsymbol{\delta}_{\perp}, \quad \boldsymbol{n}_h = \boldsymbol{n}-\boldsymbol{n}_v \tag{9.34}$$

Only the vector component $\boldsymbol{n}_h$ will be automatically rotated due to the particle rotation. The new vector component $\boldsymbol{n}'_h$ is obtained as

$$\boldsymbol{n}'_h = \boldsymbol{n}_h + |\boldsymbol{e}(t+\Delta t)-\boldsymbol{e}(t)|(\boldsymbol{\delta}_{\perp}\times\boldsymbol{n}_h) \tag{9.35}$$

Hence, the magnetic moment direction $\boldsymbol{n}'$ after the modification is obtained as

$$\boldsymbol{n}' = \boldsymbol{n}_v + \boldsymbol{n}'_h \tag{9.36}$$

Since the vector $\boldsymbol{n}'$ does not necessarily satisfy the requirement of a unit length, the normalization procedure may be necessary to obtain the final solution of $\boldsymbol{n}'$.

### 9.2.6 Parameters for simulations

Unless specifically noted, we adopted the following values for performing the present simulations. The volumetric fraction $\phi_V$ is defined using the volume of the solid body of the disk-like particle and is set as $\phi_V = 0.1$, the number of particles $N$ is $N = 135$, the thickness of the steric layer is $\delta^* = \delta/b_1 = 0.15$, the repulsive interaction is $\lambda_V = 150$, and the cutoff radius for calculating particle-particle interactions is $r_{coff}^* = r_{coff}/b_1 \simeq 3r_p$. The particle aspect ratio $r_p$ is defined as $r_p = d_1/b_1$ and restricted to $r_p = 5$. The non-dimensional parameters $Pe = Pe_0/D_\perp^{r*}$, $\xi$ and $\lambda$ are set within the wide ranges of $Pe = 1 \sim 10$, $\xi = 0 \sim 20$ and $\lambda = 20 \sim 50$. The total number of time steps per one simulation run, $N_{timemx}$, is taken as $N_{timemx} = 1{,}000{,}000$ and the last 80% of the data were used for the averaging procedure. It is noted that for a case with $Pe >> 1$, $\lambda >> 1$ and $\xi >> 1$, the shear flow, the magnetic particle-particle interaction and the applied magnetic field respectively are all more dominant than the Brownian motion. In a converse case with $Pe \lesssim 1$, $\lambda \lesssim 1$ and $\xi \lesssim 1$, the Brownian motion is more dominant than each of the respective factors.

The present simulations take into account both the translational and rotational Brownian motion, so that unless a sufficiently small time interval is used, an unreasonable overlap of the particles is likely to occur, which will induce an instability of the system due to a large repulsive interaction. Hence, from preliminary simulations, we adopted a sufficiently small time interval $\Delta t^*$ as $\Delta t^* = 5 \times 10^{-6}$. On the other hand, an extraordinary small time interval leads to a loss of precision during the calculation on a computer. In order to overcome this problem of a loss of precision during calculation, variables with quadruple precision may be employed in certain situations.

### 9.2.7 Results and discussion

#### (a) Dependence of aggregate structures on the particle-particle interaction strength

Figure 9.13 shows snapshots for the case of a weak shear flow $Pe = 1$ and a weak magnetic field $\xi = 1$ for the two different cases of the magnetic particle-particle interaction strength (a) $\lambda = 20$ and (b) $\lambda = 50$. In the case of a weak magnetic interaction, shown in Fig. 9.13(a), disk-like particles neither aggregate nor incline in a specific favored direction. As the magnetic interaction is increased to $\lambda = 50$, as shown in Fig. 9.13(b), disk-like particles aggregate to form long and stable column-like clusters but these clusters do not exhibit a specific orientational characteristic. In the case of a suspension composed of magnetic spherical particles, the value of $\lambda = 5$ is sufficient for the formation of stable clusters, but in the present disk-like particle suspension, stable clusters are generally not formed even in the strong interaction case $\lambda = 20$. This is because the rotational Brownian motion of the disk-like particles has a significant influence on the stability of column-like clusters. In other words, a much larger magnetic interaction strength is necessary for the particles to overcome the disturbance of the Brownian motion to form stable clusters. The threshold of this interaction strength is expected to become larger as the aspect ratio of the disk-like particles increases, and therefore we understand that the value of $\lambda = 20$ is not sufficiently strong for the formation of stable column-like clusters for the present aspect ratio of $r_p = 5$. In contrast, the value of $\lambda = 50$ is seen to be sufficiently strong for the cluster formation. A threshold value of the magnetic interaction strength for stable cluster formation may be predicted by considering the results of such quantitative features as the radial distribution function and the cluster size distribution.

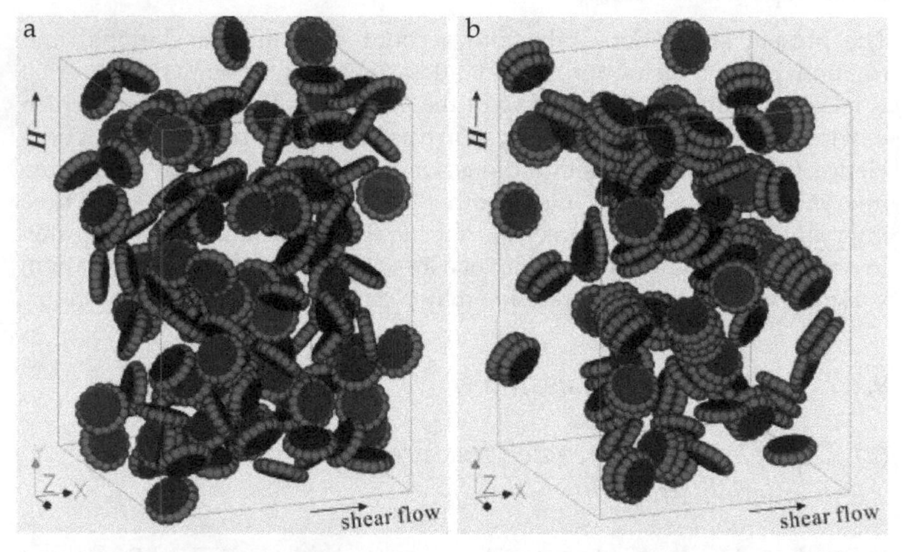


**Figure 9.13.** Appearance of column-like clusters of disk-like particles with increasing magnetic particle-particle interaction strength $\lambda$ in a weak magnetic field $\xi = 1$ and a weak shear flow $Pe = 1$: (a) $\lambda = 20$ and (b) $\lambda = 50$.

Figure 9.14 shows the radial distribution function for the same conditions of the previous case shown in Fig. 9.13 with the results of four different cases of the magnetic interaction strength, $\lambda = 20$, 30, 40 and 50. The case of $\lambda = 20$ exhibits characteristics that are different from the other cases because no peak appears at $r^* (= r/b_1) \simeq 1.3$ and from $r^* \simeq 4$ the distribution function monotonically decreases and approaches zero with decreasing radial distance. This is because the rotational Brownian motion restricts the approaching areas of the neighboring particles to the particle of interest more severely in the nearer vicinity areas. Hence, there is no cluster formation in the system for $\lambda = 20$, as shown in Fig. 9.13(a), and this is supported by the characteristics of the radial distribution function. In the other cases, a significant peak appears at $r^* \simeq 1.3$ and this peak becomes higher with increasing magnetic particle interaction strength, giving rise to 12 and 60 in height for $\lambda = 40$ and 50, respectively. Moreover, for the case of $\lambda = 50$ a second high peak and a third peak also arise at $r^* \simeq 2.7$ and 4.0, and the appearance of these peaks at the positions of approximately an integer times the particle thickness including a steric layer implies the formation of the column-like clusters shown in Fig. 9.13(b). The curve for $\lambda = 30$ has no second and third peaks, although a low peak appears at $r^*$ $\simeq 1.3$. From these characteristics of the radial distribution function for the different magnetic interaction strengths, it is expected that for the present disk-like particle with aspect ratio $r_p = 5$, stable column-like clusters start to be formed from the magnetic interaction strength $\lambda \simeq 30$ in the situation of a weak shear flow and a weak applied magnetic field. The value of this threshold magnetic interaction strength is, of course, dependent on the particle aspect ratio, the external magnetic field strength, the shear rate (Peclet number), and so forth.

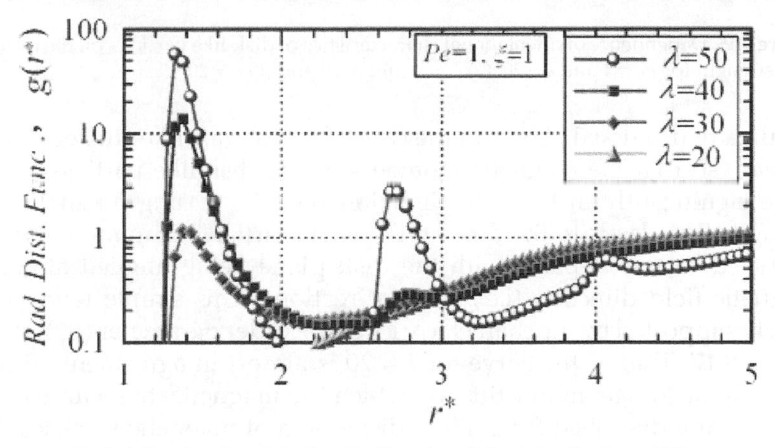


**Figure 9.14.** Radial distribution function for $Pe = 1$ and $\xi = 1$ for the four different cases of $\lambda$ = 20, 30, 40 and 50.

## (b) Dependence of aggregate structures on the magnetic field strength

Next, we consider the dependence of the aggregate structures on the magnetic field strength for a weak shear flow case $Pe = 1$: a weak shear flow enables us to capture the influence of the field strength more clearly without a significant effect arising from the shear flow. Figure 9.15 shows results of the snapshots for the magnetic interaction strength $\lambda = 20$: the three different snapshots for $\xi = 5$, 10 and 20 are shown in Figs. 9.15(a), (b) and (c), respectively. Similar results are shown in Fig. 9.16 for the strong magnetic interaction strength $\lambda = 50$. Figure 9.17 shows the order parameter regarding the alignment of the magnetic moments to the field direction, $\langle n_y \rangle$, defined by Eq. (8.15).

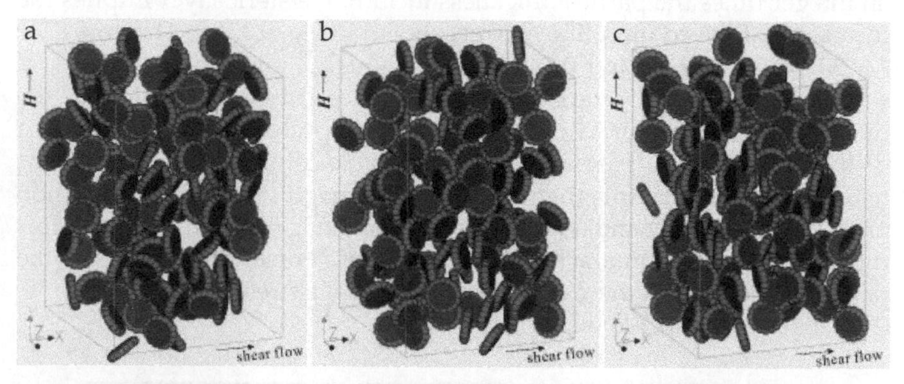


**Figure 9.15.** Dependence of orientational characteristics of disk-like particles on the magnetic field strength for $Pe = 1$ and $\lambda = 20$: (a) $\xi = 5$, (b) $\xi = 10$ and (c) $\xi = 20$.

As already discussed above, in the case of $\lambda = 20$, raft-like clusters are not formed, so that the magnetic moments of the disk-like particles incline more significantly in the field direction with increasing magnetic field strength. This leads to the direction of each particle being more strongly restricted to the $xy$-plane with the disk plane being aligned along the magnetic field direction (i.e., $y$-axis direction). This simple tendency is clearly supported by the characteristics of the order parameter $\langle n_y \rangle$, shown in Fig. 9.17. That is, the curve for $\lambda = 20$ is almost in agreement with the curve of the Langevin function, by which the magnetization curve can be theoretically described for a dilute dispersion of magnetic particles. This clearly implies that there are no clusters in the system and the particles move almost independently without significant influence from magnetic particle-particle interactions.

We now discuss the results for $\lambda = 50$ shown in Fig. 9.16. In the case of $\xi = 5$, the stable column-like clusters observed in Fig. 9.13(b) still survive and are not restricted to the magnetic field direction in orientation. In the case of $\xi = 10$, the influence of the magnetic field appears to a considerable degree, that is, many raft-like clusters are dissociated and the magnetic moments of the particles are more significantly restricted to the field direction. For the stronger magnetic field $\xi = 20$, shown in Fig. 9.16(c), any remaining column-like clusters completely disappear and the magnetic moments of the particles strongly align in the magnetic field direction. This is because the alignment of the magnetic moments to the same direction (i.e., to the magnetic field direction) induces a repulsive force between the neighboring particles in a cluster, which leads to the dissociation of the column-like clusters, as shown in Fig. 9.16(c). The order parameter $\langle n_y \rangle$ for $\lambda = 50$, shown in Fig. 9.17, clearly shows a significantly different feature from that for $\lambda = 20$. It is quite reasonable that the result for $\lambda = 50$ approaches the curve for $\lambda = 20$ in the stronger field areas with increasing magnetic field strength, because column-like clusters dissociate to give rise to a snapshot such as in Fig. 9.16(c). In the range of the field strength smaller than $\xi \simeq 10$, the values of the order parameter are much smaller than those for $\lambda = 20$, and do not exhibit the same simple monotonic increase, but seem to have a significant fluctuation in the present case between $\xi = 6$ and 8. These characteristics of the curve for $\lambda = 50$ reflect how stable and large column-like clusters are formed in the system. If the large stable column-like clusters are formed, the magnetic moments of the neighboring particles in the clusters incline in the opposite directions to each other and this leads to a lower value of the alignment of the magnetic moments to the field direction that is reflected by a smaller value of the order parameter $\langle n_y \rangle$. The second characteristic of the significant fluctuation between $\xi = 6$ and 8 is due to the size of the column-like clusters, which is strongly influenced by a variety of factors such as the shear rate, the magnetic field strength and the magnetic interaction strength. In particular, as discussed above, the increase in the field strength induces an instability in the microstructure of the column-like clusters.

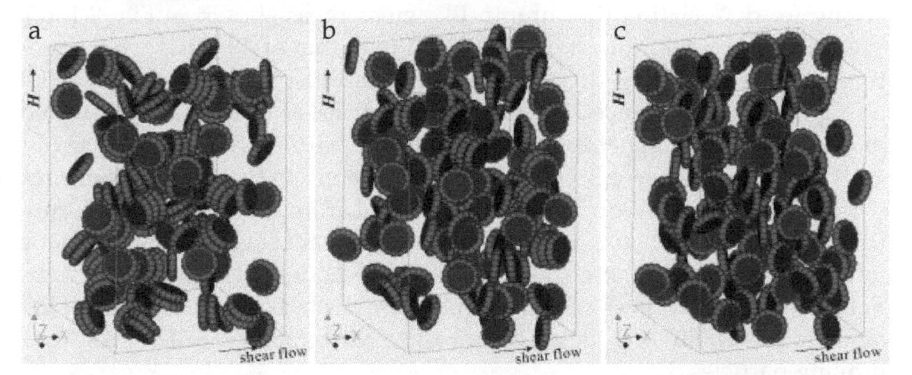


**Figure 9.16.** Collapse of column-like clusters with increasing magnetic field strength for $Pe = 1$ and $\lambda = 50$: (a) $\xi = 5$, (b) $\xi = 10$ and (c) $\xi = 20$.

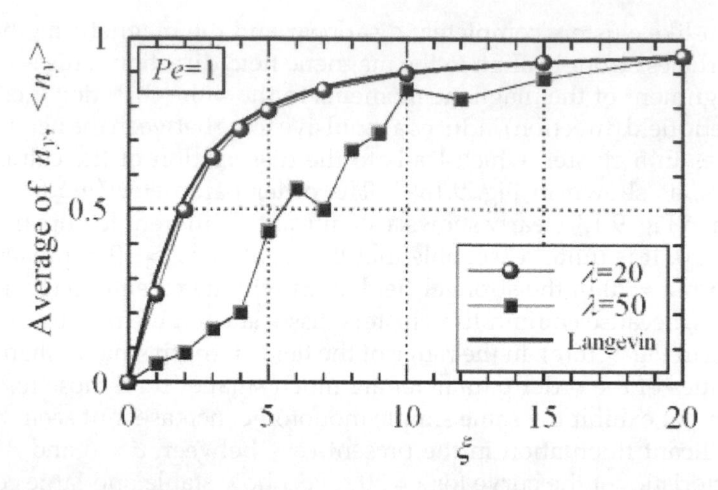


**Figure 9.17.** Characteristics of the alignment of magnetic moments to the magnetic field direction as a function of the magnetic field strength: the average of the $y$-component $n_y$ of the magnetic moment direction $\boldsymbol{n}$.

Finally, we discuss the threshold of the magnetic field strength for the phase change of the column-like clusters returning to single-moving particles, by means of the radial distribution function shown in Fig. 9.18. As illustrated in Fig. 9.18, the sharp peaks that appear at the positions of an integer times the particle thickness including the steric layer, in the present case, at $r^* \simeq 1.3, 2.6. 3.9, \ldots$, implies that long and stable column-like clusters are formed in the system. There is no peak even at $r^* \simeq 1.3$ in the case of $\xi = 20$, which implies that this field strength is sufficiently strong for dissociating the column-like clusters, as shown in Fig. 9.16(c). The curve of $\xi = 5$ is almost in agreement with that of $\xi = 1$, which implies that at these field strengths the stable column-like structures should not be significantly influenced by the magnetic field, as clearly demonstrated by the snapshot in Fig. 9.16(a). In the intermediary field strength case of $\xi = 10$, the first and second peaks have become much lower than those for $\xi = 1$ and 5, but they do not disappear, which implies that the column-like clusters are still able to survive at a significant level in this magnetic field situation. From these characteristics, we understand that a magnetic field strength between $\xi = 5$ and 10 induces the change in the phase of the aggregates structures from the column-like clusters into numerous single-moving particles.

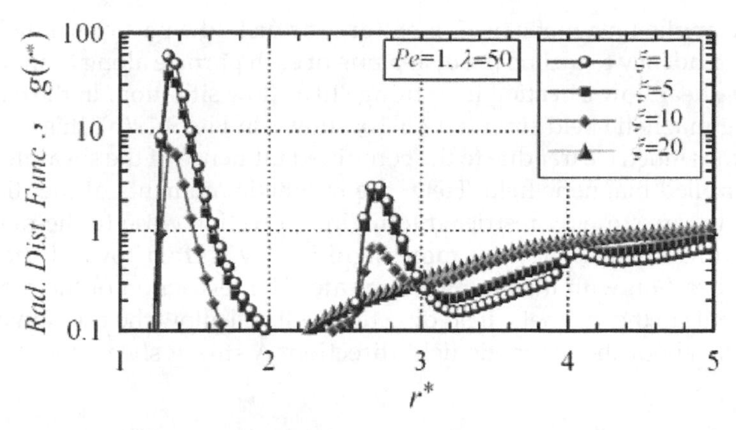


**Figure 9.18.** Radial distribution function for $Pe = 1$ and $\lambda = 50$ for the four different cases of $\xi = 1, 5, 10$ and 20.

## (c) Dependence of aggregate structures on the shear rate

We discuss the effect of the shear rate (Peclet number) on the behavior of the disk-like particles for the two typical magnetic interaction strengths $\lambda = 20$ and 50. Since we previously showed and discussed the results for a weak shear flow case $Pe = 1$, we focus here on results for a strong shear flow $Pe = 10$; the comparison between the present results and those above may enable us to clarify the effect of the shear flow strength on the behavior of the particles and the formation of the column-like clusters. Figure 9.19 shows the results in the form of snapshots for a weak magnetic particle interaction strength $\lambda = 20$, where raft-like clusters are not formed and the disk-like particles move independently. In the present situation of $\lambda = 20$, therefore, we will discuss the effect of the shear rate on the behavior of the single-moving particles. Figures 9.19(a) and (b) are for a weak $\xi = 1$ and a relatively strong magnetic field strength $\xi = 10$, respectively, where each figure has a snapshot of the front view and the side view of the system to assist the understanding of the microstructure of the main snapshot much more clearly. Similar results are shown in Fig. 9.20 for a stronger magnetic particle interaction strength $\lambda = 50$ where stable column-like clusters are able to be formed even in this strong shear flow situation. With these circumstances, therefore, we will discuss the dependence of the column-like clusters on the shear flow strength.

We first discuss the results for a weak magnetic interaction strength $\lambda = 20$, shown in Fig. 9.19. In the case of a weak magnetic field $\xi = 1$, shown in Fig. 9.19(a), the random motion is much more dominant than the effect of the applied magnetic field, therefore the disk-like particles only have a weak tendency to incline the disk plane of each particle along or in parallel to the shear flow direction in a strong shear flow situation. In the case of a strong magnetic field situation $\xi = 10$, shown in Fig. 9.19(b), this tendency becomes much clearer due to the combined influence of the shear flow and the applied magnetic field. Here, the magnetic moments of the disk-like particles are strongly restricted to lie in a direction close to the magnetic field direction but they are more significantly shifted toward the shear flow direction with increasing shear rate. The restriction of the magnetic moments to the magnetic field direction will still allow the rotation of each particle about the magnetic field direction. A strong shear flow tends to

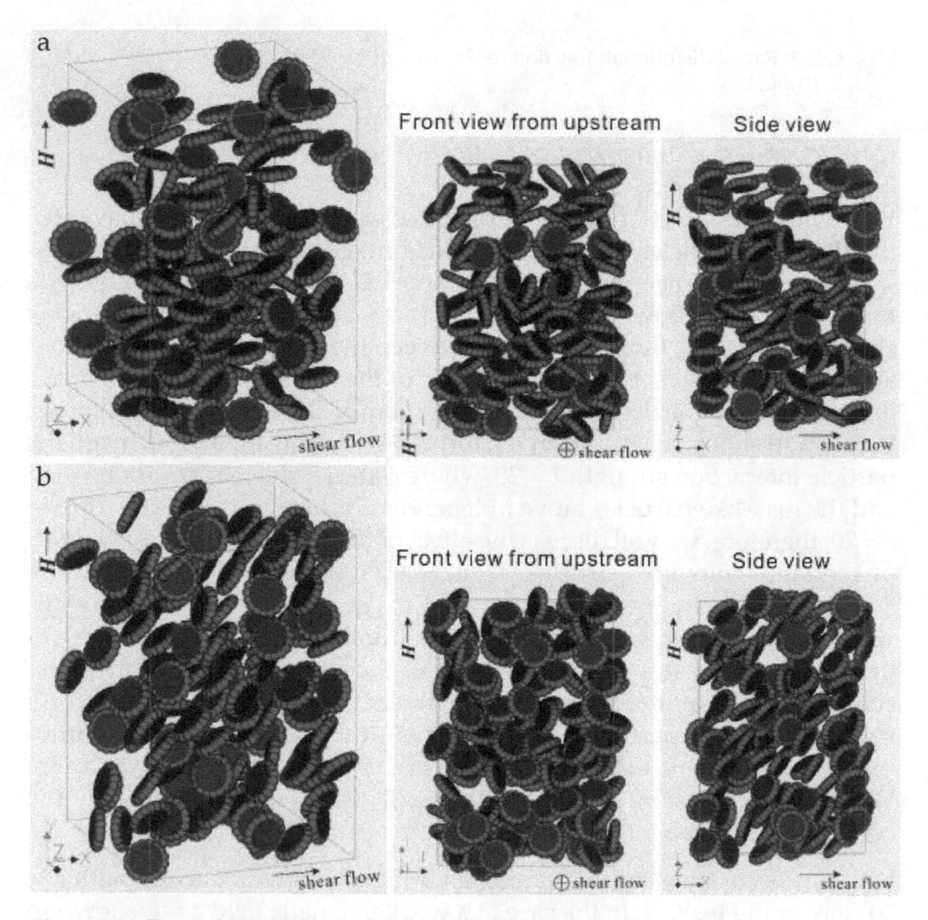


**Figure 9.19.** Dependence of orientational characteristics of disk-like particles on the magnetic field strength for $Pe = 10$ and $\lambda = 20$: (a) $\xi = 1$ and (b) $\xi = 10$.

further restrict this rotational motion in such a way that the particles are no longer able to freely rotate about that axis because the direction of each particle is strongly restricted to the plane normal to the angular velocity vector of the present shear flow. This is illustrated by the snapshots of the front and the side view of Fig. 9.19(b) where the disk-like particles are largely observed in a disk plane view and in a spherocylinder-shape view in the front and the side view, respectively. From these results, we understand that in the case of a weak magnetic interaction, a strong shear flow induces the restriction on the disk-like particles to lie in a certain direction of the plane normal to the angular velocity vector of the simple shear flow, and this direction tends to be shifted toward the shear flow direction more significantly with increasing shear rate.

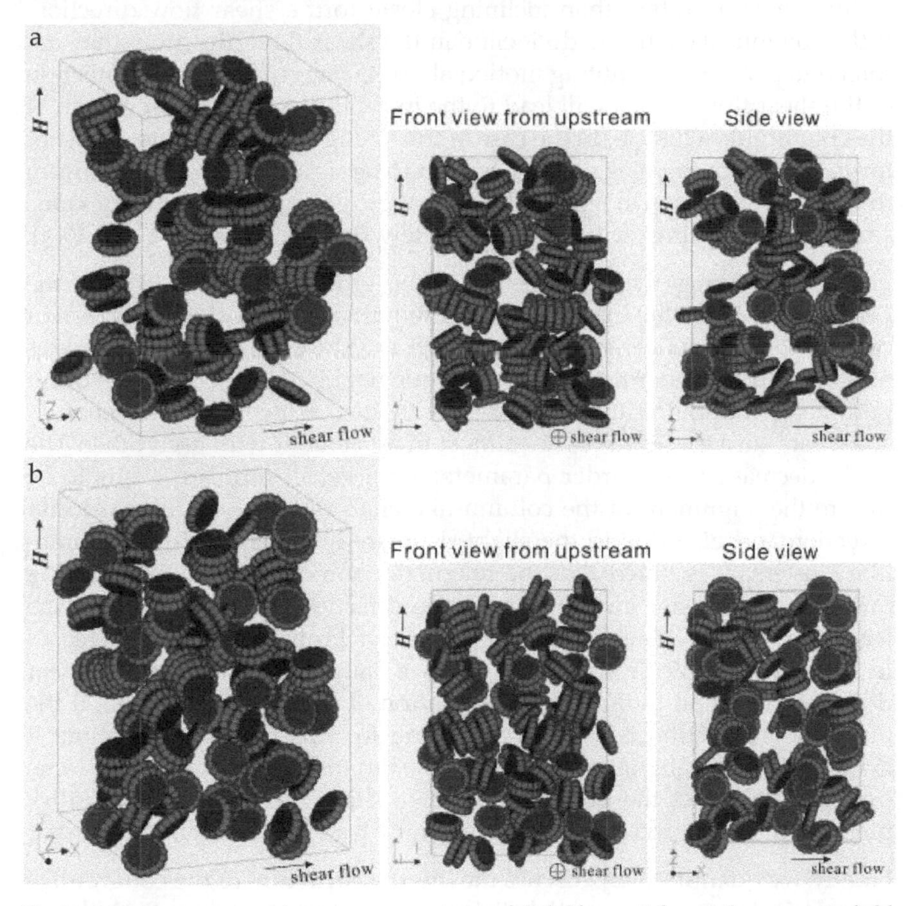


**Figure 9.20.** Dependence of aggregate structures of disk-like particles on the magnetic field strength for $Pe = 10$ and $\lambda = 50$: (a) $\xi = 1$ and (b) $\xi = 10$.

Next, we discuss the results for a strong magnetic interaction $\lambda$ = 50, shown in Fig. 9.20, where stable column-like clusters are formed in the system. Even if the shear rate is increased from $Pe$ = 1 to $Pe$ = 10 where the corresponding snapshots are Fig. 9.13(b) and Fig. 9.20(a), the internal structure of the column-like clusters does not seem to be significantly influenced by this change in the shear rate. On the other hand, the column-like clusters have now come to exhibit a characteristic orientational feature due to the strong shear flow. That is, it is seen from the front and the side views shown in Fig. 9.20(a) that the column-like clusters have a tendency to align along the angular velocity vector of the shear flow, i.e., in the $z$-axis direction, due to the effect of the stronger shear flow. This is because the column-like clusters can survive in a stronger shear flow by inclining in this direction rather than inclining closer to the shear flow direction. If the column-like clusters do incline in the shear flow direction, they are forced to perform a tumbling motion about the angular velocity direction of the shear flow, which will lead to the instability and the dissociation of the column-like clusters. In the case of the strong field $\xi$ = 10, the snapshot for $Pe$ = 10, shown in Fig. 9.20(b), is seen to be significantly different from that for $Pe$ = 1, shown in Fig. 9.16(b), although quite similar to the previous snapshots, including the front and side view, for $\xi$ = 1 shown in Fig. 9.20(a).

Figure 9.21 shows the dependence of the order parameter $\langle n_y \rangle$ on the Peclet number for the magnetic field strength $\xi$ = 10 for the two different cases of the magnetic interaction strength $\lambda$ = 20 and 50. For both the cases of $\lambda$= 20 and 50 under the strong magnetic field strength $\xi$ = 10, the order parameter monotonically decreases with increasing Peclet number or shear rate, and these curves are almost in agreement with each other. The slight decrease of the order parameter with Peclet number seems to be due to the alignment of the column-like clusters to the angular velocity direction and also due to the slight decrease in the size of the column-like clusters. It is noted that the magnetic moments of the neighboring particles in a cluster incline in opposite directions to each other, which leads to a smaller value of the order parameter. From these slight decreases in the order parameter with shear rate, we understand that a strong shear flow may function to induce the formation of column-like clusters if the magnetic interaction is sufficiently strong for this particle formation. It is noted that the alignment of the short column-like clusters in the same direction enhances the probability of two short clusters being in a disk plane-to-plane contact situation, which may lead to form a longer cluster.

Finally, we quantitatively discuss the internal structure of the column-like clusters in terms of the radial distribution function. Figure 9.22 shows results of the radial distribution function for a strong magnetic interaction strength $\lambda = 50$ in a strong magnetic field $\xi = 10$ where the three different cases are addressed for the shear rates, $Pe = 1$, 5 and 10. It is seen from this figure that in all three cases the results show the first and the second peak at positions of an integer times the particle thickness including the steric layer, at $r^* \simeq 1.3$ and 2.6, and then converge to unity with increasing radial distance. For each of the three cases, the height of the first peak is very similar and not significantly influenced by the change in the shear rate. The second peak is also almost constant in height. The lack of a third peak at $r^* \simeq 3.9$ in these curves implies that the column-like clusters formed in the system are quite short and expected to be composed of three or maybe four particles on average. From these characteristics of the radial distribution function, we understand that the shear rate does not have such a decisive effect on the internal structure of the column-like clusters that a drastic change in the phase is induced from stable column-like clusters to single-moving particles and vice versa.

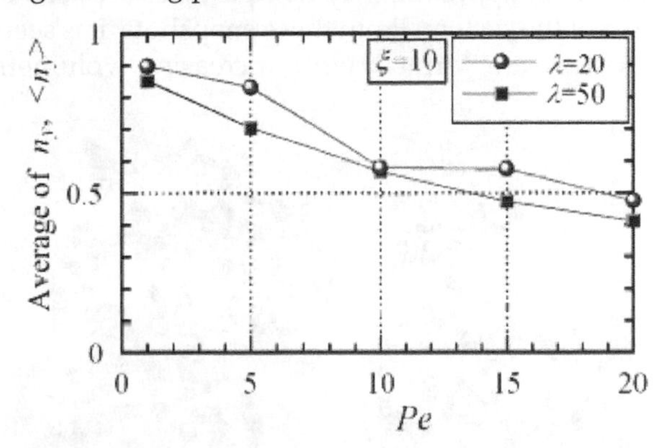


**Figure 9.21.** Characteristics of alignment of magnetic moments to the magnetic field direction as a function of the Peclet number: the ordinate is the average of the $y$-component $n_y$ of the magnetic moment direction $\boldsymbol{n}$.

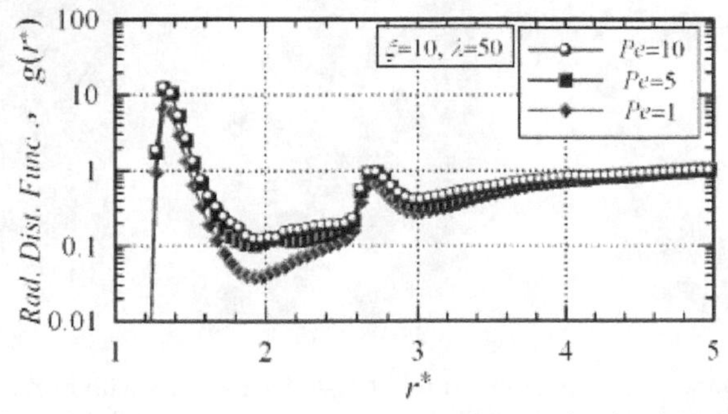


**Figure 9.22.** Radial distribution function for $\xi = 10$ and $\lambda = 50$ for the three different cases of $Pe = 1$, 5 and 10.

### *(d) Dependence of aggregate structures on the volumetric fraction of magnetic particles*

Finally, we discuss the influence of the volumetric fraction of the system particles on the column-like clusters for a strong magnetic interaction strength $\lambda = 50$. Figure 9.23 shows results of the snapshots for a weak shear flow $Pe = 1$ in a weak magnetic field $\xi = 1$ for five different cases of the volumetric fraction, $\phi_V = 0.01$, 0.02, 0.05, 0.07 and 0.2; it is noted that you should refer to the result in Fig. 9.13(b) for the case of $\phi_V = 0.1$. Figure 9.24 shows results of the radial distribution function for the typical cases, $\phi_V = 0.01, 0.05$ and 0.2. In the case of a dilute system $\phi_V = 0.01$, shown in Fig. 9.23(a), only column-like clusters composed of two particles are observed to be formed, whereas in contrast, for the case of $\phi_V = 0.2$, shown in Fig. 9.23(e), there are numerous large column-like clusters formed and these stable clusters of approximately five particles are seen to be formed in the whole area of the system. From these snapshots, it is seen that longer column-like clusters are formed with increasing volumetric fraction.

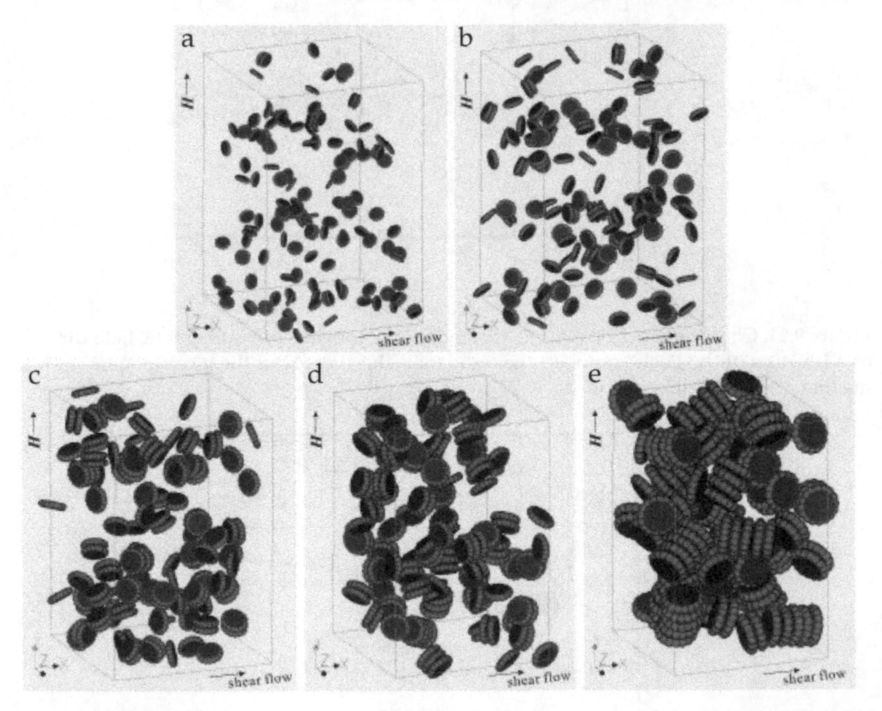


**Figure 9.23.** Change in aggregate structures of disk-like particles with increasing volumetric fraction of particles for $Pe = 1$, $\lambda = 50$ and $\xi = 1$: (a) $\phi_V = 0.01$, (b) $\phi_V = 0.02$, (c) $\phi_V = 0.05$, (d) $\phi_V = 0.07$ and (e) $\phi_V = 0.2$.

This may be reasonable because a larger volumetric fraction can provide a greater probability of the particles to approach each other and form longer clusters if the magnetic interaction strength is sufficiently strong for the formation of aggregates. On the other hand, the results of the radial distribution function, shown in Fig. 9.24, do not exhibit any significant differences even in a quantitative meaning, although the height of the first peak slightly decreases with increasing volumetric fraction and the third peak at $r^* \simeq 4$ appears for $\phi_V = 0.2$. The decrease in the height of the first peak with volumetric fraction is expected to be due to the influence of the magnetic interactions among the neighboring column-like clusters, which becomes more significant in a denser particle system.

From these characteristics of the column-like clusters, we understand that a larger volumetric fraction of particle significantly enhances the length of the column-like clusters, while on the other hand, the internal structure of the column-like clusters is not significantly influenced by the change in the volumetric fraction.

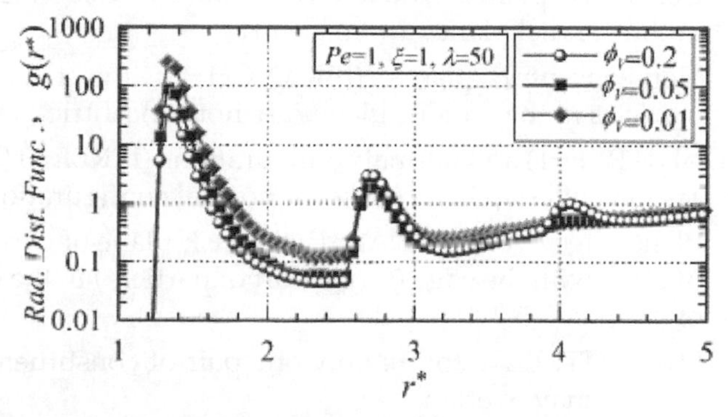


**Figure 9.24.** Radial distribution function for $Pe = 1$, $\xi = 1$ and $\lambda = 50$ for the three different cases of the volumetric fraction, $\phi_V = 0.01$, 0.05 and 0.2.

### 9.2.8 *Sample simulation program*

We show a sample simulation program, written in FORTRAN language, that was used to obtain all the present results. In this program several subroutines relating to quantities that were not treated in the previous discussion, such as for the evaluation of the orientational distribution function, the pair correlation function, several order parameters and the viscosity, have been removed from the original version in order to provide the reader with a more straightforward understanding of the simulation code. A more academic-oriented simulation code, therefore, may be developed by adding other subroutines for evaluating such quantities to the present simplified simulation code without any important modifications.

Perhaps the most important and complex subroutine in the program is the subroutine FORCECAL that is used for the assessment of the overlap of the steric layers of two disk-like particles. This subroutine may be relatively difficult for readers to understand, and therefore we offer the reader several explanations regarding the logical flow of the subroutine. Situations of the particle overlap can be classified into the several representative cases by assessing whether or not the position of $\boldsymbol{r}_{i(j)}^{Q}$ in Eq. (5.31) is situated inside the disk plane of particle $i$, and also whether or not the value of $k_j^s$ is larger than the value of $d/2$.

(1) One plane (ITREE=2): the two disk-like particles are in the same plane.

(2) Two parallel planes (ITREE=3 or 4): the two disk-like planes are in two different but parallel planes.

  (2)-1 Concentric planes (IBRANCH=0): the two disk-like particles in the parallel planes are concentric.

  (2)-2 Non-concentric planes (IBRANCH=1): the two disk-like particles in the parallel planes are not concentric.

(3) General (ITREE=1) and normal configurations (ITREE=0 (or 5)): the particles may overlap in a general or normal configuration.

  (3)-1 Plane-edge overlap (IWAY=1): the disk-plane of particle $i$ may overlap with the edge (torus) part of particle $j$ in the condition of $k_j^s$ (in Eq. (5.29)) $> d/2$.

    (3)-1-1 ITREE=0 (or 5): only one pair of constituent spheres may overlap.

    (3)-1-2 ITREE=1: all the pairs of constituent spheres may overlap.

  (3)-2 Edge-edge overlap (IWAY=2): the edge part of particle $i$ may overlap with the edge part of particle $j$ in the condition of $k_j^s > d/2$.

    (3)-2-1 ITREE=0 (or 5): only one pair of constituent spheres may overlap.

    (3)-2-2 ITREE=1: all the pairs of constituent spheres may overlap.

  (3)-3 Edge-edge overlap (IWAY=3): the edge part of particle $i$ may overlap with the edge part of particle $j$ in the condition of $k_j^s \leq d/2$. In this case, only one pair of constituent spheres may overlap.

In the cases of IWAY=2 and 3, the positions, giving rise to the minimum distance between particles $i$ and $j$, first have to be evaluated in program function RIJMNFUN before conducting the assessment procedure for particle overlap. The program function RIJMNFUN is straightforwardly understood, including the usage of the symbols, by referring to the explanation in Section 9.2.4.

It is quite essential for readers to confirm the meaning of the important variables that are defined in the first comment area before the start of the specification of parameters. As usual in simulations, a non-dimensionalized system is used in the program, previously explained in detail in Section 9.2.3, where for instance, distances and time are non-dimensionalized by the particle thickness $b_1$ and the inverse of the shear rate $1/\dot{\gamma}$, respectively. The important variables used in the program are explained as follows:

| | |
|---|---|
| RX(I),RY(I),RZ(I) | : $(x,y,z)$ coordinates of the position vector of magnetic particle $i$ |
| EX(I),EY(I),EZ(I) | : $(x,y,z)$ coordinates of the particle direction $\boldsymbol{e}_i$ of particle $i$ |
| NX(I),NY(I),NZ(I) | : $(x,y,z)$ coordinates of the magnetic moment direction $\boldsymbol{n}_i$ |
| RXB(I),RYB(I),RZB(I) | : translational random displacements in the $(x,y,z)$-axis direction |
| EXB(I),EYB(I),EZB(I) | : rotational random displacements of the particle direction |
| NXB(I),NYB(I),NZB(I) | : rotational random displacements of the magnetic moment direction |
| FX(I),FY(I),FZ(I) | : force acting on particle $i$ exerted by the ambient particles |
| TX(I),TY(I),TZ(I) | : torque acting on particle $i$ exerted by the ambient particles and the applied magnetic field |
| N,NDENS,VDENS | : number of particles $N$, number density $n^*$ and volumetric fraction $\phi_V$ |
| (HX,HY,HZ) | : unit vector denoting the magnetic field direction (= (0,1,0) in the present case) |
| (XL,YL,ZL) | : side lengths of the simulation box in the $(x, y, z)$-directions |
| RA,KU | : non-dimensional parameters $\lambda$ and $\xi$ |
| PETLD | : Peclet number based on the oblate spheroid |

| | | |
|---|---|---|
| VP | : | volume of the disk-like particle |
| DT1 | : | translational diffusion coefficient in the particle axis direction |
| DT2 | : | translational diffusion coefficient in a direction normal to the particle axis direction |
| DR1 | : | rotational diffusion coefficient for the spin rotational motion about the particle axis direction |
| DR2 | : | rotational diffusion coefficient for the usual rotational motion about a line, through the particle center, normal to the particle axis direction |

As an aid to understanding the program, explanatory descriptions have been added to the important procedures but note that the line numbers are only for the convenience of the reader and are not necessary for executing the FORTRAN program.

```
00010 C********************************************************************
00020 C*                         bddisk1.f                                 *
00030 C*                                                                   *
00040 C*     OPEN(9, FILE='@bbb1.dat', STATUS='UNKNOWN'); para,results     *
00050 C*     OPEN(11,FILE='bbb11.dat', STATUS='UNKNOWN'); parameters       *
00060 C*     OPEN(14,FILE='bbb41.mgf', STATUS='UNKNOWN'); ANIME,MicroAVS   *
00070 C*     OPEN(21,FILE='bbb001.dat',STATUS='UNKNOWN'); PTCL POS.VEL.ORI *
00080 C*     OPEN(22,FILE='bbb011.dat',STATUS='UNKNOWN'); PTCL POS.VEL.ORI *
00090 C*     OPEN(23,FILE='bbb021.dat',STATUS='UNKNOWN'); PTCL POS.VEL.ORI *
00100 C*     OPEN(24,FILE='bbb031.dat',STATUS='UNKNOWN'); PTCL POS.VEL.ORI *
00110 C*     OPEN(25,FILE='bbb041.dat',STATUS='UNKNOWN'); PTCL POS.VEL.ORI *
00120 C*                                                                   *
00130 C*     -----------    BROWNIAN DYNAMICS SIMULATIONS   -----------    *
00140 C*         THREE-DIMENSIONAL BROWNIAN DYNAMICS SIMULATIONS OF        *
00150 C*         A DISPERSION COMPOSED OF HEMATITE DISK-LIKE PARTICLES     *
00160 C*         IN A SIMPLE SHEAR FLOW.                                   *
00170 C*                                                                   *
00180 C*         1. DISK-LIKE MODEL WITH ARBITRARY ASPECT RATIO.           *
00190 C*         2. NO HYDRODYNAMIC INTERACTIONS AMONG PARTICLES.          *
00200 C*         3. HEMATITE PARTICLES ARE CONSIDERED.                     *
00210 C*         4. STERIC LAYER IS CONSIDERED.                            *
00220 C*         5. RIJ_MINIMUM IS EVALUATED FOR ASSESSMENT.               *
00230 C*                                                                   *
00240 C*                                  VER.1  BY A.SATOH , '15 10/23    *
00250 C********************************************************************
00260 C     N     : NUMBER OF PARTICLES (N=INISQUX*INISQUY*INISQUZ)
00270 C     D1    : DIAMETER OF OUTER CIRCLE OF A DISK-LIKE PARTICLE
00280 C     D     : DIAMETER OF THE PART OF CYLINDER
00290 C     B1    : THICKNESS OF PARTICLE (=1  FOR THIS CASE)
00300 C     RP    : ASPECT RATIO (=D1/B1) (=D1 FOR THIS CASE)
00310 C     RP1   : = RP-1
00320 C     DEL   : THICKNESS OF STERIC LAYER (=0.15)
00330 C     TD    : =DEL/(B1/2)=0.3
00340 C     VP    : VOLUME OF THE PARTICLE
00350 C     NDENS : NUMBER DENSITY
00360 C     VDENS : VOLUMETRIC FRACTION
00370 C     RA    : NONDIMENSIONAL PARAMETER OF PARTICLE-PARTICLE INTERACT
00380 C     RA0   : =RA/RP**3
00390 C     KU    : NONDIMENSIONAL PARAMETER OF PARTICLE-FIELD INTERACTION
00400 C     RAV   : NONDIMENSIONAL PARAMETER OF STERIC REPULSION
00410 C     HX,HY,HZ : MAGNETIC FIELD DIRECTION (UNIT VECTOR)
00420 C     RCOFF : CUTOFF RADIUS FOR CALCULATION OF INTERACTION ENERGIES
00430 C     XL,YL,ZL : DIMENSIONS OF SIMULATION REGION
00440 C              (XL,YL,ZL)=(INIPX, INIPY*BETA, INIPZ*RP) *ALPHA
00450 C     BETA  : ASPECT RATIO OF SIMULATION BOX
00460 C
00470 C     PE    : PECLET NUMBER BASED ON SPHERE
00480 C     PETLD : PECLET NUMBER BASED ON OBLATE SPHEROIDAL PARTICLE
```

• Positions and distances are non-dimensionalized by the particle thickness $b_1$; B1=1 in the present non-dimensional system.

• RA, KU and RAV imply $\lambda$, $\xi$ and $\lambda_V$, respectively.

```
00490 C       RM,RH,RV : RM=RA/PE , RH=KU/PE , RV=RAV/(3*DEL*PE)
00500 C
00510 C       DT1    : DIFFUSION COEFF. OF TRANSLATIONAL MOTION (PARALLEL)
00520 C       DT2    : DIFFUSION COEFF. OF TRANSLATIONAL MOTION (NORMAL)
00530 C       DR1    : DIFFUSION COEFF. OF ROTATIONAL MOTION (PARALLEL)
00540 C       DR2    : DIFFUSION COEFF. OF ROTATIONAL MOTION (NORMAL)
00550 C
00560 C       RX(I),RY(I),RZ(I)     : PARTICLE POSITION
00570 C       EX(N),EY(N),EZ(N)     : DIRECTION OF PARTICLE AXIS
00580 C       NX(N),NY(N),NZ(N)     : DIRECTION OF MAGNETIC MOMENT
00590 C       RXB(I),RYB(I),RZB(I)  : RANDOM DISPLACE. DUE TO BROWNIAN MOTION
00600 C       EXB(I),EYB(I),EZB(I)  : RAN.ROT.DISP. PTCL DIREC. DUE TO B.M.
00610 C       NXB(I),NYB(I),NZB(I)  : RAN.ROT.DISP. MAG.MOM.DIREC.DUE TO B.M.
00620 C       FX(I),FY(I),FZ(I)     : FORCES ACTING ON PARTICLE I
00630 C       TX(I),TY(I),TZ(I)     : TORQUES ACTING ON PARTICLE I
00640 C       MOMX(**),MOMY(**)     : MAG. MOMENT OF SYSTEM AT EACH TIME STEP
00650 C       MOMZ(**)
00660 C
00670 C       RMAT(3,3,N)        : ROTATION MATRIX
00680 C
00690 C       H       : INTERVAL OF TIME STEP
00700 C       GAMDOT  : SHEAR RATE (=1 FOR THIS CASE)
00710 C       NTIMEMX : MAXIMUM NUMBER OF TIME STEP
00720 C
00730 C       RIJMN : DISTANCE BETWEEN TWO SPHERICAL PARTICLES (CONSTITUTING
00740 C               TWO DISK-LIKE PARTICLES) IN THE CASE OF SURFACE-SURFACE
00750 C               OVERLAP SITUATION
00760 C       RMNMN : MINIMUM DISTANCE BETWEEN TWO SPHERICAL PARTICLES
00770 C               (CONSTITUTING TWO DISK-LIKE PARTICLES) AT EACH TIME STEP
00780 C       RIJMNAV : AVERAGE OF RIJMN IN A SIMULATION RUN
00790 C       RMNMNAV : AVERAGE OF RMNMN IN ONE TIME STEP
00800 C
00810 C       ++++ -XL/2<RX(I)<XL/2 , -YL/2< RY(I)<YL/2, -ZL/2<RZ(I)<ZL/2 ++++
00820 C-----------------------------------------------------------------------
00830         IMPLICIT REAL*8 (A-H,O-Z), INTEGER (I-N)
00840 C
00850         COMMON /BLOCK1/  RX    , RY    , RZ
00860         COMMON /BLOCK2/  EX    , EY    , EZ
00870         COMMON /BLOCK3/  NX    , NY    , NZ
00880         COMMON /BLOCK4/  FX    , FY    , FZ
00890         COMMON /BLOCK5/  TX    , TY    , TZ
00900         COMMON /BLOCK6/  RXB   , RYB   , RZB
00910         COMMON /BLOCK7/  EXB   , EYB   , EZB
00920         COMMON /BLOCK8/  NXB   , NYB   , NZB
00930         COMMON /BLOCK11/ XL    , YL    , ZL
00940         COMMON /BLOCK12/ HX    , HY    , HZ
00950         COMMON /BLOCK13/ RA    , KU    , RAV
00960         COMMON /BLOCK14/ RM    , RH    , RV
00970         COMMON /BLOCK15/ N     , NDENS, VDENS
00980         COMMON /BLOCK16/ D, D1, D2, B, B1, B2, RP, RP1, VP
00990         COMMON /BLOCK17/ DEL, TD, RCOFF
01000 C
01010         COMMON /BLOCK18/ INISQUX , INISQUY , INISQUZ , BETA
01020         COMMON /BLOCK19/ RA0
01030         COMMON /BLLOC20/ RCHCK , RCHCK1 , RCHCK2
01040         COMMON /BLOCK21/ RMAT
01050 C
01060         COMMON /BLOCK25/ RXSTRCG , RYSTRCG , RDSTRCG
01070         COMMON /BLOCK26/ NPROWALL, NPROW   , NPROWST
01080         COMMON /BLOCK27/ NPSTRC  , RXSTRC  , RYSTRC
01090 C
01100         COMMON /BLOCK37/ PE    , PETLD
01110         COMMON /BLOCK38/ DT1   , DT2   , DR1  , DR2  , YHYC
01120         COMMON /BLOCK39/ CRNDMT1, CRNDMT2, CRNDMR1, CRNDMR2
01130         COMMON /BLOCK41/ H , GAMDOT   , DX   , CORY
01140         COMMON /BLOCK42/ NRAN  , RAN    , IXRAN
01150         COMMON /BLOCK43/ MOMX , MOMY , MOMZ
01160 C
01170         COMMON /WORK11/  XRXI , YRYI , ZRZI , XRXJ , YRYJ , ZRZJ
01180         COMMON /WORK12/  FXIJS, FYIJS, FZIJS, FXJIS, FYJIS, FZJIS
01190         COMMON /WORK13/  TQXIJS,TQYIJS,TQZIJS,TQXJIS,TQYJIS,TQZJIS
01200         COMMON /WORK14/  RCOFF2 , RP1D2
01210         COMMON /WORK15/  RTRANMX,EROTMX,NROTMX, RTRANAV,EROTAV,NROTAV
01220         COMMON /WORK17/  NRIJMN    , RIJMN  , RIJMNAV , RIJMNSUM
01230         COMMON /WORK18/  NRMNMN    , RMNMN  , RMNMNAV , RMNMNSUM
01240 C
01250         COMMON /WORK25/  DSQ, D1SQ, D2SQ, BSQ, B1SQ, B2SQ
01260         COMMON /WORK26/  RCHCKSQ , RCHCK1SQ , RCHCK2SQ
01270         COMMON /WORK27/  RPXIMN, RPYIMN, RPZIMN, RPXJMN, RPYJMN, RPZJMN
01280 C       ---------------------------------------------------------------
01290 C
01300         INTEGER    NN , NNS
```

• $\boldsymbol{r}_i$, $\boldsymbol{e}_i$ and $\boldsymbol{n}_i$ are expressed by (RX(I),RY(I),RZ(I)), (EX(I),EY(I),EZ(I)), (NX(I),NY(I),NZ(I)), respectively.

• These variables are used for calculating steric forces using the plain sphere-constituting model for the disk-like particle.

```
01310       PARAMETER( NN=1000 , NNS=5000000 )
01320       PARAMETER( NRANMX=1000000 , PI=3.141592653589793D0 )
01330 C
01340       REAL*8   KU      , NDENS
01350       REAL*8   RX(NN) , RY(NN) , RZ(NN)
01360       REAL*8   EX(NN) , EY(NN) , EZ(NN) , NX(NN) , NY(NN) , NZ(NN)
01370       REAL*8   FX(NN) , FY(NN) , FZ(NN) , TX(NN) , TY(NN) , TZ(NN)
01380
01390       REAL*8   RMAT(3,3,NN)
01400
01410       REAL*8   RXB(NN), RYB(NN), RZB(NN)
01420       REAL*8   EXB(NN), EYB(NN), EZB(NN), NXB(NN), NYB(NN), NZB(NN)
01430       REAL     MOMX(NNS), MOMY(NNS) , MOMZ(NNS)
01440 C     --------------------------
01450       PARAMETER( NNSTRC=100 )
01460 C
01470       INTEGER  NPROWALL, NPROW( NNSTRC ) , NPROWST( NNSTRC )
01480       REAL*8   RXSTRC( NNSTRC ) , RYSTRC( NNSTRC )
01490 C     --------------------------
01500       INTEGER    TT
01510       PARAMETER( TT=1000 )
01520 C
01530       INTEGER  NRIJMN , NRMNMN
01540       REAL*8   RIJMN  , RIJMNAV , RIJMNSUM
01550       REAL*8   RMNMN  , RMNMNAV , RMNMNSUM
01560 C     --------------------------
01570       REAL     RAN(NRANMX)
01580       INTEGER  NRAN , IXRAN , NRANCHK
01590 C     --------------------------
01600       REAL*8   BETA
01610       REAL*8   RXI , RYI , RZI
01620       REAL*8   EXI , EYI , EZI , NXI , NYI , NZI
01630       REAL*8   FXI , FYI , FZI , TXI , TYI , TZI
01640       REAL*8   FXIP , FYIP , FZIP ,  FXIN , FYIN , FZIN
01650       REAL*8   TXIP , TYIP , TZIP ,  TXIN , TYIN , TZIN
01660       REAL*8   SUMORI00
01670       REAL*8   C1   , C2   , C3   ,  C0   , C00
01680       REAL*8   C0X  , C0Y  , C0Z
01690       REAL*8   C1X  , C1Y  , C1Z  ,  C2X  , C2Y  , C2Z
01700       REAL*8   C1XX , C1YY , C1ZZ ,  C2XX , C2YY , C2ZZ
01710       REAL*8   C3XX , C3YY , C3ZZ
01720       REAL*8   EXIOLD  , EYIOLD  , EZIOLD
01730       REAL*8   NXIOLD  , NYIOLD  , NZIOLD
01740       REAL*8   EXIHAT  , EYIHAT  , EZIHAT
01750       REAL*8   DELNORMX, DELNORMY, DELNORMZ
01760       REAL*8   NXIH    , NYIH    , NZIH    , NXIV , NYIV , NZIV
01770       REAL*8   NXIHHAT, NYIHHAT, NZIHHAT
01780       REAL*8   RTRANMX, EROTMX, NROTMX, RTRANAV, EROTAV, NROTAV
01790       INTEGER  NTIME , NTIMEMX , NGRAPH , DNSMPL , NP  ,  NOPT
01800       INTEGER  NANIME , NANMCTR
01810       INTEGER  DN
01820       INTEGER  NSMPL, NSMPL1, NSMPL2
01830 C     ---------------------------------------------------------------
01840              OPEN(9, FILE='@rp5pelkulra30a1.dat', STATUS='UNKNOWN')
01850              OPEN(11,FILE='rp5pelkulra30a11.dat', STATUS='UNKNOWN')
01860              OPEN(14,FILE='rp5pelkulra30a41.mgf', STATUS='UNKNOWN')
01870              OPEN(21,FILE='rp5pelkulra30a001.dat',STATUS='UNKNOWN')
01880              OPEN(22,FILE='rp5pelkulra30a011.dat',STATUS='UNKNOWN')
01890              OPEN(23,FILE='rp5pelkulra30a021.dat',STATUS='UNKNOWN')
01900              OPEN(24,FILE='rp5pelkulra30a031.dat',STATUS='UNKNOWN')
01910              OPEN(25,FILE='rp5pelkulra30a041.dat',STATUS='UNKNOWN')
01920                                                                    NP=9
01930 C                   ---------------------------------------------------
01940 C                   BE CAREFUL IN SETTING N, INISQUX, ..., INISQUZ !
01950 C                   ---------------------------------------------------
01960 C                                                --- PARAMETER (1) ---
01970       PETLD    =  1.0D0
01980       KU       =  1.0D0
01990       RA       = 30.0D0
02000       VDENS    =  0.05D0
02010       RAM      =  RA
02020 C                                                --- PARAMETER (2) ---
02030       RP       = 5.D0
02040       B1       = 1.D0
02050       D1       = RP
02060       D        = D1 - 1.D0
02070       VP       = (PI/24.D0)*(6.D0*(RP-1.D0)**2+3.D0*PI*(RP-1.D0)+4.D0)
```

- KU and RA are set as $\xi$=1 and $\lambda$=30, respectively.
- PETLD is set as $\widetilde{Pe}$=1.
- The volumetric fraction is set as $\phi_V$=0.05.

- The particle aspect ratio is $r_p$=5. The particle volume is denoted by VP.

```
02080         RP1      = RP - 1.D0
02090         RP102    = RP1/2.D0
02100 C
02110         INISQUX = 3
02120         INISQUZ = INISQUX * INT( RP+0.0001D0 )
02130         INISQUY = INISQUX
02140         N       = INISQUX * INISQUZ * INISQUY
02150         BETA    = 1.5D0
02160 C
02170         NDENS    = VDENS/VP
02180         RA0      = RA/RP**3
02190         RAV      = 150.D0
02200         DEL      = 0.15D0
02210         TD       = 0.3D0
02220 C                                               --- PARAMETER (5) ---
02230         D2       = D1 + 2.D0*DEL
02240         B2       = B1 + 2.D0*DEL
02250         RCHCK    = D/2.D0
02260         RCHCK1   = D1/2.D0
02270         RCHCK2   = D2/2.D0
02280         DSQ      = D**2
02290         D1SQ     = D1**2
02300         D2SQ     = D2**2
02310         B1SQ     = B1**2
02320         B2SQ     = B2**2
02330         RCHCKSQ = RCHCK**2
02340         RCHCK1SQ= RCHCK1**2
02350         RCHCK2SQ= RCHCK2**2
02360 C                                               --- PARAMETER (6) ---
02370         CS       = DSQRT( 1.D0 - 1.D0/RP**2 )
02380         QCS      = DATAN( CS / DSQRT(1.D0-CS**2) )
02390         C1       = RP**3 * (2.D0/3.D0) * ( CS**3 *(2.D0-CS**2) )
02400      &                 / (  CS*DSQRT(1.D0-CS**2)-(1.D0-2.D0*CS**2)*QCS  )
02410         PE       = PETLD / C1
02420         RM       = RA /PE
02430         RH       = KU /PE
02440         RV       = RAV/(3.D0*DEL*PE)
02450 C                                               --- PARAMETER (7) ---
02460         H        = 0.000005D0
02470         GAMDOT = 1.D0
02480         HX       = 0.D0
02490         HY       = 1.D0
02500         HZ       = 0.D0
02510         DX       = 0.D0
02520 C
02530         RCOFF    = 5.D0*D1
02540         RCOFF2 = RCOFF**2
02550 C                                               --- PARAMETER (9) ---
02560         NTIMEMX = 1000000
02570         NGRAPH  = NTIMEMX/5
02580         NANIME  = NTIMEMX/100
02590 C
02600         DNSMPL   = 1
02610         DN       = 10
02620         NOPT     = 20
02630 C                                               --- PARAMETER (10) ---
02640         IXRAN = 0
02650         CALL RANCAL( NRANMX, IXRAN, RAN )
02660         NRAN     = 1
02670         NRANCHK = NRANMX - 16*N
02680 C                                               --- PARAMETER (11) ---
02690         DT1 = 3.D0/(4.D0*RP)*
02700      &        ( (2.D0*CS**2-1.D0)*QCS + CS*DSQRT(1.D0-CS**2) ) / CS**3
02710         DT2 = 3.D0/(8.D0*RP)*
02720      &        ( (2.D0*CS**2+1.D0)*QCS - CS*DSQRT(1.D0-CS**2) ) / CS**3
02730         DR1 = 3.D0/(2D0*RP**3)*
02740      &        ( QCS - CS*DSQRT(1.D0-CS**2) ) / CS**3
02750         DR2 = 3.D0/(2D0*RP**3)*
02760      &        ( CS*DSQRT(1.D0-CS**2) - (1.D0-2.D0*CS**2)*QCS )
02770      &                                / ( CS**3*(2.D0-CS**2) )
02780         CRNDMT1 = (1.D0/(3.D0*PE)) * (2.D0*DT1) * H
02790         CRNDMT2 = (1.D0/(3.D0*PE)) * (2.D0*DT2) * H
02800         CRNDMR1 = (1.D0/PE)          * (2.D0*DR1) * H
02810         CRNDMR2 = (1.D0/PE)          * (2.D0*DR2) * H
02820 C                                               --- PARAMETER (12) ---
02830         YHYC = - CS**2 / (2.D0-CS**2 )
02840 C
02850 C     -------------------------------------------------------------
02860 C     -----------------   INITIAL CONFIGURATION   -----------------
02870 C     -------------------------------------------------------------
02880 C                                           --- SET INITIAL CONFIG. ---
```

• INISQUX should be set as a sufficiently large value such as 5 for a research objective.
• The size in the $y$-axis direction is (BETA=)1.5 times that in the $x$-axis direction. The bottom area of the simulation box is set as a square parallel to the $xz$-plane.

• A smaller time interval H should be adopted for a smaller Peclet number.
• The shear rate is unity in the present non-dimensional system.
• The cutoff distance $r_{coff}$ is denoted by RCOFF.

• DNSMPL or DN implies that data sampling is carried out every DNSMPL or DN time steps.

• DT1 and DT2 imply the translational diffusion coefficients $D_{\parallel}^{T}$ and $D_{\perp}^{T}$, respectively.

• DR1 implies the rotational diffusion coefficient $D_{\parallel}^{R}$.

• DR2 implies the rotational diffusion coefficient $D_{\perp}^{R}$.

• $Y^H/Y^C$ is denoted by YHYC.

```
02890 CCCC  OPEN(19,FILE='aaa141.dat',STATUS='OLD')
02900 CCCC  READ(19,472)  N , XL , YL , ZL , D , DT , RP , RP1 , DX
02910 CCCC  READ(19,474)  (RX(I),I=1,N),(RY(I),I=1,N),(RZ(I),I=1,N),
02920 CCCC &              (EX(I),I=1,N),(EY(I),I=1,N),(EZ(I),I=1,N),
02930 CCCC &              (NX(I),I=1,N),(NY(I),I=1,N),(NZ(I),I=1,N)
02940 CCCC  CLOSE(19,STATUS='KEEP')
02950 CCCC  GOTO 7
02960 C
02970       CALL INITIAL
02980 C
02990     7 IF( XL .LE. YL ) THEN
03000         IF( RCOFF .GE. XL/2.D0 ) THEN
03010           RCOFF = XL/2.D0 - 0.00001D0
03020         END IF
03030       ELSE
03040         IF( RCOFF .GE. YL/2.D0 ) THEN
03050           RCOFF = YL/2.D0 - 0.00001D0
03060         END IF
03070       END IF
03080 C
03090       RCOFF2 = RCOFF**2
03100 C     -------------------
03110 C
03120       CALL PTCLMDL( RP )
03130 C
03140       NTIME   = 0
03150       CALL FORCECAL( DX, NP, NTIME )
03160 C                                        --- CAL
03170       CALL RANDISP( N, H, NTIME )
03180 C
03190 C     ------------------------------------------------ PRINT OUT ---
03200       WRITE(NP,12) N, VDENS, NDENS, RA, KU, RAV, RM, RH, RV
03210       WRITE(NP,14) PE, PETLD
03220       WRITE(NP,16) D, RP, TD, DEL, D1, RP1, PV
03230       WRITE(NP,17) D1, D2, B1, B2, RCHCK, RCHCK1, RCHCK2
03240       WRITE(NP,18) H, GAMDOT, HX, HY, HZ, RCOFF, BETA, XL,YL, ZL
03250       WRITE(NP,20) DT1, DT2, DR1, DR2, YHYC
03260       WRITE(NP,21) NTIMEMX, NGRAPH, NANIME
03270       WRITE(NP,22) DNSMPL
03280       WRITE(NP,25) CRNDMT1, CRNDMT2, CRNDMR1, CRNDMR2
03290       WRITE(NP,28) INISQUX , INISQUY , INISQUZ
03300 C     ------------------------------------------ INITIALIZATION ---
03310       NSMPL   = 0
03320       NANMCTR = 0
03330 C
03340       RTRANMX = 0.D0
03350       EROTMX  = 0.D0
03360       NROTMX  = 0.D0
03370       RTRANAV = 0.D0
03380       EROTAV  = 0.D0
03390       NROTAV  = 0.D0
03400 C
03410       NRIJMN   = 0
03420       RIJMNSUM = 0.D0
03430       NRMNMN   = 0
03440       RMNMNSUM = 0.D0
03450 C     -------------------------------------------------------------
03460 C     -------------------------------------------------------------
03470 C     ------------------    START OF MAIN LOOP    --------------------
03480 C     -------------------------------------------------------------
03490 C
03500       DO 1000 NTIME = 1, NTIMEMX
03510 C
03520         DX = GAMDOT*YL*H*DBLE(NTIME)
03530         DX = DMOD( DX, XL )
03540 C
03550         DO 100 I = 1,N
03560 C
03570           EXI = EX(I)
03580           EYI = EY(I)
03590           EZI = EZ(I)
03600           NXI = NX(I)
03610           NYI = NY(I)
03620           NZI = NZ(I)
03630           FXI = FX(I)
03640           FYI = FY(I)
03650           FZI = FZ(I)
03660           TXI = TX(I)
03670           TYI = TY(I)
03680           TZI = TZ(I)
03690           EXIOLD = EX(I)
03700           EYIOLD = EY(I)
```

• Subroutine INITIAL sets the initial position and orientation of each particle.

• The cutoff distance $r_{coff}$ should be shorter than the half side lengths.

• The position of the constituent spheres of the plain sphere-constituting model is specified in subroutine PTCLMDL. These variables are used in evaluating the steric interaction forces.

• Forces and torques are calculated.

• Translational and rotational random displacements are set.

• These quantities are used for confirming the validity of the choice of the time interval. RTRANMX, EROTMX and NROTMX imply the maximum displacements of the translational position, the particle direction and the magnetic moment direction, respectively.

• RMNMNSUM is used for evaluating the average of the minimum distance between two particles in one time step, RMNMNAV.

```
03710          EZIOLD = EZ(I)
03720 C                                       ----------------------------------
03730 C        -------------------------------- (1) TRANSLATIONAL MOTION ---
03740 C                                       ----------------------------------
03750 C
03760          C00    = FXI*EXI + FYI*EYI + FZI*EZI
03770          FXIP   = C00*EXI
03780          FYIP   = C00*EYI
03790          FZIP   = C00*EZI
03800          FXIN   = FXI - FXIP
03810          FYIN   = FYI - FYIP
03820          FZIN   = FZI - FZIP
03830 C
03840          COX    = H*( DT1*FXIP + DT2*FXIN ) + RXB(I) + RY(I)*GAMDOT*H
03850          COY    = H*( DT1*FYIP + DT2*FYIN ) + RYB(I)
03860          COZ    = H*( DT1*FZIP + DT2*FZIN ) + RZB(I)
03870                 C0     = DSQRT( COX**2 + COY**2 + COZ**2 )
03880                 IF( C0 .GT. RTRANMX )  RTRANMX = C0
03890                 RTRANAV = RTRANAV + C0
03900          IF( C0 .GT. 0.05D0 ) THEN
03910            WRITE(6,*) NTIME,'--- R_TRANS IS TOO LARGE ---', REAL(C0)
03920            C0  = 0.05D0/C0
03930            COX = C0 * COX
03940            COY = C0 * COY
03950            COZ = C0 * COZ
03960          END IF
03970          RXI    = RX(I) + COX
03980          RYI    = RY(I) + COY
03990          RZI    = RZ(I) + COZ
04000          CORY   = DNINT( RYI/YL )
04010          RXI    = RXI - CORY*DX
04020          RX(I) = RXI - DNINT( RXI/XL )*XL
04030          RY(I) = RYI - CORY*YL
04040          RZ(I) = RZI - DNINT( RZI/ZL )*ZL
04050 C
04060 C                                       ----------------------------------
04070 C        ---------------------------------- (2) ROTATIONAL MOTION ---
04080 C                                       ----------------------------------
04090 C                                       ++++++++++++++++++++++++++++++++++
04100 C                                       ++++++++++++++ (2)-1 (EX,EY,EZ) +++
04110 C                                       ++++++++++++++++++++++++++++++++++
04120          C00  = TXI*EXI + TYI*EYI + TZI*EZI
04130          TXIP = C00*EXI
04140          TYIP = C00*EYI
04150          TZIP = C00*EZI
04160          TXIN = TXI - TXIP
04170          TYIN = TYI - TYIP
04180          TZIN = TZI - TZIP
04190 C
04200          C00     = -0.5D0*EZI
04210          OMEXIP = C00*EXI
04220          OMEYIP = C00*EYI
04230          OMEZIP = C00*EZI
04240          OMEXIN =          - OMEXIP
04250          OMEYIN =          - OMEYIP
04260          OMEZIN = -0.5D0 - OMEZIP
04270 C
04280          C1X = DR2*TXIN
04290          C1Y = DR2*TYIN
04300          C1Z = DR2*TZIN
04310          C1XX = C1Y*EZI - C1Z*EYI
04320          C1YY = C1Z*EXI - C1X*EZI
04330          C1ZZ = C1X*EYI - C1Y*EXI
04340 C
04350          C2X = OMEXIN
04360          C2Y = OMEYIN
04370          C2Z = OMEZIN
04380          C2XX = C2Y*EZI - C2Z*EYI
04390          C2YY = C2Z*EXI - C2X*EZI
04400          C2ZZ = C2X*EYI - C2Y*EXI
04410 C
04420          C3XX = 0.5D0*EYI - (EXI*EYI)*EXI
04430          C3YY = 0.5D0*EXI - (EXI*EYI)*EYI
04440          C3ZZ =            - (EXI*EYI)*EZI
04450          C3XX = YHYC*C3XX
04460          C3YY = YHYC*C3YY
04470          C3ZZ = YHYC*C3ZZ
04480 C
04490          COX    = H*( C1XX + C2XX + C3XX ) + EXB(I)
04500          COY    = H*( C1YY + C2YY + C3YY ) + EYB(I)
```

• $\boldsymbol{F}_{\|}$ and $\boldsymbol{F}_{\perp}$ are denoted by (FXIP,FYIP,FZIP) and (FXIN,FYIN,FZIN), respectively. These quantities are obtained from Eq. (6.51).

• The basic equations are Eqs. (9.16) and (9.17).

• If too large displacement happens, this message is printed out on the display.

• The Lees-Edwards boundary condition has to be applied.

• $\boldsymbol{T}_{\|}$ and $\boldsymbol{T}_{\perp}$ are denoted by (TXIP,TYIP,TZIP) and (TXIN,TYIN,TZIN), respectively. These quantities are obtained from Eq. (6.51).

• $\boldsymbol{\Omega}_{\|}$ and $\boldsymbol{\Omega}_{\perp}$ are denoted by (OMEXIP,OMEYIP,OMEZIP) and (OMEXIN,OMEYIN,OMEZIN), respectively. These quantities are obtained from Eq. (6.51).

• (C1XX,C1YY,C1ZZ) denotes the third term on the right-hand side of Eq. (9.18) without factor $\Delta t$.

• (C2XX,C2YY,C2ZZ) denotes the second term on the right-hand side of Eq. (9.18) without factor $\Delta t$.

• (C3XX,C3YY,C3ZZ) denotes the fourth term on the right-hand side of Eq. (9.18) without factor $\Delta t$.

• EXB(*), EYB(*) and EZB(*) are random displacements.

```
04510          C0Z    = H*( C1ZZ + C2ZZ + C3ZZ ) + EZB(I)
04520                 C0     = DSQRT( C0X**2 + C0Y**2 + C0Z**2 )
04530                 IF( C0 .GT. EROTMX )  EROTMX = C0
04540                 EROTAV  = EROTAV + C0
04550          IF( C0*RCHCK .GT. 0.05D0 ) THEN
04560            WRITE(6,*) NTIME,'--- E_ROT IS TOO LARGE ---', REAL(C0)
04570            C0  = 0.05D0/(C0*RCHCK)
04580            C0X = C0 * C0X
04590            C0Y = C0 * C0Y
04600            C0Z = C0 * C0Z
04610          END IF
04620          EXI    = EXI + C0X
04630          EYI    = EYI + C0Y
04640          EZI    = EZI + C0Z
04650          C00    = DSQRT( EXI**2 + EYI**2 + EZI**2 )
04660          EXI    = EXI/C00
04670          EYI    = EYI/C00
04680          EZI    = EZI/C00
04690          EX(I)  = EXI
04700          EY(I)  = EYI
04710          EZ(I)  = EZI
04720 C
04730 C                        +++ (NX,NY,NZ) IS MODIFIED IN DIRECTION +++
04740          EXIHAT = EXI - EXIOLD
04750          EYIHAT = EYI - EYIOLD
04760          EZIHAT = EZI - EZIOLD
04770 C
04780          DELNORMX = EYI*EZIHAT - EZI*EYIHAT
04790          DELNORMY = EZI*EXIHAT - EXI*EZIHAT
04800          DELNORMZ = EXI*EYIHAT - EYI*EXIHAT
04810 C
04820          C0 = DELNORMX**2 + DELNORMY**2 + DELNORMZ**2
04830          IF( C0 .LT. 1.D-30 )  GOTO 70
04840 C
04850          C0 = DSQRT( C0 )
04860          DELNORMX = DELNORMX /C0
04870          DELNORMY = DELNORMY /C0
04880          DELNORMZ = DELNORMZ /C0
04890          C0 = NXI*DELNORMX + NYI*DELNORMY + NZI*DELNORMZ
04900          NXIV = C0 * DELNORMX
04910          NYIV = C0 * DELNORMY
04920          NZIV = C0 * DELNORMZ
04930          NXIH = NXI - NXIV
04940          NYIH = NYI - NYIV
04950          NZIH = NZI - NZIV
04960          C0 = DSQRT( EXIHAT**2 + EYIHAT**2 + EZIHAT**2 )
04970          NXIHHAT = ( DELNORMY*NZIH - DELNORMZ*NYIH ) * C0
04980          NYIHHAT = ( DELNORMZ*NXIH - DELNORMX*NZIH ) * C0
04990          NZIHHAT = ( DELNORMX*NYIH - DELNORMY*NXIH ) * C0
05000          NXI = NXI + NXIHHAT
05010          NYI = NYI + NYIHHAT
05020          NZI = NZI + NZIHHAT
05030 C                     +++ (NX,NY,NZ) MUST BE NORMAL TO (EX,EY,EZ) +++
05040          C0     = NXI*EXI + NYI*EYI + NZI*EZI
05050          NXI    = NXI - C0*EXI
05060          NYI    = NYI - C0*EYI
05070          NZI    = NZI - C0*EZI
05080          C00    = DSQRT( NXI**2 + NYI**2 + NZI**2 )
05090          NXI    = NXI/C00
05100          NYI    = NYI/C00
05110          NZI    = NZI/C00
05120 C                            ++++++++++++++++++++++++++++++++++++
05130 C                            ++++++++++++++ (2)-2 (NX,NY,NZ) +++
05140 C                            ++++++++++++++++++++++++++++++++++++
05150     70   C1X = DR1*TXIP
05160          C1Y = DR1*TYIP
05170          C1Z = DR1*TZIP
05180          C1XX = C1Y*NZI - C1Z*NYI
05190          C1YY = C1Z*NXI - C1X*NZI
05200          C1ZZ = C1X*NYI - C1Y*NXI
05210 C
05220          C2X = OMEXIP
05230          C2Y = OMEYIP
05240          C2Z = OMEZIP
05250          C2XX = C2Y*NZI - C2Z*NYI
05260          C2YY = C2Z*NXI - C2X*NZI
05270          C2ZZ = C2X*NYI - C2Y*NXI
05280 C
05290          C0X    = H*( C1XX + C2XX ) + NXB(I)
05300          C0Y    = H*( C1YY + C2YY ) + NYB(I)
05310          C0Z    = H*( C1ZZ + C2ZZ ) + NZB(I)
05320                 C0     = DSQRT( C0X**2 + C0Y**2 + C0Z**2 )
```

- If too large displacement happens, this message is printed out on the display.
- The basic equation is Eq. (9.18).
- The following procedure is for modifying the magnetic moment direction. This is because the change in the particle direction naturally induces the change in the moment direction.
- The vector $\hat{\boldsymbol{e}}$ is defined as $\hat{\boldsymbol{e}} = \boldsymbol{e}(t+\Delta t) - \boldsymbol{e}(t)$, denoted by (EXIHAT,EYIHAT, EZIHAT).
- The vector $\boldsymbol{\delta}_\perp$ is defined as $\boldsymbol{\delta}_\perp = (\boldsymbol{e}\times\hat{\boldsymbol{e}})/|\boldsymbol{e}\times\hat{\boldsymbol{e}}|$, denoted by (DELNORMX, DELNORMY, DELNORMZ).
- The vector $\boldsymbol{n}$ is decomposed into $\boldsymbol{n}_h$ and $\boldsymbol{n}_v$ which are the components in the directions normal and parallel to the vector $\boldsymbol{\delta}_\perp$, respectively.
- The vector $\hat{\boldsymbol{n}}_h$ is defined as $\hat{\boldsymbol{n}}_h = |\hat{\boldsymbol{e}}| (\boldsymbol{\delta}_\perp \times \boldsymbol{n}_h)$.
- The vector $\boldsymbol{n}_v$ is not changeable for the particle rotation.
- The modified $\boldsymbol{n}$ is obtained from $\boldsymbol{n} = \boldsymbol{n}_v + \boldsymbol{n}_h + \hat{\boldsymbol{n}}_h$ ; $\boldsymbol{n}$ has to be a unit vector .
- Also, the modified $\boldsymbol{n}$ has to be normal to the particle direction $\boldsymbol{e}$.
- (C1XX,C1YY,C1ZZ) denotes the third term on the right-hand side of Eq. (9.19) without factor $\Delta t$.
- (C2XX,C2YY,C2ZZ) denotes the second term on the right-hand side of Eq. (9.19) without factor $\Delta t$.
- NXB(*), NYB(*) and NZB(*) are random displacements.

```
05330                IF( C0 .GT. NROTMX )  NROTMX = C0
05340                NROTAV  = NROTAV + C0
05350          IF( C0 .GT. 0.05D0 ) THEN
05360            WRITE(6,*) NTIME,'--- N_ROT IS TOO LARGE ---', REAL(C0)
05370            C0  = 0.05D0/C0
05380            C0X = C0 * C0X
05390            C0Y = C0 * C0Y
05400            C0Z = C0 * C0Z
05410          END IF
05420          NXI   = NXI + C0X
05430          NYI   = NYI + C0Y
05440          NZI   = NZI + C0Z
05450 C                  +++ (NX,NY,NZ) MUST BE NORMAL TO (EX,EY,EZ) +++
05460          C0    = NXI*EXI + NYI*EYI + NZI*EZI
05470          NXI   = NXI - C0*EXI
05480          NYI   = NYI - C0*EYI
05490          NZI   = NZI - C0*EZI
05500          C00   = DSQRT( NXI**2 + NYI**2 + NZI**2 )
05510          NX(I) = NXI/C00
05520          NY(I) = NYI/C00
05530          NZ(I) = NZI/C00
05540 C
05550   100   CONTINUE
05560 C
05570 C                                     ---------------- CAL FORCES ---
05580         CALL FORCECAL( DX, NP, NTIME )
05590 C                                     --- CAL RANDOM DISPLACEMENT ---
05600         CALL RANDISP( N , H , NTIME )
```

• If too large displacement happens, this message is printed out on the display.

• The basic equation is Eq. (9.19).

• Forces and torques are calculated.

• Translational and rotational random displacements are set.

```
05610 C
05620 C       -----------------------------------------------------------
05630 C
05640 C                                          --- MOMENT OF SYSTEM ---
05650         IF( MOD(NTIME,DNSMPL) .EQ. 0 ) THEN
05660           C1 = 0.D0
05670           C2 = 0.D0
05680           C3 = 0.D0
05690           DO 150 J=1,N
05700             C1 = C1 + NX(J)
05710             C2 = C2 + NY(J)
05720             C3 = C3 + NZ(J)
05730   150     CONTINUE
05740           NSMPL        = NSMPL + 1
05750           MOMX(NSMPL)  = REAL(C1)/REAL(N)
05760           MOMY(NSMPL)  = REAL(C2)/REAL(N)
05770           MOMZ(NSMPL)  = REAL(C3)/REAL(N)
05780         END IF
05790 C
05800 C       ---------------------------- DATA OUTPUT (1) FOR GRAPHICS ---
05810         IF( MOD(NTIME,NGRAPH) .EQ. 0 ) THEN
05820           NOPT = NOPT + 1
05830           WRITE(NOPT,472)  N , XL , YL , ZL , D , DT , RP , RP1 , DX
05840           WRITE(NOPT,474)  (RX(I),I=1,N),(RY(I),I=1,N),(RZ(I),I=1,N),
05850      &                     (EX(I),I=1,N),(EY(I),I=1,N),(EZ(I),I=1,N),
05860      &                     (NX(I),I=1,N),(NY(I),I=1,N),(NZ(I),I=1,N)
05870                                           CLOSE(NOPT,STATUS='KEEP')
05880         END IF
05890 C                                     --- DATA OUTPUT
05900         IF( MOD(NTIME,NANIME) .EQ. 0 ) THEN
05910           NANMCTR = NANMCTR + 1
05920           NOPT1   = 14
05930           CALL ANIMEOPT( NOPT1, NANMCTR, NTIMEMX, NANIME, N )
05940         END IF
05950 C                                           --- NORMALIZATION ---
05960         IF( MOD(NTIME,DN) .EQ. 0 ) THEN
05970           DO 490 I=1,N
05980             C1 = DSQRT( NX(I)**2 + NY(I)**2 + NZ(I)**2 )
05990             NX(I)  = NX(I)/C1
06000             NY(I)  = NY(I)/C1
06010             NZ(I)  = NZ(I)/C1
06020             C1 = DSQRT( EX(I)**2 + EY(I)**2 + EZ(I)**2 )
06030             EX(I)  = EX(I)/C1
06040             EY(I)  = EY(I)/C1
06050             EZ(I)  = EZ(I)/C1
06060   490     CONTINUE
06070         END IF
06080 C
06090 C
06100  1000 CONTINUE
06110 C
```

• Sampling is conducted every DNSMPL steps for evaluating characteristics of the system.

• Data output for making snapshots and animations by means of a comercial software MicroAVS.

• Modification is necessary for keeping the unit length of the vectors $\boldsymbol{e}_i$ and $\boldsymbol{n}_i$.

```
06120 C       ---------------------------------------------------------------
06130 C       -------------------- END OF BROWNIAN DYNAMICS     --------------
06140 C       ---------------------------------------------------------------
06150 C
06160 C                                              --- PRINT OUT (2) ---
06170         WRITE(NP,1011)
06180         NSMPL1 = 1
06190         NSMPL2 = NSMPL
06200         CALL PRNTDATA( NSMPL1 , NSMPL2 , NP )
06210         WRITE(NP,1013) NSMPL1 , NSMPL2
06220 C                                              --- DATA OUTPUT (3) ---
06230         WRITE(11,1111) N, VDENS, NDENS, RA, KU, RAV, RM, RH, RV
06240         WRITE(11,1113) PE, PETLD
06250         WRITE(11,1114) D, RP, TD, DEL, D1, RP1, PV
06260         WRITE(11,1115) D1, D2, B1, B2, RCHCK, RCHCK1, RCHCK2
06270         WRITE(11,1116) H, GAMDOT, HX, HY, HZ, RCOFF, BETA, XL,YL, ZL
06280         WRITE(11,1118) DT1, DT2, DR1, DR2, YHYC
06290         WRITE(11,1121) NTIMEMX, NGRAPH, NANIME
06300         WRITE(11,1123) DNSMPL
06310         WRITE(11,1128) INISQUX , INISQUY , INISQUZ
06320         WRITE(11,1130) NSMPL
06330         WRITE(11,1132) ( MOMX(I),I=NSMPL1, NSMPL2)
06340      &                ,( MOMY(I),I=NSMPL1, NSMPL2)
06350      &                ,( MOMZ(I),I=NSMPL1, NSMPL2)
06360 C
06370 C
06380         RIJMNAV = RIJMNSUM     / DBLE(NRIJMN)
06390         RMNMNAV = RMNMNSUM     / DBLE(NRMNMN)
06400         CNRIJMN = DBLE(NRIJMN) / DBLE(NTIMEMX)
```

• The following quantities are used for confirming the validity of the choice of the time interval. RTRANMX, EROTMX and NROTMX imply the maximum displacements of the translational position, the particle direction and the magnetic moment direction, respectively.

```
06410         WRITE(NP,1365) NRIJMN , NRMNMN , RIJMNAV , RMNMNAV , CNRIJMN
06420 C
06430         RTRANAV = RTRANAV/( DBLE(NTIMEMX*N) )
06440         EROTAV  = EROTAV /( DBLE(NTIMEMX*N) )
06450         NROTAV  = NROTAV /( DBLE(NTIMEMX*N) )
06460         WRITE(NP,1371) RTRANMX, EROTMX, NROTMX, RTRANAV, EROTAV, NROTAV
06470 C
06480                                             CLOSE(9, STATUS='KEEP')
06490                                             CLOSE(11,STATUS='KEEP')
06500                                             CLOSE(14,STATUS='KEEP')
06510 C       -------------------------- FORMAT ----------------------------
06520    12 FORMAT(/1H ,'-----------------------------------------------------'
06530      &        /1H ,'-    BROWNIAN DYNAMICS SIMULATIONS OF DISK-LIKE  -'
06540      &        /1H ,'-    PARTICLES IN A SIMPLE SHEAR FLOW            -'
06550      &        /1H ,'-----------------------------------------------------'
06560      &       //1H ,'N=',I4, 2X, 'VDENS=',F8.5, 2X ,'NDENS=',F9.6
06570      &        /1H ,'RAM=',F6.2, 1X, 'KU=',F9.2, 1X ,'RAV=',F6.2, 1X,
06580      &             'RM=',F10.3, 1X, 'RH=',F8.2, 1X ,'RV=',F9.2)
06590    14 FORMAT(/1H ,'PE=',F8.3, 2X, 'PETLD=',F8.2)
06600    16 FORMAT(/1H ,'D=',F5.2, 2X ,'RP=',F5.2, 2X, 'TD=',F5.2, 2X,
06610      &             'DEL=',F5.2, 2X ,'D1=',F5.2, 2X, 'RP1=',F5.2, 2X,
06620      &             'PV=',F7.3)
06630    17 FORMAT(/1H , 'D1=',F8.3,2X, 'D2=',F8.3,2X, 'B1=',F8.3,2X,
06640      &             'B2=',F8.3,2X, 'RCHCK=',F8.3,2X, 'RCHCK1=',F8.3,2X,
06650      &             'RCHCK2=',F8.3)
06660    18 FORMAT(/1H ,'H=',F10.7, 2X, 'GAMDOT=',F6.2, 2X,'HX=',F5.2, 2X,
06670      &             'HY=',F5.2, 2X, 'HZ=',F5.2, 2X,'RCOFF=',F6.2, 2X,
06680      &             'BETA=',F5.2
06690      &        /1H ,'XL=',F8.3, 2X, 'YL=',F8.3, 2X, 'ZL=',F8.3)
06700    20 FORMAT(/1H ,'DT1=',F8.4, 2X ,'DT2=',F8.4, 2X ,'DR1=',F8.4, 2X,
06710      &             'DR2=',F8.4, 2X ,'YHYC=',F8.4)
06720    21 FORMAT(/1H ,'NTIMEMX=',I8, 2X, 'NGRAPH=',I8,2X,'NANIME=',I8)
06730    22 FORMAT(/1H ,'DNSMPL=',I4 )
06740    25 FORMAT(/1H ,'CRNDMT1=',E12.4, 2X, 'CRNDMT2=',E12.4,
06750      &        /1H ,'CRNDMR1=',E12.4, 2X, 'CRNDMR2=',E12.4)
06760    28 FORMAT(/1H ,'INISQUX=',I3,2X, 'INISQUY=',I3,2X, 'INISQUZ=',I3/)
06770 C
06780   472 FORMAT( I5 , 3F9.4 , 4F8.4 , E16.8 )
06790   474 FORMAT( (5F16.10) )
06800  1011 FORMAT(/1H ,'++++++++++++++++++++++++++++++'
06810      &        /1H ,'        BD SIMULATIONS        '
06820      &        /1H ,'++++++++++++++++++++++++++++++'/)
06830  1013 FORMAT(///1H ,18X, 'START OF BD SAMPLING STEP=',I7
06840      &        /1H ,18X, 'END   OF BD SAMPLING STEP=',I7/)
06850  1111 FORMAT( I5 , 2F7.4 , 3F9.3 , F8.4 , 2F11.3 )
06860  1113 FORMAT( 2E12.4 )
06870  1114 FORMAT( 7F10.4 )
06880  1115 FORMAT( 7F8.4 )
06890  1116 FORMAT( E12.4 , F6.2 , 3F6.2 , 2F7.2, 3F7.2 )
06900  1118 FORMAT( 5E12.4 )
```

• RMNMNAV is the average of the minimum distance between two particles in one time step.

```
06910  1121 FORMAT( 3I9 )
06920  1123 FORMAT( I6 )
06930  1128 FORMAT( 3I8 )
06940  1130 FORMAT( I9 )
06950  1132 FORMAT( (10F8.5) )
06960  1365 FORMAT(/1H ,'NRIJMN=', I8, 1X,'NRMNMN=', I8, 1X, 'RIJMNAV=',
06970      &              F6.3, 1X, 'RMNMNAV=', F6.3, 1X, 'CNRIJMN=', F6.3/)
06980  1371 FORMAT(/1H ,'R_TRAN_MAX=',E10.3, 1X, 'E_ROT_MAX=',E10.3, 1X,
06990      &             'N_ROT_MAX= ',E10.3
07000      &         /1H ,'R_TRAN_AV =',E10.3, 1X, 'E_ROT_AV =',E10.3, 1X,
07010      &             'N_ROT_AV = ',E10.3       )
07020                                                                     STOP
07030                                                                     END
07040 C*********************************************************************
07050 C**************************    SUBROUTINE    *************************
07060 C*********************************************************************
07070 C
07080 C**** SUB INITIAL ****
07090       SUBROUTINE INITIAL
07100 C
07110       IMPLICIT REAL*8 (A-H,O-Z), INTEGER (I-N)
07120 C
07130       COMMON /BLOCK1/  RX    , RY    , RZ
07140       COMMON /BLOCK2/  EX    , EY    , EZ
07150       COMMON /BLOCK3/  NX    , NY    , NZ
```

· An initial configuration regarding the particle positions and the particle and magnetic moment directions is assigned.

```
07160       COMMON /BLOCK6/  RXB   , RYB   , RZB
07170       COMMON /BLOCK7/  EXB   , EYB   , EZB
07180       COMMON /BLOCK8/  NXB   , NYB   , NZB
07190       COMMON /BLOCK11/ XL    , YL    , ZL
07200       COMMON /BLOCK12/ HX    , HY    , HZ
07210       COMMON /BLOCK13/ RA    , KU    , RAV
07220       COMMON /BLOCK14/ RM    , RH    , RV
07230       COMMON /BLOCK15/ N     , NDENS, VDENS
07240       COMMON /BLOCK16/ D, D1, D2, B, B1, B2, RP, RP1, VP
07250       COMMON /BLOCK17/ DEL, TD, RCOFF
07260 C
07270       COMMON /BLOCK18/ INISQUX , INISQUY , INISQUZ , BETA
07280       COMMON /BLOCK21/ RMAT
07290 C   --------------------------
07300       INTEGER    NN , NNS
07310       PARAMETER( NN=1000 , NNS=5000000 )
07320       PARAMETER( NRANMX=1000000 , PI=3.141592653589793D0 )
07330 C
07340       REAL*8    KU      , NDENS
07350       REAL*8    RX(NN) , RY(NN) , RZ(NN)
07360       REAL*8    EX(NN) , EY(NN) , EZ(NN) , NX(NN) , NY(NN) , NZ(NN)
07370
07380       REAL*8    RMAT(3,3,NN)
07390
07400       REAL*8    RXB(NN), RYB(NN), RZB(NN)
07410       REAL*8    EXB(NN), EYB(NN), EZB(NN), NXB(NN), NYB(NN), NZB(NN)
07420 C
07430       INTEGER   PTCL , ICNTR
07440       REAL*8    XLUNT , YLUNT , ZLUNT,  RAN1 , RAN2 , RAN3
07450       REAL*8    VDENSMX , CRATIO , C0 , C1 , C2 , C3 , CNXB , CNYB
07460 C   ------------------------------------------------------------------
07470       INIPX = INISQUX
07480       INIPY = INISQUY
07490       INIPZ = INISQUZ
07500 C
07510       CL = (  DBLE(N)*VP/(BETA*VDENS)  )**(1./3.)
07520 C
07530       XLUNT    = CL / DBLE(INIPX )
07540       ZLUNT    = CL / DBLE(INIPZ )
07550       YLUNT    = CL*BETA / DBLE(INIPY )
07560       XL       = CL
07570       ZL       = CL
07580       YL       = CL*BETA
07590 C                                                    ----- POSITION -----
07600       RAN1 = DSQRT( 2.D0 )
07610       RAN2 = DSQRT( 7.D0 )
07620       RAN3 = DSQRT( 11.D0 )
07630       C0   = 1.D-4
07640       PTCL = 0
07650       DO 10 K=0, INIPZ-1
07660       DO 10 J=0, INIPY-1
07670       DO 10 I=0, INIPX-1
07680         PTCL = PTCL + 1
07690         C1 = RAN1*DBLE(PTCL)
07700         C1 = C1 - DINT(C1)
07710         C1 = C1 - 0.5D0
```

· An initial position of each particle is assigned using quasi-random numbers ranging from 0 to 1.

```
07720            C2 = RAN2*DBLE(PTCL)
07730            C2 = C2 - DINT(C2)
07740            C2 = C2 - 0.5D0
07750            C3 = RAN3*DBLE(PTCL)
07760            C3 = C3 - DINT(C3)
07770            C3 = C3 - 0.5D0
07780            RX(PTCL) = DBLE(I)*XLUNT + C1*C0 + D1/2.D0 - XL/2.D0
07790            RY(PTCL) = DBLE(J)*YLUNT + C2*C0 + D1/2.D0 - YL/2.D0
07800            RZ(PTCL) = DBLE(K)*ZLUNT + C3*C0 + B1/2.D0 - ZL/2.D0
07810     10 CONTINUE
07820         N = PTCL
07830 C                                                   ----- DIRECTION -----
07840         RAN1 = DSQRT( 2.D0 )
07850         RAN2 = DSQRT( 3.D0 )
07860         DO 80 I=1,N
07870           C1 = 0.D0
07880           C2 = 0.D0
07890           EX(I) = DSIN(C1)*DCOS(C2)
07900           EY(I) = DSIN(C1)*DSIN(C2)
07910           EZ(I) = DCOS(C1)
07920 C
07930           RMAT(1,1,I) =  DCOS(C1)*DCOS(C2)
07940           RMAT(2,1,I) =  DCOS(C1)*DSIN(C2)
07950           RMAT(3,1,I) = -DSIN(C1)
07960           RMAT(1,2,I) = -DSIN(C2)
07970           RMAT(2,2,I) =  DCOS(C2)
07980           RMAT(3,2,I) =  0.D0
07990           RMAT(1,3,I) =  DSIN(C1)*DCOS(C2)
08000           RMAT(2,3,I) =  DSIN(C1)*DSIN(C2)
08010           RMAT(3,3,I) =  DCOS(C1)
08020     80 CONTINUE
08030 C
08040         RAN1 = DSQRT( 2.D0 )
08050         DO 90 I=1,N
08060           C1      = RAN1*DBLE(I)
08070           C1      = C1 - DINT(C1)
08080           CNPSI   = 2.D0*PI*C1
08090           CNXB    = DCOS(CNPSI)
08100           CNYB    = DSIN(CNPSI)
08110           NX(I)   = RMAT(1,1,I)*CNXB + RMAT(1,2,I)*CNYB
08120           NY(I)   = RMAT(2,1,I)*CNXB + RMAT(2,2,I)*CNYB
08130           NZ(I)   = RMAT(3,1,I)*CNXB + RMAT(3,2,I)*CNYB
08140     90 CONTINUE
08150                                                           RETURN
08160                                                           END
08170 C**** SUB PTCLMDL *****
08180         SUBROUTINE PTCLMDL( RP )
08190 C
08200         IMPLICIT REAL*8 (A-H,O-Z), INTEGER (I-N)
08210 C
08220         COMMON /BLOCK25/  RXSTRCG , RYSTRCG , RDSTRCG
08230         COMMON /BLOCK26/  NPROWALL, NPROW   , NPROWST
08240         COMMON /BLOCK27/  NPSTRC  , RXSTRC  , RYSTRC
08250 C
08260         PARAMETER( NNSTRC=100 )
08270 C
08280         INTEGER  NPROWALL, NPROW( NNSTRC ) , NPROWST( NNSTRC )
08290         REAL*8   RXSTRC( NNSTRC ) , RYSTRC( NNSTRC )
08300 C
08310         REAL*8   CX , CY
08320         INTEGER  IPTCL , IPTCL0 , IROW , IIEND
08330 C       -------------------------------------------------------------
08340         NPROWALL = INT( RP + 0.0001D0)
08350         IIEND    = NPROWALL
08360 C
08370         RXSTRCG    = 0.D0
08380         RYSTRCG    = (RP-1.D0)/2.D0
08390         RDSTRCG    = (RP-1.D0)/2.D0 + 0.0001D0
08400 C
08410         NPROW(1)    = 1
08420         NPROWST(1)  = 1
08430         RXSTRC(1)   = 0.D0
08440         RYSTRC(1)   = 0.D0
08450 C
08460         IPTCL = 1
08470         DO 100 IROW = 2, NPROWALL
08480           IPTCL = IPTCL + 1
08490           NPROWST(IROW)  = IPTCL
08500           RXSTRC(IPTCL)  = 0.D0
08510           RYSTRC(IPTCL)  = DBLE(IROW-1)
08520           IPTCL0 = IPTCL
08530           III = 1
08540           DO 50 II=1, IIEND
```

· The origin of the absolute coordinate system is at the center of the simulation box.

· An initial direction of each particle direction is (0,0,1).

· The matrix RMAT(*,+,#) can relate the body-fixed to the absolute coordinate systems.

· An initial direction of each magnetic moment is randomly assigned in the range of 0° to 360° in the body-fixed coordinate system.

· Transfer to the absolute coordinate system is conducted using the orientation matrix.

· The position of the constituent spheres of the plain sphere-constituting model is specified in subroutine PTCLMDL. These variables are used in evaluating the steric interaction forces.

· The first criterion sphere is located at the bottom position in an inscribed manner. Then, the *XY*-coordinate system is assigned along the disk surface: the *X*-axis and *Y*-axis are set in the right and in the upward direction, respectively.

· The particles are located in the shell area with the same radial distance. This procedure is sequentially performed with increasing radial distance by particle diameter in increments.

· The position of each sphere is specified by (RXSTRC(*), RYSTRC(*)).

```
08550              CX =  RXSTRC(IPTCL0) - DBLE(II)*( DSQRT(3.D0)/2.D0 )
08560              CY =  RYSTRC(IPTCL0) - DBLE(II)*( 1.D0/2.D0 )
08570              IF( DSQRT( (CX-RXSTRCG)**2+(CY-RYSTRCG)**2 )
08580     &                                        .LE. RDSTRCG ) THEN
08590                IPTCL = IPTCL + 1
08600                RXSTRC(IPTCL) =-CX
08610                RYSTRC(IPTCL) = CY
08620                IPTCL = IPTCL + 1
08630                RXSTRC(IPTCL) = CX
08640                RYSTRC(IPTCL) = CY
08650                III = III + 2
08660              END IF
08670    50      CONTINUE
08680            NPROW(IROW) = III
08690   100    CONTINUE
08700 C
08710          NPSTRC = IPTCL
08720                                                              RETURN
08730                                                              END
```

• The particles are located in a contact situation.

• The number of spheres in each shell area is specified by NPROW(*).
• The number of the shells is NPROWALL.
• The number of constituent spheres forming the disk-like particle is denoted by NPSTRC.

```
08740 C**** SUB PRNTDATA ****
08750          SUBROUTINE PRNTDATA( MCSST, MCSMX, NP )
08760 C
08770          IMPLICIT REAL*8 (A-H,O-Z), INTEGER (I-N)
08780 C
08790          COMMON /BLOCK43/ MOMX , MOMY , MOMZ
08800 C
08810          INTEGER    NN , NNS
08820          PARAMETER( NN=1000 , NNS=5000000 )
08830          PARAMETER( NRANMX=1000000 , PI=3.14159265358979300 )
08840 C
08850          INTEGER  MCSST     , MCSMX      , NP
08860 C
08870          REAL     MOMX(NNS), MOMY(NNS) , MOMZ(NNS)
08880 C
08890          REAL     AMOMX(10) , AMOMY(10)  , AMOMZ(10)  , C0
08900          REAL     AVIS(10)  , AVISF(10)  , AVISTF(10) , AVISTH(10)
08910          INTEGER  IC , IMC(0:10) , JS , JE
08920 C        ---------------------------------------------------------------
08930 C                                               ----- INSTANT VALUES -----
08940          IC = ( MCSMX-MCSST+1 )/50
08950          DO 20 I= MCSST-1+IC , MCSMX , IC
08960            WRITE(NP,10) I, MOMX(I), MOMY(I), MOMZ(I)
08970     20 CONTINUE
08980 C                                             ----- TIME STEP AVERAGE -----
08990          IC = ( MCSMX-MCSST+1 )/10
09000          DO 30 I=0,10
09010            IMC(I) = MCSST - 1 + IC*I
09020            IF( I .EQ. 10 ) IMC(I) =MCSMX
09030     30 CONTINUE
09040 C
09050 C
09060          DO 35 I=1,10
09070            AMOMX( I) = 0.
09080            AMOMY( I) = 0.
09090            AMOMZ( I) = 0.
09100     35 CONTINUE
09110 C
09120          DO 50 I=1,10
09130            JS = IMC(I-1) + 1
09140            JE = IMC(I)
09150            DO 40 J=JS,JE
09160              AMOMX(I)   = AMOMX(I)   + MOMX( J)
09170              AMOMY(I)   = AMOMY(I)   + MOMY( J)
09180              AMOMZ(I)   = AMOMZ(I)   + MOMZ( J)
09190     40     CONTINUE
09200     50 CONTINUE
09210 C
09220          DO 70 I=1,10
09230            C0        = REAL( IMC(I)-IMC(I-1) )
09240            AMOMX(I)   = AMOMX(I) /C0
09250            AMOMY(I)   = AMOMY(I) /C0
09260            AMOMZ(I)   = AMOMZ(I) /C0
09270     70 CONTINUE
09280 C                                     ----- PRINT OUT STEP AVERAGE -----
09290          WRITE(NP,75)
09300          DO 90 I=1,10
09310           WRITE(NP,80) I, IMC(I-1)+1, IMC(I), AMOMX(I),AMOMY(I),AMOMZ(I)
09320     90 CONTINUE
09330 C        ---------------------------------------------------------------
09340     10 FORMAT(1H ,'SMPL=',I7, 1X ,'NX=',F7.4, 1X,'NY=',F7.4,
09350     &                                          1X,'NZ=',F7.4)
09360     75 FORMAT(//1H ,'-----------------------------------------'
```

• In subroutine PRNTDATA, the ten sub-averages of MOMY(*) are calculated and printed out.

```
09370      &       /1H ,'                         TIME AVERAGE                    '
09380      &       /)
09390   80 FORMAT(1H ,'I=',I2, 2X ,'SMPLMN=',I8, 2X ,'SMPLMX=',I8
09400      &       /1H ,5X ,'NX=',F7.4, 2X,'NY=',F7.4, 2X,'NZ=',F7.4 )
09410                                                                       RETURN
09420                                                                       END
09430 C**** SUB ANIMEOPT ****
09440       SUBROUTINE ANIMEOPT( NOPT1, NANMCTR, MCSMPLMX, NANIME, N )
09450 C
09460       IMPLICIT REAL*8 (A-H,O-Z), INTEGER (I-N)
09470 C
09480       COMMON /BLOCK1/  RX   , RY   , RZ
09490       COMMON /BLOCK2/  EX   , EY   , EZ
09500       COMMON /BLOCK3/  NX   , NY   , NZ
09510       COMMON /BLOCK11/ XL   , YL   , ZL
09520       COMMON /BLOCK12/ HX   , HY   , HZ
09530       COMMON /BLOCK16/ D, D1, D2, B, B1, B2, RP, RP1, VP
09540       COMMON /BLOCK17/ DEL, TD, RCOFF
09550 C
09560       INTEGER    NN , NNS
09570       PARAMETER( NN=1000 , NNS=5000000 )
09580       PARAMETER( NRANMX=1000000 , PI=3.141592653589793D0 )
09590 C
09600       REAL*8   RX(NN) , RY(NN) , RZ(NN)
09610       REAL*8   EX(NN) , EY(NN) , EZ(NN)
09620       REAL*8   NX(NN) , NY(NN) , NZ(NN)
09630       REAL*8   D02 , D102 , D202 , CX1 , CY1 , CZ1 , CX2 , CY2 , CZ2
09640       REAL*8   C0 , CR
09650       REAL*8   CNX(50) , CNY(50) , CNZ(50)
09660 C     -------------------------------------------------------------------
09670       D02  = D/2.D0
09680       D102 = D1/2.D0
09690       D202 = D2/2.D0
09700 C
09710       IF( NANMCTR .EQ. 1 ) THEN
09720         WRITE(NOPT1,181) (MCSMPLMX/NANIME)
09730       END IF
09740 C
09750       IF( (NANMCTR.GE.1) .AND. (NANMCTR.LE.9) ) THEN
09760         WRITE(NOPT1,183) NANMCTR
09770       ELSE IF( (NANMCTR.GE.10) .AND. (NANMCTR.LE.99) ) THEN
09780         WRITE(NOPT1,184) NANMCTR
09790       ELSE IF( (NANMCTR.GE.100) .AND. (NANMCTR.LE.999) ) THEN
09800         WRITE(NOPT1,185) NANMCTR
09810       ELSE IF( (NANMCTR.GE.1000) .AND. (NANMCTR.LE.9999) ) THEN
09820         WRITE(NOPT1,186) NANMCTR
09830       END IF
09840 C     ---------------------------------------------------- CYLINDER (1) ---
09850       WRITE(NOPT1,211)  N
09860       DO 250 I=1,N
09870 CCCCC   C0 = 0.5D0
09880         C0 = B2/2.D0
09890         CX1 = RX(I) - EX(I)*C0
09900         CY1 = RY(I) - EY(I)*C0
09910         CZ1 = RZ(I) - EZ(I)*C0
09920         CX2 = RX(I) + EX(I)*C0
09930         CY2 = RY(I) + EY(I)*C0
09940         CZ2 = RZ(I) + EZ(I)*C0
09950         WRITE(NOPT1,248) CX1,CY1,CZ1, CX2,CY2,CZ2, D02, 1.0, 0.0, 0.0
09960   250 CONTINUE
09970 C     ------------------------------------------------------ SPHERE (1) ---
09980 C                                          --- FOR MAKING OUTER SHAPE ---
09990       WRITE(NOPT1,311)  N*16
10000       DO 350 I=1,N
10010         CNX(1) = NX(I)
10020         CNY(1) = NY(I)
10030         CNZ(1) = NZ(I)
10040 C
10050         C1X    = EY(I)*NZ(I) - EZ(I)*NY(I)
10060         C1Y    = EZ(I)*NX(I) - EX(I)*NZ(I)
10070         C1Z    = EX(I)*NY(I) - EY(I)*NX(I)
10080         C1     = DSQRT( C1X**2 + C1Y**2 + C1Z**2 )
10090         CNX(2) = C1X/C1
10100         CNY(2) = C1Y/C1
10110         CNZ(2) = C1Z/C1
10120         CNX(3) = - CNX(2)
10130         CNY(3) = - CNY(2)
```

• For a commercial software MicroAVS which makes snapshots and animations.

```
10140          CNZ(3) = - CNZ(2)
10150 C
10160          CNX(4) =  ( CNX(1) + CNX(2) )/1.4142D0
10170          CNY(4) =  ( CNY(1) + CNY(2) )/1.4142D0
10180          CNZ(4) =  ( CNZ(1) + CNZ(2) )/1.4142D0
10190          CNX(5) =  ( CNX(1) + CNX(3) )/1.4142D0
10200          CNY(5) =  ( CNY(1) + CNY(3) )/1.4142D0
10210          CNZ(5) =  ( CNZ(1) + CNZ(3) )/1.4142D0
10220 C
10230          CNX(6) =  ( CNX(1) + CNX(4) )/1.8478D0
10240          CNY(6) =  ( CNY(1) + CNY(4) )/1.8478D0
10250          CNZ(6) =  ( CNZ(1) + CNZ(4) )/1.8478D0
10260          CNX(7) =  ( CNX(2) + CNX(4) )/1.8478D0
10270          CNY(7) =  ( CNY(2) + CNY(4) )/1.8478D0
10280          CNZ(7) =  ( CNZ(2) + CNZ(4) )/1.8478D0
10290          CNX(8) =  ( CNX(1) + CNX(5) )/1.8478D0
10300          CNY(8) =  ( CNY(1) + CNY(5) )/1.8478D0
10310          CNZ(8) =  ( CNZ(1) + CNZ(5) )/1.8478D0
10320          CNX(9) =  ( CNX(3) + CNX(5) )/1.8478D0
10330          CNY(9) =  ( CNY(3) + CNY(5) )/1.8478D0
10340          CNZ(9) =  ( CNZ(3) + CNZ(5) )/1.8478D0
10350 C
10360          CNX(10) = - CNX(1)
10370          CNY(10) = - CNY(1)
10380          CNZ(10) = - CNZ(1)
10390          CNX(11) =- CNX(4)
10400          CNY(11) = - CNY(4)
10410         CNZ(11) = - CNZ(4)
10420         CNX(12) = - CNX(5)
10430         CNY(12) = - CNY(5)
10440         CNZ(12) = - CNZ(5)
10450         CNX(13) = - CNX(6)
10460         CNY(13) = - CNY(6)
10470         CNZ(13) = - CNZ(6)
10480         CNX(14) = - CNX(7)
10490         CNY(14) = - CNY(7)
10500         CNZ(14) = - CNZ(7)
10510         CNX(15) = - CNX(8)
10520         CNY(15) = - CNY(8)
10530         CNZ(15) = - CNZ(8)
10540         CNX(16) = - CNX(9)
10550         CNY(16) = - CNY(9)
10560         CNZ(16) = - CNZ(9)
10570 C
10580         DO 340 J=1,16
10590           CX1 = RX(I) + CNX(J)*D02
10600           CY1 = RY(I) + CNY(J)*D02
10610           CZ1 = RZ(I) + CNZ(J)*D02
10620 CCCCC     CR = 0.499D0
10630           CR = B2/2.D0 - 0.001D0
10640           WRITE(NOPT1,348) CX1, CY1, CZ1, CR , 1.0, 0.2, 0.2
10650   340   CONTINUE
10660 C
10670   350 CONTINUE
10680 C      ---------------------------------------------------- SPHERE (2) ---
10690 C                                                --- FOR MAG MOMENT ---
10700       WRITE(NOPT1,311)  N
10710       DO 450 I=1,N
10720         CX1 = RX(I) + NX(I)*D202
10730         CY1 = RY(I) + NY(I)*D202
10740         CZ1 = RZ(I) + NZ(I)*D202
10750         WRITE(NOPT1,348) CX1, CY1, CZ1, 0.12 , 0.0, 0.8, 1.0
10760   450 CONTINUE
10770 C      -------------------------------------- SIM.REGEON LINES  (3) ---
10780       XL2 = XL/2.0
10790       YL2 = YL/2.0
10800       ZL2 = ZL/2.0
10810       WRITE(NOPT1,648)  17
10820       WRITE(NOPT1,649)  0.-XL2 ,  0.-YL2 ,  0.-ZL2
10830       WRITE(NOPT1,649)  XL-XL2 ,  0.-YL2 ,  0.-ZL2
10840       WRITE(NOPT1,649)  XL-XL2 ,  YL-YL2 ,  0.-ZL2
10850       WRITE(NOPT1,649)  0.-XL2 ,  YL-YL2 ,  0.-ZL2
10860       WRITE(NOPT1,649)  0.-XL2 ,  0.-YL2 ,  0.-ZL2
10870       WRITE(NOPT1,649)  0.-XL2 ,  0.-YL2 ,  ZL-ZL2
10880       WRITE(NOPT1,649)  XL-XL2 ,  0.-YL2 ,  ZL-ZL2
10890       WRITE(NOPT1,649)  XL-XL2 ,  YL-YL2 ,  ZL-ZL2
10900       WRITE(NOPT1,649)  0.-XL2 ,  YL-YL2 ,  ZL-ZL2
10910       WRITE(NOPT1,649)  0.-XL2 ,  0.-YL2 ,  ZL-ZL2
10920       WRITE(NOPT1,649)  0.-XL2 ,  0.-YL2 ,  0.-ZL2
10930       WRITE(NOPT1,649)  0.-XL2 ,  YL-YL2 ,  0.-ZL2
```

```
10940        WRITE(NOPT1,649)  0.-XL2 ,  YL-YL2 ,  ZL-ZL2
10950        WRITE(NOPT1,649)  XL-XL2 ,  YL-YL2 ,  ZL-ZL2
10960        WRITE(NOPT1,649)  XL-XL2 ,  YL-YL2 ,  0.-ZL2
10970        WRITE(NOPT1,649)  XL-XL2 ,  0.-YL2 ,  0.-ZL2
10980        WRITE(NOPT1,649)  XL-XL2 ,  0.-YL2 ,  ZL-ZL2
10990 C
11000 C      ---------------------------- FORMAT ---------------------------
11010    181 FORMAT('# Micro AVS Geom:2.00'
11020      &        /'# Animation of MC simulation results'
11030      &        /I4)
11040    183 FORMAT('step',I1)
11050    184 FORMAT('step',I2)
11060    185 FORMAT('step',I3)
11070    186 FORMAT('step',I4)
11080    211 FORMAT( 'column'/'cylinder'/'dvertex_and_color'/'32'/I7 )
11090    248 FORMAT( 6F10.3 , F6.2 , 3F4.1)
11100    311 FORMAT( 'sphere'/'sphere_sample'/'color'/I7 )
11110    348 FORMAT( 3F10.3 , F6.2 , 3F5.2 )
11120    648 FORMAT( 'polyline'/'pline_sample'/'vertex'/I3 )
11130    649 FORMAT( 3F10.3 )
11140                                                                 RETURN
11150                                                                 END
11160 C**** SUB FORCECAL ****
11170        SUBROUTINE FORCECAL( DX, NP, NTIME )
11180 C
11190        IMPLICIT REAL*8 (A-H,O-Z), INTEGER (I-N)
11200 C
11210          COMMON /BLOCK1/  RX    , RY    , RZ
11220          COMMON /BLOCK2/  EX    , EY    , EZ
11230          COMMON /BLOCK3/  NX    , NY    , NZ
11240          COMMON /BLOCK4/  FX    , FY    , FZ
11250          COMMON /BLOCK5/  TX    , TY    , TZ
11260          COMMON /BLOCK11/ XL    , YL    , ZL
11270          COMMON /BLOCK12/ HX    , HY    , HZ
11280          COMMON /BLOCK13/ RA    , KU    , RAV
11290          COMMON /BLOCK14/ RM    , RH    , RV
11300          COMMON /BLOCK15/ N     , NDENS, VDENS
11310          COMMON /BLOCK16/ D, D1, D2, B, B1, B2, RP, RP1, VP
11320          COMMON /BLOCK17/ DEL, TD, RCOFF
11330          COMMON /BLOCK18/ INISQUX , INISQUY , INISQUZ , BETA
11340          COMMON /BLOCK19/ RA0
11350          COMMON /BLLOC20/ RCHCK , RCHCK1 , RCHCK2
11360 C
11370          COMMON /BLOCK25/ RXSTRCG , RYSTRCG , RDSTRCG
11380          COMMON /BLOCK26/ NPROWALL, NPROW   , NPROWST
11390          COMMON /BLOCK27/ NPSTRC  , RXSTRC  , RYSTRC
11400 C
11410          COMMON /BLOCK37/ PE    , PETLD
11420          COMMON /BLOCK38/ DT1   , DT2   , DR1 , DR2 , YHYC
11430 C
11440          COMMON /WORK11/  XRXI , YRYI , ZRZI , XRXJ , YRYJ , ZRZJ
11450          COMMON /WORK12/  FXIJS, FYIJS, FZIJS, FXJIS, FYJIS, FZJIS
11460          COMMON /WORK13/  TQXIJS,TQYIJS,TQZIJS,TQXJIS,TQYJIS,TQZJIS
11470          COMMON /WORK14/  RCOFF2 , RP102
11480          COMMON /WORK17/  NRIJMN   , RIJMN  , RIJMNAV , RIJMNSUM
11490          COMMON /WORK18/  NRMNMN   , RMNMN  , RMNMNAV , RMNMNSUM
11500 C
11510          COMMON /WORK25/  DSQ, D1SQ, D2SQ, BSQ, B1SQ, B2SQ
11520          COMMON /WORK26/  RCHCKSQ , RCHCK1SQ , RCHCK2SQ
11530          COMMON /WORK27/  RPXIMN, RPYIMN, RPZIMN, RPXJMN, RPYJMN, RPZJMN
11540 C
11550 C        ----------------
11560          INTEGER     NN , NNS
11570          PARAMETER( NN=1000 , NNS=5000000 )
11580          PARAMETER( NRANMX=1000000 , PI=3.141592653589793D0 )
11590 C
11600          REAL*8    KU      , NDENS
11610          REAL*8    RX(NN) , RY(NN) , RZ(NN)
11620          REAL*8    EX(NN) , EY(NN) , EZ(NN) , NX(NN) , NY(NN) , NZ(NN)
11630          REAL*8    FX(NN) , FY(NN) , FZ(NN) , TX(NN) , TY(NN) , TZ(NN)
11640 C
11650          INTEGER  NRIJMN , NRMNMN
11660          REAL*8    RIJMN  , RIJMNAV , RIJMNSUM
11670          REAL*8    RMNMN  , RMNMNAV , RMNMNSUM
11680 C        ---------------
11690          PARAMETER( NNSTRC=100 )
11700 C
```

・Forces and torques are calculated in this subroutine.

```
11710        INTEGER  NPROWALL, NPROW( NNSTRC ) , NPROWST( NNSTRC )
11720        REAL*8   RXSTRC( NNSTRC ) , RYSTRC( NNSTRC )
11730 C      ---------------
11740        REAL*8   CORY
11750        REAL*8   RXI  , RYI  , RZI  , RXJ  , RYJ  , RZJ
11760        REAL*8   RXIJ , RYIJ , RZIJ , RXJI , RYJI , RZJI
11770        REAL*8   RIJ  , RIJSQ, RIJ3
11780        REAL*8   NXI  , NYI   , NZI  , NXJ  , NYJ  , NZJ
11790        REAL*8   EXI  , EYI   , EZI  , EXJ  , EYJ  , EZJ
11800        REAL*8   FXI , FYI , FZI , TXI  , TYI  , TZI
11810        REAL*8   FXJ , FYJ , FZJ , TXJ  , TYJ  , TZJ
11820        REAL*8   FXIJ   , FYIJ  , FZIJ  , FXJI  , FYJI  , FZJI
11830        REAL*8   RTXIJ  , RTYIJ , RTZIJ , RTXJI , RTYJI , RTZJI
11840        REAL*8   TXIJ0  , TYIJ0 , TZIJ0
11850        REAL*8   XI , YI , ZI , XJ , YJ , ZJ
11860 C
11870        REAL*8   TXIJS, TYIJS, TZIJS
11880        REAL*8   EXIS , EYIS , EZIS , EXJS , EYJS , EZJS
11890        REAL*8   KIS  , KJS  , KIJS , KIJQ
11900 C
11910        REAL*8   RRXI  , RRYI  , RRZI  , RRXJ  , RRYJ  , RRZJ
11920        REAL*8   RRXIJ , RRYIJ , RRZIJ , RRXJI , RRYJI , RRZJI
11930        REAL*8   TTXIJS, TTYIJS, TTZIJS
11940        REAL*8   EEXI  , EEYI  , EEZI  , EEXJ  , EEYJ  , EEZJ
11950        REAL*8   NNXI  , NNYI  , NNZI  , NNXJ  , NNYJ  , NNZJ
11960        REAL*8   EEXIS , EEYIS , EEZIS , EEXJS , EEYJS , EEZJS
11970        REAL*8   KKIS  , KKJS  , KKIJS
11980 C
11990        REAL*8   RIJMNFUN , RCOFF2 , D02 , CHCK0 , CHCK1
12000 C
12010        REAL*8   RPXI , RPYI , RPZI , RPXJ , RPYJ , RPZJ
12020        REAL*8   RPXIJ, RPYIJ, RPZIJ, RPXJI, RPYJI, RPZJI
12030        REAL*8   EPXI , EPYI , EPZI , EPXJ , EPYJ , EPZJ
12040        REAL*8   DPXXI, DPXYI, DPXZI, DPYXJ, DPYYJ, DPYZJ
12050        REAL*8   RXJQ , RYJQ , RZJQ , RXIJQ, RYIJQ, RZIJQ
12060        REAL*8   RXSTRC0 , RYSTRC0 , CRISQ , CRJSQ
12070        REAL*8   RORGNXI , RORGNYI , RORGNXZI
12080        REAL*8   RORGNXJ , RORGNYJ , RORGNXZJ
12090 C
12100        REAL*8   HIQ   , HJQ
12110        REAL*8   RRXIQ , RRYIQ , RRZIQ , RRXJQ , RRYJQ , RRZJQ
12120        REAL*8   RRMDLX, RRMDLY, RRMDLZ
12130 C
12140        REAL*8   C0  , C1  , C2   , C00 , C01 , C02 , C03
12150        REAL*8   C11 , C12 , C21  , C22
12160        REAL*8   C11X, C11Y, C22X, C22Y, C22Z, C33X, C33Y, C33Z
12170        REAL*8   C44X, C44Y, C44Z, C55X, C55Y, C55Z
12180        REAL*8   C1X , C1Y , C1Z  , C1SQ
12190        REAL*8   CIX , CIY , CIZ  , CJX , CJY , CJZ
12200        REAL*8   CEIEJ , CEIRIJ , CEJEIX , CEJEIY  , CEJEIZ , CRIJSQ
12210        REAL*8   CRIJ  , CRIJ3  , CRIJ4
12220        REAL*8   C1XI , C1YI , C1ZI , C1XJ , C1YJ , C1ZJ
12230        REAL*8   CC1X , CC1Y , CC1Z , CC1SQ
12240 C
12250        INTEGER  ITREE, IPATH, JPTCL, ISKIP, IBRANCH, IWAY
12260        INTEGER  ICNTR, NPROWAL9
12270 C
12280        LOGICAL  KEEP , IJOVRLAP , IOVRLAP , LGCWALL , KEEP1 , SKIPTRQ
12290 C      ----------------------------------------------------------------
12300        DO 10 I=1,N
12310          FX(I) =  0.D0
12320          FY(I) =  0.D0
12330          FZ(I) =  0.D0
12340          TX(I) =  0.D0
12350          TY(I) =  0.D0
12360          TZ(I) =  0.D0
12370     10 CONTINUE
12380 C
12390        D02     = D/2.D0
12400        RMNMN   = 1.0D30
12410        IOVRLAP= .FALSE.
12420 C
12430 C      ---------------------------------------------- MAIN LOOP START
12440        DO 2000 I=1,N-1
12450 C
12460          RXI  = RX(I)
12470          RYI  = RY(I)
12480          RZI  = RZ(I)
```

```
12490          EXI  = EX(I)
12500          EYI  = EY(I)
12510          EZI  = EZ(I)
12520          NXI  = NX(I)
12530          NYI  = NY(I)
12540          NZI  = NZ(I)
12550          FXI  = FX(I)
12560          FYI  = FY(I)
12570          FZI  = FZ(I)
12580          TXI  = TX(I)
12590          TYI  = TY(I)
12600          TZI  = TZ(I)
12610 C
12620        DO 1000 JPTCL=I+1,N
12630 C
12640          J = JPTCL
12650          IF( J .EQ. I )     GOTO 1000
12660 C
12670          RIJMN     = 1.D30
12680          IJOVRLAP = .FALSE.
12690 C
12700          RXJ  = RX(J)
12710          RYJ  = RY(J)
12720          RZJ  = RZ(J)
12730          EXJ  = EX(J)
12740          EYJ  = EY(J)
12750          EZJ  = EZ(J)
12760          NXJ  = NX(J)
12770          NYJ  = NY(J)
12780          NZJ  = NZ(J)
12790 C
12800          RZIJ = RZI  - RZJ
12810          IF( RZIJ .GT. ZL/2.D0 ) THEN
12820            RZIJ = RZIJ - ZL
12830            RZJ  = RZJ  + ZL
12840          ELSE IF( RZIJ .LT. -ZL/2.D0 ) THEN
12850            RZIJ = RZIJ + ZL
12860            RZJ  = RZJ  - ZL
12870          END IF
12880          IF( DABS(RZIJ) .GE. RCOFF )       GOTO 1000
12890 C
12900          RYIJ  = RYI  - RYJ
12910          CORY  = DNINT( RYIJ/YL)
12920          RYIJ  = RYIJ - CORY*YL
12930          RYJ   = RYJ  + CORY*YL
12940          IF( DABS(RYIJ) .GE. RCOFF )       GOTO 1000
12950 C
12960          RXIJ  = RXI  - RXJ
12970          RXIJ  = RXIJ - CORY*DX
12980          RXJ   = RXJ  + CORY*DX
12990          IF( RXIJ .GT. XL/2.D0 ) THEN
13000            RXIJ = RXIJ - XL
13010            RXJ  = RXJ  + XL
13020          ELSE IF( RXIJ .LT. -XL/2.D0 ) THEN
13030            RXIJ = RXIJ + XL
13040            RXJ  = RXJ  - XL
13050          END IF
13060          IF( DABS(RXIJ) .GE. RCOFF )       GOTO 1000
13070 C
13080          RIJSQ= RXIJ**2 + RYIJ**2 + RZIJ**2
13090          IF( RIJSQ .GE. RCOFF2 )           GOTO 1000
13100          RIJ  = DSQRT(RIJSQ)
13110 C
13120 C          --- FOR TREATMENT OF EVALUATION OF MAGNETIC INTERACTIONS ---
13130          IF( RIJ .GT. B1 ) THEN
13140            CRIJ   = RIJ
13150            CRIJSQ = RIJSQ
13160            CRXIJ  = RXIJ
13170            CRYIJ  = RYIJ
13180            CRZIJ  = RZIJ
13190          ELSE
13200            CRIJ   = B1
13210            CRIJSQ = B1**2
13220            CRXIJ  = RXIJ*(B1/RIJ)
13230            CRYIJ  = RYIJ*(B1/RIJ)
13240            CRZIJ  = RZIJ*(B1/RIJ)
13250          END IF
13260 C
13270 C        ---------------------------------- START OF MAGNETIC FORCES ---
13280          CRIJ3 = CRIJSQ * CRIJ
13290          CRIJ4 = CRIJ3  * CRIJ
13300          RTXIJ = RXIJ/RIJ
13310          RTYIJ = RYIJ/RIJ
```

· Particles I and J are treated for calculation.

· Treatment of the periodic boundary condition.

· Treatment of the Lees-Edwards boundary condition.

· If the two magnetic moments are separated at a distance larger than the particle thickness $b_1$, ordinary treatment is conducted.

· In order to prevent the system from diverging, the magnetic forces and torques are calculated by regarding the two magnetic moments as being separated by the particle thickness $b_1$.

· The force $\boldsymbol{F}_{ij}$ in Eq. (2.29) exerted by particle $j$ is calculated.

```
13320             RTZIJ = RZIJ/RIJ
13330             C00    = NXI*NXJ   + NYI*NYJ   + NZI*NZJ
13340             C11    = NXI*RTXIJ + NYI*RTYIJ + NZI*RTZIJ
13350             C22    = NXJ*RTXIJ + NYJ*RTYIJ + NZJ*RTZIJ
13360             C0     = RM/CRIJ4
13370 C
13380             FMXIJ = -C0 * (  (-C00 + 5.D0*C11*C22)*RTXIJ
13390        &                         -(C22*NXI + C11*NXJ)  )
13400             FMYIJ = -C0 * (  (-C00 + 5.D0*C11*C22)*RTYIJ
13410        &                         -(C22*NYI + C11*NYJ)  )
13420             FMZIJ = -C0 * (  (-C00 + 5.D0*C11*C22)*RTZIJ
13430        &                         -(C22*NZI + C11*NZJ)  )
13440             FXI = FXI + FMXIJ
13450             FYI = FYI + FMYIJ
13460             FZI = FZI + FMZIJ
13470             FX(J) =  FX(J) - FMXIJ
13480             FY(J) =  FY(J) - FMYIJ
13490             FZ(J) =  FZ(J) - FMZIJ
13500 C                                              --- MAGNETIC TORQUES ---
13510             C33X = NYI*NZJ - NZI*NYJ
13520             C33Y = NZI*NXJ - NXI*NZJ
13530             C33Z = NXI*NYJ - NYI*NXJ
13540             C44X = NYI*RTZIJ - NZI*RTYIJ
13550             C44Y = NZI*RTXIJ - NXI*RTZIJ
13560             C44Z = NXI*RTYIJ - NYI*RTXIJ
13570             C55X = NYJ*RTZIJ - NZJ*RTYIJ
13580             C55Y = NZJ*RTXIJ - NXJ*RTZIJ
13590             C55Z = NXJ*RTYIJ - NYJ*RTXIJ
13600 C
13610             C1    = RM/CRIJ3
13620 C
13630             TMXIJ = -C1*(  C33X - 3.D0*C22*C44X )
13640             TMYIJ = -C1*(  C33Y - 3.D0*C22*C44Y )
13650             TMZIJ = -C1*(  C33Z - 3.D0*C22*C44Z )
13660             TMXJI = -C1*( -C33X - 3.D0*C11*C55X )
13670             TMYJI = -C1*( -C33Y - 3.D0*C11*C55Y )
13680             TMZJI = -C1*( -C33Z - 3.D0*C11*C55Z )
13690 C
13700             TXI   = TXI   + TMXIJ
13710             TYI   = TYI   + TMYIJ
13720             TZI   = TZI   + TMZIJ
13730             TX(J) = TX(J) + TMXJI
13740             TY(J) = TY(J) + TMYJI
13750             TZ(J) = TZ(J) + TMZJI
13760 C           ---------------------------------- END OF MAGNETIC FORCES ---
13770 C
13780             IF( RIJ .GE. D2 ) THEN
13790               GOTO 1000
13800             END IF
13810 C
13820 C           ***************************
13830 C           ****************************************************************
13840 C           *********      CHECK THE OVERLAP OF PARTICLES I AND J     *******
13850 C           ****************************************************************
13860 C           ****************************************************************
13870 C
13880             RXJI = -RXIJ
13890             RYJI = -RYIJ
13900             RZJI = -RZIJ
13910 C
13920             CEIEJ = EXI*EXJ + EYI*EYJ + EZI*EZJ
13930             RTXIJ = RXIJ/RIJ
13940             RTYIJ = RYIJ/RIJ
13950             RTZIJ = RZIJ/RIJ
13960             RTXJI = -RTXIJ
13970             RTYJI = -RTYIJ
13980             RTZJI = -RTZIJ
13990             C11    = RTXIJ*EXJ + RTYIJ*EYJ + RTZIJ*EZJ
14000 C
14010             IF( DABS(CEIEJ) .LT. DCOS(PI*89.9D0/180.D0) ) THEN
14020               ITREE = 0
14030             ELSE IF( DABS(CEIEJ) .GT. DCOS(PI*0.5D0/180.D0) ) THEN
14040               IF(  DABS(C11) .LT. DCOS(PI*89.9D0/180.D0) )THEN
14050                 ITREE = 2
14060               ELSE
14070                 ITREE = 3
14080               END IF
14090             ELSE
14100               IF( DABS(CEIEJ) .GT. DCOS(PI*1.D0/180.D0) ) THEN
14110                 ITREE = 4
14120               ELSE IF( DABS(CEIEJ) .LE. DCOS(PI*40.D0/180.D0) ) THEN
```

· The torque $\boldsymbol{T}_{ij}$ in Eq. (2.32) exerted by particle $j$ is calculated.

· In the following, the steric repulsive interactions are evaluated by classifying the situation into several typical overlap cases of the steric layers.

· (RTXIJ,RTYIJ,RTZIJ) is the unit vector defined as $\boldsymbol{t}_{ij}=\boldsymbol{r}_{ij}/r_{ij}$, where $\boldsymbol{r}_{ij}=\boldsymbol{r}_i-\boldsymbol{r}_j$ and $r_{ij}=|\boldsymbol{r}_{ij}|$.

```
14130              ITREE = 5
14140            ELSE
14150              ITREE = 1
14160            END IF
14170          END IF
14180 C
14190 C                                    ITREE=0: NORMAL
14200 C                                    ITREE=1: GENERAL
14210 C                                    ITREE=2: ONE PLANE
14220 C                                    ITREE=3: TWO PARALLEL PLANES
14230 C                                    ITREE=4: NEARLY TWO PARALLEL
14240 C                                             PLANES
14250 C                                    ITREE=5: SIMILAR TO NORMAL
14260 C                                   ------------------------------
14270 C        ***********************************************************
14280 C        *********************************** (1) ITREE=2: ONE PLANE ***
14290 C        ***********************************************************
14300 C
14310          IF( ITREE .EQ. 2 ) THEN
14320 C
14330            IF( RIJ .GE. D2 ) THEN
14340              GOTO 1000
14350            ELSE
14360 C
14370              XJ    = RXJ + RTXIJ*RCHCK
14380              YJ    = RYJ + RTYIJ*RCHCK
14390              ZJ    = RZJ + RTZIJ*RCHCK
14400              XI    = RXI - RTXIJ*RCHCK
14410              YI    = RYI - RTYIJ*RCHCK
14420              ZI    = RZI - RTZIJ*RCHCK
14430 C
14440              KEEP1    = .TRUE.
14450              SKIPTRQ = .TRUE.
14460 C                                        ++ FORCES ALONE ARE USED ++
14470              CALL STEIJCAL( XI,YI,ZI, XJ,YJ,ZJ, RXI,RYI,RZI,
14480      &                      RXJ, RYJ, RZJ, RV, IJOVRLAP, IOVRLAP,
14490      &                      KEEP1, FXI,FYI,FZI, TXI,TYI,TZI, J, SKIPTRQ )
14500 C
14510              GOTO 900
14520            END IF
14530 C
14540          END IF
14550 C
14560 C        ***********************************************************
14570 C        ***************** (2) ITREE=3: TWO PARALLEL PLANES   OR    ***
14580 C        *****************     ITREE=4: NEARLY TWO PARALLEL PLANES ***
14590 C        ***********************************************************
14600 C
14610          IF(  (ITREE.EQ.3) .OR. (ITREE.EQ.4)   ) THEN
14620 C
14630            CEJRIJ = EXJ*RXIJ + EYJ*RYIJ + EZJ*RZIJ
14640            C1     = CEJRIJ/RIJ
14650            IF( DABS(C1) .GT. DCOS(PI*3.0D0/180.D0) ) THEN
14660 C                            +++ IBRANCH=0: CONCENTRIC CIRCLES +++
14670              IBRANCH = 0
14680            ELSE
14690 C                        +++ IBRANCH=1: NOT CONCENTRIC CIRCLES +++
14700              IBRANCH = 1
14710            END IF
14720 C
14730            IF( CEJRIJ .GE. 0.0D0 ) THEN
14740              EEXJ =  EXJ
14750              EEYJ =  EYJ
14760              EEZJ =  EZJ
14770            ELSE
14780              EEXJ = -EXJ
14790              EEYJ = -EYJ
14800              EEZJ = -EZJ
14810              CEJRIJ = -CEJRIJ
14820            END IF
14830 C
14840            CEIRJI = EXI*(-RXIJ) + EYI*(-RYIJ) + EZI*(-RZIJ)
14850            IF( CEIRJI .GE. 0.0D0 ) THEN
14860              EEXI =  EXI
14870              EEYI =  EYI
14880              EEZI =  EZI
14890            ELSE
14900              EEXI = -EXI
14910              EEYI = -EYI
14920              EEZI = -EZI
```

• The case of ITREE=5 is treated in a similar manner to the normal (vertical) situation of ITREE=0.
• The value of 40° is here employed in the present simulation.

• The case of ITREE=2 is treated inside IF( ITREE. EQ.2) THEN … END IF. This is the same for the other cases of IREE=0 to ITREE=5.

• The nearest points are (XJ,YJ,ZJ) and (XI,YI,ZI) for particles *j* and *i*, respectively; these points are on the circle with radius *d*/2 of each disk-like particle, as shown in Fig. 5.6.

• The forces due to the overlap of the steric layers of particles *i* and *j* are calculated in the subroutine STEIJCAL.

• If the two parallel disk-like particles are in the concentric location, the treatment is straightforward.

• The direction of particle *j* is necessarily defined toward the disk surface of particle *i* for the following procedure; this direction is now denoted by (EEXJ, EEYJ, EEZJ).

• Similarly, the direction of particle *i* is necessarily defined toward the disk surface of particle *j* for the following procedure; this direction is now denoted by (EEXI, EEYI, EEZI).

```
14930               CEIRJI = -CEIRJI
14940             END IF
14950 C
14960             IF( ITREE.EQ.3 ) THEN
14970               IF( DABS(CEJRIJ) .GE. B2 ) THEN
14980                 GOTO 1000
14990               END IF
15000             END IF
15010 C                                  +++ THIS CONDITON MUST BE CAREFUL +++
15020             IF( ITREE.EQ.4 ) THEN
15030               IF( DABS(CEIRJI) .GE. (B2+D*DSIN(PI*3.0D0/180.D0)) ) THEN
15040                 GOTO 1000
15050               END IF
15060             END IF
15070 C                                               ++ RX(I),RY(I),RZ(I) ++
15080 C                                               ++ RPXI,RPYI,RPZI    ++
15090 C                                               ++ EPXI,EPYI,EPZI    ++
15100 C                                               ++ EEXI,EEYI,EEZI    ++
15110 C                                               ++ EEXJ,EEYJ,EEZJ    ++
15120             RPXIJ = RXIJ - CEJRIJ*EEXJ
15130             RPYIJ = RYIJ - CEJRIJ*EEYJ
15140             RPZIJ = RZIJ - CEJRIJ*EEZJ
15150             RPXJI = -RPXIJ
15160             RPYJI = -RPYIJ
15170             RPZJI = -RPZIJ
```

• $\boldsymbol{r}_{ij}^P$ is the vector on the disk plane of particle $j$, which is obtained by projecting the vector $\boldsymbol{r}_{ij}$ on that plane, defined as $\boldsymbol{r}_{ij}^P=\boldsymbol{r}_{ij}-(\boldsymbol{e}_j\cdot\boldsymbol{r}_{ij})\boldsymbol{e}_j$.

```
15180             C0   = DSQRT( RPXIJ**2 + RPYIJ**2 + RPZIJ**2 )
15190             EPXJ = RPXIJ/C0
15200             EPYJ = RPYIJ/C0
15210             EPZJ = RPZIJ/C0
```

• $\boldsymbol{e}_j^P$ is the unit vector defined as $\boldsymbol{e}_j^P=\boldsymbol{r}_{ij}^P/|\boldsymbol{r}_{ij}^P|$.

```
15220             IF( ITREE.EQ.3 ) THEN
15230               EPXI = -EPXJ
15240               EPYI = -EPYJ
15250               EPZI = -EPZJ
```

• $\boldsymbol{e}_i^P$ is the unit vector defined as $\boldsymbol{e}_i^P=-\boldsymbol{e}_j^P$ in the parallel situation.

```
15260             ELSE IF( ITREE.EQ.4 ) THEN
15270               RPXJI = (-RXIJ) - CEIRJI*EEXI
15280               RPYJI = (-RYIJ) - CEIRJI*EEYI
15290               RPZJI = (-RZIJ) - CEIRJI*EEZI
```

• $\boldsymbol{r}_{ji}^P$ is the projected vector on the disk plane of particle $i$, defined as $\boldsymbol{r}_{ji}^P=\boldsymbol{r}_{ji}-(\boldsymbol{e}_i\cdot\boldsymbol{r}_{ji})\boldsymbol{e}_i$.

```
15300               C0    = DSQRT( RPXJI**2 + RPYJI**2 + RPZJI**2 )
15310               EPXI = RPXJI/C0
15320               EPYI = RPYJI/C0
15330               EPZI = RPZJI/C0
```

• $\boldsymbol{e}_i^P$ is the unit vector defined as $\boldsymbol{e}_i^P=\boldsymbol{r}_{ji}^P/|\boldsymbol{r}_{ji}^P|$.

```
15340               C1 = EPXI*EPXJ + EPYI*EPYJ + EPZI*EPZJ
15350               IF( C1 .GT. 0.D0 ) THEN
15360                 EPXI = -EPXI
15370                 EPYI = -EPYI
15380                 EPZI = -EPZI
15390               END IF
15400             END IF
```

• $\boldsymbol{e}_i^P$ is taken in the opposite direction to $\boldsymbol{e}_j^P$.

```
15410 C
15420             IF( C0 .GE. D2 ) THEN
15430               GOTO 1000
15440             END IF
15450 C
15460             IF( (C0 .LT. D2) .AND. (C0 .GE. D) ) THEN
15470 C                                        ++ EDGE-TO-EDGE INTERACTION ++
15480               IPATH = 1
15490               XI    = RXI + RCHCK*EPXI
15500               YI    = RYI + RCHCK*EPYI
15510               ZI    = RZI + RCHCK*EPZI
15520               XJ    = RXJ + RCHCK*EPXJ
15530               YJ    = RYJ + RCHCK*EPYJ
15540               ZJ    = RZJ + RCHCK*EPZJ
```

• In this situation, only two spheres located at the torus part of each disk-like particle may possibly overlap in the steric layers.

• The nearest points are (XJ,YJ,ZJ) and (XI,YI,ZI) for particles $j$ and $i$, respectively.

```
15550             ELSE
15560 C                                        ++ PLANE-TO-PLANE INTERACTION ++
15570               IPATH = 2
15580               RPXJ = RXJ + RPXIJ/2.D0
15590               RPYJ = RYJ + RPYIJ/2.D0
15600               RPZJ = RZJ + RPZIJ/2.D0
15610               RPXI = RXI + RPXJI/2.D0
15620               RPYI = RYI + RPYJI/2.D0
15630               RPZI = RZI + RPZJI/2.D0
15640             END IF
```

• In this situation, spheres located at the disk part (i.e., at the body without the torus part) have to be taken into account for calculation of forces and torques due to the overlap of the steric layers.

```
15650 C                                ++++++++++ EDGE-TO-EDGE INTERACTION ++
15660 C                                               +++++ IPATH = 1 +++++
15670             IF( IPATH .EQ. 1 ) THEN
15680 C
15690               KEEP1    = .TRUE.
15700               SKIPTRQ  = .FALSE.
15710               CALL STEIJCAL( XI,YI,ZI, XJ,YJ,ZJ, RXI,RYI,RZI,
15720      &                       RXJ, RYJ, RZJ, RV, IJOVRLAP, IOVRLAP,
15730      &                       KEEP1, FXI,FYI,FZI, TXI,TYI,TZI, J, SKIPTRQ )
```

• The forces and torques due to the overlap of the steric layers of particles $i$ and $j$ are calculated.

```
15740 C
15750           GOTO 900
15760         END IF
```

• Since particles $i$ and $j$ are not exchanged for treatment, the logical variable KEEP1 is ".TRUE."

```
15770 C                        +++++++ PLANE-TO-PLANE INTERACTION ++
15780 C                                  +++++ IPATH = 2 +++++
15790         IF( IPATH .EQ. 2 ) THEN
15800 C                     ++ RX(I),RY(I),RZ(I), RX(J),RY(J),RZ(J) ++
15810 C                     ++ EEXI ,EEYI ,EEZI , EEXJ ,EEYJ ,EEZJ  ++
15820 C                     ++ RPXI ,RPYI ,RPZI , RPXJ ,RPYJ ,RPZJ  ++
15830 C                     ++ EPXI ,EPYI ,EPZI , EPXJ ,EPYJ ,EPZJ  ++
15840 C                     ++ DPXXI,DPXYI,DPXZI, DPYXI,DPYYI,DPYZI ++
15850 C                     ++ DPXXJ,DPXYJ,DPXZJ, DPYXJ,DPYYJ,DPYZJ ++
15860 C                     ++ RORGNXI , RORGNYI , RORGNXZI         ++
15870 C                     ++ RORGNXJ , RORGNYJ , RORGNXZJ         ++
15880           IF( IBRANCH .EQ. 1 ) THEN
15890 C                       +++ IBRANCH=1: NOT CONCENTRIC CIRCLES +++
15900             DPYXJ =  -EPXJ
15910             DPYYJ =  -EPYJ
15920             DPYZJ =  -EPZJ
```

• Preparation for using the plain sphere-constituting model that was prescribed in subroutine PTCLMDL.

```
15930             DPXXJ =  DPYYJ*EEZJ - DPYZJ*EEYJ
15940             DPXYJ =  DPYZJ*EEXJ - DPYXJ*EEZJ
15950             DPXZJ =  DPYXJ*EEYJ - DPYYJ*EEXJ
```

• $\boldsymbol{\delta}_j^{pX}$ is defined by $\boldsymbol{\delta}_j^{pY} \times \boldsymbol{e}_j$.

```
15960             DPYXI =   EPXI
15970             DPYYI =   EPYI
15980             DPYZI =   EPZI
15990             DPXXI =   DPXXJ
16000             DPXYI =   DPXYJ
```

• $\boldsymbol{\delta}_j^{pX}$ and $\boldsymbol{\delta}_j^{pY}$ are the unit vectors in the $X$- and $Y$-axis directions of the plane of the sphere-constituting model of particle $j$, explained in Section 9.2.4.
• Similar symbols $\boldsymbol{\delta}_i^{pX}$ and $\boldsymbol{\delta}_i^{pY}$ are used for particle $i$.

```
16010             DPXZI =   DPXZJ
16020           ELSE IF( IBRANCH .EQ. 0 ) THEN
16030 C                         +++ IBRANCH=0: CONCENTRIC CIRCLES +++
16040             IF( ITREE .EQ. 3 ) THEN
16050               EPXJ =   NXJ
16060               EPYJ =   NYJ
16070               EPZJ =   NZJ
16080               EPXI =  -NXJ
16090               EPYI =  -NYJ
16100               EPZI =  -NZJ
```

• $\boldsymbol{e}_j^p$ is defined as $\boldsymbol{e}_j^p = \boldsymbol{n}_j$. Similarly, $\boldsymbol{e}_i^p$ is defined as $\boldsymbol{e}_i^p = -\boldsymbol{n}_j$.

```
16110             ELSE IF( ITREE .EQ. 4 ) THEN
16120               CEJEIX = EEYJ*EEZI - EEZJ*EEYI
16130               CEJEIY = EEZJ*EEXI - EEXJ*EEZI
16140               CEJEIZ = EEXJ*EEYI - EEYJ*EEXI
16150               C1     = DSQRT( CEJEIX**2 + CEJEIY**2 + CEJEIZ**2 )
16160               TXIJS  = CEJEIX / C1
16170               TYIJS  = CEJEIY / C1
16180               TZIJS  = CEJEIZ / C1
```

• $\boldsymbol{t}_{ij}^s$ is defined as $\boldsymbol{t}_{ij}^s = \boldsymbol{e}_j \times \boldsymbol{e}_i / |\boldsymbol{e}_j \times \boldsymbol{e}_i|$.

```
16190               EXJS   =  ( EEYJ*TZIJS - EEZJ*TYIJS )
16200               EYJS   =  ( EEZJ*TXIJS - EEXJ*TZIJS )
16210               EZJS   =  ( EEXJ*TYIJS - EEYJ*TXIJS )
16220               EPXJ   =  EXJS
16230               EPYJ   =  EYJS
16240               EPZJ   =  EZJS
```

• $\boldsymbol{e}_j^p$ is defined as $\boldsymbol{e}_j^p = \boldsymbol{e}_j \times \boldsymbol{t}_{ij}^s$.

```
16250               EPXI   = -EXJS
16260               EPYI   = -EYJS
16270               EPZI   = -EZJS
```

• $\boldsymbol{e}_i^p$ is defined as $\boldsymbol{e}_i^p = -\boldsymbol{e}_j^p$. This vector is not exactly on the disk plane of particle $i$.

```
16280             END IF
16290             C0    =   EEXI*EPXI + EEYI*EPYI + EEZI*EPZI
16300             EPXI  =   EPXI - C0*EEXI
16310             EPYI  =   EPYI - C0*EEYI
16320             EPZI  =   EPZI - C0*EEZI
```

• $\boldsymbol{e}_i^p$ is modified to make it being on the disk plane of particle $i$.

```
16330             C0    =   DSQRT( EPXI**2 + EPYI**2 + EPZI**2 )
16340             EPXI  =   EPXI/C0
16350             EPYI  =   EPYI/C0
16360             EPZI  =   EPZI/C0
```

• That is, $\boldsymbol{e}_i^p = \boldsymbol{e}_i^p - (\boldsymbol{e}_i \cdot \boldsymbol{e}_i^p)\boldsymbol{e}_i$, and then it is normalized.

```
16370             DPYXJ =  -EPXJ
16380             DPYYJ =  -EPYJ
16390             DPYZJ =  -EPZJ
```

• $\boldsymbol{\delta}_j^{pY}$ is defined by $\boldsymbol{\delta}_j^{pY} = -\boldsymbol{e}_j^p$.

```
16400             DPXXJ =  DPYYJ*EEZJ - DPYZJ*EEYJ
16410             DPXYJ =  DPYZJ*EEXJ - DPYXJ*EEZJ
16420             DPXZJ =  DPYXJ*EEYJ - DPYYJ*EEXJ
```

• $\boldsymbol{\delta}_j^{pX}$ is defined by $\boldsymbol{\delta}_j^{pY} \times \boldsymbol{e}_j$.

```
16430             DPYXI =   EPXI
16440             DPYYI =   EPYI
16450             DPYZI =   EPZI
```

• $\boldsymbol{\delta}_i^{pY}$ is defined by $\boldsymbol{\delta}_i^{pY} = \boldsymbol{e}_i^p$.

```
16460             DPXXI =   DPXXJ
16470             DPXYI =   DPXYJ
16480             DPXZI =   DPXZJ
```

• $\boldsymbol{\delta}_i^{pX}$ is defined by $\boldsymbol{\delta}_i^{pX} = \boldsymbol{\delta}_j^{pX}$.

```
16490           END IF
16500 C                         +++ FOR FINDING THE STARTING POINT +++
16510           ICNTR = 1
16520           DO 250 JJJ=1,INT(RP+0.001D0)
16530             ICNTR = ICNTR + 1
16540             C1XJ = RPXJ + DBLE(JJJ)*EPXJ
```

• $\boldsymbol{r}_j^p$ was defined as $\boldsymbol{r}_j^p = \boldsymbol{r}_j + \boldsymbol{r}_{ij}^p/2$. Similarly, $\boldsymbol{r}_i^p$ was defined as $\boldsymbol{r}_i^p = \boldsymbol{r}_i + \boldsymbol{r}_{ji}^p/2$.

```
16550              C1YJ = RPYJ + DBLE(JJJ)*EPYJ
16560              C1ZJ = RPZJ + DBLE(JJJ)*EPZJ
16570              CC1X = C1XJ - RXJ
16580              CC1Y = C1YJ - RYJ
16590              CC1Z = C1ZJ - RZJ
16600              CC1SQ = CC1X**2 + CC1Y**2 + CC1Z**2
16610              C1XI = RPXI + DBLE(JJJ)*(-EPXI)
16620              C1YI = RPYI + DBLE(JJJ)*(-EPYI)
16630              C1ZI = RPZI + DBLE(JJJ)*(-EPZI)
16640              IF( CC1SQ .GT. RCHCKSQ+0.02D0 ) THEN
16650                ICNTR    = ICNTR - 1
16660                RORGNXJ =  C1XJ -    EPXJ
16670                RORGNYJ =  C1YJ -    EPYJ
16680                RORGNZJ =  C1ZJ -    EPZJ
16690                RORGNXI =  C1XI - (-EPXI)
16700                RORGNYI =  C1YI - (-EPYI)
16710                RORGNZI =  C1ZI - (-EPZI)
16720                GOTO 255
16730            END IF
16740   250      CONTINUE
16750   255      NPROWAL9 = 1 + 2*(ICNTR-1)
16760 C
16770 C            +++ STARTING POINT (RORGNXI,...,RORGNZI) FOR PTCL I +++
16780 C            +   STARTING POINT (RORGNXJ,...,RORGNZJ) FOR PTCL J   +
16790 C            +   NUMBER OF MAX ROW IS NPROWAL9 FOR CHECKING FORCES +
16800 C
```

• The location at the nearest point to the torus part in the direction $e_j^p$ is the starting position for evaluation, where the first sphere is located.

• This position is denoted by $r_j^{p(orgn)}$, (RORGNXJ, RORGNYJ, RORGNZJ).

• Similar symbols are used for particle $i$.

• The plain sphere-constituting model was explained in subroutine PTCLMDL.

```
16810            DO 500 IROW =1, NPROWAL9
16820              IPTCL0  = NPROWST(IROW)
16830              LGCWALL = .FALSE.
16840            DO 480 III=1, NPROW(IROW)
16850              IPTCL   = IPTCL0 + (III-1)
16860              RXSTRC0 = RXSTRC(IPTCL)
16870              RYSTRC0 = RYSTRC(IPTCL)
16880              XJ = RORGNXJ + RXSTRC0*DPXXJ + RYSTRC0*DPYXJ
16890              YJ = RORGNYJ + RXSTRC0*DPXYJ + RYSTRC0*DPYYJ
16900              ZJ = RORGNZJ + RXSTRC0*DPXZJ + RYSTRC0*DPYZJ
16910              XI = RORGNXI + RXSTRC0*DPXXI + RYSTRC0*DPYXI
16920              YI = RORGNYI + RXSTRC0*DPXYI + RYSTRC0*DPYYI
16930              ZI = RORGNZI + RXSTRC0*DPXZI + RYSTRC0*DPYZI
16940 C
16950              CRISQ = (XI-RXI)**2 + (YI-RYI)**2 + (ZI-RZI)**2
16960              CRJSQ = (XJ-RXJ)**2 + (YJ-RYJ)**2 + (ZJ-RZJ)**2
16970 C
16980              IF(  (CRISQ.GT.RCHCKSQ+0.02D0).OR.
16990      &            (CRJSQ.GT.RCHCKSQ+0.02D0)        ) GOTO 480
17000 C
17010              LGCWALL = .TRUE.
17020 C
17030              KEEP1    = .TRUE.
17040              SKIPTRQ = .FALSE.
17050              CALL STEIJCAL( XI,YI,ZI, XJ,YJ,ZJ, RXI,RYI,RZI,
17060      &                      RXJ, RYJ, RZJ, RV, IJOVRLAP, IOVRLAP,
17070      &                      KEEP1, FXI,FYI,FZI, TXI,TYI,TZI, J, SKIPTRQ )
17080 C
17090   480      CONTINUE
17100 CCCCCCC    IF( .NOT. LGCWALL ) GOTO 900
17110 C
17120   500      CONTINUE
17130 C
17140            GOTO 900
17150 C
17160          END IF
17170 C
17180        END IF
```

• IPTCL0 denotes the name of the first sphere of the shell area IROW; see Fig. 9.12.

• The number of spheres belonging to the shell area IROW is NPROW(IROW).

• The positions of the two spheres are denoted by (XJ,YJ,ZJ) and (XI,YI,ZI) for particles $j$ and $i$, respectively.

• The point (XJ,YJ,ZJ) has to be within the disk plane of particle $j$. This has to be satisficd also by the point (XI, YI, ZI) of particle $i$.

• The forces and torques due to the overlap of the steric layers of particles $i$ and $j$ are calculated.

• Since particles $i$ and $j$ are not exchanged for treatment, the logical variable KEEP1 is ".TRUE."

• Preparation for the following assessment for the other cases.

```
17190 C
17200 C      ////////////////////////////////////////////////////////////////
17210 C      ///////////   FOR ITREE=1:GENERAL , ITREE=0:NORMAL     //////////
17220 C      ///////////             ITREE=5:SIMILAR TO NORMAL      //////////
17230 C      ////////////////////////////////////////////////////////////////
17240 C
17250 C      ---------------------------------------------------- TIJS ---
17260        CEJEIX = EYJ*EZI - EZJ*EYI
17270        CEJEIY = EZJ*EXI - EXJ*EZI
17280        CEJEIZ = EXJ*EYI - EYJ*EXI
17290        C1     = DSQRT( CEJEIX**2 + CEJEIY**2 + CEJEIZ**2 )
17300        TXIJS  = CEJEIX / C1
17310        TYIJS  = CEJEIY / C1
17320        TZIJS  = CEJEIZ / C1
17330 C
17340 C      ---------------------------------------------- EIS , EJS ---
```

• $t_{ij}^s$ is defined as $t_{ij}^s = e_j \times e_i / | e_j \times e_i |$ in Eq. (5.27).

```
17350         EXIS = -( EYI*TZIJS - EZI*TYIJS )
17360         EYIS = -( EZI*TXIJS - EXI*TZIJS )
17370         EZIS = -( EXI*TYIJS - EYI*TXIJS )
17380         EXJS =  ( EYJ*TZIJS - EZJ*TYIJS )
17390         EYJS =  ( EZJ*TXIJS - EXJ*TZIJS )
17400         EZJS =  ( EXJ*TYIJS - EYJ*TXIJS )
17410 C
17420 C       ---------------------------------------- KIS , KJS , KIJS ---
17430         KIS  = -(EXJ*RXIJ + EYJ*RYIJ + EZJ*RZIJ)/
17440      &          (EXJ*EXIS + EYJ*EYIS + EZJ*EZIS)
17450         KJS  =  (EXI*RXIJ + EYI*RYIJ + EZI*RZIJ)/
17460      &          (EXI*EXJS + EYI*EYJS + EZI*EZJS)
17470         KIJS = RXIJ*TXIJS  + RYIJ*TYIJS + RZIJ*TZIJS
17480 C
17490 C       ------------------------ REPLACEMENT OF PARTICLES I AND J ---
17500         IF( DABS(KJS) .GE. DABS(KIS) ) THEN
17510           KEEP   = .TRUE.
17520           II     = I
17530           JJ     = J
17540           RRXI   = RXI
17550           RRYI   = RYI
17560           RRZI   = RZI
17570           RRXJ   = RXJ
17580           RRYJ   = RYJ
17590           RRZJ   = RZJ
17600           RRXIJ  = RXIJ
17610           RRYIJ  = RYIJ
17620           RRZIJ  = RZIJ
17630           RRXJI  = RXJI
17640           RRYJI  = RYJI
17650           RRZJI  = RZJI
17660           TTXIJS= TXIJS
17670           TTYIJS= TYIJS
17680           TTZIJS= TZIJS
17690           EEXI   = EXI
17700           EEYI   = EYI
17710           EEZI   = EZI
17720           EEXJ   = EXJ
17730           EEYJ   = EYJ
17740           EEZJ   = EZJ
17750           EEXIS  = EXIS
17760           EEYIS  = EYIS
17770           EEZIS  = EZIS
17780           EEXJS  = EXJS
17790           EEYJS  = EYJS
17800           EEZJS  = EZJS
17810           KKIS   = KIS
17820           KKJS   = KJS
17830           KKIJS  = KIJS
17840         ELSE
17850           KEEP   = .FALSE.
17860           II     = J
17870           JJ     = I
17880           RRXI   = RXJ
17890           RRYI   = RYJ
17900           RRZI   = RZJ
17910           RRXJ   = RXI
17920           RRYJ   = RYI
17930           RRZJ   = RZI
17940           RRXIJ  = -RXIJ
17950           RRYIJ  = -RYIJ
17960           RRZIJ  = -RZIJ
17970           RRXJI  = -RXJI
17980           RRYJI  = -RYJI
17990           RRZJI  = -RZJI
18000           TTXIJS= -TXIJS
18010           TTYIJS= -TYIJS
18020           TTZIJS= -TZIJS
18030           EEXI   = EXJ
18040           EEYI   = EYJ
18050           EEZI   = EZJ
18060           EEXJ   = EXI
18070           EEYJ   = EYI
18080           EEZJ   = EZI
18090           EEXIS  = EXJS
18100           EEYIS  = EYJS
18110           EEZIS  = EZJS
18120           EEXJS  = EXIS
18130           EEYJS  = EYIS
18140           EEZJS  = EZIS
```

• $\boldsymbol{e}_i^s$ is $-\boldsymbol{e}_i \times \boldsymbol{t}_{ij}^s$ and $\boldsymbol{e}_j^s$ is $\boldsymbol{e}_j \times \boldsymbol{t}_{ij}^s$ in Eq. (5.27).

• $k_i^s = -\boldsymbol{e}_j \cdot \boldsymbol{r}_{ij} / (\boldsymbol{e}_j \cdot \boldsymbol{e}_i^s)$, $k_j^s = \boldsymbol{e}_i \cdot \boldsymbol{r}_{ij} / (\boldsymbol{e}_i \cdot \boldsymbol{e}_j^s)$ and $k_{ij}^s = \boldsymbol{r}_{ij} \cdot \boldsymbol{t}_{ij}^s$ in Eq. (5.29).

• In the assessment of the overlap of the steric layers, the particle that has a longer distance from the particle center to the line of intersection, shown in Fig. 5.5, is treated as particle JJ.

• In this case, exchange of particles $i$ and $j$ is not necessary.

• The logical variable KEEP is set as ".TRUE."

• In this case, exchange of particles $i$ and $j$ is necessary.

• The logical variable KEEP is set as ".FALSE."

```
18150           KKIS   = KJS
18160           KKJS   = KIS
18170           KKIJS  = KIJS
18180         END IF
18190 C
18200 C       ------------------- REPLACEMENT OF DIRECTIONS OF EI AND EJ ---
18210         CHCK0 = RRXJI*EEXI + RRYJI*EEYI + RRZJI*EEZI
18220         IF( CHCK0 .LT. 0.D0 ) THEN
18230           EEXI    = -EEXI
18240           EEYI    = -EEYI
18250           EEZI    = -EEZI
18260         END IF
18270 C
```

• The direction of particle $i$ is necessarily defined toward the disk surface of particle $j$ for the following procedure; this direction is now denoted by (EEXI, EEYI, EEZI).

```
18280         CEIEJ = EEXI*EEXJ + EEYI*EEYJ + EEZI*EEZJ
18290         IF( CEIEJ .LT. 0.D0 ) THEN
18300           EEXJ    = -EEXJ
18310           EEYJ    = -EEYJ
18320           EEZJ    = -EEZJ
18330           CEIEJ   = -CEIEJ
18340         END IF
18350 C
```

• Similarly, the direction of particle $j$ is necessarily defined toward the disk surface of particle $i$ for the following procedure; this direction is now denoted by (EEXJ, EEYJ, EEZJ).

```
18360 C       ------------------------- REPLACEMENT OF DIRECTION OF TIJS ---
18370         CHCK0 = TTXIJS*RRXIJ + TTYIJS*RRYIJ + TTZIJS*RRZIJ
18380         IF( CHCK0 .LT. 0.D0 ) THEN
18390           TTXIJS = -TTXIJS
18400           TTYIJS = -TTYIJS
18410           TTZIJS = -TTZIJS
18420         END IF
18430 C
```

• $t_{ij}^s$ has to be in the direction from particles JJ to II.

```
18440 C       -------- REPLACEMENT OF DIRECTIONS OF EIS,EJS,KIS,KJS,KIJS ---
18450         IF( KKIS .LT. 0.D0 ) THEN
18460           KKIS   = -KKIS
18470           EEXIS  = -EEXIS
18480           EEYIS  = -EEYIS
18490           EEZIS  = -EEZIS
18500         END IF
```

• $k_i^s$ has to be positive (or equal to zero).

```
18510         IF( KKJS .LT. 0.D0 ) THEN
18520           KKJS   = -KKJS
18530           EEXJS  = -EEXJS
18540           EEYJS  = -EEYJS
18550           EEZJS  = -EEZJS
18560         END IF
```

• $k_j^s$ has to be positive (or equal to zero).

```
18570         IF( KKIJS .LT. 0.D0 ) THEN
18580           KKIJS  = -KKIJS
18590         END IF
```

• $k_{ij}^s$ has to be positive (or equal to zero).

```
18600 C                   +++++++++++++++++++++++++++++++++++++++++++++++++++++
18610 C                   II , JJ
18620 C                   RRXI , RRYI , RRZI , RRXJ , RRYJ , RRZJ
18630 C                   RRXIJ , RRYIJ , RRZIJ , RRXJI , RRYJI , RRZJI
18640 C                   TTXIJS, TTYIJS, TTZIJS
18650 C                   EEXI , EEYI , EEZI , EEXJ , EEYJ , EEZJ
18660 C                   EEXIS, EEYIS, EEZIS, EEXJS, EEYJS, EEZJS
18670 C                   KKIS , KKJS , KKIJS
18680 C                   +++++++++++++++++++++++++++++++++++++++++++++++++++++
18690 C       *************************************************************
18700 C       ***********  (3) ITREE=1:GENERAL , ITREE=0:NORMAL    **********
18710 C       ***********          ITREE=5:SIMILAR TO NORMAL       **********
18720 C       *************************************************************
18730 C
18740         IF( (ITREE.EQ.1) .OR. (ITREE.EQ.0) .OR. (ITREE.EQ.5) ) THEN
18750 C
18760           IF( KKJS .GT. D02 ) THEN
18770 C
18780             C0 = DABS( EEXJS*EEXI + EEYJS*EEYI + EEZJS*EEZI )
18790             KIJQ   = ( KKJS - D02 )*C0
18800             IF( KIJQ .GT. B2 ) THEN
18810               GOTO 1000
18820             END IF
18830 C
```

• $k_{i(j)}^Q$ is defined as $k_{i(j)}^Q = (k_j^s - d/2)|e_j^s \cdot e_i|$ in Eq. (5.32).

```
18840             RXJQ   = RRXJ + D02*EEXJS
18850             RYJQ   = RRYJ + D02*EEYJS
18860             RZJQ   = RRZJ + D02*EEZJS
18870             RXIJQ  = RXJQ - KIJQ*EEXI
18880             RYIJQ  = RYJQ - KIJQ*EEYI
18890             RZIJQ  = RZJQ - KIJQ*EEZI
```

• $\boldsymbol{r}_j^Q$ is defined as $\boldsymbol{r}_j^Q = \boldsymbol{r}_j + (d/2)\boldsymbol{e}_j^s$.

• $\boldsymbol{r}_{i(j)}^Q$ is defined as $\boldsymbol{r}_{i(j)}^Q = \boldsymbol{r}_j^Q - k_{i(j)}^Q \boldsymbol{e}_i$ in Eq. (5.31).

```
18900 C                         +++++++++++++++++++++++++++++++++++++++
18910 C                          IWAY=1: II-PLANE TO JJ-EDGE (KKJS>D/2)
18920 C                          IWAY=2: II-EDGE  TO JJ-EDGE (KKJS>D/2)
18930 C                          IWAY=3: II-EDGE  TO JJ-EDGE (KKJS<D/2)
18940 C                         +++++++++++++++++++++++++++++++++++++++
18950             CHCK1    = DSQRT( (RXIJQ-RRXI)**2 + (RYIJQ-RRYI)**2
18960      &                                        + (RZIJQ-RRZI)**2 )
18970             IF( CHCK1 .LE. D02 ) THEN
```

```
18980               IWAY = 1
18990             ELSE
19000               IWAY = 2
19010             END IF
19020           ELSE
19030               IWAY = 3
19040           END IF
19050 C
19060 C                                   --------------------------------------
19070 C         ------------------------------------ (3)-1 INNER CIRCLE ---
19080 C                                   --------------------------------------
19090 C
19100           IF( IWAY .EQ. 1 ) THEN
19110 C
19120             CEIEJ = EEXI*EEXJ + EEYI*EEYJ + EEZI*EEZJ
19130 C
19140             IF( (ITREE.EQ.0) .OR. (ITREE.EQ.5) ) THEN
19150 C
19160 C                                 ++++++++++++++++++++++++ (3)-1-1 +++
19170 C                                 +++++++++  NORMAL AND NEARLY NORMAL +
19180 C
19190 C                                             ++ RXJQ  , RYJQ  , RZJQ  ++
19200 C                                             ++ RXIJQ, RYIJQ, RZIJQ ++
19210               XI = RXIJQ
19220               YI = RYIJQ
19230               ZI = RZIJQ
19240               XJ = RXJQ
19250               YJ = RYJQ
19260               ZJ = RZJQ
```

· The nearest points are (XJ,YJ,ZJ) and (XI,YI,ZI) for particles $j$ and $i$, respectively.
· A pair of spheres at these positions are sufficient for assessment in a normal or nearly normal situation.

```
19270 C
19280               SKIPTRQ = .FALSE.
19290               CALL STEIJCAL( XI, YI, ZI, XJ, YJ, ZJ, RRXI, RRYI, RRZI,
19300      &                  RRXJ, RRYJ, RRZJ, RV, IJOVRLAP, IOVRLAP,
19310      &                  KEEP, FXI, FYI, FZI, TXI, TYI, TZI, J, SKIPTRQ )
19320 C
19330               GOTO 900
19340 C
19350             ELSE IF( ITREE .EQ. 1 ) THEN
19360 C                                 ++++++++++++++++++++++++ (3)-1-2 +++
19370 C                                 ++++++++++++++++++++   GENERAL CASE  +
19380 C
19390 C                   ----------------------------------------------------
19400 C                   II , JJ
19410 C                   RRXI , RRYI , RRZI , RRXJ , RRYJ , RRZJ
19420 C                   RRXIJ , RRYIJ , RRZIJ , RRXJI , RRYJI , RRZJI
19430 C                   TTXIJS, TTYIJS, TTZIJS
19440 C                   EEXI , EEYI , EEZI , EEXJ , EEYJ , EEZJ
19450 C                   EEXIS, EEYIS, EEZIS, EEXJS, EEYJS, EEZJS
19460 C                   KKIS , KKJS , KKIJS
19470 C                   ----------------------------------------------------
19480 C                           ++ RX(I),RY(I),RZ(I), RX(J),RY(J),RZ(J) ++
19490 C                           ++ EPXI ,EPYI ,EPZI , EPXJ ,EPYJ ,EPZJ  ++
19500 C                           ++ DPXXI,DPXYI,DPXZI, DPYXJ,DPYYJ,DPYZJ ++
19510 C                           ++ RORGNXI , RORGNYI , RORGNXZI         ++
19520 C                           ++ RORGNXJ , RORGNYJ , RORGNXZJ         ++
19530 C                                             ++ RXJQ  , RYJQ  , RZJQ  ++
19540 C                                             ++ RXIJQ, RYIJQ, RZIJQ ++
19550               RORGNXJ = RXJQ
19560               RORGNYJ = RYJQ
19570               RORGNZJ = RZJQ
19580               RORGNXI = RXIJQ
19590               RORGNYI = RYIJQ
19600               RORGNZI = RZIJQ
```

· In this situation, spheres located at the disk part (i.e., at the body without the torus part) has to be taken into account for calculation of forces and torques due to the overlap of the steric layers.

```
19610 C
19620               SKIPTRQ = .FALSE.
19630               CALL STEALLCA( RRXI, RRYI, RRZI, RRXJ, RRYJ, RRZJ,
19640      &          RORGNXI,RORGNYI,RORGNZI, RORGNXJ,RORGNYJ,RORGNZJ,
19650      &          EEXIS, EEYIS, EEZIS, EEXJS, EEYJS, EEZJS,
19660      &          EEXJ, EEYJ, EEZJ, LGCWALL, RCHCKSQ, RV, IJOVRLAP,
19670      &          IOVRLAP, KEEP, FXI,FYI,FZI, TXI,TYI,TZI, J, SKIPTRQ )
19680 C
19690               GOTO 900
19700 C
19710             END IF
19720 C
19730           END IF
19740 C
19750 C         ------------------------------------ (3)-2 OUTER CIRCLE ---
19760 C                                   --------------------------------------
19770 C
```

· The starting positions for evaluation are denoted by $r_j^{orgn}$ and $r_i^{orgn}$, where the spheres are possibly located at the nearest positions.

· According to the plain sphere-constituting model for the disk-like particle, the forces and torques are evaluated in the subroutine STEALLCA.

```
19780           IF( IWAY .EQ. 2 ) THEN
19790 C
```

· $\boldsymbol{r}_{i(j)}^{Q}$ was defined as $\boldsymbol{r}_{i(j)}^{Q} = \boldsymbol{r}_j^{Q} - k_{i(j)}^{Q}\,\boldsymbol{e}_i$ in Eq. (5.31).

```
19800 C                                   --- SEARCH FOR STARTING POINTS ---
19810          C1X = RXIJQ - RRXI
19820          C1Y = RYIJQ - RRYI
19830          C1Z = RZIJQ - RRZI
19840 C
19850          C0  = DSQRT( C1X**2 + C1Y**2 + C1Z**2 )
19860          HIQ = D02/C0
19870 C
19880          RRXIQ = RRXI + HIQ * C1X
19890          RRYIQ = RRYI + HIQ * C1Y
19900          RRZIQ = RRZI + HIQ * C1Z
19910 C
19920          RRXJQ = RRXJ + D02 * EEXJS
19930          RRYJQ = RRYJ + D02 * EEYJS
19940          RRZJQ = RRZJ + D02 * EEZJS
19950 C
19960 C                                +++++++++++++++++++++++++++++++++++
19970 C                                 STARTING POINTS :
19980 C                                    PTCL II : (RRXIQ, RRYIQ, RRZIQ)
19990 C                                    PTCL JJ : (RRXJQ, RRYJQ, RRZJQ)
20000 C
```

• $r_i^Q$ is defined as $r_i^Q = r_i + (d/2)\,(r_{i(j)}^Q - r_i)/|r_{i(j)}^Q - r_i|$; this vector is along the disk plane.

• $r_j^Q$ is defined as $r_j^Q = r_j + (d/2)e_j^s$.

• These points $r_i^Q$ and $r_j^Q$ are used as starting points for searching for the minimum distance positions of two spheres on each disk-like particle.

```
20010 C
20020          RIJMN = RIJMNFUN( RRXI, RRYI, RRZI, RRXJ, RRYJ, RRZJ,
20030     &                      EEXI, EEYI, EEZI, EEXJ, EEYJ, EEZJ,
20040     &                    RRXIQ, RRYIQ, RRZIQ, RRXJQ, RRYJQ, RRZJQ )
20050 C
20060 C                             ++++++++++++++++++++++++++++++++++++++
20070 C                              POINTS YIELDING THE MINIMUM DISTANCE:
20080 C                                    PTCL II : (RPXIMN,RPYIMN,RPZIMN
20090 C                                    PTCL JJ : (RPXJMN,RPYJMN,RPZJMN
20100 C                              MINIMUM DISTANCE : RIJMN
20110 C                             ++++++++++++++++++++++++++++++++++++++
20120 C
20130          IF( RIJMN .GE. B2 ) THEN
20140            GOTO 1000
20150          END IF
20160 C
20170 C
```

• The positions of two spheres, which give rise to a minimum distance, are obtained in the subroutine RIJMNFUN.

```
20180          IF( (ITREE.EQ.0) .OR. (ITREE.EQ.5) ) THEN
20190 C
20200 C                            ++++++++++++++++++++++++ (3)-2-1 +++
20210 C                            +++++++++  NORMAL AND NEARLY NORMAL +
20220            XI = RPXIMN
20230            YI = RPYIMN
20240            ZI = RPZIMN
20250            XJ = RPXJMN
20260            YJ = RPYJMN
20270            ZJ = RPZJMN
20280 C
20290            SKIPTRQ = .FALSE.
20300            CALL STEIJCAL( XI, YI, ZI, XJ, YJ, ZJ, RRXI, RRYI, RRZI,
20310     &                RRXJ, RRYJ, RRZJ, RV, IJOVRLAP, IOVRLAP,
20320     &                KEEP, FXI, FYI, FZI, TXI, TYI, TZI, J, SKIPTRQ )
20330 C
20340            GOTO 900
20350 C
```

• The nearest points are (XJ,YJ,ZJ) and (XI,YI,ZI) for particles *j* and *i*, respectively.

• A pair of spheres at these positions are sufficient for assessment in a normal or nearly normal situation.

```
20360          ELSE IF( ITREE.EQ.1 ) THEN
20370 C
20380 C                            ++++++++++++++++++++++++ (3)-2-2 +++
20390 C                            +++++++++++++++++++++++++   GENERAL +
20400            RORGNXJ = RPXJMN
20410            RORGNYJ = RPYJMN
20420            RORGNZJ = RPZJMN
20430            RORGNXI = RPXIMN
20440            RORGNYI = RPYIMN
20450            RORGNZI = RPZIMN
20460 C
20470            SKIPTRQ = .FALSE.
20480            CALL STEALLCA( RRXI, RRYI, RRZI, RRXJ, RRYJ, RRZJ,
20490     &        RORGNXI,RORGNYI,RORGNZI, RORGNXJ,RORGNYJ,RORGNZJ,
20500     &        EEXIS, EEYIS, EEZIS, EEXJS, EEYJS, EEZJS,
20510     &        EEXJ, EEYJ, EEZJ, LGCWALL, RCHCKSQ, RV, IJOVRLAP,
20520     &        IOVRLAP, KEEP, FXI,FYI,FZI, TXI,TYI,TZI, J, SKIPTRQ )
20530 C
20540            GOTO 900
20550 C
20560          END IF
20570 C
20580        END IF
```

• In this situation, spheres located at the disk part (i.e., at the body without the torus part) has to be taken into account for calculation of forces and torques due to the overlap of the steric layers.

• The starting positions for evaluation are denoted by $r_j^{orgn}$ and $r_i^{orgn}$, where the spheres are located at the nearest positions.

```
20590 C                                   --------------------------------------
20600 C          --------------------------------------- (3)-3 OUTER CIRCLE ---
20610 C                                   --
20620          IF( IWAY .EQ. 3 ) THEN
20630 C                                   --- SEARCH FOR STARTING POINTS ---
20640            HIQ    = DSQRT( D02**2 - KKIS**2 )
20650            RRXIQ = RRXI + KKIS*EEXIS - HIQ * TTXIJS
20660            RRYIQ = RRYI + KKIS*EEYIS - HIQ * TTYIJS
20670            RRZIQ = RRZI + KKIS*EEZIS - HIQ * TTZIJS
20680 C
20690            HJQ    = DSQRT( D02**2 - KKJS**2 )
20700            RRXJQ = RRXJ + KKJS*EEXJS + HJQ * TTXIJS
20710            RRYJQ = RRYJ + KKJS*EEYJS + HJQ * TTYIJS
20720            RRZJQ = RRZJ + KKJS*EEZJS + HJQ * TTZIJS
20730 C
20740 C                                 +++++++++++++++++++++++++++++++++++
20750 C                                  STARTING POINTS :
20760 C                                      PTCL II : (RRXIQ, RRYIQ, RRZIQ)
20770 C                                      PTCL JJ : (RRXJQ, RRYJQ, RRZJQ)
20780 C                                 +++++++++++++++++++++++++++++++++++
20790 C
20800            RIJMN = RIJMNFUN( RRXI, RRYI, RRZI, RRXJ, RRYJ, RRZJ,
20810     &                   EEXI, EEYI, EEZI, EEXJ, EEYJ, EEZJ,
20820     &                 RRXIQ, RRYIQ, RRZIQ, RRXJQ, RRYJQ, RRZJQ )
20830 C
20840 C                               +++++++++++++++++++++++++++++++++++++
20850 C                                POINTS YIELDING THE MINIMUM DISTANCE:
20860 C                                       PTCL II : (RPXIMN,RPYIMN,RPZIMN
20870 C                                       PTCL JJ : (RPXJMN,RPYJMN,RPZJMN
20880 C                                MINIMUM DISTANCE : RIJMN
20890 C                               +++++++++++++++++++++++++++++++++++++
20900            IF( RIJMN .GE. B2 ) THEN
20910              GOTO 1000
20920            END IF
20930 C
20940            XI = RPXIMN
20950            YI = RPYIMN
20960            ZI = RPZIMN
20970            XJ = RPXJMN
20980            YJ = RPYJMN
20990            ZJ = RPZJMN
21000 C
21010            SKIPTRQ = .FALSE.
21020            CALL STEIJCAL( XI, YI, ZI, XJ, YJ, ZJ, RRXI, RRYI, RRZI,
21030     &               RRXJ, RRYJ, RRZJ, RV, IJOVRLAP, IOVRLAP,
21040     &               KEEP, FXI, FYI, FZI, TXI, TYI, TZI, J, SKIPTRQ )
21050 C
21060            GOTO 900
21070          END IF
21080 C
21090        END IF
21100 C
21110 C      ///////////////////////////////////////////////////////////////////
21120 C      /////////////  END OF FORCE and TORQUE DUE TO STERIC INER. ///
21130 C      ///////////////////////////////////////////////////////////////////
21140 C
21150   900  IF( IJOVRLAP ) THEN
21160          RIJMNSUM = RIJMNSUM + RIJMN
21170        END IF
21180 C
21190 C
21200  1000 CONTINUE
21210 C
21220       FX(I) = FXI
21230       FY(I) = FYI
21240       FZ(I) = FZI
21250       TX(I) = TXI
21260       TY(I) = TYI
21270       TZ(I) = TZI
21280 C
21290 C
21300  2000 CONTINUE
21310 C
21320 C      ---------------------------------------------------------------
21330       IF( IOVRLAP ) THEN
21340         RMNMNSUM = RMNMNSUM + RMNMN
21350       END IF
21360 C
21370       DO 2010 I=1,N
21380         TX(I) = TX(I) + ( NY(I)*HZ - NZ(I)*HY )*RH
21390         TY(I) = TY(I) + ( NZ(I)*HX - NX(I)*HZ )*RH
21400         TZ(I) = TZ(I) + ( NX(I)*HY - NY(I)*HX )*RH
```

· This case corresponds to the situation in Fig. 5.6(a).

· $\boldsymbol{r}_i^Q$ is defined as $\boldsymbol{r}_i^Q = \boldsymbol{r}_i^s - h_i^Q \boldsymbol{t}_{ij}^s$, where $h_i^Q$ is expressed as $h_i^Q = ((d/2)^2 - (k_i^s)^2)^{1/2}$.

· $\boldsymbol{r}_j^Q$ is defined as $\boldsymbol{r}_j^Q = \boldsymbol{r}_j^s + h_j^Q \boldsymbol{t}_{ij}^s$, where $h_j^Q$ is expressed as $h_j^Q = ((d/2)^2 - (k_j^s)^2)^{1/2}$.

· The nearest points are (XJ,YJ,ZJ) and (XI,YI,ZI) for particles $j$ and $i$, respectively.

· A pair of spheres at these positions are sufficient for assessment in this edge-to-edge contact situation.

· The torques due to the applied magnetic field are added to the corresponding variables.

```
21410   2010 CONTINUE
21420                                                                    RETURN
21430                                                                    END
21440 C**** SUB STEIJCAL ****
21450       SUBROUTINE STEIJCAL( XI, YI, ZI, XJ, YJ, ZJ, RRXI, RRYI, RRZI,
21460      &                     RRXJ, RRYJ, RRZJ, RV, IJOVRLAP, IOVRLAP,
21470      &                     KEEP, FXI, FYI, FZI, TXI, TYI, TZI, JPTCL, SKIPTRQ )
21480 C
21490       IMPLICIT REAL*8 (A-H,O-Z), INTEGER (I-N)
21500 C
21510       COMMON /BLOCK4/  FX    , FY    , FZ
21520       COMMON /BLOCK5/  TX    , TY    , TZ
21530 C
21540       COMMON /WORK11/  XRXI , YRYI , ZRZI , XRXJ , YRYJ , ZRZJ
21550       COMMON /WORK12/  FXIJS, FYIJS, FZIJS, FXJIS, FYJIS, FZJIS
21560       COMMON /WORK13/  TQXIJS,TQYIJS,TQZIJS,TQXJIS,TQYJIS,TQZJIS
21570 C     -----------------
21580       INTEGER       NN , NNS
21590       PARAMETER( NN=1000 , NNS=5000000 )
21600       PARAMETER( NRANMX=1000000 , PI=3.141592653589793D0 )
21610 C
21620       REAL*8    FX(NN) , FY(NN) , FZ(NN) , TX(NN) , TY(NN) , TZ(NN)
21630       LOGICAL   KEEP , IJOVRLAP , IOVRLAP , SKIPTRQ
21640 C
21650       REAL*8    RRIJ , TXIJ0 , TYIJ0 , TZIJ0 , C1 , C2 , C3
21660       INTEGER   ISKIP
21670 C     ---------------------------------------------------------------------
21680       RRIJ = DSQRT( (XI-XJ)**2 + (YI-YJ)**2 + (ZI-ZJ)**2 )
21690       XRXI = XI - RRXI
21700       YRYI = YI - RRYI
21710       ZRZI = ZI - RRZI
21720       XRXJ = XJ - RRXJ
21730       YRYJ = YJ - RRYJ
21740       ZRZJ = ZJ - RRZJ
21750       IF( .NOT.SKIPTRQ ) THEN
21760         TXIJ0= (XI-XJ)/RRIJ
21770         TYIJ0= (YI-YJ)/RRIJ
21780         TZIJ0= (ZI-ZJ)/RRIJ
21790       END IF
21800 C                                          ++ SKIP=0 : CAL. FORCES AND TORQUES ++
21810 C                                          ++ SKIP=1 : CAL. FORCES ALONE       ++
21820       ISKIP = 0
21830       CALL STEFORCE( RRIJ, RV, ISKIP,
21840      &                    TXIJ0, TYIJ0, TZIJ0, IJOVRLAP, IOVRLAP )
21850       IF( .NOT.KEEP ) THEN
21860         C1     = FXIJS
21870         C2     = FYIJS
21880         C3     = FZIJS
21890         FXIJS  = FXJIS
21900         FYIJS  = FYJIS
21910         FZIJS  = FZJIS
21920         FXJIS  = C1
21930         FYJIS  = C2
21940         FZJIS  = C3
21950         C1      = TQXIJS
21960         C2      = TQYIJS
21970         C3      = TQZIJS
21980         TQXIJS = TQXJIS
21990         TQYIJS = TQYJIS
22000         TQZIJS = TQZJIS
22010         TQXJIS = C1
22020         TQYJIS = C2
22030         TQZJIS = C3
22040       END IF
22050 C
22060       FXI   = FXI   + FXIJS
22070       FYI   = FYI   + FYIJS
22080       FZI   = FZI   + FZIJS
22090       FX(JPTCL) = FX(JPTCL) + FXJIS
22100       FY(JPTCL) = FY(JPTCL) + FYJIS
22110       FZ(JPTCL) = FZ(JPTCL) + FZJIS
22120       IF( .NOT.SKIPTRQ ) THEN
22130         TXI   = TXI   + TQXIJS
22140         TYI   = TYI   + TQYIJS
22150         TZI   = TZI   + TQZIJS
22160         TX(JPTCL) = TX(JPTCL) + TQXJIS
22170         TY(JPTCL) = TY(JPTCL) + TQYJIS
22180         TZ(JPTCL) = TZ(JPTCL) + TQZJIS
22190       END IF
22200                                                                    RETURN
22210                                                                    END
22220 C**** SUB STEALLCA ****
```

• The forces and torques due to steric interactions are calculated in this subroutine for particles I and JPTCL.

• The "arm" position vector (XRXI,YRYI,ZRZI) is used for evaluating torques by cross product of the arm position vector and the force vector. Similar symbols are used for particle JPTCL.

• The vector (TXIJ0,TYIJ0,TZIJ0) denotes the relative vector from particles JPTCL to I.

• The forces and torques due to steric interactions are calculated in subroutine STEFORCE.

• If the particles were exchanged in the variables before entrance into this subroutine, the calculated forces and torques are saved in the appropriate variables; in this case, the logical variable KEEP was set as ".FALSE." before entrance into this subroutine.

• According to the plain sphere-constituting model for the disk-like particle, the forces and torques are evaluated in subroutine STEALLCA.

```
22230        SUBROUTINE STEALLCA( RRXI, RRYI, RRZI, RRXJ, RRYJ, RRZJ,
22240      &        RORGNXI, RORGNYI, RORGNZI, RORGNXJ, RORGNYJ, RORGNZJ,
22250      &        EEXIS, EEYIS, EEZIS, EEXJS, EEYJS, EEZJS,
22260      &        EEXJ, EEYJ, EEZJ, LGCWALL, RCHCKSQ, RV, IJOVRLAP,
22270      &        IOVRLAP, KEEP, FXI, FYI, FZI, TXI, TYI, TZI, J, SKIPTRQ )
22280 C
22290        IMPLICIT REAL*8 (A-H,O-Z), INTEGER (I-N)
22300 C
22310        COMMON /BLOCK26/ NPROWALL, NPROW   , NPROWST
22320        COMMON /BLOCK27/ NPSTRC  , RXSTRC  , RYSTRC
22330 C      ---------------
22340        PARAMETER( NNSTRC=100 )
22350 C
22360        INTEGER  NPROWALL, NPROW( NNSTRC ) , NPROWST( NNSTRC )
22370        REAL*8   RXSTRC( NNSTRC ) , RYSTRC( NNSTRC )
22380 C      ---------------
22390        LOGICAL  KEEP , IJOVRLAP , IOVRLAP , LGCWALL , SKIPTRQ
22400 C      ---------------
22410        REAL*8   DPXXI, DPXYI, DPXZI, DPYXI, DPYYI, DPYZI
22420        REAL*8   DPXXJ, DPXYJ, DPXZJ, DPYXJ, DPYYJ, DPYZJ
22430        REAL*8   RXSTRC0, RYSTRC0, XI , YI , ZI , XJ , YJ , ZJ
22440        REAL*8   CRISQ, CRJSQ, C0
22450        INTEGER  IROW, IPTCL0, IPTCL
22460 C      ------------------------------------------------------------------------------
22470        DPYXJ =  -EEXJS
22480        DPYYJ =  -EEYJS
22490        DPYZJ =  -EEZJS
22500        DPXXJ =  DPYYJ*EEZJ - DPYZJ*EEYJ
22510        DPXYJ =  DPYZJ*EEXJ - DPYXJ*EEZJ
22520        DPXZJ =  DPYXJ*EEYJ - DPYYJ*EEXJ
22530 C
22540        DPYXI =  EEXIS
22550        DPYYI =  EEYIS
22560        DPYZI =  EEZIS
22570        C0    =  DPYXJ*DPYXI +  DPYYJ*DPYYI +  DPYZJ*DPYZI
22580        IF( C0 .LT. 0.D0 ) THEN
22590          DPYXI = -DPYXI
22600          DPYYI = -DPYYI
22610          DPYZI = -DPYZI
22620        END IF
22630        DPXXI =  DPXXJ
22640        DPXYI =  DPXYJ
22650        DPXZI =  DPXZJ
22660 C               +++ STARTING POINT (RORGNXI,...,RORGNZI) FOR PTCL I +++
22670 C               +  STARTING POINT (RORGNXJ,...,RORGNZJ) FOR PTCL J   +
22680 C               +  NUMBER OF MAX ROW IS NPROWALL FOR CHECKING FORCES +
22690 C
22700        DO 800 IROW =1, NPROWALL
22710          IPTCL0  = NPROWST(IROW)
22720          LGCWALL = .FALSE.
22730          DO 780 III=1, NPROW(IROW)
22740            IPTCL   = IPTCL0 + (III-1)
22750            RXSTRC0 = RXSTRC(IPTCL)
22760            RYSTRC0 = RYSTRC(IPTCL)
22770            XJ = RORGNXJ + RXSTRC0*DPXXJ + RYSTRC0*DPYXJ
22780            YJ = RORGNYJ + RXSTRC0*DPXYJ + RYSTRC0*DPYYJ
22790            ZJ = RORGNZJ + RXSTRC0*DPXZJ + RYSTRC0*DPYZJ
22800            XI = RORGNXI + RXSTRC0*DPXXI + RYSTRC0*DPYXI
22810            YI = RORGNYI + RXSTRC0*DPXYI + RYSTRC0*DPYYI
22820            ZI = RORGNZI + RXSTRC0*DPXZI + RYSTRC0*DPYZI
22830 C
22840            IF( IROW .EQ. 1 )   GOTO 760
22850 C
22860            CRISQ = (XI-RRXI)**2 + (YI-RRYI)**2 + (ZI-RRZI)**2
22870            CRJSQ = (XJ-RRXJ)**2 + (YJ-RRYJ)**2 + (ZJ-RRZJ)**2
22880 C
22890            IF(  (CRISQ.GT.RCHCKSQ+0.02D0).OR.
22900      &                (CRJSQ.GT.RCHCKSQ+0.02D0)       ) GOTO 780
22910 C
22920   760      LGCWALL = .TRUE.
22930 C
22940 CCCC       SKIPTRQ = .FALSE.
22950            CALL STEIJCAL( XI,YI,ZI, XJ,YJ,ZJ, RRXI,RRYI,RRZI,
22960      &                    RRXJ, RRYJ, RRZJ, RV, IJOVRLAP, IOVRLAP,
22970      &                    KEEP, FXI,FYI,FZI, TXI,TYI,TZI, J, SKIPTRQ)
22980 C
22990 C
23000   780   CONTINUE
23010 C
23020   800 CONTINUE
23030                                                                   RETURN
23040                                                                   END
```

• Preparation for using the plain sphere-constituting model that was explained in subroutine PTCLMDL.

• $\boldsymbol{\delta}_j^{pX}$ is defined by $\boldsymbol{\delta}_j^{pY}\times\boldsymbol{e}_j$.

• $\boldsymbol{\delta}_j^{pX}$ and $\boldsymbol{\delta}_j^{pY}$ are the unit vectors in the *X*- and *Y*-axis directions of the plane of the sphere-constituting model of particle *j*.

• Similar symbols $\boldsymbol{\delta}_i^{pX}$ and $\boldsymbol{\delta}_i^{pY}$ are used for particle *i*.

• $\boldsymbol{\delta}_i^{pY}$ is set as $\boldsymbol{\delta}_i^{pY}=\boldsymbol{e}_i^{s}$.

• $\boldsymbol{\delta}_i^{pX}$ is set as $\boldsymbol{\delta}_i^{pX}=\boldsymbol{\delta}_j^{pX}$.

• The plain sphere-constituting model.

• This starting position is denoted by $\boldsymbol{r}_j^{orgn}$, (RORGNXJ, RORGNYJ, RORGNZJ). Similar symbols are used for particle *i*.

• IPTCL0 denotes the name of the first sphere of the shell area IROW.

• The positions of the two spheres are denoted by (XJ,YJ,ZJ) and (XI,YI,ZI) for particles *j* and *i*, respectively.

• The point (XJ,YJ,ZJ) has to be within the disk plane of particle *j*. This has to be valid also by the point (XI, YI, ZI) of particle *i*.

• The forces and torques due to the overlap of the steric layers of particles *i* and *j* are calculated.

```
23050 C#### FUN RIJMNFUN ####
23060       DOUBLE PRECISION FUNCTION  RIJMNFUN( RRXI, RRYI, RRZI,
23070      &                                    RRXJ, RRYJ, RRZJ,
23080      &                      EEXI, EEYI, EEZI, EEXJ, EEYJ, EEZJ,
23090      &                      RRXIQ, RRYIQ, RRZIQ, RRXJQ, RRYJQ, RRZJQ )
23100 C
23110       IMPLICIT REAL*8 (A-H,O-Z), INTEGER (I-N)
23120 C
23130       COMMON /BLOCK16/ D, D1, D2, B, B1, B2, RP, RP1, VP
23140       COMMON /WORK27/  RPXIMN, RPYIMN, RPZIMN, RPXJMN, RPYJMN, RPZJMN
23150 C
23160       PARAMETER( PI=3.141592653589793D0 )
23170 C
23180       REAL*8   NXIQ , NYIQ , NZIQ , NXJQ , NYJQ , NZJQ
23190       REAL*8   C1X  , C1Y  , C1Z  , C2X  , C2Y  , C2Z , C0 , C1 , D02
23200       REAL*8   DTHETAMX , RAN1 , RAN2
23210       REAL*8   DTMX1   , DTMX2
23220       REAL*8   RRXIQ0 , RRYIQ0 , RRZIQ0 , RRXJQ0 , RRYJQ0 , RRZJQ0
23230       REAL*8   NXIQ0  , NYIQ0  , NZIQ0  , NXJQ0  , NYJQ0  , NZJQ0
23240       REAL*8   RRXIC0 , RRYIC0 , RRZIC0 , RRXJC0 , RRYJC0 , RRZJC0
23250       REAL*8   RRDIST , RRIJMN0(20)
23260       REAL*8   RRXIMN(20) , RRYIMN(20) , RRZIMN(20)
23270       REAL*8   RRXJMN(20) , RRYJMN(20) , RRZJMN(20)
23280       REAL*8   CMN
23290       REAL*8   CRRXIMN , CRRYIMN , CRRZIMN
23300       REAL*8   CRRXJMN , CRRYJMN , CRRZJMN
23310 C
23320       INTEGER  IREPEAT , III , IIIEND , ICMN
23330 C     -------------------------------------------------------------
23340 C                                   ++++++++++++++++
23350 C                                   STARTING POINTS :
23360 C                                      PTCL II : (RRXIQ, RRYIQ, RRZIQ)
23370 C                                      PTCL JJ : (RRXJQ, RRYJQ, RRZJQ)
23380 C                                   ++++++++++++++++++++++++++++++++++++
23390       D02     = D/2.D0
23400 C
```

· In this function, the minimum distance and the particle positions giving rise to this value are solved by means of scanning of ambient positions using quasi-random numbers.

```
23410       C1X = RRXIQ - RRXI
23420       C1Y = RRYIQ - RRYI
23430       C1Z = RRZIQ - RRZI
23440       C2X = EEYI * C1Z - EEZI * C1Y
23450       C2Y = EEZI * C1X - EEXI * C1Z
23460       C2Z = EEXI * C1Y - EEYI * C1X
23470       C0  = DSQRT( C2X**2 + C2Y**2 + C2Z**2 )
23480       NXIQ = C2X / C0
23490       NYIQ = C2Y / C0
23500       NZIQ = C2Z / C0
23510 C
```

· $\boldsymbol{n}_i^Q$ is defined as $\boldsymbol{n}_i^Q = \boldsymbol{e}_i \times (\boldsymbol{r}_i^Q - \boldsymbol{r}_i) / |\boldsymbol{e}_i \times (\boldsymbol{r}_i^Q - \boldsymbol{r}_i)|$ in Eq. (5.33).

```
23520       C1X = RRXJQ - RRXJ
23530       C1Y = RRYJQ - RRYJ
23540       C1Z = RRZJQ - RRZJ
23550       C2X = EEYJ * C1Z - EEZJ * C1Y
23560       C2Y = EEZJ * C1X - EEXJ * C1Z
23570       C2Z = EEXJ * C1Y - EEYJ * C1X
23580       C0  = DSQRT( C2X**2 + C2Y**2 + C2Z**2 )
23590       NXJQ = C2X / C0
23600       NYJQ = C2Y / C0
23610       NZJQ = C2Z / C0
23620 C
```

· $\boldsymbol{n}_j^Q$ is defined as $\boldsymbol{n}_j^Q = \boldsymbol{e}_j \times (\boldsymbol{r}_j^Q - \boldsymbol{r}_j) / |\boldsymbol{e}_j \times (\boldsymbol{r}_j^Q - \boldsymbol{r}_j)|$ in Eq. (5.33).

```
23630       DTHETAMX = (30.D0/180.D0)*PI * D02
23640       RAN1 = DSQRT( 2.D0 )
23650       RAN2 = DSQRT( 7.D0 )
23660 C
23670       DO 500 IREPEAT = 1, 3
23680 C
23690         IF( IREPEAT.EQ.1 ) THEN
23700           RRXIQ0 = RRXIQ
23710           RRYIQ0 = RRYIQ
23720           RRZIQ0 = RRZIQ
23730           RRXJQ0 = RRXJQ
23740           RRYJQ0 = RRYJQ
23750           RRZJQ0 = RRZJQ
23760           DTMX1  = DTHETAMX
23770           DTMX2  = DTHETAMX
23780           NXIQ0  = NXIQ
23790           NYIQ0  = NYIQ
23800           NZIQ0  = NZIQ
23810           NXJQ0  = NXJQ
23820           NYJQ0  = NYJQ
23830           NZJQ0  = NZJQ
23840         END IF
23850 C
23860         IIIEND = 10
23870         DO 100 III= 1, IIIEND
```

· Scanning area (angle) is denoted by DTHETAMX; in this simulation, ±30° is adopted in the first (beginning) stage.

· IREPEAT=1,2 and 3 denote the first, second and third scanning stage, respectively; a scanning angle is diminished with advancing scanning stage. In this simulation, the final minimum distance and positions are determined after three scanning stages.

· In each stage, ten number of scanning procedures are conducted.

· $\boldsymbol{n}_{i0}^Q$ is used as $\boldsymbol{n}_i^Q$ in this stage. Similarly, $\boldsymbol{n}_{j0}^Q$, $\boldsymbol{r}_{i0}^Q$, $\boldsymbol{r}_{j0}^Q$, $\Delta\theta_1^{(max)}$ and $\Delta\theta_2^{(max)}$ are used as the corresponding quantities in this stage.

· $\Delta\theta_1^{(max)}$ and $\Delta\theta_2^{(max)}$ are scanning areas (angles) for particles $i$ and $j$, respectively.

```
23880 C                                                  --- FOR PTCL II ---
23890          C1 = RAN1 * DBLE(III*IREPEAT)
23900          C1 = C1 - DINT(C1)
23910          C1 = (2.D0*C1 - 1.D00) * DTMX1
23920 C
23930          RRXIC0 = RRXIQ0 + C1*NXIQ0
23940          RRYIC0 = RRYIQ0 + C1*NYIQ0
23950          RRZIC0 = RRZIQ0 + C1*NZIQ0
23960 C
23970          C1X = RRXIC0 - RRXI
23980          C1Y = RRYIC0 - RRYI
23990          C1Z = RRZIC0 - RRZI
24000 C
24010          C0  = DSQRT( C1X**2 + C1Y**2 + C1Z**2 )
24020          C1  = D02 / C0
24030 C
24040          RRXIC0 = RRXI + C1 * C1X
24050          RRYIC0 = RRYI + C1 * C1Y
24060          RRZIC0 = RRZI + C1 * C1Z
24070 C                                                  --- FOR PTCL JJ ---
24080          C1 = RAN2 * DBLE(III*IREPEAT)
24090          C1 = C1 - DINT(C1)
24100          C1 = (2.D0*C1 - 1.D00) * DTMX2
24110 C
24120          RRXJC0 = RRXJQ0 + C1*NXJQ0
24130          RRYJC0 = RRYJQ0 + C1*NYJQ0
24140          RRZJC0 = RRZJQ0 + C1*NZJQ0
24150 C
24160          C1X = RRXJC0 - RRXJ
24170          C1Y = RRYJC0 - RRYJ
24180          C1Z = RRZJC0 - RRZJ
24190 C
24200          C0  = DSQRT( C1X**2 + C1Y**2 + C1Z**2 )
24210          C1  = D02 / C0
24220 C
24230          RRXJC0 = RRXJ + C1 * C1X
24240          RRYJC0 = RRYJ + C1 * C1Y
24250          RRZJC0 = RRZJ + C1 * C1Z
24260 C
24270          RRDIST = DSQRT(  (RRXIC0-RRXJC0)**2 + (RRYIC0-RRYJC0)**2
24280      &                                        + (RRZIC0-RRZJC0)**2  )
24290 C
24300          RRIJMN0(III) = RRDIST
24310          RRXIMN( III) = RRXIC0
24320          RRYIMN( III) = RRYIC0
24330          RRZIMN( III) = RRZIC0
24340          RRXJMN( III) = RRXJC0
24350          RRYJMN( III) = RRYJC0
24360          RRZJMN( III) = RRZJC0
24370 C
24380   100  CONTINUE
24390 C                                             --- SET THE DATA IN ORDER ---
24400        DO 120 JJ=1,IIIEND
24410 C
24420          CMN  = RRIJMN0(JJ)
24430          ICMN = JJ
24440          DO 115 JJJ=JJ+1,IIIEND
24450            IF( RRIJMN0(JJJ) .LT. CMN ) THEN
24460              CMN  = RRIJMN0(JJJ)
24470              ICMN = JJJ
24480            END IF
24490   115    CONTINUE
24500          RRIJMN0(ICMN) = RRIJMN0(JJ)
24510          RRIJMN0(JJ)   = CMN
24520 C
24530          CRRXIMN = RRXIMN(ICMN)
24540          CRRYIMN = RRYIMN(ICMN)
24550          CRRZIMN = RRZIMN(ICMN)
24560          CRRXJMN = RRXJMN(ICMN)
24570          CRRYJMN = RRYJMN(ICMN)
24580          CRRZJMN = RRZJMN(ICMN)
24590          RRXIMN(ICMN) = RRXIMN(JJ)
24600          RRYIMN(ICMN) = RRYIMN(JJ)
24610          RRZIMN(ICMN) = RRZIMN(JJ)
24620          RRXJMN(ICMN) = RRXJMN(JJ)
24630          RRYJMN(ICMN) = RRYJMN(JJ)
24640          RRZJMN(ICMN) = RRZJMN(JJ)
24650          RRXIMN(JJ)   = CRRXIMN
24660          RRYIMN(JJ)   = CRRYIMN
24670          RRZIMN(JJ)   = CRRZIMN
24680          RRXJMN(JJ)   = CRRXJMN
24690          RRYJMN(JJ)   = CRRYJMN
24700          RRZJMN(JJ)   = CRRZJMN
```

• $\boldsymbol{r}_{i(try)}^{(min)}$ is evaluated from Eq. (5.34), denoted by (RRXIC0, RRYIC0, RRZIC0).

• Modification is conducted according to Eq. (5.35).

• $\boldsymbol{r}_{j(try)}^{(min)}$ is evaluated from Eq. (5.34), denoted by (RRXJC0, RRYJC0, RRZJC0).

• Modification is conducted according to Eq. (5.35).

• The distance and the particle positions are saved in these variables.

• This procedure is a typical routine for setting the data in order; smaller values are saved in earlier variables.

```
24710 C
24720   120    CONTINUE
24730 C                          --- SET THE DATA FOR THE NEXT NARROW RANGE ---
24740 C                                                   --- FOR PTCL II ---
24750          RRXIQ0 = RRXIMN(1)
24760          RRYIQ0 = RRYIMN(1)
24770          RRZIQ0 = RRZIMN(1)
24780          C1X = RRXIMN(3)- RRXIMN(1)
24790          C1Y = RRYIMN(3)- RRYIMN(1)
24800          C1Z = RRZIMN(3)- RRZIMN(1)
```

- $r_{i1}^{(min)}$ is now adopted as $r_{i0}^{Q}$ for the next scanning stage. Also, $|r_{i3}^{(min)} - r_{i1}^{(min)}|$ is adopted as $\Delta\theta_1^{(max)}$. $(r_{i3}^{(min)} - r_{i1}^{(min)}) / |r_{i3}^{(min)} - r_{i1}^{(min)}|$ is used as $n_{i0}^{Q}$ in the next scanning stage.
- Similar quantities are adopted for particle *j*.

```
24810          C0 = DSQRT( C1X**2 + C1Y**2 + C1Z**2 )
24820          DTMX1  = C0
24830          NXIQ0  = C1X /C0
24840          NYIQ0  = C1Y /C0
24850          NZIQ0  = C1Z /C0
24860 C                                                   --- FOR PTCL JJ ---
24870          RRXJQ0 = RRXJMN(1)
24880          RRYJQ0 = RRYJMN(1)
24890          RRZJQ0 = RRZJMN(1)
24900          C1X = RRXJMN(3)- RRXJMN(1)
24910          C1Y = RRYJMN(3)- RRYJMN(1)
24920          C1Z = RRZJMN(3)- RRZJMN(1)
24930          C0 = DSQRT( C1X**2 + C1Y**2 + C1Z**2 )
24940          DTMX2  = C0
24950          NXJQ0  = C1X /C0
24960          NYJQ0  = C1Y /C0
24970          NZJQ0  = C1Z /C0
24980 C
24990   500 CONTINUE
25000 C
25010       RIJMNFUN = RRIJMN0(1)
25020 C                                  ++++++++++++++++++++++++++++++++++++++++
25030 C                                   POINTS YIELDING THE MINIMUM DISTANCE:
25040 C                                         PTCL II : (RPXIMN,RPYIMN,RPZIMN
25050 C                                         PTCL JJ : (RPXJMN,RPYJMN,RPZJMN
25060 C                                  MINIMUM DISTANCE : RIJMNFUNC
25070 C                                  ++
25080       RPXIMN = RRXIMN(1)
25090       RPYIMN = RRYIMN(1)
25100       RPZIMN = RRZIMN(1)
25110       RPXJMN = RRXJMN(1)
25120       RPYJMN = RRYJMN(1)
25130       RPZJMN = RRZJMN(1)
25140                                                                   RETURN
25150                                                                   END
```

- RIJMNFUN denotes the minimum distance.
- (RPXIMN,RPYIMN,RPZIMN) denotes the position of sphere of particle *i*, which gives rise to the minimum distance; similar position is obtained for particle *j*.

```
25160 C**** SUB STEFORCE ****
25170       SUBROUTINE STEFORCE( RRIJ, RV, ISKIP, TXIJ, TYIJ, TZIJ,
25180      &                     IJOVRLAP, IOVRLAP )
25190 C
25200       IMPLICIT REAL*8 (A-H,O-Z), INTEGER (I-N)
25210 C
25220       COMMON /BLOCK16/ D, D1, D2, B, B1, B2, RP, RP1, VP
25230       COMMON /BLOCK17/ DEL, TD, RCOFF
25240       COMMON /WORK11/  XRXI , YRYI , ZRZI , XRXJ , YRYJ , ZRZJ
25250       COMMON /WORK12/  FXIJS, FYIJS, FZIJS, FXJIS, FYJIS, FZJIS
25260       COMMON /WORK13/  TQXIJS,TQYIJS,TQZIJS,TQXJIS,TQYJIS,TQZJIS
25270       COMMON /WORK14/  RCOFF2 , RP102
25280       COMMON /WORK17/  NRIJMN    , RIJMN    , RIJMNAV  , RIJMNSUM
25290       COMMON /WORK18/  NRMNMN    , RMNMN    , RMNMNAV  , RMNMNSUM
25300 C
25310       INTEGER    NN , NNS
25320       PARAMETER( NN=1000 , NNS=5000000 )
25330       PARAMETER( NRANMX=1000000 , PI=3.141592653589793D0 )
25340 C
25350       REAL*8   FXIJ , FYIJ , FZIJ , C0
25360       LOGICAL  IJOVRLAP,IOVRLAP
25370 C
25380       INTEGER  NRIJMN , NRMNMN
25390       REAL*8   RIJMN  , RIJMNAV , RIJMNSUM
25400       REAL*8   RMNMN  , RMNMNAV , RMNMNSUM
25410 C     -----------------------------------------------------------------
25420 C                                              --- STERIC REPULSION ---
25430       FXIJ   = 0.D0
25440       FYIJ   = 0.D0
25450       FZIJ   = 0.D0
25460 C
25470       IF( RRIJ .LT. B2 ) THEN
25480         IF( RRIJ .LT.  RIJMN )   RIJMN = RRIJ
25490         IF( .NOT. IJOVRLAP ) THEN
```

- The forces and torques due to the overlap of the steric layers of particles *i* and *j* are calculated.

```
25500             NRIJMN    = NRIJMN + 1
25510             IJOVRLAP = .TRUE.
25520           END IF
25530 C
25540           IF( RRIJ .LT.  RMNMN )    RMNMN = RRIJ
25550           IF( .NOT. IOVRLAP ) THEN
25560             NRMNMN    = NRMNMN + 1
25570             IOVRLAP = .TRUE.
25580           END IF
25590         END IF
25600 C
25610         IF( RRIJ .LT. B2 ) THEN
25620           IF( RRIJ .LE. 1.D0 ) RRIJ = 1.0001D0
25630           CO    = DLOG( B2 / RRIJ )
25640           FXIJ = TXIJ*C0
25650           FYIJ = TYIJ*C0
25660           FZIJ = TZIJ*C0
25670         END IF
25680 C
25690         FXIJS = FXIJ*RV
25700         FYIJS = FYIJ*RV
25710         FZIJS = FZIJ*RV
25720         FXJIS = - FXIJS
25730         FYJIS = - FYIJS
25740         FZJIS = - FZIJS
25750         IF( ISKIP .EQ. 1 )   RETURN
25760 C                                                        --- TORQUES ---
25770         TQXIJS  = ( YRYI*FZIJS - ZRZI*FYIJS )*3.D0
25780         TQYIJS  = ( ZRZI*FXIJS - XRXI*FZIJS )*3.D0
25790         TQZIJS  = ( XRXI*FYIJS - YRYI*FXIJS )*3.D0
25800         TQXJIS  = ( YRYJ*FZJIS - ZRZJ*FYJIS )*3.D0
25810         TQYJIS  = ( ZRZJ*FXJIS - XRXJ*FZJIS )*3.D0
25820         TQZJIS  = ( XRXJ*FYJIS - YRYJ*FXJIS )*3.D0
25830                                                                  RETURN
25840                                                                  END
25850 C**** SUB RANDISP *****
25860         SUBROUTINE RANDISP( N, H, NTIME )
25870 C                                              -----------------------------
25880 C                                               CTRAB1 FOR PRAPALLEL (TBM)
25890 C                                               CTRAB2 FOR NORMAL 1  (TBM)
25900 C                                               CTRAB3 FOR NORMAL 2  (TBM)
25910 C                                               CROTB1 FOR PRAPALLEL (RBM)
25920 C                                               CROTB2 FOR NORMAL 1  (RBM)
25930 C                                               CROTB3 FOR NORMAL 2  (RBM)
25940 C                                              -----------------------------
25950         IMPLICIT REAL*8 (A-H,O-Z), INTEGER (I-N)
25960 C
25970         COMMON /BLOCK2/  EX    , EY    , EZ
25980         COMMON /BLOCK3/  NX    , NY    , NZ
25990         COMMON /BLOCK6/  RXB   , RYB   , RZB
26000         COMMON /BLOCK7/  EXB   , EYB   , EZB
26010         COMMON /BLOCK8/  NXB   , NYB   , NZB
26020         COMMON /BLOCK14/ RM    , RH    , RV
26030         COMMON /BLOCK38/ DT1   , DT2   , DR1 , DR2 , YHYC
26040         COMMON /BLOCK39/ CRNDMT1, CRNDMT2, CRNDMR1, CRNDMR2
26050         COMMON /BLOCK42/ NRAN , RAN   , IXRAN
26060 C
26070         INTEGER     NN , NNS
26080         PARAMETER( NN=1000 , NNS=5000000 )
26090         PARAMETER( NRANMX=1000000 , PI=3.141592653589793D0 )
26100 C
26110         REAL*8    EX(NN) , EY(NN) , EZ(NN) , NX(NN) , NY(NN) , NZ(NN)
26120         REAL*8    RXB(NN), RYB(NN), RZB(NN)
26130         REAL*8    EXB(NN), EYB(NN), EZB(NN), NXB(NN), NYB(NN), NZB(NN)
26140 C
26150         REAL      RAN(NRANMX)
26160         INTEGER   NRAN , IXRAN , NRANCHK
26170 C
26180         REAL*8    EXI  , EYI  , EZI , NXI  , NYI  , NZI
26190         REAL*8    RXBI , RYBI , RZBI
26200         REAL*8    RXBI1, RYBI1, RZBI1
26210         REAL*8    RXBI2, RYBI2, RZBI2
26220         REAL*8    RXBI3, RYBI3, RZBI3
26230         REAL*8    EXBI , EYBI , EZBI , NXBI , NYBI , NZBI
26240         REAL*8    EXBI1, EYBI1, EZBI1
26250         REAL*8    EXBI2, EYBI2, EZBI2
26260         REAL*8    EXBI3, EYBI3, EZBI3
26270         REAL*8    CS1  , SN1  , CS2  , SN2
26280         REAL*8    RAN1 , RAN2
26290         REAL*8    CTRAB1 , CTRAB2 , CTRAB3 , CROTB1 , CROTB2 , CROTB3
```

• Treatment for the overlap of the solid parts of spheres in order to prevent the system from diverging, just in case.

• The forces due to the overlap of the steric layers of particles *i* and *j* are calculated according to Eq. (3.1).

• The torques due to the overlap of the steric layers are calculated by cross product of the "arm" position vector and the force vector.

• The translational and rotational random displacements are generated for Brownian motion of disk-like particles.

```
26300        REAL*8   CTRAB1MX , CTRAB2MX , CTRAB3MX
26310        REAL*8   CROTB1MX , CROTB2MX , CROTB3MX
26320        REAL*8   C00 , XI , YI , XXI , YYI , ZZI
26330 C -----------------------------------------------
26340        CTRAB1MX = DSQRT( CRNDMT1 ) * 3.5D0
26350        CTRAB2MX = DSQRT( CRNDMT2 ) * 3.5D0
26360        CTRAB3MX = DSQRT( CRNDMT2 ) * 3.5D0
26370        CROTB1MX = DSQRT( CRNDMR1 ) * 3.5D0
26380        CROTB2MX = DSQRT( CRNDMR2 ) * 3.5D0
26390        CROTB3MX = DSQRT( CRNDMR2 ) * 3.5D0
26400 C
```

· The maximum displacements may be necessitated to be set for preventing the system from diverging. CTRAB1MX is for translational displacement in the particle direction, and CTRAB2MX and CTRAB3MX are for that in the directions normal to the particle direction.

```
26410        DO 100 I=1,N
26420 C
26430          RXBI = 0.D0
26440          RYBI = 0.D0
26450          RZBI = 0.D0
26460          EXBI = 0.D0
26470          EYBI = 0.D0
26480          EZBI = 0.D0
26490          NXBI = 0.D0
26500          NYBI = 0.D0
26510          NZBI = 0.D0
26520 C
26530          EXI  = EX(I)
26540          EYI  = EY(I)
26550          EZI  = EZ(I)
26560          C0   = DSQRT( EXI**2 + EYI**2 + EZI**2 )
26570          EXI  = EXI/C0
26580          EYI  = EYI/C0
26590          EZI  = EZI/C0
26600          NXI  = NX(I)
26610          NYI  = NY(I)
26620          NZI  = NZ(I)
26630          C0   = DSQRT( NXI**2 + NYI**2 + NZI**2 )
26640          NXI  = NXI/C0
26650          NYI  = NYI/C0
26660          NZI  = NZI/C0
26670 C                             ----------------------------------------
26680 C                               (1)  TRANSLATIONAL BROWNIAN MOTION
26690 C                             ----------------------------------------
26700          CS1  = EZI
26710          SN1  = DSQRT( 1.D0 - CS1**2 )
26720          IF( SN1 .LT. 0.00001D0 ) THEN
26730            CS2 = 1.D0
26740            SN2 = 0.D0
26750 CCCCCC     CS2  = EXI/SN1
26760 CCCCCC     SN2  = EYI/SN1
26770            WRITE(6,*) '++++ in sub RANDSIP ++++ NTIME,I=',NTIME,I
26780          ELSE
26790            CS2  = EXI/SN1
26800            SN2  = EYI/SN1
26810          END IF
26820 C
26830          IF ( (NRAN+14) .GE. NRANMX ) THEN
26840            CALL RANCAL( NRANMX, IXRAN, RAN )
26850            NRAN = 1
26860          END IF
26870 C                                     --- CTRAB1 FOR PRAPALLEL ---
26880          NRAN = NRAN+1
26890          RAN1 = DBLE( RAN(NRAN) )
26900          NRAN = NRAN+1
26910          RAN2 = DBLE( RAN(NRAN) )
26920          CTRAB1 = ( -2.D0*CRNDMT1*DLOG(RAN1) )**0.5
26930       &              * DCOS( 2.D0*PI*RAN2 )
26940          IF( DABS(CTRAB1) .GT. CTRAB1MX ) CTRAB1=DSIGN(CTRAB1MX,CTRAB1)
26950 C                                     --- CTRAB2 FOR NORMAL 1 ---
26960          NRAN = NRAN+1
26970          RAN1 = DBLE( RAN(NRAN) )
26980          NRAN = NRAN+1
26990          RAN2 = DBLE( RAN(NRAN) )
27000          CTRAB2 = ( -2.D0*CRNDMT2*DLOG(RAN1) )**0.5
27010       &              * DCOS( 2.D0*PI*RAN2 )
27020          IF( DABS(CTRAB2) .GT. CTRAB2MX ) CTRAB2=DSIGN(CTRAB2MX,CTRAB2)
27030 C                                     --- CTRAB3 FOR NORMAL 2 ---
27040          NRAN = NRAN+1
27050          RAN1 = DBLE( RAN(NRAN) )
27060          NRAN = NRAN+1
27070          RAN2 = DBLE( RAN(NRAN) )
27080          CTRAB3 = ( -2.D0*CRNDMT2*DLOG(RAN1) )**0.5
27090       &              * DCOS( 2.D0*PI*RAN2 )
```

· Similar maximum displacements are set for spin and ordinary rotational displacements,CROTB1MX, CROTB2MX and CROTB3MX.

· Caution is necessary for this situation.

· The translational random displacement $\Delta r_{\parallel}^{B}$ in the particle axis direction in Eq. (9.20).

· The translational random displacement $\Delta r_{\perp 1}^{B}$ in a direction normal to the particle axis direction in Eq. (9.20).

· The translational random displacement $\Delta r_{\perp 2}^{B}$ in a direction normal to the particle axis direction in Eq. (9.20).

```
27100          IF( DABS(CTRAB3) .GT. CTRAB3MX ) CTRAB3=DSIGN(CTRAB3MX,CTRAB3)
27110 C                                       --- FOR PARALLEL MOVEMENT ---
27120          RXBI1= CTRAB1*EXI
27130          RYBI1= CTRAB1*EYI
27140          RZBI1= CTRAB1*EZI
27150          RXBI2= CTRAB2*( CS1*CS2 )
27160          RYBI2= CTRAB2*( CS1*SN2 )
27170          RZBI2= CTRAB2*(   -SN1 )
27180          RXBI3= CTRAB3*(   -SN2 )
27190          RYBI3= CTRAB3*(    CS2 )
27200          RZBI3= 0.D0
```

• The random displacements in the particle-fixed coordinate are transferred into those in the absolute coordinate system in terms of the rotation matrix.

```
27210 C
27220          RXBI = RXBI1 + RXBI2 + RXBI3
27230          RYBI = RYBI1 + RYBI2 + RYBI3
27240          RZBI = RZBI1 + RZBI2 + RZBI3
27250 C
27260          RXB(I) = RXBI
27270          RYB(I) = RYBI
27280          RZB(I) = RZBI
```

• The net random displacements vector $\boldsymbol{r}_i^B$ is obtained by summing the components as $\boldsymbol{r}_i^B = \Delta r_{\parallel}{}^B \boldsymbol{e}_i + \Delta r_{\perp 1}{}^B \boldsymbol{e}_{i\perp 1} + \Delta r_{\perp 2}{}^B \boldsymbol{e}_{i\perp 2}$ in Eqs. (6.38) and (6.39).

```
27290 C                                 ++++++++++++++++++++++++++++++++++++++
27300 C                                   (2)  ROTATIONAL BROWNIAN MOTION
27310 C                                 ++++++++++++++++++++++++++++++++++++++
27320 C                                 -----------  FOR PARALLEL MOVEMENT ---
27330 C                                      ---  CROTB1 FOR PRAPALLEL ---
27340          NRAN = NRAN+1
27350          RAN1 = DBLE( RAN(NRAN) )
27360          NRAN = NRAN+1
27370          RAN2 = DBLE( RAN(NRAN) )
27380          CROTB1 = ( -2.D0*CRNDMR1*DLOG(RAN1) )**0.5
27390     &                   * DCOS( 2.D0*PI*RAN2 )
27400          IF( DABS(CROTB1) .GT. CROTB1MX ) CROTB1=DSIGN(CROTB1MX,CROTB1)
```

• The rotational random displacement $\Delta\varphi_{\parallel}{}^B$ about the particle axis direction in Eq. (6.50).

```
27410 C
27420          NXBI = CROTB1*( EYI*NZI - EZI*NYI )
27430          NYBI = CROTB1*( EZI*NXI - EXI*NZI )
27440          NZBI = CROTB1*( EXI*NYI - EYI*NXI )
27450          NXB(I) = NXBI
27460          NYB(I) = NYBI
27470          NZB(I) = NZBI
```

• The magnetic moment randomly orients in the direction of $\boldsymbol{e}_i \times \boldsymbol{n}_i$ in Eq. (6.49).

```
27480 C                                 -------------  FOR NORMAL MOVEMENT ---
27490 C                                      ---  CROTB2 FOR NORMAL 1 ---
27500          NRAN = NRAN+1
27510          RAN1 = DBLE( RAN(NRAN) )
27520          NRAN = NRAN+1
27530          RAN2 = DBLE( RAN(NRAN) )
27540          CROTB2 = ( -2.D0*CRNDMR2*DLOG(RAN1) )**0.5
27550     &                   * DCOS( 2.D0*PI*RAN2 )
27560          IF( DABS(CROTB2) .GT. CROTB2MX ) CROTB2=DSIGN(CROTB2MX,CROTB2)
```

• The rotational random displacement $\Delta\varphi_{\perp 1}{}^B$ about a line normal to the particle axis direction in Eq. (6.46).

```
27570 C                                      ---  CROTB3 FOR NORMAL 2 ---
27580          NRAN = NRAN+1
27590          RAN1 = DBLE( RAN(NRAN) )
27600          NRAN = NRAN+1
27610          RAN2 = DBLE( RAN(NRAN) )
27620          CROTB3 = ( -2.D0*CRNDMR2*DLOG(RAN1) )**0.5
27630     &                   * DCOS( 2.D0*PI*RAN2 )
27640          IF( DABS(CROTB3) .GT. CROTB3MX ) CROTB3=DSIGN(CROTB3MX,CROTB3)
```

• The rotational random displacement $\Delta\varphi_{\perp 2}{}^B$ about a line normal to the particle axis direction in Eq. (6.46).

```
27650 C
27660          EXBI2= CROTB2*( CS1*CS2 )
27670          EYBI2= CROTB2*( CS1*SN2 )
27680          EZBI2= CROTB2*(   -SN1 )
27690          EXBI3= CROTB3*(   -SN2 )
27700          EYBI3= CROTB3*(    CS2 )
27710          EZBI3= 0.D0
```

• The random displacements in the particle-fixed coordinate are transferred into those in the absolute coordinate system in terms of the rotation matrix.

```
27720 C
27730          EXB(I) = EXBI2 + EXBI3
27740          EYB(I) = EYBI2 + EYBI3
27750          EZB(I) = EZBI2 + EZBI3
27760 C
27770   100 CONTINUE
27780                                                                  RETURN
27790                                                                  END
```

• The net random displacements vector $\boldsymbol{e}_i^B$ is obtained by summing the components as $\boldsymbol{e}_i^B = \Delta\varphi_{\perp 1}{}^B \boldsymbol{e}_{i\perp 1} + \Delta\varphi_{\perp 2}{}^B \boldsymbol{e}_{i\perp 2}$ in Eq. (6.48).

```
27800 C*******************************************************************
27810 C   THIS SUBROUTINE IS FOR GENERATING UNIFORM RANDOM NUMBERS         *
27820 C   (SINGLE PRECISION) FOR 64-BIT COMPUTER.                          *
27830 C       N       : NUMBER OF RANDOM NUMBERS TO GENERATE               *
27840 C       IXRAN   : INITIAL VALUE OF RANDOM NUMBERS (POSITIVE INTEGER) *
27850 C               : LAST GENERATED VALUE IS KEPT                       *
27860 C       X(N)    : GENERATED RANDOM NUMBERS (0<X(N)<1)                *
27870 C*******************************************************************
27880 C**** SUB RANCAL ****
27890       SUBROUTINE RANCAL( N, IXRAN, X )
27900 C
27910       IMPLICIT REAL*8 (A-H,O-Z), INTEGER (I-N)
```

• Generation of a series of pseudo-random numbers ranging from 0 to 1.

```
27920 C
27930         REAL      X(N)
27940         INTEGER INTEGMX, INTEG64, INTEGST, INTEG
27950 C
27960         DATA INTEGMX/2147483647/
27970 CCC     DATA INTEG64/2147483648/
27980         DATA INTEGST,INTEG/584287,48828125/
27990 C
28000         AINTEGMX = REAL( INTEGMX )


28010 CCC     AINTEGMX = REAL( INTEG64 )
28020 C
28030         IF ( IXRAN.LT.0 ) STOP
28040         IF ( IXRAN.EQ.0 ) IXRAN = INTEGST
28050         DO 30 I=1,N
28060    10     IXRAN = IXRAN*INTEG
28070 CCC       IXRAN = KMOD(IXRAN,INTEG64)
28080           IF (IXRAN .EQ. 0 ) THEN
28090             IXRAN   = INTEGST
28100             GOTO 10
28110           ELSE IF (IXRAN .LT. 0 ) THEN
28120             IXRAN   = (IXRAN+INTEGMX)+1
28130             X(I)    = REAL(IXRAN)/AINTEGMX
28140           ELSE
28150             X(I)    = REAL(IXRAN)/AINTEGMX
28160           END IF
28170    30   CONTINUE
28180         RETURN
28190         END
```

• This subroutine is for a 32-bit computer, but simple replacement of "CCC" makes the subroutine for a 64-bit computer.

• By printing out X(*) before starting the present simulation program, it is necessary to confirm that random values ranging from 0 to 1 are surely generated.

# Bibliography


[1] Aoshima, M., Ozaki, M. and Satoh, A. 2012. Structural analysis of self-assembled lattice structures composed of cubic hematite particles. J. Phys. Chem. 116: 17862–17871.

[2] Coverdale, G. N., Chantrell, R. W., Hart, A. and Parker, D. 1993. A 3-D simulation of a particulate dispersion. J. Magn. Magn. Mater. 120: 210–212.

[3] Coverdale, G. N., Chantrell, R. W., Hart, A. and Parker, D. 1994. A computer simulation of the microstructure of a particulate dispersion. J. Appl. Phys. 75: 5574–5576.

[4] Coverdale, G. N., Chantrell, R. W., Martin, G. A. R., Bradbury, A., Hart, A. and Parker, D. 1998. Cluster analysis of the microstructure of colloidal dispersions using the maximum entropy technique. J. Magn. Magn. Mater. 188: 41–51.

[5] Ozaki, M., Ookoshi, N. and Matijević, E. 1990. Preparation and magnetic properties of uniform hematite platelets. J. Colloid Interface Sci. 137: 546–549.

CHAPTER 10

# Topics of Current Applications

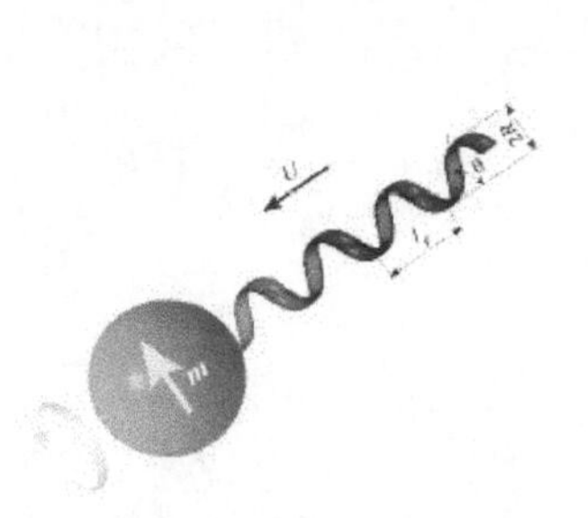

In this final chapter, we highlight several topics of current applications of magnetic particle suspensions. As already described in Chapter 1, the studies on the application in the field of biomedical engineering have been growing and expanding sufficiently to successfully lead to clinical practices which are relatively of gentle-burden to patients compared with standard surgery. Although remarkable advances in the synthesis and characterization of magnetic particles have been made for their use in the biomedical application, we focus here on the physical aspect of some of the current applications of magnetic particle suspensions in this field. These research subjects will surely stimulate and challenge students and young researchers who will pioneer new fields of research in magnetic particles or materials in the future. In the following we address the two topics of magnetic particle hyperthermia and magnetically-propelled swimmers, where the former application makes use of the heat generated by the interaction between a magnetic particle and an alternating magnetic field, and the latter application makes use of the locomotive mechanism controlled by a time-dependent external magnetic field such as an alternating or rotational magnetic field.

## 10.1 Magnetic particle hyperthermia

The concept of a magnetic particle hyperthermia treatment is to make use of the heat of magnetic particles that is generated by applying an alternating magnetic field in order to destroy a targeted cancer or tumor tissue without causing any serious damage to the surrounding healthy tissue [1–4]. The cancer tissue is more fragile in response to heat than ordinary tissue and therefore if the temperature of the target tissue can be controlled in an accurate manner, only the target cancer tissue will be destroyed by this treatment. This remarkable characteristic may yield a significant advantage to the magnetic particle hyperthermia treatment compared with an ordinary surgery approach.

The mechanism for an alternating magnetic field to induce the heating effect of magnetic particles is both the relaxation effect of magnetic moments and the hysteresis loss of the magnetization of magnetic moments [4]. Hence, the heating ability is dependent on a variety of factors such as the magnetic characteristics of particles, the base liquid characteristics such as viscosity and the frequency and strength of an alternating magnetic field. We consider here the heating characteristics of a suspension composed of single domain magnetic particles and in this case the heat generation is accomplished by the two mechanisms of Néel and Brownian relaxation of magnetic particles, as was explained in Chapter 1. If the magnetic moment of each particle rotates against an energy barrier within the particle body while the particle itself remains fixed, then the particle undergoes Néel relaxation where the thermal energy is dissipated during the rearrangement of the atomic dipole moment within the particle body. This is the heating mechanism due to the motion of the Néel relaxation of the magnetic moment aligning to the magnetic field direction. If the magnetic particle itself rotates with the magnetic moment direction fixed at the particle body, then the particle undergoes Brownian relaxation and in this case the thermal energy will be dissipated through the friction motion against the ambient fluid. In the case of magnetic particles of size in the range of nano-order, the combination of these two heating mechanisms may simultaneously contribute to the heat generation and, as will be shown in the following, the Brownian relaxation effect becomes a main factor in the range of larger particles, whereas the Néel relaxation effect becomes more dominant with decreasing particle diameter.

### *10.1.1 Relaxation of magnetic moment*

The relaxation phenomenon of the magnetic moment of magnetic particles has already been referred to in Chapter 1 and therefore here we briefly summarize the relaxation times for the Brownian and Néel relaxation modes. In the following discussion, we assume that the magnetic particles have only one magnetic domain or a single-domain within the particle body. A suspension composed of these single-domain particles will exhibit a superparamagnetic feature.

In the Néel relaxation mode, the magnetic moment is forced to reorient in the direction of an alternating magnetic field and will exhibit a delay of certain period known as the relaxation time. This relaxation is characterized by the Néel relaxation time $\tau_N$, expressed as [5]

$$\tau_N = \tau_0 \exp\left(\frac{K_a V_c}{k_B T}\right) \tag{10.1}$$

or the modified version is written as [6]

$$\tau_N = \frac{\sqrt{\pi}}{2}\tau_0 \left(\frac{K_a V_c}{k_B T}\right)^{-\frac{1}{2}} \exp\left(\frac{K_a V_c}{k_B T}\right) \tag{10.2}$$

in which $\tau_0$ is a constant that is generally in the order of $10^{-9}$ s, $V_c$ is the magnetic core volume of the particle, $K_a$ is the anisotropy constant that generally depends on the crystalline or shape effects of the particle, $k_B$ is Boltzmann's constant and $T$ is the liquid temperature. The quantity $K_a V_c$ is the activation energy which is the effective energy barrier inhibiting the change in the orientation of the magnetic moment. From this expression of the relaxation time, it is clear that the Néel relaxation strongly depends on the particle diameter and the anisotropy constant in an exponential manner.

In the Brownian relaxation mode, the particle itself rotates in a carrier liquid in order for the magnetic moment to follow the change in the direction of an applied time-dependent magnetic field. In this relaxation mode, the phenomenon is characterized by the rotational diffusion time or the Brownian relaxation time $\tau_B$. This relaxation time is expressed, for instance, in the case of a spherical particle as [7]

$$\tau_B = \frac{3\eta V_h}{k_B T} \tag{10.3}$$

in which $\eta$ is the viscosity of a carrier liquid and $V_h$ is the hydrodynamic volume of the particle. The expression in Eq. (10.3) implies that this relaxation time $\tau_B$ mainly depends on the carrier liquid viscosity and the size of the particle.

In an actual magnetic particle suspension, the relaxation of the magnetic moment aligning to the magnetic field direction of an alternating magnetic field is accomplished by the combination of the above-mentioned two relaxation mechanisms with an effective relaxation time $\tau$, which is expressed as

$$\frac{1}{\tau}=\frac{1}{\tau_B}+\frac{1}{\tau_N} \tag{10.4}$$

Figure 10.1 illustrates the dependence of the above relaxation times on the diameter of spherical magnetite particles in the typical nano-size range [4, 8]. It is seen that the Néel relaxation $\tau_N$ is much shorter than the Brownian relaxation time $\tau_B$ in the range of the smaller nano-size magnetic particles, which implies that in this particle size range the effect of the Néel relaxation strongly dominates the heating efficiency in a suspension of single-domain nano-particles. In the larger nano-particle size range, the Brownian relaxation time is much shorter than the Néel relaxation time, and therefore the effect of the Brownian relaxation is much more dominant in the mechanism of the alignment of the magnetic moments to the field direction. There is an intermediate range of nano-size magnetic particles for which $\tau_N \simeq \tau_B$ where both the Néel and Brownian relaxation mechanism will contribute to the heating efficiency of a magnetic particle suspension.

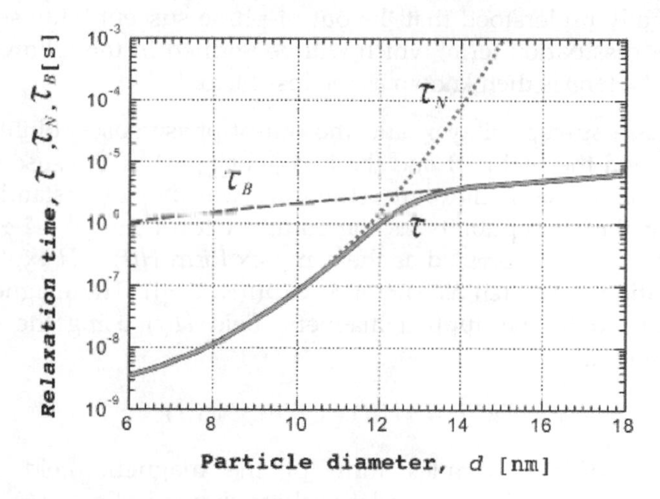


**Figure 10.1.** Typical example of relaxation times for magnetic particles: $\tau_N$ is the Néel relaxation time, $\tau_B$ is the Brownian relaxation time, and $\tau$ is the effective relaxation time.

## *10.1.2 In-phase and out-of-phase susceptibility components*

In this section we address the in-phase and the out-of-phase components of the susceptibility of a suspension of single-domain nano-particles in order to evaluate the power dissipation that is induced by the relaxation of the alignment of the magnetic moment of the particles to the direction of an alternating magnetic field [4, 5]. If an alternating magnetic field is applied to this magnetic particle suspension, the magnetization of this system responds to the magnetic field with a phase delay, which is induced by the relaxation that has been explained above.

If the alternating applied field $H(t)$ is expressed as $H(t) = H_0\cos(\omega t)$, then the magnetization $M(t)$ is written with a phase delay $\delta$ as

$$M(t) = M_0 \cos(\omega t - \delta) = (M_0 \cos\delta)\cos(\omega t) + (M_0 \sin\delta)\sin(\omega t) \quad (10.5)$$

In the above expression, the first term on the far right-hand side is known as the in-phase component and the second term is the out-of-phase component. The magnetic susceptibility of the system, $\chi$, is therefore expressed by both the in-phase susceptibility $\chi'$ and the out-of-phase susceptibility $\chi''$. These susceptibility components are expressed as

$$\chi' = \frac{M_0}{H_0}\cos\delta, \quad \chi'' = \frac{M_0}{H_0}\sin\delta \quad (10.6)$$

It is generally understood that the out-of-phase susceptibility serves for the power dissipation term, which will be verified in the following. The ratio $\chi''/\chi' = \tan\delta$ is then known as the loss factor.

The in-phase susceptibility $\chi'$ and the out-of-phase susceptibility $\chi''$ are usually called the real part and the imaginary part of the susceptibility, respectively. These names are straightforwardly understandable by expressing the susceptibility as the complex form $\chi = \chi' - \mathrm{i}\,\chi''$. If the alternating field is expressed as the complex form $H(t) = H_0\exp(\mathrm{i}\omega t)$, the magnetization is written as $M(t) = M_0\exp(\mathrm{i}(\omega t - \delta))$. The magnetization $M(t)$ is related to the applied magnetic field $H(t)$ using the complex susceptibility as

$$M(t) = \chi H(t) = (\chi' - \mathrm{i}\,\chi'')H(t) \quad (10.7)$$

Substitution of the complex form of the magnetic field and the magnetization gives rise to the relationships shown in Eq. (10.6).

From this consideration, the out-of-phase component or imaginary part $\chi''$ is expected to contribute to the heat generation in a magnetic particle suspension by an alternating magnetic field, which will be shown below.

In general, the magnetization $M$ responds to the input magnetic field $H$, which is schematically shown in Fig. 10.2 where a time-dependent magnetic field is applied in a step function manner. The response of the magnetization may be generally expressed in the following relaxation equation [5]:

$$M(t) = M_{sat}\left(1 - \exp(-t/\tau)\right) \tag{10.8}$$

in which $M_{sat}$ is the saturation magnetization and $\tau$ is the relaxation time that implies the index of the relaxation speed. Differentiating the expression in Eq. (10.8) leads to the following equation:

$$\frac{dM(t)}{dt} = \frac{1}{\tau}(M_{sat} - M(t)) \tag{10.9}$$

We now apply this equation to the present alternating magnetic field. As defined previously, the alternating applied field $H(t)$ is expressed as $H(t) = H_0\cos(\omega t)$, and the induced magnetization with no delay is assumed to be expressed as $M_{nodelay}(t) = \chi_0 H_0\cos(\omega t)$, where $\chi_0$ is the equilibrium susceptibility. In the present phenomenon, this quantity $M_{nodelay}(t)$ corresponds to the quantity $M_{sat}$ in Eq. (10.9). Since the magnetization $M(t)$ is expressed in Eq. (10.5), substitution of the expressions of $M(t)$ and $M_{nodelay}(t)$ into Eq. (10.9) leads to the solution of the in-phase and out-of-phase susceptibilities as

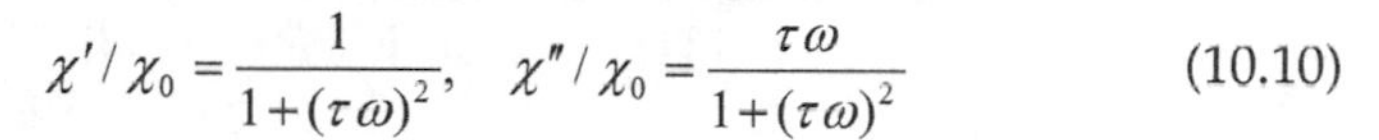

$$\chi'/\chi_0 = \frac{1}{1+(\tau\omega)^2}, \quad \chi''/\chi_0 = \frac{\tau\omega}{1+(\tau\omega)^2} \tag{10.10}$$

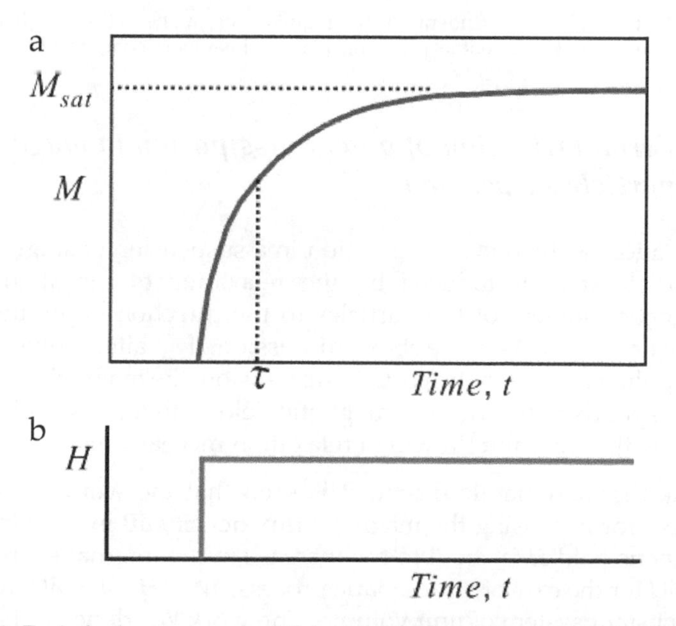


**Figure 10.2.** Typical relaxation curve of the magnetization after application of a step-type external magnetic field: (a) magnetization and (b) applied magnetic field.

Figure 10.3 shows characteristics of the in-phase and out-of-phase susceptibilities expressed in Eq. (10.10) as a function of the frequency factor $\tau\omega$. It is seen that the out-of-phase susceptibility has a maximum value $\chi''/\chi_0 = 0.5$ at $\tau\omega = 1$, and in the low frequency range, the magnetization can approximately respond to the alternating magnetic field without a phase delay.

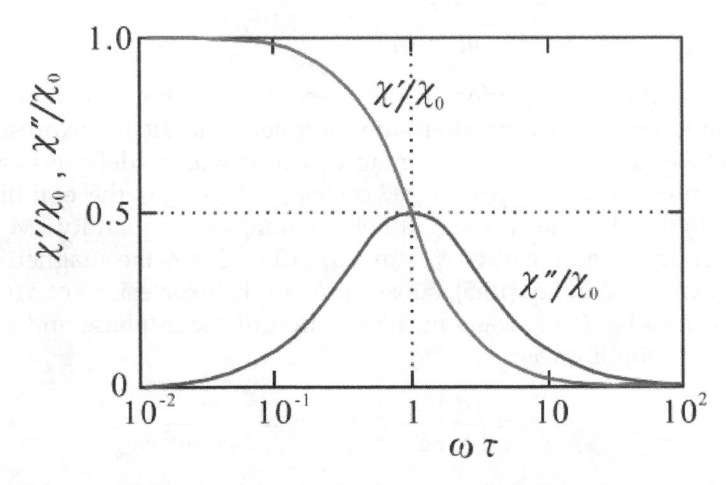


**Figure 10.3.** Susceptibility dependent on the frequency of an alternating magnetic field: $\chi'$ is the real component (synchronous part) and $\chi''$ is the imaginary component (out-of-phase part) of the complex susceptibility.

### *10.1.3 Formularization of power dissipation in magnetic particle suspension*

Here we address the power dissipation in a suspension of single-domain nano-particles that is induced by the relaxation of the alignment of the magnetic moment of the particles to the direction of an alternating magnetic field [4, 5]. As in the above discussion, if an alternating magnetic field is applied to a magnetic particle suspension, the magnetization of the system responds to the varying magnetic field with a phase delay that is induced by the Néel and Brownian relaxation mechanisms.

From the thermodynamic theory, it is seen that the work $\delta W$ done on the system for increasing the magnetic flux density $\delta \boldsymbol{B}$ in the situation of the magnetic field $\boldsymbol{H}$ is equal to the increase in the internal energy of the system $\delta U$ for the case of an adiabatic process, $\delta W = \boldsymbol{H} \cdot \delta \boldsymbol{B} = \delta U$, where we treat a constant system of unit volume. The work $W_{21}$ done on the system for increasing the magnetic flux density from $\boldsymbol{B}_1$ to $\boldsymbol{B}_2$ is written as

$$W_{21} = \int_{B_1}^{B_2} \boldsymbol{H} \cdot d\boldsymbol{B} \tag{10.11}$$

If the work is used for inducing the magnetization alone and is treated per one cycle of the alternating magnetic field, then the work done on the system, $W_{cycl}$, is expressed as

$$W_{cycle} = \mu_0 \oint \boldsymbol{H} \cdot d\boldsymbol{M} \tag{10.12}$$

in which $\mu_0$ is the permeability of free space. Substitution of the magnetic field $H(t) = H_0\cos(\omega t)$ and the magnetization $M(t)$ shown in Eq. (10.5) with the relationship shown in Eq. (10.6) into Eq. (10.12) gives rise to the following equation:

$$W_{cycle} = \mu_0 \oint \boldsymbol{H} \cdot d\boldsymbol{M} = \mu_0 \int_0^{2\pi/\omega} H \frac{dM}{dt} dt = \mu_0 H_0^2 \int_0^{2\pi/\omega} \cos\omega t \frac{d}{dt}\left(\chi' \cos\omega t + \chi'' \sin\omega t\right) dt$$

$$= \mu_0 H_0^2 \omega \chi'' \int_0^{2\pi/\omega} \cos^2 \omega t dt = \mu_0 \pi H_0^2 \chi'' \tag{10.13}$$

This expression clearly shows that the power dissipation is related only to the out-of-phase susceptibility and not to the in-phase susceptibility.

Hence, the power dissipation per unit time, $P$, is expressed as

$$P = \frac{\omega}{2\pi} U_{cycle} = \frac{\omega}{2\pi} W_{cycle} = \frac{\mu_0 H_0^2 \omega \chi''}{2} \tag{10.14}$$

Using this expression for the power dissipation, the increase in temperature $\Delta T$ in the suspension during time interval $\Delta t$ is obtained as $\Delta T = P\Delta t/c$, where $c$ is the specific heat of a suspension.

This power dissipation will be used as the mechanism for heat generation in a magnetic particle suspension in the situation of an alternating magnetic field. It is therefore important to elucidate the magnetic characteristics of the magnetic suspension of interest in order to apply the magnetic suspension to hyperthermia treatments.

If we take into account the particle size distribution of the magnetic particles in a suspension, the power dissipation $P_{multi}$ is expressed using the particle size distribution function $g(R)$ as [4, 9]

$$P_{multi} = \int_0^{\infty} P g(R) dR \tag{10.15}$$

in which $R$ is the particle diameter and the distribution $g(R)$ has to satisfy the following stochastic condition:

$$\int_0^{\infty} g(R)dR = 1 \tag{10.16}$$

## 10.2 Magnetically-propelled microswimmer

Microswimmers in a liquid may be a hopeful technique for delivering the magnetic materials loaded with anti-cancer drugs to the site of the targeted tissue by the use of a time-dependent magnetic field such as a rotational magnetic field [10–15]. In Chapter 1, we referred to a magnetically-controlled drug delivery system, where magnetic composite materials designed for delivering drugs are captured and transported to the target site by a non-uniform applied magnetic field. In this concept, one of key techniques to be developed for a successful application, on the physical side, is the generation of an effective non-uniform magnetic field by using an arrangement of a combination of magnets in order to transport the drugs to the target tissue in a variety of complex circumstances. In contrast to this approach, a magnetic microswimmer designed for loading drugs will actively perform the locomotive motion to move itself toward the target site in a liquid or liquid-like environment. In this application, the motion of microswimmers will be controlled by the use of a time-dependent applied magnetic field, and a rotational magnetic field is typically used for this objective. Here we concentrate on several representative magnetically-propelled microswimmers from the viewpoint of applying them to the drug delivery system or in the development of a micromachine that may be used in the field of medical treatments in the future, such as for instance, destroying kidney stones.

Figures 10.4(a) and (b) show microswimmers due to the more conventional propulsion mechanism of the rotational motion and Fig. 10.4(c) shows a microswimmer with propulsion due to cooperative reciprocal motion. The power inducing the locomotive motion of a microswimmer in the case of a magnetically-propelled system is supplied by a time-dependent applied magnetic field because, as already discussed in Sections 2.1 and 2.6, a steady uniform applied magnetic field does not induce a body force acting on the magnetic material. In the rotational propulsion mechanism, a time-dependent or rotational magnetic field induces the rotational motion and this rotational motion is transferred into the translational or unidirectional motion of the magnetic swimmer. The unidirectional motion may be accomplished by the geometrical shape of a non-magnetic material, for instance, in the case of Figs. 10.4(a) and 10.4(b), the helical body will play this role. In the reciprocal propulsion mechanism, a time-dependent magnetic field induces a different translational motion in each of the two magnetic particles shown in Fig. 10.4(c), and the combination of these motions leads to the unidirectional reciprocal motion of the whole swimmer system. In the following we discuss the physical aspects of these magnetically-propelled microswimmers in more detail.

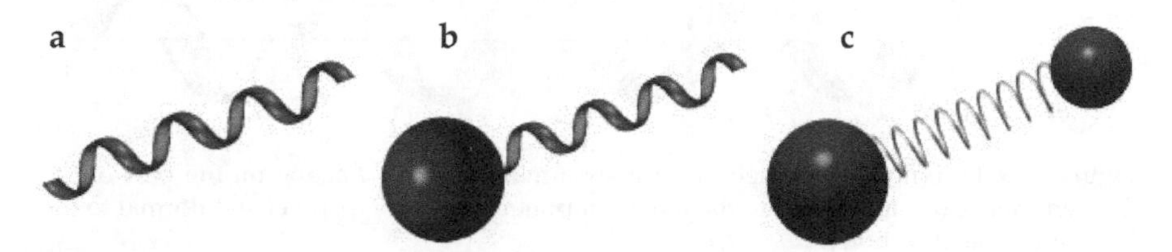


**Figure 10.4.** Typical magnetically-propelled microswimmers: (a) magnetic helix, (b) magnetic particle head with a helical tail and (c) two-magnetic particles connected by an elastic element (polymer).

## *10.2.1 Microswimmer of a magnetic helix*

Figure 10.4(a) shows a magnetic microswimmer with the simple shape of a spiral or helical structure [10, 11, 16, 17]. In a magnetically-propelled system, this helical body is a thin magnetic wire such as a curved nickel nanowire or a helix covered by a magnetic thin film, therefore, the helical microswimmer can respond to a rotational magnetic field, which leads to the rotation about the helical axis and finally leads to the locomotive motion in the helical axis direction. The speed of the swimmer will be dependent on a variety of factors such as characteristics of the helical structures themselves (i.e., the helical radius, pitch angle and length), the magnetic characteristics, the properties of the rotational magnetic field (i.e., the amplitude and frequency), and the characteristics of the ambient fluid (i.e., the viscosity). In general, the swimming speed increases with the increasing frequency of a rotational magnetic field.

We characterize the geometrical properties of the helix by using the notation $R$ for the radius, θ for the pitch angle and $l_p$ for the pitch, as shown in Figs. 10.5 and 10.6. If the thin or slender helix is rotating with angular velocity $\omega$ in a Newtonian fluid in a force-free situation, then the locomotion speed $U_{loco}$ is expressed as [10, 11],

$$U_{loco} = R\omega \frac{\sin 2\theta}{2(1+\sin^2\theta)} \tag{10.17}$$

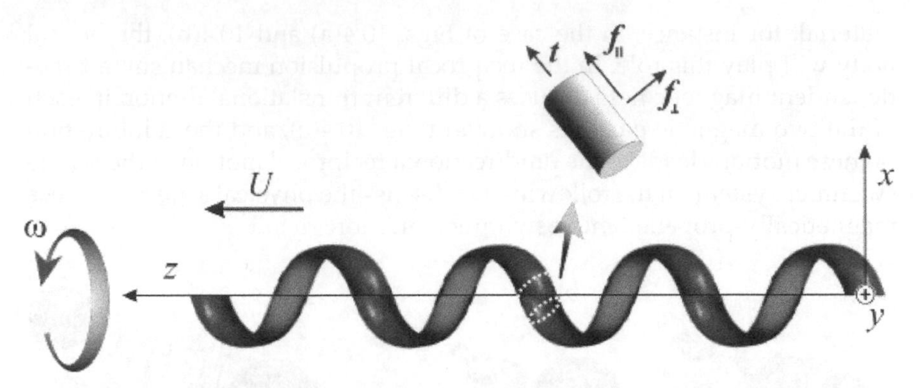


**Figure 10.5.** Locomotion of a helical microswimmer: the force $f$ acting on the part of the filament can be decomposed into the force components $f_{\parallel}$ and $f_{\perp}$ parallel and normal to the local tangential direction $\boldsymbol{t}$.

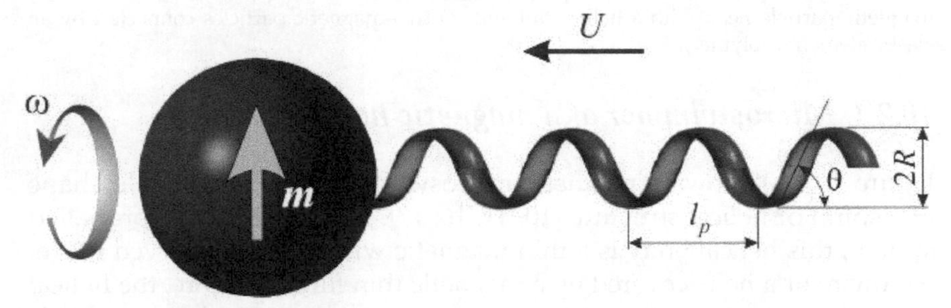


**Figure 10.6.** Magnetically-propelled microswimmer: this microswimmer is composed of a head of magnetic particle and a non-magnetic helical tail.

It is noted that in this assumption, the locomotion speed is independent of the viscosity of the ambient fluid.

We now consider a general helix made from a filament with a certain diameter $d = 2a$ where the assumption of a slender body is not applicable. If the $z$-axis is taken in the helical axis direction denoted by the unit vector $\boldsymbol{\delta}_z$ and the origin of the orthogonal coordinate system is taken at one end of the helix on the helical axis line, as shown in Fig. 10.5, then the center-line of the body of the helical filament, $\boldsymbol{r}(s,t)$, is expressed as

$$\boldsymbol{r}(s,t) = \left(R\cos(2\pi s/s_p + \omega t),\ R\sin(2\pi s/s_p + \omega t),\ bs + Ut\right) \tag{10.18}$$

in which s is the length along the helical filament from the origin, two constants are evaluated as $s_p = 2\pi R/\sin\theta$ and $b = \cos\theta$, $\omega$ is the angular velocity of the rotation of the helix, and $U$ is the locomotion speed to be determined.

As shown in Fig. 10.5, the force $f$ acting on a segment of the filament, which is partially cut from the filament and regarded as a cylinder, can be decomposed into the force components $f_{||}$ and $f_{\perp}$ parallel and normal to the local tangential direction $\boldsymbol{t}$, respectively, where $\boldsymbol{t}$ is expressed as $\boldsymbol{t} = \partial \boldsymbol{r}/\partial s$. Employing the friction coefficients of the short cylinder, $\xi_{||}$ and $\xi_{\perp}$ parallel and normal to the direction $\boldsymbol{t}$, respectively, the force components $f_{||}$ and $f_{\perp}$ are expressed as

$$f_{||} = \xi_{||}(\boldsymbol{u}\cdot\boldsymbol{t})\boldsymbol{t}, \quad f_{\perp} = \xi_{\perp}(\boldsymbol{u}-(\boldsymbol{u}\cdot\boldsymbol{t})\boldsymbol{t}) \tag{10.19}$$

in which $\boldsymbol{u}$ is the velocity of the filament segment or the short cylinder, expressed as $\boldsymbol{u} = \partial \boldsymbol{r}/\partial t$. The force $f$ will then be obtained as $f = f_{||} + f_{\perp}$. In the force-free situation, there is no resultant force acting on the whole body of the helix in the locomotion direction $\boldsymbol{\delta}_z$, that is,

$$\int_0^L f(s)\cdot\boldsymbol{\delta}_z\,ds = \int_0^L \left(f_{||}(s) + f_{\perp}(s)\right)\cdot\boldsymbol{\delta}_z\,ds = 0 \tag{10.20}$$

in which $L$ is the length of the filament line forming the helix. Employing the notation $\boldsymbol{u}_{tan}$ for the local velocity component normal to the helical axis and tangential to the circular segment, the velocity of the filament segment, $\boldsymbol{u}$, is expressed as $\boldsymbol{u} = U\boldsymbol{\delta}_z + \boldsymbol{u}_{tan}$. Substitution of this local velocity and the relationships in Eqs. (10.19) into Eq. (10.20) gives rise to the expression for the locomotion speed $U$ as

$$U = -(\xi_{||}-\xi_{\perp})\int_0^L (\boldsymbol{t}\cdot\boldsymbol{u}_{tan})(\boldsymbol{t}\cdot\boldsymbol{\delta}_z)ds \Bigg/ \int_0^L \left\{\xi_{\perp} + (\xi_{||}-\xi_{\perp})(\boldsymbol{t}\cdot\boldsymbol{\delta}_z)^2\right\}ds \tag{10.21}$$

The expressions of the local velocity $\boldsymbol{u}$ of the filament segment and the local tangential direction $\boldsymbol{t}$ are obtained from Eq. (10.18) after a relatively complex mathematical manipulation, and substitution of these expressions into Eq. (10.21) finally yields the expression of the locomotion speed as [10, 11]

$$U = R\omega\frac{(\hat{\xi}-1)\sin 2\theta}{2\{1+(\hat{\xi}-1)\sin^2\theta\}} \tag{10.22}$$

in which $\hat{\xi}$ is the ratio of the friction coefficients expressed as $\hat{\xi} = \xi_{\perp}/\xi_{||}$. It is noted that the expression in Eq. (10.22) is valid for the force-free motion of a helical swimmer which is made of a cylindrical filament rotating about the helical axis with angular velocity $\omega$. In the limit of $s_p/a$ approaching infinity, the ratio $\hat{\xi}$ converges to $\hat{\xi} = 2$ and consequently the expression reduces to the solution with the assumption of a slender helical body shown in Eq. (10.17).

Employing the above mathematical description for a microswimmer, the viscous power dissipation per unit time due to the locomotion of the swimmer, $P_{loss}$, is evaluated as

$$P_{loss} = \int_{o}^{L} \boldsymbol{f}(s) \cdot \boldsymbol{u}(s) ds = \frac{\hat{\xi}}{1 + (\hat{\xi} - 1)\sin^2\theta} \xi_{||} (R\omega)^2 L \tag{10.23}$$

## *10.2.2 Microswimmer of a magnetic particle head with a helical tail*

Figure 10.6 shows a magnetic microswimmer that is composed of a magnetic particle head with a helical tail [12, 18, 19]. In this type of swimmer, the magnetic head functions as a driving energy source and the helical tail converts this energy into the locomotive motion of the swimmer toward the target site. In contrast to the helical microswimmers explained in Section 10.2.1, the helical tail of this microswimmer is not necessarily made of a magnetic material. In application for a drug delivery system, for instance, the drugs may be loaded in either a magnetic head formed with composite materials or in a non-magnetic helical tail that may be made of polymeric materials. The mechanism of the locomotive motion is essentially the same as for the previous helical microswimmers but there is an important difference where the head body now becomes a resistance against the translational locomotive motion of the swimmer.

In Stokesian regime of a slow flow problem, the torque $\boldsymbol{T}_{flow}$ acting on the ambient fluid by the rotational motion of the angular velocity $\boldsymbol{\omega}$ of a sphere with radius $a$ is written as

$$\boldsymbol{T}_{flow} = 8\pi\eta a^3 \boldsymbol{\omega} \tag{10.24}$$

in which $\eta$ is the viscosity of ambient liquid. If the inertia moment of the swimmer and the viscous friction loss of the helix tail are negligible, this torque is equal to the input magnetic torque $\boldsymbol{T}_{in}$ by an applied rotational magnetic field, then $\boldsymbol{T}_{in} = \boldsymbol{T}_{flow}$. It is therefore understood that a larger angular velocity is obtained with decreasing viscosity of the ambient liquid and also with decreasing particle size for a given input of magnetic torque.

As the rotational motion of the microswimmer is generated by a time-dependent or rotational applied magnetic field $\boldsymbol{H}$, the torque $\boldsymbol{T}$ acting on the magnetic particle head with magnetic moment $\boldsymbol{m}$ is expressed as

$$\boldsymbol{T} = \mu_0 \boldsymbol{m} \times \boldsymbol{H} \tag{10.25}$$

Hence, it is seen that if a rotational magnetic field is applied normal to the helical axis and the head body is also magnetized in this direction, as shown in Fig. 10.6, then the driving energy that is input by a magnetic field is converted into the rotational motion about the helical axis in the most effective manner, whereas if the head body is magnetized in the helical axis direction, the rotational field cannot induce the cork-screw motion that is required of the magnetic microswimmer.

For the successful application of this type of microswimmer, it is necessary to optimize the helical tail characteristics in regard to the helical radius, pitch angle and length [18], and to develop a magnetic head body that can respond to an applied rotational magnetic field in an effective manner, which leads to a high performance in the transfer of the driving energy by the magnetic field into the rotational motion or the locomotive motion of the microswimmers.

### *10.2.3 Microswimmer of two-magnetic particles connected by an elastic element*

Figure 10.4(c) shows a magnetic microswimmer that is composed of two magnetic particles with different magnetic characteristics combined by an elastic element [20, 21]. A time-dependent external magnetic field induces an attractive or repulsive force acting between the two magnetic particles that results from the difference in the magnetic characteristics of the particles. In a time-dependent external magnetic field, these attractive and repulsive forces are moderated by an elastic connection element that leads to the locomotive motion or reciprocal motion in this type of swimmer system. The connection element might be a polymer or protein fiber if the objective application is a drug delivery system and therefore has the possibility to include the drugs to be transported to the site of a specific tumor or cancer tissue. Hence, a successful application of this magnetic microswimmer to the biomedical engineering field requires the development of an accurate control technique for a persistent unidirectional swimming motion toward a specific site whilst minimizing any unpleasant behavior such as a tumbling motion.

The locomotive motion of this two-magnetic particle swimmer may generally require magnetic particles that exhibit soft and hard magnetic characteristics, which lead to the different responses to the change in the direction of an alternating magnetic field. In general, the particle with soft magnetic characteristics changes its magnetic moment direction in a much shorter time than the particle with hard magnetic characteristics, and this tends to induce an alternating force of attraction and repulsion between a pair of magnetic particles. This force arises from the interaction between the magnetic moment and the local non-uniform magnetic field induced by the moment of the other particle. Ideally, the combination of these alternating forces acting between the two magnetic particles will generate the translational locomotive motion in a direction along the connecting line.

As previously discussed in Chapter 2, the application of a uniform applied magnetic field does not induce a body force acting on a particle with a dipole moment. However, in the above situation of a microswimmer, the magnetic particle induces a non-uniform magnetic field at the position of the other magnetic particle. If the second particle with a point dipole moment $\boldsymbol{m}_2$ is located at the relative position $\boldsymbol{r}_{21}$ from the first particle, then a magnetic field $\boldsymbol{H}_{21}$ induced at the position by the first particle with a point dipole moment $\boldsymbol{m}_1$ is expressed as

$$\boldsymbol{H}_{21} = \frac{1}{4\pi|\boldsymbol{r}_{21}|^3}\left\{-\boldsymbol{m}_1 + \frac{3}{|\boldsymbol{r}_{21}|^2}(\boldsymbol{r}_{21}\cdot\boldsymbol{m}_1)\boldsymbol{r}_{21}\right\} \tag{10.26}$$

The interaction of this induced magnetic field $\boldsymbol{H}_{21}$ and the magnetic moment $\boldsymbol{m}_2$ of the second particle yields a magnetic body force $\boldsymbol{F}_{21}$ acting on the second particle, expressed as

$$\boldsymbol{F}_{21} = \mu_0(\boldsymbol{m}_2\cdot\nabla)\boldsymbol{H}_{21} \tag{10.27}$$

in which $\mu_0$ is the permeability of free space. A similar force $\boldsymbol{F}_{12}$ acts on the first particle due to the non-uniform magnetic field $\boldsymbol{H}_{12}$ induced by the magnetic moment $\boldsymbol{m}_2$ of the second particle.

Independent of a uniform or non-uniform nature of the applied magnetic field, a magnetic torque $\boldsymbol{T}$ acts on the particle with a point dipole moment $\boldsymbol{m}$, expressed as

$$\boldsymbol{T} = \mu_0\boldsymbol{m}\times\boldsymbol{H} \tag{10.28}$$

It is seen that this torque acts in such a way that the magnetic moment will tend to align to the magnetic field direction.

There is a possibility for a reciprocal swimmer to be composed of more than two head spheres combined in various ways by elastic segments, but in this case the reciprocal motion will be significantly more complex in order to generate the cooperatively locomotive motion of the swimmer system as a whole [21].

# Bibliography


[1] Schmidt, A. M. 2007. Thermoresponsive magnetic colloids. Colloid Polymer Sci. 285: 953–966.

[2] Pankhurst, Q. A., Connolly, J., Jones, S. K. and Dobson, J. 2003. Applications of magnetic nanoparticles in biomedicine. J. Phys. D: Appl. Phys. 36: R167.

[3] Obaidat, I. M., Issa, B. and Haik, Y. 2015. Magnetic properties of magnetic nanoparticles for efficient hyperthermia. Nanomater. 5: 63–89.

[4] Rosensweig, R. E. 2002. Heating magnetic fluid with alternating magnetic field. J. Magn. Magn. Mater. 252: 370–374.

[5] Chikazumi, S. 1997. Physics of Ferromagnetism, 2nd ed., Oxford Science Publications, London.

[6] Brown, W. F. 1963. Thermal fluctuations of a single-domain particle. Phys. Rev. 130: 1677–1686.

[7] Frenkel, J. 1955. Kinetic Theory of Liquids, Dover, New York.

[8] Deatsch, A. E. and Evans, B. A. 2014. Heating efficiency in magnetic nanoparticle hyperthermia. J. Magn. Magn. Mater. 354: 163–172.

[9] Gonzales-Weimuller, M., Zeisberger, M. and Krishnan, K. M. 2009. Size-dependant heating rates of iron oxide nanoparticles for magnetic fluid hyperthermia. J. Magn. Magn. Mater. 321: 1947–1950.

[10] Leshansky, A. M. 2009. Enhanced low-Reynolds-number propulsion in heterogeneous viscous environments. Phys. Rev. E. 80: 051911.

[11] Wada, H. and Netz, R. R. 2009. Hydrodynamics of helical-shaped bacterial motility. Phys. Rev. E. 80: 021921.

[12] Ghosh, A., Paria, D., Rangarajan, G. and Ghosh, A. 2014. Velocity fluctuations in helical propulsion: How small can a propeller be. J. Phys. Chem. Lett. 5: 62–68.

[13] Vach, P.J., Fratzl, P., Klumpp, S. and Faivre, D. 2015. Fast magnetic micropropellers with random shapes. Nano Lett. 15: 7064–7070.

[14] Roper, M., Dreyfus, R., Baudry, J., Fermigier, M., Bibette, J. and Stone, H. A. 2008. Do magnetic micro-swimmers move like eukaryotic cells?. Proc. R. Soc. A. 464: 877–904.

[15] Gilbert, A. D., Ogrin, F. Y., Petrov, P. G. and Winlove, C. P. 2011. Theory of ferromagnetic microswimmers. Quarterly J. Mech. Appl. Math. 64: 239–263.

[16] Morozov, K. I. and Leshansky, A. M. 2014. The chiral magnetic nanomotors. Nanoscale. 6: 1580–1588.

[17] Yamazaki, A., Sendoh, M., Ishiyama, K., Arai, K., Kato, R., Nakano, M. and Fukunaga, H. 2004. Wireless micro swimming machine with magnetic thin film. J. Magn. Magn. Mater. 272–276: E1741–E1742.

[18] Eric E. Keaveny, Shawn W. Walker and Michael J. Shelley. 2013. Optimization of chiral structures for microscale propulsion. Nano Lett. 13: 531–537.

[19] Julian Espinosa-Garcia, Eric Lauga and Roberto Zenit. 2013. Fluid elasticity increases the locomotion of flexible swimmers. Phys. Fluids 25: 031701.
[20] Ogrin, F. Y., Petrov, P. G. and Winlove, C. P. 2008. Ferromagnetic microswimmers. Phys. Rev. Lett. 100: 218102.
[21] Earl, D. J., Pooley, C. M., Ryder, J. F., Bredberg, I. and Yeomans, J. M. 2007. Modeling microscopic swimmers at low Reynolds number. J. Chem. Phys. 126: 064703.

# How to Acquire the Sample Simulation Programs

A copy of the sample simulation programs that are shown in Chapter 9 can be requested directly from the author at the e-mail address below:

e-mail address: asatoh_book2016@excite.co.jp

Please note that the following information is required;

(1) the purchase date,
(2) the number of purchased copies,
(3) the profession of the purchaser.

In addition the request by a university library may be acceptable from an educational point of view.

It is noted that the sample simulation programs in this book can be freely used for educational purposes in an academic environment, but they are not permitted to be used for commercial purposes. Moreover, the user should be solely responsible for all the results that are obtained from using the sample simulation programs.

The author would deeply appreciate the report of any bugs in the programs, but regrets that he is unable to accept any enquiry concerning the content of the simulation programs.

# Index